Process Technology Equipment

Second Edition

Technical Editor

Gayle Cannon
Middlesex County College
Edison, NJ

330 Hudson Street, NY NY 10013

Managing Director, Career Development and Employability: Leah Jewell
Director, Alliance/Partnership Management: Andrew Taylor
Director, Learning Solutions: Kelly Trakalo
Managing Content Producer: Laura Burgess
Associate Content Producer: Shannon Stanton
Media Producer: Jose Carchi
Development Editor: Rachel Bedard
Instructor and Student Supplement Development: Perci LLC dba/Publisher's Resource Center
Executive Marketing Manager: Brian Hoehl
Product Marketing Manager: Rachele Strober
Manufacturing Buyer: Maura Zaldivar-Garcia, LSC Communications
Cover Designer: Laurie Entringer
Cover Image Credit: Lonnie Duka/Getty Images
Editorial and Full-Service Production and Composition Services: Pearson CSC
Editorial Project Manager: Susan Hannahs
Full-Service Project Manager: Billu Suresh
Printer/Bindery: LSC Communications
Cover Printer: LSC Communications

Library of Congress Cataloging-in-Publication Data

Names: North American Process Technology Alliance.
Title: Process technology equipment / Technical editor, Gayle Cannon, Middlesex County College,
 technical editor.
Description: Second edition. | New York : Pearson Education, [2018]
Identifiers: LCCN 2018028641 | ISBN 9780134891262
Subjects: LCSH: Chemical processes--Equipment and supplies.
Classification: LCC TP155.7 .P763 2018 | DDC 660/.283--dc23 LC record available at https://lccn.loc.gov/2018028641

www.pearsonhighered.com

ISBN-10: 0-13-489126-0
ISBN-13: 978-0-13-489126-2

31 2023

Preface

The Process Industries Challenge

In the early 1990s, the process industries recognized that they would face a major staffing shortage because of the large number of "baby boomer" employees who would be retiring. Industry partnered with community colleges, technical colleges, and universities to remedy this situation. Together, they developed this series, which provides consistent curriculum content and exit competencies for process technology graduates to ensure a knowledgeable and competent staff that is ready to take over the demands of the field. The collaborators in education and industry also recognized that training for process technicians would benefit industry by reducing the costs associated with training and traditional hiring methods. This was how the NAPTA series for Process Technology was born.

To achieve consistency of exit competencies among graduates from different schools and regions, the North American Process Technology Alliance identified a core technical curriculum for the Associate Degree in Process Technology. This core consists of eight technical courses and is taught in alliance member institutions throughout the United States. Instructors who teach the process technology core curriculum, and who are recognized in industry for their years of experience and depth of subject matter expertise, requested that a textbook be developed to match the standardized curriculum. A broad range of reviewers from process industries and educational institutions participated in the production of these materials so that the presentation of content would address the widest audience possible. This textbook is intended to provide a common national standard reference for the *Process Technology Equipment* course in the Process Technology degree program.

This textbook is intended for use in high schools, community colleges, technical colleges, universities, and any corporate setting in which process technology is taught. Current and future process technicians will use the information within this textbook as their foundation for work in the process industries. This knowledge will make them better prepared to meet the ever-changing roles and responsibilities within their specific process industry.

What's New!

The second edition has been thoroughly updated and revised.

- **New** Learning Outcome alignment with NAPTA core objectives, with links from objective to text page provided.

- **All New** Full Color Art Program with more than 450 fully revised drawings and photos, including new visuals of equipment.

- **New** Key term definitions on text pages where content appears, as well as at beginning of each chapter.

- **Updated Review and New Answers Appendix!** Checking Your Knowledge questions have been updated to meet new chapter objectives and now include an Appendix with answers.

- **Thoroughly Updated Content** including more in-depth content, formulas for use with different processes, and more information about standards, materials of construction, sizing, and testing.

- **Safety** emphasized strongly throughout.

- **Expanded Content** about the Process Technician's role, piping and associated components, materials of construction, filters and dryers, federal regulations, and more, including tables for reference use in the field.

- **New Content** about lockout-tagout procedures, including a generic LOTO procedure; piston compressors, reactive and impulse turbines; bearings, seals, and gaskets; cold box and air cooled heat exchangers, rules of thumb for heat exchangers; and metric conversions to prepare students for working with international partners in industry.

- **ALL NEW Instructor Resource Package** including lesson plans, test banks, review questions, PowerPoints, and a correlation guide to NAPTA curriculum.

Organization of the Textbook

This book has been divided into five parts with 20 chapters. Part 1 (Chapters 1–5) provides an overview of process technology equipment and types of tools. Part 2 (Chapters 6–10) describes the movers and drivers of process equipment, plus the elements of power transmission and lubrication. Part 3 (Chapters 11–14) focuses on heating and cooling equipment. Part 4 (Chapters 15–17) addresses common types of vessels used in process industries. Part 5 (Chapters 18–20) covers important miscellaneous equipment the process technician will encounter in the workplace.

Each chapter has the same organization.

- **Objectives** for each chapter are aligned with the revised NAPTA curriculum and can cover one or more sessions in a course.

- **Key Terms** list important words or phrases and their respective definitions, which students should know and understand before proceeding to the next chapter.
- The **Introduction** might be a simple introductory paragraph or might introduce concepts necessary to the development of the chapter's content.
- Any of the **Key Topics** can have several subtopics. Topics and subtopics address the objectives stated at the beginning of each chapter.
- The **Summary** is a restatement of important points in the chapter.
- **Checking Your Knowledge** questions are designed to help students do self-testing on essential information in the chapter.
- **Student Activities** provide opportunities for individual students or small groups to apply some of the knowledge they have gained from the chapter. These activities generally should be performed with instructor involvement.

Acknowledgements

The second edition of this series would not have been possible without the support of the entire NAPTA Board and, in particular, without the leadership and dedication of Executive Director Eric Newby. A particular and special thank you also goes to Gayle Cannon for her key role in the revision of this title. Her clear writing style and careful attention to detail are greatly appreciated.

Contributors

Gayle Cannon, Middlesex County College Edison, NJ (Art program, Text revisions)

Cleve Fontenot, BASF Corporation (retired), Baton Rouge, LA (Instructor materials)

Ronald Gamble, Remington College, Mobile, AL (Instructor materials)

Jeffrey Laube, Kenai Peninsula College Soldotna, AK (Art program, Text revisions)

Martha McKinley, McKinley Consulting Longview, TX (Art program, Text revisions)

Mike Tucker, Eastman Chemical Company, Texas Operations, Texas (Art program)

Reviewers

Ammar Alkhawaldeh, PhD, HCC Global Energy Institute, Houston, TX

Bobby Ray Player, Eastman Chemical Company, Longview, TX

Cleve Fontenot, Equipment Training Facility Developer, BASF Corporation (retired), Baton Rouge, LA

Lance Harlan, Formosa Plastics, Point Comfort, TX

Curtis Briggs, Development Coordinator for INEOS, League City, TX

David Hendrix, Brazosport College, Lake Jackson, TX

David W. T. King, BSC Chemical Engineering, League City, TX

Frank Huckabee, Remington College-Mobile, Mobile, AL

Jeffrey T. Key, Shell Chemical, Pearland, TX

Jennifer FIllinger, Nunez Community College, Chalmette, LA

Regina Cooper, Marathon Petroleum, Texas City, TX

Shawn Wehmeyer, Formosa Plastics, Point Comfort, TX

Tommie Ann Broome, Mississippi Gulf Coast Community College, Gautier, MS

Walbert Schulpen, University of Alaska Anchorage, Anchorage, AK

The following organizations and their dedicated personnel supported the development of the first edition of this textbook. Their contributions set the foundation for this revision and continue to be greatly appreciated.

Alaska Process Industry Careers Consortium

Anchorage Water and Wastewater Utility

Basell USA

BASF

Bayport Technical

BP

Chevron Texaco

ConocoPhillips

Dow Chemical Company

Eastman Chemical Company

Equistar Chemicals

ExxonMobil

Formosa Plastics Corp.

HeadsUp Systems

Huish Detergents, Inc.

Ingenious, Inc.

John Zink

Kraft Foods

Marathon

Mississippi Power

Novartis

Pasadena Refining System, Inc.

Shell Chemical LP

Shell Oil Products

Sherwin Alumina Company

TAP Safety Services

Union Carbide Corp.

Valero

Westlake Chemical Corp.

Industry Content Developers and Reviewers

Chuck Baukal, John Zink Company LLC, Oklahoma

Ted Borel, Equistar, Texas

Linda Brown, Pasadena Refining System, Inc., Texas

Gayle Cannon, ConocoPhillips, New Jersey

Candy Carrigan, ConocoPhillips, New Jersey
Karl Diederich, Sterling Solutions, Inc., Texas
Larry Ely, ConocoPhillips, California
Steve Erickson, Gulf Coast Process Technology Alliance, Texas
Jimmy Greene, Eastman Chemical Company, Texas Operations, Texas
Debera Hanrahan, British Petroleum, Washington
Richard Honea, The Dow Chemical Company, Texas
Leslie Hunt, TailorMade Training, Tennessee
Glenn E. Johnson, The Sun Products Corporation, Texas
Alex Kharazi, The Dow Chemical Company, New Jersey
Susanne Kolodzy, Troubleshooting Resources, Texas
Steve Lagger, Citgo Petroleum Corporation, Illinois
John Leedy, The Dow Chemical Company, Texas
Diane McGinn, INEOS, Texas
Walter Eric Newby, BASF, Texas
Don Parsley, Valero Refining Corporation, Texas
Ray Player, Eastman Chemical Company, Texas Operations, Texas
Lyndon Pousson, Independent Reviewer, Louisiana
Brian Smith, ChevronTexaco, Louisiana
Barbara Tracy, ConocoPhillips, New Jersey
Mike Tucker, Eastman Chemical Company, Texas Operations, Texas
Roy Viator, ChevronTexaco, Louisiana
Norris Watt, British Petroleum, Louisiana
John E. Wilson, Training & Development Systems, Inc. (TDS), Texas

Education Content Developers and Reviewers

Chuck Beck, Red Rocks Community College, Colorado (formerly of Coors Brewing Company)
Tommie Ann Broome, Mississippi Gulf Coast Community College, Mississippi
Tom Carleson, Bellingham Technical College, Washington
Mike Cobb, College of the Mainland, Texas

Mike Connella, McNeese State University, Louisiana
David Corona, College of the Mainland, Texas
Mary Darden, Independent Reviewer, Texas
John Dees, d3 Consulting, Texas
Lisa Arnold Diederich, Independent Reviewer, Texas
Mark Demark, Alvin Community College, Texas
Eric Douglas, University of the Virgin Islands
Jerry Duncan, College of the Mainland, Texas
Jim Forthman, Calhoun Community College, Alabama (formerly with Monsanto/Solutia)
Gary Hicks, Brazosport College, Texas (formerly of The Dow Chemical Company)
Lauren Hightower, Baylor University, Texas
Jerry Layne, Baton Rouge Community College, Louisiana
Linton Lecompte, Independent Reviewer, Louisiana
Mike Link, Delaware Technical and Community College, Delaware
Jim Lockett, Lee College, Texas
Derrill Mallett, College of the Mainland, Texas
Martha McKinley, Texas State Technical College, Marshall, Texas (formerly of Eastman Chemical Company)
Larry Perswell, College of the Mainland, Texas
Kelsey Rexroat, Baylor University, Texas
Paul Rodriguez, Lamar Institute of Technology, Texas
Vicki Rowlett, Lamar Institute of Technology, Beaumont, Texas
Pete Rygaard, College of the Mainland, Texas
Dan Schmidt, Bismarck State College, North Dakota
Dale Smith, Alabama Southern Community College, Alabama
Robert (Bobby) Smith, Texas State Technical College, Marshall, Texas (formerly of Eastman Chemical Company)
Walter Tucker, Lamar Institute of Technology, Texas
Steve Wethington, College of the Mainland, Texas

This material is based upon work supported, in part, by the National Science Foundation under Grant No. DUE 0202400. Any opinions, findings, and conclusions or recommendations expressed in this material are those of the author(s) and do not necessarily reflect the views of the National Science Foundation.

Contents

Part 2 Movers and Drivers

Part 4 Vessels

Chapter 1

Introduction to Process Equipment

 Objectives

After completing this chapter, you will be able to:

1.1 Describe the process industries and what they produce. (NAPTA Intro History 1*) p. 2

1.2 Describe the process technician's role in operations and maintenance. (NAPTA Intro to Tools and Equipment 5, 10–12) p. 3

1.3 Describe the types of common equipment used in the process industries. (NAPTA Intro to Tools and Equipment 7, 8) p. 5

1.4 Describe safety and environmental hazards associated with equipment usage in the process industries. (NAPTA Intro to Tools and Equipment 9) p. 12

*North American Process Technology Alliance (NAPTA) developed curriculum to ensure that Process Technology courses will produce knowledgeable graduates to become entry level employees in process technology. Objectives from that curriculum are named here in abbreviated form. For example, "(NAPTA Intro Tools and Equipment 5)" means that this chapter's objective relates to objective 5 of NAPTA's curriculum about tools and equipment.

Key Terms

Boiler—a device in which water is boiled and converted into steam under controlled conditions, **p. 10.**

Compressor—a mechanical device used to increase the pressure of a gas or vapor, **p. 8.**

Cooling tower—a structure designed to lower the temperature of water using latent heat of evaporation, **p. 9.**

Dryer—a device used to remove moisture from a process stream, **p. 11.**

Engine—a machine that converts chemical (fuel) energy into mechanical force, **p. 9.**

Filter—device that removes particles from a process, allowing the clean product to pass through, **p. 11.**

Fitting—system component used to connect two or more pieces of piping, tubing, or other equipment, **p. 6.**

Furnace—a piece of equipment that burns fuel in order to generate heat that can be transferred to process fluids flowing through tubes; also referred to as a process heater, **p. 9.**

Heat exchanger—a device used to transfer heat from one substance to another without physical contact between the two, **p. 9.**

Hose—flexible tube that carries fluids; can be made of plastic, rubber, fiber, metal, or a combination of materials, **p. 6.**

Lubrication—the application of a substance between moving surfaces in order to reduce friction and minimize heating, **p. 11.**

Motor—a mechanical driver that converts electrical energy into useful mechanical work and provides power for rotating equipment, **p. 8.**

Pipe—long, hollow cylinder through which fluids are transmitted; primarily made of metal, but also can be made of glass, plastic, or plastic-lined material, **p. 6.**

Process—the conversion of raw materials into a finished or intermediate product, **p. 2.**

Process drawing—illustration that provides a visual description and explanation of the processes' equipment, flows, and other important items in a facility, **p. 5.**

Process industries—a broad term for industries that convert raw materials, using a series of actions or operations, into products for consumers, **p. 2.**

Process technician—a worker in a process facility who monitors and controls mechanical, physical, and/or chemical changes throughout a process in order to create a product from raw materials, **p. 3.**

Pump—a mechanical device that transfers energy to move materials through piping systems, **p. 7.**

Reactor—a vessel in which a controlled chemical reaction is initiated and takes place either continuously or as a batch operation, **p. 10.**

Solids handling equipment—equipment that is used to process and transfer solid materials from one location to another in a process facility; it also might provide storage for those materials, **p. 11.**

Tank—a vessel in which a feedstock or product (intermediate or finished) is stored; might be classified as atmospheric or pressurized, aboveground or underground, fixed or floating roof, **p. 7.**

Troubleshooting—the systematic search for the source of a problem so that it can be solved, **p. 4.**

Tubing—hose or pipe of small diameter (typically less than 1 inch [2.5 cm]) used to transport fluids, **p. 6.**

Turbine—a machine that is used to produce power and rotate shaft-driven equipment such as pumps, compressors, and generators, **p. 8.**

Unit—an integrated group of process equipment used to produce a specific product; it might be referred to by the processes it performs or be named after its end products, **p. 3.**

Valve—piping system component used to control, throttle, or stop the flow of fluids through a pipe, **p. 6.**

Vessel—an enclosed process container such as a tank, drum, tower, filter, or reactor, **p. 7.**

1.1 Introduction

This chapter provides an overview of the process industries, the role of the process technician, the types of equipment a process technician might encounter, and safety and environmental hazards.

Throughout this textbook, the term **process industries** is used broadly to describe industries that use processes to create products for consumers. A **process** is the conversion of raw materials into a finished or intermediate product. Process industries are some of the world's largest industries, with hundreds of thousands of workers in virtually every country. Products from these industries directly or indirectly affect the daily lives of almost everyone on the planet.

Generally speaking, process industries involve technologies that take specific quantities of raw materials and safely transform them into other products. The result might be an end product for a consumer or an intermediate product that is later converted to a different end product.

Process industries a broad term for industries that convert raw materials, using a series of actions or operations, into products for consumers.

Process the conversion of raw materials into a finished or intermediate product.

Companies in the process industries use a system of people, methods, equipment, and procedures to create products. A **process technician** is a worker in a process facility who continually monitors and controls mechanical, physical, and/or chemical changes in order to create a product from raw materials, while maintaining a safe work environment.

Although processes and products vary, process industries share some basic equipment. For example, pumps are used to move liquids through piping systems, motors are used to drive equipment, and furnaces are used to generate heat.

Process technician a worker in a process facility who monitors and controls mechanical, physical, and/or chemical changes throughout a process in order to create a product from raw materials.

How Process Industries Operate

Various industries are classified as process industries. These include oil and gas exploration and production, petroleum refining, chemical manufacturing, mining, power generation, water and wastewater treatment, food and beverage production, pharmaceutical manufacturing, and paper and pulp processing. While a wide range of processes and products are associated with each industry, they all share some common operations:

1. Raw materials, sometimes called *input* or *feedstock*, are made available to a process facility or plant unit. Most facilities have distinct units that perform different processes. A **unit** is an integrated group of process equipment used to produce a specific product or carry out a specific activity. Units can be referred to by the processes they perform (e.g., reforming unit) or by their end products (e.g., olefin unit).

2. The raw materials are sorted by process requirements.

3. The raw materials are processed. Process technicians monitor and control the mechanical, physical, and/or chemical changes that occur during the process, while maintaining safety, health, environmental, quality, and efficiency standards.

4. A product (output), or desired component, is the result of a particular process. The product can be either an end product for consumers or an intermediate product used as part of another process to make a different end product. The end product of one process unit or process industry can become the feedstock of another.

5. The product then is distributed to consumers, including other businesses that use the product of one industry's process as the feedstock for their own (e.g., ethylene as the feedstock for polyethylene).

Unit an integrated group of process equipment used to produce a specific product; it might be referred to by the processes it performs or be named after its end products.

1.2 Process Technician's Role in Operations and Maintenance

A process technician is a key member of a team responsible for safe planning, analyzing, and controlling of process operations. The production process includes everything from the acquisition of raw materials through the production and distribution of products to customers.

The job duties of a process technician include a wide variety of tasks, which can include monitoring and controlling processes from a control room or in the field; assisting with equipment maintenance; communicating and working with others; performing administrative duties; troubleshooting; maintaining environmental and quality standards; and working in teams. Safety must always be the first consideration when performing any process duty.

While monitoring and controlling a process (Figure 1.1), the process technician might be required to sample processes; inspect equipment; start, stop, and regulate equipment; view instrumentation readouts; analyze data; evaluate processes for improvement; make process adjustments; maintain area housekeeping; respond to changes, emergencies, and abnormal operations; and document activities, issues, and changes.

Process technicians may be required to perform minor maintenance tasks, including predictive/preventive maintenance, on the equipment for which they are responsible. For example, they might need to change or clean filters or strainers, repair minor leaks, lubricate

A.

B.

equipment, or monitor or analyze equipment performance using infrared or ultrasonic testing equipment. They also might be required to conduct regulatory required testing for fugitive emissions (very small leaks), using a portable volatile organic compound (VOC) analyzer.

Process technicians may be responsible for plant maintenance tasks beyond the scope of "minor" maintenance. Some facilities train technicians in one of the maintenance crafts (instrument technician, electrician, machinist, etc.), in addition to their operations training, and they rotate employees between the operations and maintenance roles.

Process technicians also will be responsible for preparing equipment for mechanical work. This preparation includes removing the equipment from service, isolating and draining it, and following the appropriate hazardous energy isolation procedures. When mechanical work has been completed, process technicians are responsible for returning process equipment to service.

One of the most valuable skills a process technician can master is the ability to communicate and work with others. This is especially important when the issue being communicated pertains to health, environmental impact, and/or safety of both plant personnel and the surrounding community.

Communication involves sending and receiving information through written reports, documenting equipment analysis, writing and reviewing procedures, and documenting incidents. It also includes listening to and training others, learning new skills and information, and working as part of a team. Strong computer, oral, and written communication skills are essential for process technicians operating within the organizational structure of a company. These skills are used on a daily basis when describing activities for relief personnel, maintaining data logs, and preparing reports. Good communication between field and control room operators is essential to the safe, efficient running of any unit. It is normally accomplished with the use of intrinsically safe two-way radios. Each plant will have established protocols for radio use.

Process technicians also might be required to perform additional duties, which include housekeeping and performing safety and environmental checks. *Safety*, *health*, and *environment* are typically the most spoken words in the process industry. Because of this, process technicians must keep safety, health, and environmental regulations and procedures in mind at all times. They must look for unsafe or abnormal conditions and watch for signs of potentially hazardous situations.

Equipment troubleshooting is a skill that is both taught in the classroom and learned through on-the-job training. **Troubleshooting** is the systematic search for the source of a problem, taking the steps necessary to eliminate the problem, and returning the process to its normal operating conditions. A process technician working on a process unit may be required to apply troubleshooting techniques and principles to identify an operational issue with a piece of equipment or an entire process. Equipment procedures are available, and a list of potential problems and solutions is often provided. If an item is not listed, a process technician might be required to conduct a basic investigation and apply troubleshooting skills to determine the cause of the problem.

Troubleshooting the systematic search for the source of a problem so that it can be solved.

In many facilities, the maintenance department is responsible for repairing most problems associated with equipment. However, it is helpful when process technicians are able to tell maintenance personnel what they observed prior to the equipment stoppage. For example, if a compressor shuts down, a technician might have heard a rattling sound prior to shutdown. Observations and feedback about equipment behavior can help the maintenance technician identify the cause of the problem and return the system to normal operation more quickly.

Process technicians must have a basic understanding of quality and how it can affect a company's reputation and production. Quality of a product or service has two major elements: (1) being free of deficiencies; and (2) satisfying stated or implied needs. Without quality measures, products and services could be deficient or unsatisfactory. Unsatisfactory products lead to loss of customers, increased waste, inefficiencies, increased costs, reduced profits, and inability to maintain a competitive edge.

Process technicians are expected to work effectively in a team-based environment. People are selected for teams because they have skills that complement those of other team members. Everyone on a team shares a common purpose (successful accomplishment of a goal). All team members hold each other mutually accountable for their success.

When working as part of a team, process technicians must understand diversity and practice its principles. Process technicians must recognize and appreciate others for their contributions and perspectives, rather than disregard people because of their differences.

The life of a process technician must be flexible because in many cases, it involves a rotating shift work schedule. This career provides a variety of experiences for an individual looking for a challenging occupation. Employees in the process industries generally are rewarded for job excellence through salary increases, promotions, and bonuses. Job benefits usually include health and dental insurance, profit sharing, and retirement plans.

1.3 Equipment

As part of their daily tasks, process technicians routinely work with many different types of equipment on a process unit. It is critical for a process technician to have a clear understanding of the various types of equipment, their components, and how they work. Types of equipment that are common in process industries include:

- Movers (pumps, compressors, conveyor belts, elevators, etc.)
- Power sources (electric, steam, fuels, hydraulic, pneumatic, etc.)
- Processors (distillation towers, reactors, filters, strippers, etc.)
- Heating/cooling units (furnaces, boilers, cooling towers, heat exchangers, etc.)

In addition to knowing the types of equipment used in the various industries, process technicians also must know how to read process diagrams and have a basic understanding of equipment standards. The chapters in this textbook provide a detailed understanding of various types of equipment and how they are used in the process industries. The following is a brief explanation of the various pieces of equipment described in this textbook.

Process Drawings and Equipment Standards

Process drawings refer to the various types of drawings and diagrams such as plot plans, block flow diagrams (BFDs), process flow diagrams (PFDs), and piping and instrumentation diagrams (P&IDs) that are used by process technicians. These drawings provide a visual description of the equipment or process and help a process technician understand the process operation and the relationship of various pieces of equipment. Process drawings also help in work planning and troubleshooting activities. Most drawings include a legend, title block, and application block. On each drawing are several symbols that represent the various components of a process. Each of these symbols is standardized to allow people

Process drawing illustration that provides a visual description and explanation of the processes, equipment, flows, and other important items in a facility.

from different facilities to read the process diagrams. Standardization reduces confusion and mistakes.

Some institutions that are responsible for creating these standardized symbols are:

- International Society of Automation (ISA); see Figure 1.2
- American National Standards Institute (ANSI)
- American Petroleum Institute (API)
- American Society of Mechanical Engineers (ASME)
- National Electric Code (NEC).

Figure 1.2 ISA logo.
CREDIT: Courtesy of ISA.

ISA
67 Alexander Drive
Research Triangle Park, NC 27709
www.isa.org

Pipes, Tubing, Hoses, and Fittings

Pipes, tubing, hoses, and fittings are critical to the process industries and account for a significant portion of the initial investment when creating a new facility. **Pipes** are long, hollow cylinders through which fluids or solids are transmitted. They are primarily made of metal, but they also can be made of glass, plastic, or plastic-lined material.

Tubing is small-diameter hose or pipe that can be used to transport fluids or solids. Tubing is made from a variety of materials and is used in many applications, including sample systems and instrumentation.

Hoses are larger than tubing and less permanent than piping. **Hoses** are flexible tubes that carry fluids; they can be made of plastic, rubber, fiber, metal, or a combination of materials. They are used to make temporary connections, as in the case where facility air is used to drive equipment or when a rail car or tank truck is hooked into the piping system.

Fittings are system components used to connect two or more pieces of pipe, tubing, or other equipment (Figure 1.3). Fittings are selected based on the demands of the process. Fittings should be used only in the manner they were intended and for the service or application for which they were designed. Use of improper or modified fittings could result in significant safety or environmental incidents.

Pipe long, hollow cylinder through which fluids are transmitted; primarily made of metal, but also can be made of glass, plastic, or plastic-lined material.

Tubing hose or pipe of small diameter (typically less than 1 inch [2.5 cm]) used to transport fluids.

Hose flexible tube that carries fluids; can be made of plastic, rubber, fiber, metal, or a combination of materials.

Fitting system component used to connect two or more pieces of piping, tubing, or other equipment.

Figure 1.3 Sample fittings used in process industries. A. Flange. B. Plug C. 45-degree elbow. D. Cross. E. Union.

A. Flange   **B.** Plug   **C.** 45° Elbow   **D.** Cross   **E.** Union

Valves

Valves are piping system components used to control, throttle, or stop the flow of fluids through a pipe. Valve types include:

- Gate
- Ball

Valve piping system component used to control, throttle, or stop the flow of fluids through a pipe.

- Plug
- Butterfly
- Globe
- Diaphragm
- Relief and safety
- Multiport
- Check
- Control valve

Valves can be operated either manually or automatically. The most commonly used type of valve in the process industries is the gate valve (Figure 1.4). Main valve components include the hand wheel, packing, bonnet, valve body, valve disc, and valve seat.

Figure 1.4 Gate valve.
CREDIT: Courtesy of Design Assistance Corp.

Process Vessels

Tanks are vessels in which material such as feedstock, intermediate products, or finished products are stored. They can operate under conditions at or close to atmospheric pressure or be maintained at higher pressure conditions. The construction and repair of atmospheric tanks are governed by API and NFPA codes and standards. API, NFPA, and ASME codes and standards govern the operation of high-pressure vessels.

Towers, drums, filters, and reactors are other types of process vessels. These **vessels** are enclosed containers that can operate at pressures ranging from full vacuum to very high pressure. They are governed by API, NFPA, and ASME codes and standards.

Tanks and other process vessels allow companies to store large amounts of products more effectively and efficiently. They also can act as intermediate storage between processing steps, provide residence time for reactions or fractionation to occur, and provide settling and filtering for process fluids.

Companies use mobile tanks and vessels to store and transport products throughout the United States and to other countries.

Tank a vessel in which a feedstock or product (intermediate or finished) is stored; might be classified as atmospheric or pressurized, aboveground or underground, fixed or floating roof.

Vessel an enclosed process container such as a tank, drum, tower, filter, or reactor.

Pumps

Pumps are mechanical devices that provide the energy required to move liquids through a piping system. Most process operations would not be able to function without pumps. Pumps are used in many applications, including transferring material between process units, filling or emptying tanks, providing water to boilers, supplying fire control water, and lubricating equipment. There are two main pump categories: dynamic and positive displacement.

Dynamic pumps convert centrifugal force to dynamic pressure to move liquids. Dynamic pumps are classified as either centrifugal or axial. Figure 1.5 shows a centrifugal pump.

Pump a mechanical device that transfers energy to move materials through piping systems.

Positive displacement pumps use pistons, diaphragms, gears, vanes, lobes, or screws to deliver a constant volume of fluid with each rotation or stroke (see Chapter 6). Positive displacement pumps deliver the same amount of liquid regardless of the discharge pressure. Positive displacement pumps are classified as either rotary or reciprocating.

Compressors

Compressor   a mechanical device used to increase the pressure of a gas or vapor.

A **compressor** is a mechanical device used to increase the pressure of a gas or vapor. There are two types of compressors: dynamic and positive displacement. Dynamic compressors use a centrifugal or rotational force to move gases and are classified as either centrifugal or axial. Positive displacement compressors use screws, sliding vanes, lobes, gears, a diaphragm, or pistons to deliver a set volume of gas. Positive displacement compressors can be classified as either reciprocating or rotary.

Turbines

Turbine   a machine that is used to produce power and rotate shaft-driven equipment such as pumps, compressors, and generators.

Turbines are machines that are used in place of electric drivers to produce power and rotate shaft-driven equipment such as pumps, compressors, and generators. Turbines also can be used to provide auxiliary power back to the process units they service. Figure 1.6 shows a steam turbine.

Turbines convert the kinetic and potential energy of a motive fluid into the mechanical energy required to drive a piece of equipment. Common types of turbines are steam, gas, hydraulic, and wind.

Motors

Motor   a mechanical driver that converts electrical energy into useful mechanical work and provides power for rotating equipment.

Motors are mechanical drivers that convert electrical energy into useful mechanical work and provide power for rotating equipment. Power distribution is crucial to the function of process industries. Electricity is the most common source of energy for driving motors, and motors are typically the largest electrical loads in an industrial manufacturing facility. Thus, the efficiency of the motors and their operation directly affects the cost structure of the process plant.

Motors and equipment use two forms of electricity: alternating current (AC) and direct current (DC). AC uses a changing flow of electrons in a conductor, while DC flows in a single direction. A wide range of voltages is encountered in a typical plant (such as 120 V, 480 V, 4160 V, etc.). The connection and disconnection of motors introduce significant safety hazards and usually are performed by trained and qualified electrical technicians.

Engine a machine that converts chemical (fuel) energy into mechanical force.

Heat exchanger a device used to transfer heat from one substance to another without physical contact between the two.

Engines

Engines are machines that convert chemical (fuel) energy into mechanical force. Types of engines include internal combustion, such as a gasoline or diesel engine, and external combustion, like a steam engine.

Heat Exchangers

Heat exchangers are devices used to transfer heat from one substance to another, usually without the two substances physically contacting each other. Types of heat exchangers include double-pipe, shell and tube, air cooled, spiral, and plate and frame. The shell and tube design is the most common in the process industry. Flow through a heat exchanger can be characterized as laminar or turbulent. Flow paths can be defined as countercurrent, parallel, or cross flow. These terms will be described more fully later in this book. Figure 1.7 shows a shell and tube heat exchanger.

Did You Know?

Alternating current (AC) is the form of electricity that is delivered to businesses and residences.

Direct current (DC) is the form of electricity that is used in flashlights.

CREDIT: Africa Studios / Shutterstock

CREDIT: Akkalak / Shutterstock

Figure 1.7 Shell and tube heat exchanger.
CREDIT: Photo smile / Shutterstock.

Cooling Towers

A **cooling tower** is another type of heat exchanger, but in this case, the two fluids, water and air, do contact each other. Cooling towers are designed to lower the temperature of water using evaporative cooling. Cooling towers are classified based on how air flow moves through them and whether the air movement is natural, induced, or forced. Air flow inside a cooling tower can be either cross flow or counterflow. A small amount of heat is transferred directly by heating the air (through sensible heat). Most of the heat is transferred through evaporation. As the water evaporates (changes from liquid into a vapor), it absorbs a substantial amount of heat from the remaining water. This heat is called latent heat of vaporization and the process is similar to the way sweat cools your body on a hot summer day. A small portion of the warm water evaporates into the air, while the remaining water gives up heat energy and is thereby cooled.

Cooling tower a structure designed to lower the temperature of water using latent heat of evaporation.

Furnaces

Furnaces (also referred to as *process heaters*) are pieces of equipment that burn fuel to generate heat that can be transferred to process fluids flowing through tubes. Furnace designs vary according to the function, heating duty, type of fuel, and method of introducing *combustion*

Furnace a piece of equipment that burns fuel in order to generate heat that can be transferred to process fluids flowing through tubes; also referred to as a process heater.

air (air needed to burn a fuel). Furnace designs include box, vertical cylindrical, and cabin. Furnaces can be natural draft, forced draft, induced draft, or balanced draft type. Furnace components include firebox, main burners, pilot burners, radiant tubes, refractory lining, shock bank, convection tubes, soot blowers, stack, and damper.

Boilers

Boiler a device in which water is boiled and converted into steam under controlled conditions.

Boilers are devices in which water is converted into steam under controlled conditions. In process industries, steam is used to heat and cool process fluids, to fight fires, to strip light ends, to purge equipment, and to promote reactions. Two main types of boilers are water tube and fire tube. The main components of boilers are the firebox, burners, downcomer tubes, riser tubes, mud drum, steam drum, economizer, superheater, steam distribution system, fuel system, and boiler feedwater system.

Auxiliary Equipment

Auxiliary equipment includes items such as mixers, agitators, eductors, centrifuges, and hydrocyclones used to support process operations. Although they are not the main pieces of equipment in process operations, they are vital to the processes in which they are used.

Tools

The selection, knowledge, and use of hand and power tools are essential to the safety and well-being of the process technician. Process technicians use basic hand tools to perform various work functions. It is crucial to select the proper tool to prevent injuries to the employee or damage to the equipment. For example, lifting equipment often is used to raise heavy items to elevated locations. Types of lifting equipment include hoists, cranes, forklifts, and personnel lifts. Improper usage of this equipment or using it outside of its design limitations can result in serious injury to personnel. Likewise, electric, pneumatic, hydraulic, and powder-actuated power tools should be used only after proper training and selection based on the specific requirements of the task.

Process technicians must know which tools can be used on each piece of equipment. If there is any doubt about which tool should be used, the process technician should review the standard operating procedures, consult with a supervisor, or consult with the safety department. As a general rule, a process technician should not operate any tools or equipment without proper training and authorization.

Separation Equipment

Separation is often required to convert raw materials into products for sale and distribution to consumers. Processes such as distillation systems are some of the most widely used in process industries to separate these raw materials. A few types of distillation are binary, azeotropic, multicomponent, and extractive. Separation also can be carried out using other processes, such as extraction, evaporation, crystallization, adsorption, absorption, and stripping.

Reactors

Reactor a vessel in which a controlled chemical reaction is initiated and takes place either continuously or as a batch operation.

Reactors are types of vessels in which a controlled chemical reaction is initiated. Within a reactor, raw materials are combined at various flow rates, pressures, and temperatures and react with each other to form a product. Reactor processes can be batch or continuous, which refers to how the reactants are added and products are removed. In batch processes, all material is added at one time and the product is removed after the reaction has completed. In continuous reaction processes, raw materials are fed into the reactor continuously, and the products are continuously being formed and removed as raw materials are fed into the reactor. Reactors are normally controlled to precise operating conditions for pressure, temperature, flow, and composition to optimize quality and output. Types of reactors include stirred tank, tubular, fixed bed, fluidized bed, hot wall, cold wall, and nuclear.

Filters and Dryers

Filters and dryers are used in the process industries to remove undesirable components from process streams. **Filters** remove particles from a process, allowing the clean product to pass through; **dryers** are typically used to remove moisture from a process stream. Types of filters include cartridge, bag, leaf, fixed bed, strainers, gravity, rotary drum, and plate and frame. Types of dryers include fixed bed, fluid bed, flash, rotary, and cyclone. Figure 1.8 shows a fluidized bed dryer.

Filter device that removes particles from a process, allowing the clean product to pass through.

Dryer a device used to remove moisture from a process stream.

Figure 1.8 Fluidized bed dryer.

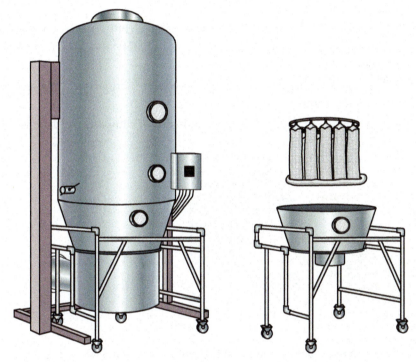

Solids Handling Equipment

Solids handling equipment is used to process and transfer solid materials from one location to another in a process facility. It also can be used to provide additives and colorants to materials such as plastics to enhance their properties. Types of solids handling equipment include flow inducers, conveyors, feeders, extruders, elevators (e.g., bucket elevators), cyclones, screening systems, bins, silos, hoppers, dryers, trickle valves, bulk bag stations, hopper cars, and blowers.

Solids handling equipment equipment that is used to process and transfer solid materials from one location to another in a process facility; it also might provide storage for those materials.

Environmental Control Equipment

Environmental control equipment is used to maintain the operation of process equipment and piping in order to meet environmental standards and government regulations. Types of air pollution control equipment include baghouses and precipitators, vapor recovery systems, scrubbers, incinerators, and flare systems. Types of water and soil pollution controls include the activated sludge process, clarifiers, dikes, separators, settling ponds, and landfills.

Mechanical Power Transmission and Lubrication

Mechanical power transmission transfers rotational energy from a driver to driven equipment with minimal loss of energy from friction. **Lubrication** is the application of a lubricant between moving surfaces to reduce friction and wear, and remove heat. Mechanical transmission includes the following components: couplings, belts, gearboxes, chains, magnetic or hydraulic drives, gears, and bearings.

Lubrication the application of a substance between moving surfaces in order to reduce friction and wear, and remove heat.

1.4 Safety and Environmental Hazards

When working with process equipment, process technicians must be aware of the factors that can have an impact on worker safety or the environment:

- Abnormal sounds coming from equipment
- Excessive vibration
- Leaks around equipment
- Faulty equipment gauges
- Use of incorrect tools
- Lack of proper personal protective equipment (PPE) such as hearing protection, safety glasses, goggles, hard hats, work gloves, safety shoes, or flame-retardant clothing (Figure 1.9)
- Not following standard operating procedures (normal operating conditions of pressure and temperatures)
- Not complying with government, industry, or company regulations.

Figure 1.9 Examples of PPE used in industry.

CREDIT: L Barnwell / Shutterstock.

Any of these factors could lead to significant personal injury and injury to others, as well as extensive damage to process equipment. Each of these factors needs to be taken seriously and addressed immediately. Procedures need to be followed explicitly, the specified PPE must be worn, and all safety rules must be followed.

Process technicians also are expected to:

- Comply with all environmental regulations
- Work smart and focus on safety and business goals
- Look for ways to safely reduce waste and improve efficiency
- Stay current with industry trends and continue to improve skills.

Summary

A process technician is a worker in a process facility who safely monitors and controls mechanical, physical, and/or chemical changes throughout a process to produce either a final product or an intermediate product made from raw materials. Protecting the environment while creating these products is an important goal.

Numerous industries are classified as process industries, including oil and gas exploration and production, petroleum refining, chemical manufacturing, mining, power generation, waste and water treatment, food and beverage production, pharmaceutical manufacturing, and paper and pulp processing.

Process technicians are responsible for planning, analyzing, and controlling the production and distribution of products. The duties of a process technician include maintaining a safe work environment, as well as controlling, monitoring, and troubleshooting problems with equipment.

Various types of equipment are used throughout the process industries, so it is important for process technicians to have a clear understanding of equipment, its components, and its operation. Common equipment used in the process industries includes valves, pumps, compressors, turbines, motors, engines, heat exchangers, cooling towers, furnaces, boilers, reactors, and filters. With each piece of equipment, process technicians must understand potential problems, environmental and safety concerns, typical procedures, and their role based on standard operating procedures and company requirements.

Safety is a major expectation of each employee in the workforce today. It is important for all safety rules to be followed and for all employees to have a proactive attitude regarding their own safety as well as the safety of fellow employees, the community, and the environment.

Checking Your Knowledge

1. Define the following terms:
   a. Process drawings
   b. Process industries
   c. Process technician
   d. Troubleshooting
   e. Unit

2. (True or False) A process technician controls mechanical, physical, and/or chemical changes throughout many processes to produce a final product made from raw materials.

3. (True or False) Process technicians analyze data and communicate data to the appropriate employees.

4. All of the following are types of process heat exchangers except:
   a. Shell-and-tube
   b. Double-pipe
   c. Hot wall
   d. Air-cooled

5. Process drawings include (select all that apply):
   a. Plot plans
   b. Block flow diagrams
   c. Process flow diagrams
   d. System diagrams

6. In an operations role, a process technician might be required to perform the following maintenance activities (select all that apply):
   a. Lubricate equipment
   b. Replace a unit compressor
   c. Monitor and analyze equipment performance
   d. Change or clean filters

7. List five items of personal protective equipment that a process technician wears.

8. List four factors that process technicians must be aware of that can have an impact on worker safety or the environment.

9. A mechanical piece of equipment used to move liquid material through a piping system within various processes is a:
   a. Heat exchanger
   b. Compressor
   c. Burner
   d. Pump

10. Drivers that convert electrical energy into useful mechanical work and provide power for rotating equipment such as pumps, compressors, and conveyor drives are:
    a. Reactors
    b. Solids handling equipment
    c. Motors
    d. Furnaces

11. Equipment used to move nonfluid material within a process facility includes:
    a. Environmental control equipment
    b. Solids handling equipment
    c. Separation equipment
    d. Auxiliary equipment

12. How are raw materials sorted?
    a. By plant unit
    b. By their end products
    c. By process requirements
    d. By physical requirements

NOTE: Answers to Checking Your Knowledge questions are in the Appendix.

Student Activities

1. Research the process industries in your city or region that might hire you after you graduate from this program. Place a map of the area in the front of the classroom. Share names, locations, and information about the various types of process industries with your classmates.

2. Work with a classmate to draw a flow sequence of a simple process and describe the basic steps associated with the process.

Chapter 2
Process Drawings and Industry Standards

Objectives

After completing this chapter, you will be able to:

2.1 Explain the purpose of diagrams, including why, when, and where they are used. (NAPTA Diagrams 1*) p. 16

2.2 Identify the major unit sections in block flow diagrams. (NAPTA Diagrams 2) p. 17

2.3 Identify components on a typical process flow diagram (PFD). (NAPTA Diagrams 4) p. 18

2.4 Identify components on a typical piping and instrumentation diagram (P&ID). (NAPTA Diagrams 5) p. 20

2.5 Identify a plot plan and explain the purpose of equipment layout drawings (plot plans). (NAPTA Diagrams 1) p. 23

2.6 Identify symbols and common elements on drawings used for process equipment and instrumentation. (NAPTA Diagrams 3) p. 24

2.7 Explain the purpose of industry standards. (NAPTA Diagrams 3) p. 35

*North American Process Technology Alliance (NAPTA) developed curriculum to ensure that Process Technology courses will produce knowledgeable graduates to become entry level employees in process technology. Objectives from that curriculum are named here in abbreviated form. For example, "(NAPTA Diagrams 1)" means that this chapter's objective relates to objective 1 of NAPTA's course content on diagrams.

Key Terms

ANSI—American National Standards Institute, an organization that oversees and coordinates the voluntary standards in the United States, **p. 35.**

API—American Petroleum Institute, a trade association that represents the oil and gas industry in the areas of advocacy, research, standards, certification, and education, **p. 36.**

Application block—the main part of a drawing that contains symbols and defines elements such as relative position, types of materials, equipment descriptions, flows, and functions, **p. 27.**

ASME—American Society of Mechanical Engineers, an organization that specifies requirements and standards for pressure vessels, piping, and their fabrication, **p. 36.**

Block flow diagram (BFD)—a simple illustration that shows a general overview of a process, indicating its parts and their interrelationships, **p. 17.**

Electrical diagram—illustration showing power transmission and how it relates to the process, **p. 21.**

Electrical schematic—a drawing that shows the direction of electrical current flow in a circuit, typically beginning at the power source, **p. 22.**

ISA—The International Society of Automation (originally known as the Instrument Society of America), a global, nonprofit technical society that develops standards for automation, instrumentation, control, and measurement, **p. 35.**

Isometric drawing—an illustration showing objects as they would appear to the viewer (similar to a three-dimensional drawing that appears to come off the page), **p. 23.**

Legend—the section of a drawing that explains or defines the information or symbols contained within the drawing, **p. 24.**

NEC—National Electric Code; standard established by the National Fire Protection Agency (NFPA), which promotes the safe installation of electrical wiring and equipment, **p. 36.**

OSHA—Occupational Safety and Health Administration, a U.S. government agency created to establish and enforce workplace safety and health standards, conduct workplace inspections, provide worker training and education, and investigate serious workplace incidents, **p. 36.**

Piping and instrumentation diagram (P&ID)—detailed illustration that graphically represents the relationship of equipment, piping, instrumentation, and flows contained within a process in the facility, **p. 20.**

Plot plan—illustration drawn to scale, showing the layout and dimensions of equipment, units, and buildings; also called an *equipment location drawing*, **p. 23.**

Process flow diagram (PFD)—basic illustration that uses symbols and direction arrows to show the primary flow path of material through a process, **p. 18.**

Process schematic—screen display of a process in the computer control system; it shows control parameters but not piping and equipment details, **p. 24.**

Symbol—simple illustration used to represent a piece of equipment, an instrument, or other device on a PFD or P&ID, **p. 27.**

Title block—the section of a drawing (typically located in the bottom-right corner) that contains the drawing title, drawing number, revision number, sheet number, company information, process unit, and approval signatures, **p. 24.**

Utility flow diagram (UFD)—illustration that provides process technicians a PFD-type view of the utilities used for a process, **p. 21.**

2.1 Introduction

Diagrams or process drawings are used to provide process technicians with a visual description and explanation of the processes, equipment, and other important items in a facility. Process drawings for a process technician are like topographical maps for hikers in the deep woods. Process diagrams show process technicians what they will encounter in the process unit and how to navigate around the process flow of a facility.

There are many different types of drawings, each of which represents different aspects of the process and different level of detail. Looking at combinations of these drawings provides a more complete picture of the processes and the facility. Without process drawings, it would be difficult for process technicians to understand a process and how it operates.

When examining process drawings, it is important to remember that all drawings share three common functions: (1) *simplifying* (using common symbols to make processes easy to understand), (2) *explaining* (describing how all of the parts or components of a system work together), and (3) *standardizing* (using a common set of lines and symbols to represent components).

Diagrams are also used extensively by process technicians for learning a process, for troubleshooting, at startup and shutdown, and after initial commissioning.

To be considered proper industrial drawings, process drawings must meet several requirements. These include specific, universal rules about how lines are drawn, how proportions are used, what measurements are used, and what components are included.

Common Process Drawings and Their Uses

Process technicians must recognize a wide variety of drawings and understand how to use each one. The most commonly encountered drawings are:

- Block flow diagrams (BFDs)
- Process flow diagrams (PFDs)
- Piping and instrumentation diagrams (P&IDs)
- Utility flow diagrams (UFDs)
- Electrical diagrams
- Schematics
- Isometrics.

2.2 Block Flow Diagrams (BFDs)

Block flow diagrams (BFDs) are simple drawings that show a general overview of a process and contain few specifics. BFDs use blocks to represent sections of a process and use flow arrows to show the order and relationship of components.

The BFD in Figure 2.1 represents a boiler feedwater treatment process. Numbers are provided on this diagram to indicate the order of the steps in the process, but numbers are not typically found on a BFD. The list below describes what occurs at each step.

Block flow diagram (BFD) a simple illustration that shows a general overview of a process, indicating its parts and their interrelationships.

1. **Raw water**—Raw water is imported from whichever source is available at the facility (e.g., river, lake, etc.). Raw water is not clean enough to be used as boiler feed water, so it is sent to step two to be clarified.

2. **Clarification**—In the clarifier, suspended solids are removed from the raw water.

3. **Filtration**—In this phase, clarified water is passed through a set of filters to remove any remaining solid matter that could accumulate in the boiler tubes and cause plugging or possible tube failure.

4. **Demineralization**—In this phase, most of the mineral content is removed from the water.

5. **Deaeration**—This section of the boiler feedwater process removes excess air or gas that might be contained in the water. The process is accomplished by heating (flashing) the water above steam temperature and removing any entrained gases or free air.

Figure 2.1 Block flow diagram (BFD).

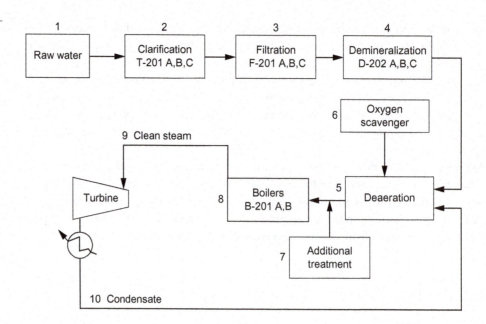

6. **Oxygen scavenging**—Because water that goes into the boiler cannot contain excess oxygen (O_2), an oxygen scavenger is used to remove oxygen from the water. This prevents problems such as bacterial growth and rust.

7. **Additional treatment**—If any additional water treatment is required, it will occur at this point.

8. **Boilers**—Water is sent to the boiler so it can be converted into steam. The amount of steam produced depends on the size and makeup of the particular boiler.

9. **Steam**—Steam feeds process units throughout the facility. The output from the units is collected and removed as condensate.

10. **Condensate**—Condensate is collected and reused. The more condensate that can be recycled, the less raw water has to be introduced and treated. This translates to an economic savings for the company.

Process technicians are exposed to different types of industrial drawings on the job. The two most common types of drawings are process flow diagrams (PFDs) and piping and instrument diagrams (P&IDs).

2.3 Process Flow Diagrams (PFDs)

Process flow diagram (PFD) basic illustration that uses symbols and direction arrows to show the primary flow path of material through a process.

Process flow diagrams (PFDs) are basic drawings that use symbols and direction arrows to show the primary flow path of material through a process. They include information such as operating conditions, location of main instruments, and major pieces of equipment. PFDs allow process technicians to trace the step-by-step flow of a process. PFDs use symbols to represent the major pieces of equipment and piping used in the process and directional arrows to show the path of the process. Figure 2.2 shows an example of a PFD.

The process flow is typically drawn from left to right, starting with feed or raw materials on the left, and ending with finished products on the right. Other information found on a PFD includes process variables, pump capacities, heat exchangers, equipment symbols, equipment designations, major process piping, and control valves.

Symbology charts are used along with PFDs and P&IDs to identify the major pieces of equipment, piping, temperatures, pressures at critical points, and the flow of the process. The use of symbology allows for the standardization of information on industrial drawings. Each industrial drawing has similar lines and symbols that represent the various components. These lines and symbols (with subtle changes) are used all over the world.

Figure 2.2 Sample process flow diagram (PFD).

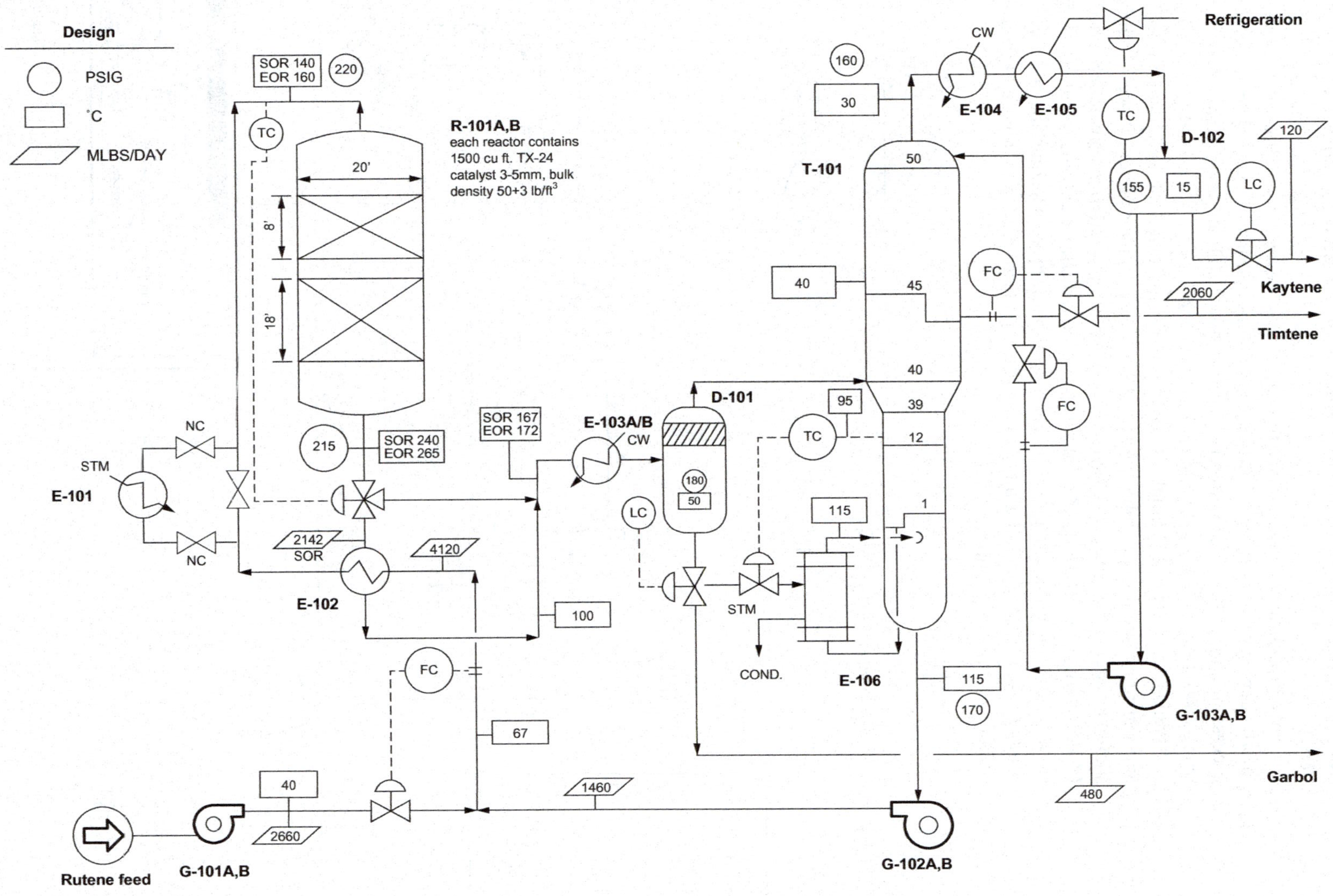

2.4 Piping and Instrumentation Diagrams (P&IDs)

Piping and instrumentation diagram (P&ID) detailed illustration that graphically represents the relationship of equipment, piping, instrumentation, and flows contained within a process in the facility.

Piping and instrumentation diagrams (P&IDs), sometimes referred to as Process and Instrument Drawings, are similar to PFDs. However, P&IDs show more detailed process information, such as equipment, piping, flow arrows, materials of construction, and insulation. Additional information on a P&ID includes equipment numbers, piping specifications, and complete and detailed instrumentation. Figure 2.3 shows an example of a P&ID.

For a process technician, a vital part of a P&ID is the instrumentation information. This information gives the technician a firm understanding about how material flows through the process and how it can be monitored and controlled. P&IDs are also critical during maintenance tasks, troubleshooting, modifications, and upgrades.

P&IDs use standard symbols of the International Society of Automation (ISA). (ISA and instrumentation tag numbers are discussed later in this chapter.) Process technicians should be able to recognize these symbols and any special lettering conventions used on a P&ID. Many symbols are standard throughout the industry, but some might be specific to the individual manufacturing or engineering company. Process technicians must recognize symbols used in their facilities and also be able to interpret process flows and instrument and equipment designations.

Figure 2.3 Sample piping and instrument diagram (P&ID).

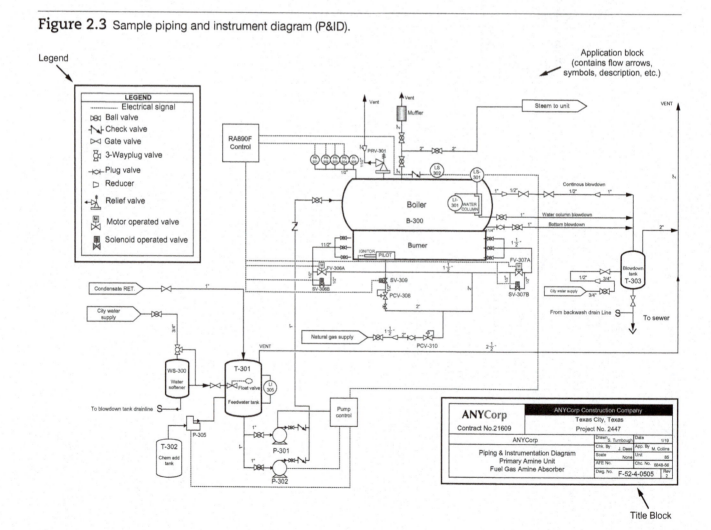

Utility Flow Diagrams (UFDs)

Utility flow diagrams (UFDs) provide process technicians a PFD-type view of the utilities used for a process. UFDs represent the way utilities connect to the process equipment, along with the piping and main instrumentation used to operate those utilities.

Typical utilities shown on a UFD include:

- Steam
- Condensate
- Cooling water
- Instrument air
- Plant air
- Nitrogen
- Fuel gas.

Figure 2.4 is an example of a utility flow diagram (UFD).

Utility flow diagram (UFD) illustration that provides process technicians a PFD-type view of the utilities used for a process.

Figure 2.4 Sample utility flow diagram (UFD).

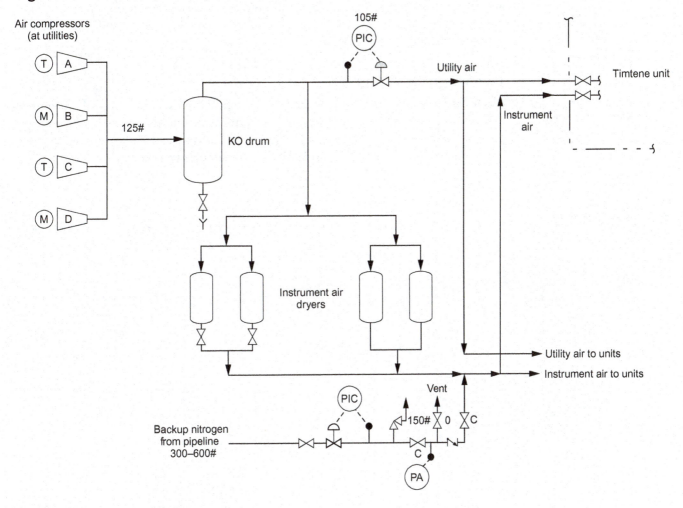

Electrical Diagrams

Since most processes rely on electricity, it is important for process technicians to understand electrical systems and how they work. **Electrical diagrams** help process technicians understand power transmission and how it relates to the process. A firm understanding of these

Electrical diagram illustration showing power transmission and how it relates to the process.

relationships is critical when performing lockout/tagout procedures (i.e., control of hazardous energy) and when monitoring various electrical measurements.

Electrical diagrams show the various electrical components and their relationships, for example:

- Switches used to stop, start, or change the flow of electricity in a circuit
- Power sources provided by transmission lines, generators, or batteries
- Loads (the components that actually use the power)
- Coils or wire used to increase the voltage of a current
- Inductors (coils of wire that generate a magnetic field and are used to create a brief current in the opposite direction of the original current) that can be used for surge protection
- Transformers (devices that take electricity of one voltage and change it into another voltage)
- Resistors (coils of wire used to provide resistance in a circuit)
- Contacts used to join two or more electrical components.

There are different types of electrical diagrams, but the most common are wiring diagrams and schematics. Wiring diagrams are used by electricians to understand the physical connections between components in an electrical circuit. They are especially useful when installing new equipment or troubleshooting.

Electrical Schematics

Electrical schematic a drawing that shows the direction of electrical current flow in a circuit, typically beginning at the power source.

Electrical schematics show the direction of current flow in a circuit, typically beginning at the power source. Process technicians use electrical schematics to visualize how current flows between two or more circuits. Electrical schematics also help electricians detect potential trouble spots in a circuit. Figure 2.5 shows an example of an electrical schematic.

Figure 2.5 Electrical diagram (schematic).

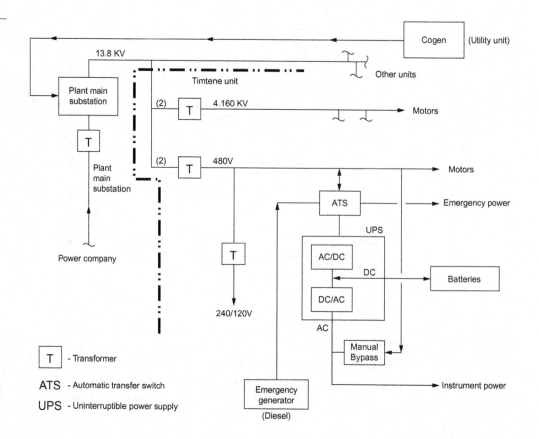

Isometric Drawings

Isometric drawings are drawings that show objects (piping, equipment, etc.) as they would appear in a three-dimensional drawing. Isometric drawings also might contain cutaway views to show the inner workings of an object. Figure 2.6 shows an example of an isometric drawing.

Isometric drawings show the three sides of the object that can be seen, with the object appearing at a 30-degree angle with respect to the viewer. Isometrics are typically used during new unit construction or unit revisions. These drawings may be useful to new process technicians as they learn to identify equipment and understand its inner workings.

Isometric drawing an illustration showing objects as they would appear to the viewer (similar to a three-dimensional drawing that appears to come off the page).

Figure 2.6 Isometric drawing.

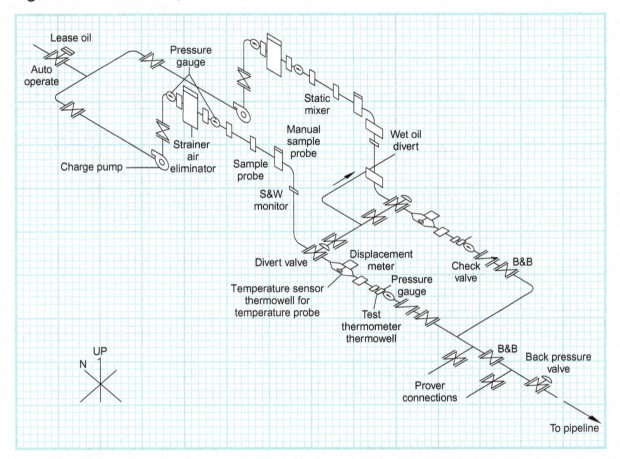

2.5 Plot Plan

Plot plans (also called *equipment location drawings*) show the layout and dimensions of equipment, units, and buildings. They are drawn to scale so that everything is of the correct relative size. For example, plot plans show the location of machinery (e.g., pumps and heat exchangers) in an equipment room. On a larger scale, a plot plan shows the location and dimensions of process units, buildings, roads, and other site constructions such as fences. A site plot plan also shows elevations and grades of the ground surface. Figure 2.7 shows an example of a plot plan.

Plot plan illustration drawn to scale, showing the layout and dimensions of equipment, units, and buildings; also called an *equipment location drawing*.

Other Drawings

Along with the drawings mentioned in the previous sections, process technicians also might encounter other types of drawings, such as elevation diagrams, process schematics, loop diagrams, and logic diagrams.

Figure 2.7 Plot plan.

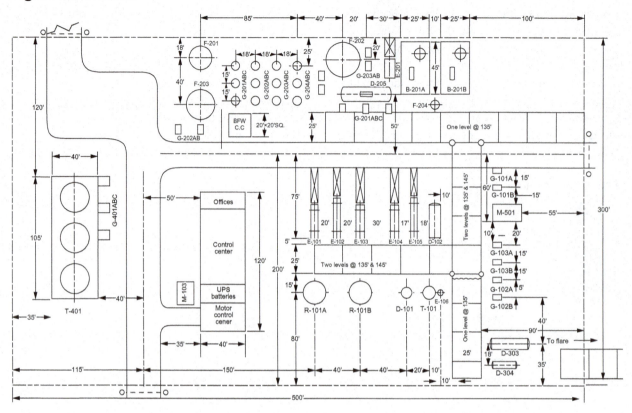

An elevation diagram shows the relationship of equipment to ground level and other structures.

Process schematic screen display of a process in the computer control system; it shows control parameters but not piping and equipment details.

A **process schematic** is the screen display used by control room operators to control unit operations.

A loop diagram shows all components and connections between instrumentation and the control room.

A logic diagram shows the sequential steps within the computer or safety system. Figure 2.8 describes logic diagrams and actions that occur.

2.6 Common Information Contained on Process Drawings

Legend

Legend the section of a drawing that explains or defines the information or symbols contained within the drawing.

A **legend** is the section of a drawing that explains or defines the information or symbols contained within the drawing (similar to a legend on a map). Legends include information such as abbreviations, numbers, symbols, and tolerances. Figure 2.9 shows an example of a legend.

Title Block

Title block the section of a drawing (typically located in the bottom-right corner) that contains the drawing title, drawing number, revision number, sheet number, company information, process unit, and approval signatures.

The **title block** is the section of a drawing (typically located in the bottom-right corner) that contains the drawing name, drawing number, revision number, sheet number, company or process unit, and approval signatures and dates. Figure 2.10 shows an example of a title block.

Figure 2.8 Logic diagrams.

Example A

GENERAL NOTES:
A. ANY INITIATOR WILL ACTIVATE THE INTERLOCK UNLESS OTHERWISE NOTED.
B. ALL ACTIONS (EFFECTS) WILL ACTIVATE ONCE THE INTERLOCK IS ACTIVATED UNLESS OTHERWISE NOTED.
C.
D.
E.

IPS CLASSIFICATION / INTERLOCK TYPE: SIL 2

EXAMPLE

NOTES:
1. THIS IS A FAIL OPEN VALVE
2. VALVE CLOSES ONLY ON A HIGH HIGH R-21 BASE LEVEL.
3.
4.
5.

INITIATORS

	IPS CLASS	VOTING	NOTES	PRIMARY INSTRUMENT TAG	INITIATOR TAG NUMBER	INITIATOR SERVICE DESCRIPTION	INTERLOCK LOGIC LOCATION	P&ID DRAWING	ELEMENTARY DRAWING	INITIATOR NORMAL LEVEL	INITIATOR ACTION LEVEL
1	SIL 2	1oo1	2	LT-722-621	LAHH-722-621	R-21 COLUMN BASE LEVEL	PLC USY-722-24	99-9T-722	99_14-3E-1241	< 50 PERCENT	> 90 PERCENT
2	SIL 2	1oo1		LT-722-621	LALL-722-621	R-21 COLUMN BASE LEVEL	PLC USY-722-24	99-9T-722	99_14-3E-1241	< 50 PERCENT	< 10 PERCENT
3	SIL 2	2oo2		PT-722-622A	PAHH-722-622A	R-21 COLUMN BASE PRESSURE	PLC USY-722-24	99-9T-722	99_14-3E-1242	< 300 PSIG	> 350 PSIG
4	SIL 2	2oo2		PT-722-622B	PAHH-722-622B	R-21 COLUMN BASE PRESSURE	PLC USY-722-24	99-9T-722	99_14-3E-1243	< 300 PSIG	>350 PSIG
5											
20											

REFERENCE DRAWING DESC	REFERENCE DRAWING #	REVISION DESC	REVISION #	LINES	DRAFTER	ENGR APPROVAL	DATE	INITIATOR ACTION LEVEL
P & ID, R-21	99-9T-722							
ELEMENTARY DWG, R-21 LEVEL	99_14-3E-1241	DESIGN AND BUILD WITH CARE						

ACTIONS (EFFECTS)

IPS CLASS	ACTION/ EFFECT	ACTION ITEM TAG NUMBER	NOTES	ACTION (EFFECT) DESCRIPTION and HOW THE ACTION IS ACCOMPLISHED	P&ID DRAWING	ELEMENTARY DRAWING
SIL 2	CLOSE	FV-722-19	1	STOP PROPANOL FLOW TO R-21 -by DENERGIZING SOLENOID FV-722-19	99-9T-722	99_12-3E-1429
SIL 2	CLOSE	PV-722-12	2	STOP AIR FLOW TO R-21 -by DENERGIZING SOLENOID PV-722-12	99-9T-722	99_12-3E-1430
SIL 2	CLOSE	TV-722-23		STOP STEAM TO R-21 BASE HEATER E-21 -by DENERGIZING SOLENOID TV-722-	99-9T-722	99_12-3E-1431

APPROVALS		OPER APPROVAL	DATE
PROJECT MGR	MRT		12/6/16
DEPT. HEAD	KGD		12/6/16
DIVISION HEAD			
WORKS MGR			

LONGVIEW OPERATIONS
XYZ Corporation

INTERLOCK NARRATIVE DIAGRAM
R-21 REACTOR INTERLOCKS
I-0001

DRAWN:	SCL	12/1/2016	DRAWING NO
JOB ENGINEER:	JWA	12/2/2016	99-9L-033

SIS INTERLOCK (SINGLE ONLY!)

A.

Example B

GENERAL NOTES:
A. ANY INITIATOR WILL ACTIVATE THE INTERLOCK UNLESS OTHERWISE NOTED.
B. ALL ACTIONS (EFFECTS) WILL ACTIVATE ONCE THE INTERLOCK IS ACTIVATED UNLESS OTHERWISE NOTED.
C.
D.
E.

IPS CLASSIFICATION / INTERLOCK TYPE: SAF

EXAMPLE

NOTES:
1. THIS IS A FAIL OPEN VALVE
2. VALVE CLOSES ONLY ON A HIGH HIGH R-12 BASE LEVEL3.
3.
4.
5.

INITIATORS

LINKS	VOTING	NOTES	PRIMARY INSTRUMENT TAG NUMBER	INITIATOR TAG NUMBER	INITIATOR SERVICE DESCRIPTION	INTERLOCK LOGIC LOCATION	P&ID DRAWING	ELEMENTARY DRAWING	INITIATOR NORMAL LEVEL	INITIATOR ACTION LEVEL
1	1oo1	2	LT-722-621	LAH-722-621	R-21 COLUMN BASE LEVEL	DCS UUC-722-30	99-9T-722	99_14-3E-1241	< 50 PERCENT	> 85 PERCENT
2 OR	1oo1		LT-722-621	LAL-722-621	R-21 COLUMN BASE LEVEL	DCS UUC-722-30	99-9T-722	99_14-3E-1241	< 50 PERCENT	< 1.5 PERCENT
3										
20										

REFERENCE DRAWING DESC	REFERENCE DRAWING #	REVISION DESC	REVISION #	LINES	DRAFTER	ENGR APPROVAL	OPER APPROVAL
P & ID, R-21	99-9T-722						
ELEMENTARY DWG, R-21 LEVEL	99_14-3E-1241	DESIGN AND BUILD WITH CARE					

ACTIONS (EFFECTS)

ACTION / EFFECT	ACTION ITEM TAG NUMBER	NOTES	ACTION (EFFECT) DESCRIPTION and HOW THE ACTION IS ACCOMPLISHED	P&ID DRAWING	ELEMENTARY DRAWING
RESET	FV-722-19	1	DECREASE PROPANOL FLOW TO R-21 -by DECREASING FIC-722-19 SETPOINT BY 10 GPM	99-9T-722	99_12-3E-1429
RESET	PV-722-12	2	INCREASE AIR FLOW TO R-21 -by INCREASING PIC-722-12 SETPOINT BY 5 PSIG	99-9T-722	99_12-3E-1430

APPROVALS		DATE
PROJECT MGR	MRT	12/6/16
DEPT. HEAD	KGD	12/6/16
DIVISION HEAD		
DIVISION SUPT		
WORKS MGR		

LONGVIEW OPERATIONS
XYZ Corporation

INTERLOCK NARRATIVE DIAGRAM
R-21 REACTOR INTERLOCK
I-2002

DRAWN:	SCL	12/1/2016	DRAWING NO
JOB ENGINEER:	JWA	12/2/2016	99-9L-034

NON SIS INTERLOCK DWG SINGLE

B.

Figure 2.8 Logic diagrams. (Continued)

EXAMPLE

GENERAL NOTES:
A. ANY INITIATOR WILL ACTIVATE THE INTERLOCK UNLESS OTHERWISE NOTED.
B. All ACTIONS (EFFECTS) WILL ACTIVATE ONCE THE INTERLOCK IS ACTIVATED UNLESS OTHERWISE NOTED.
C. EMERGENCY SHUT DOWN (ESD) INTERLOCKS ARE USED TO MANUALLY INITIATE A PROCESS SHUTDOWN DUE TO AN EXTERNAL EMERGENCY (LOSS OF STEAM OR LOSS OF FEEDSTOCK).
D.
E.

IPS CLASSIFICATION / INTERLOCK TYPE: SEE BELOW (SAF = SAFETY, PRC = PROCESS CONTROL, EPP = EQUIPMENT/PROPERTY PROTECTION, ESD = EMERGENCY SHUT DOWN)

NOTES:
1. THIS IS A FAIL OPEN VALVE
2. VALVE CLOSES ONLY ON A HIGH HIGH R-12 BASE LEVEL.
3. FLOW MUST BE LOW FOR >15 SECONDS TO TRIGGER INITIATOR
4. CONTROLLER REVERTS TO -5% OUTPUT AFTER 15 MINUTES
5.

INITIATORS

IPS CLASS / INTERLOCK TYPE	INTERLOCK NUMBER	INTERLOCK NAME	LINKS	VOTING	NOTES	PRIMARY INITIATOR TAG NUMBER	INITIATOR TAG NUMBER	INITIATOR SERVICE DESCRIPTION	INTERLOCK LOGIC LOCATION	P&ID DRAWING	ELEMENTARY DRAWING	INITIATOR NORMAL LEVEL	INITIATOR ACTION LEVEL
SAF	TX99 I-2002	R-21 LEVEL SHUT DOWN		1oo1	2	LT-722-621	LAH-722-621	R-21 COLUMN BASE LEVEL	DCS UUC-722-30	99-9T-722	99_14-3E-1241	<50 PERCENT	>85 PERCENT
			OR	1oo1		LT-722-621	LAL-722-621	R-21 COLUMN BASE LEVEL	DCS UUC-722-30	99-9T-722	99 14-3E-1241	<50 PERCENT	<15 PERCENT
PRC	TX99 I-2003	R-21 PRESSURE SHUT DOWN		2oo2		PT-722-622A	PAH-722-622A	R-21 COLUMN BASE PRESSURE	DCS UUC-722-30	99-9T-722	99 14-3E-1242	<300 PSIG	>325 PSIG
			AND	2oo2		PT-722-622B	PAH-722-622B	R-21 COLUMN BASE PRESSURE	DCS UUC-722-30	99-9T-722	99 14-3E-1243	<300 PSIG	>325 PSIG
			OR	2oo2		PT-722-622A	PAL-722-622A	R-21 COLUMN BASE PRESSURE	DCS UUC-722-30	99-9T-722	99 14-3E-1242	<250 PSIG	<200 PSIG
			AND	2oo2		PT-722-622B	PAL-722-622B	R-21 COLUMN BASE PRESSURE	DCS UUC-722-30	99-9T-722	99 14-3E-1243	<250 PSIG	<200 PSIG
			OR	1oo1		PT-722-623	PAH-722-623	R-21 COLUMN TOP PRESSURE	DCS UUC-722-30	99-9T-722	99 14-3E-1244	<275 PSIG	>200 PSIG
EPP	TX99 I-2004	R-21 BASE HEATER E-21 SHUT DOWN		1oo1	3	FT-722-624	FAL-722-624	R-21 BASE HEATER E-21 FEED FLOW	DCS UUC-722-30	99-9T-722	99_14-3E-1245	>50 GPM	<25 GPM
			OR	1oo1		TT-722-23	TAH-722-23	R-21 COLUMN BASE TEMPERATURE	DCS UUC-722-30	99-9T-722	99 14-3E-1246	<325 DEG F	>350 DEG F
ESD	TX99 I-2005	R-21 EMERGENCY SHUT DOWN		1oo1		HS-722-621A	HS-722-621A	R-21 EMERGENCY SHUT DOWN	DCS UUC-722-30	99-9T-722	99 14-3E-1247	NOT ACTIVATED	ACTIVATED
						HS-722-021B		R-21 COLUMN EMERGENCY SHUT DOWN	PANELBOARD PUSH BUTTON	99-9T-722	99 14-3E-1248	NOT ACTIVATED	ACTIVATED

ACTIONS (EFFECTS)

NOTE	ACTION/EFFECT	ACTION ITEM TAG NUMBER	ACTION (EFFECT) DESCRIPTION and HOW THE ACTION IS ACCOMPLISHED	P&ID DRAWING	ELEMENTARY DRAWING
1	RESET	FV-722-19	DECREASE PROPANOL FLOW TO R-21 -by DECREASING FIC-722-19 SETPOINT BY 10 GPM	99-9T-722	99_12-3E-1429
2	RESET	PV-722-12	INCREASE AIR FLOW TO R-21 -by INCREASING PIC-722-12 SETPOINT BY 5 PSIG	99-9T-722	99 12-3E-1430
7	CLOSE	FV-722-19	STOP PROPANOL FLOW TO R-21 -by DENERGIZING SOLENOID FY-722-19	99-9T-722	99 12-3E-1429
8	CLOSE	PV-722-12	STOP AIR FLOW TO R-21 -by DENERGIZING SOLENOID PY-722-12	99-9T-722	99 12-3E-1430
9	CLOSE	TV-722-23	STOP STEAM TO R-21 BASE HEATER E-21 -by DENERGIZING SOLENOID TY-722-23	99-9T-722	99 12-3E-1431
13	CLOSE	TV-722-23	STOP STEAM TO R-21 BASE HEATER E-21 -by DENERGIZING SOLENOID TY-722-23	99-9T-722	99 12-3E-1431
18	CLOSE	FV-722-19	STOP PROPANOL FLOW TO R-21 -by DENERGIZING SOLENOID FY-722-19	99-9T-722	99_12-3E-1429
19	CLOSE	PV-722-12	STOP AIR FLOW TO R-21 -by DENERGIZING SOLENOID PY-722-192	99-9T-722	99 12-3E-1430
20	CLOSE	TV-722-23	STOP STEAM TO R-21 BASE HEATER E-21 -by DENERGIZING SOLENOID TY-722-23	99-9T-722	99 12-3E-1431
21	CLOSE	FV-722-24	STOP PROPIONALDEHYDE FLOW FROM R-21 -by DENERGIZING SOLENOID FY-722-24	99-9T-722	99 12-3E-1432
22	1,4 OPEN	TV-722-25	FULL COOLING TO FINAL PRODUCT EXCH EX-21A -by PLACING TIC-722-25 IN MAN MODE & 100% OUTPUT	99-9T-722	99 12-3E-1431

LONGVIEW OPERATIONS
XYZ Corporation
INTERLOCK NARRATIVE DIAGRAM
R-21 REACTOR INTERLOCKS
I-2002, I-2003, I-2004, I-2005

APPROVALS		DATE
PROJECT MGR	SHIFT	12/2/16
DEPT. HEAD	KGD	12/2/16
DIVISION HEAD		
WORKS MGR	DIVISION SHIFT	

DRAWN: SCL   12/1/2016
JOB ENGINEER: JWA   12/2/2016
DRAWING NO 99-9L-035   REV 0

DESIGN AND BUILD WITH CARE	REFERENCE DRAWING DESC	REFERENCE DRAWING	REV #	JOB NUMBER	REVISION DESC	DRAFTER	LINES	ENGR APPROVAL	OPER APPROVAL	DATE
	P & ID R-21	99-9T-722								
	ELEMENTARY DWG. R-21 LEVEL	99_14-3E-1241								

NON SIS INTERLOCK MULTIPLE

c.

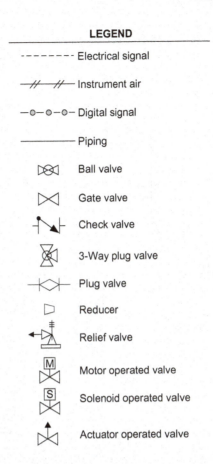

LEGEND

- - - - - - - Electrical signal

—//—//— Instrument air

—o—o—o— Digital signal

————— Piping

⋈ Ball valve

◁▷ Gate valve

Check valve

3-Way plug valve

Plug valve

Reducer

Relief valve

Motor operated valve

Solenoid operated valve

Actuator operated valve

Figure 2.9 Example of a P&ID legend.

ANYCorp	ANYCorp Construction Company				
Contract No.21609	Texas City, Texas Project No. 2447				
ANYCorp	Drawn S. Turnbough		Date 1/19		
	Chk. By J. Dees		App. By M. Collins		
Piping & Instrumentation Diagram Primary Amine Unit Fuel Gas Amine Absorber	Scale None		Unit 85		
	AFE No.		Chc. No. 6848-56		
	Dwg. No. F-52-4-0505				Rev 2

Figure 2.10 Example of a P&ID title block.

Application Block

An **application block** is the main part of a drawing that contains symbols and defines elements such as relative position, types of materials, equipment descriptions, and functions. Figure 2.11 shows an example of an application block.

Symbols

Symbols are simple illustrations used to identify types of equipment. A set of common symbols has been developed to represent actual equipment, piping, instrumentation, and other components. While some symbols differ from facility to facility, many are universal, with only subtle differences. It is critical that process technicians recognize and understand symbols used in their facilities. Figure 2.12 shows examples of some of the equipment symbols that are commonly found on process diagrams.

- Compressors (see Figure 2.12) are used to increase the pressure on gases and vapors.
- Cooling towers (see Figure 2.12) are used to lower the temperature of water using latent heat of evaporation and sensible heat loss.

Application block the main part of a drawing that contains symbols and defines elements such as relative position, types of materials, equipment descriptions, flows, and functions.

Symbol simple illustration used to represent a piece of equipment, an instrument, or other device on a PFD or P&ID.

Figure 2.11 Example of a P&ID application block.

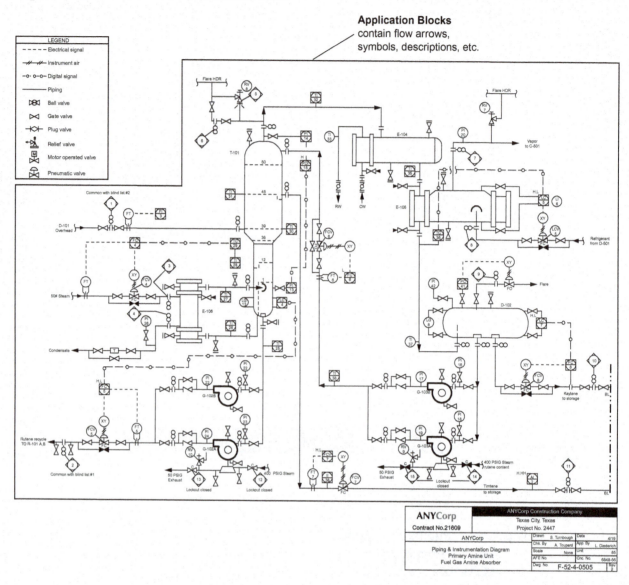

- Turbines (see Figure 2.12) produce the power necessary to drive equipment.
- Furnaces (see Figure 2.12) are used to produce heat required for processes.
- Boilers (see Figure 2.12) produce the steam required for various parts of a process.
- Reactors (see Figure 2.12) are vessels in which chemical reactions are initiated, controlled, and sustained.
- Distillation columns are used to separate the components of a liquid mixture by partially vaporizing the lighter components of the liquid and then recovering both the vapors and the bottoms.
- Pressure vessels (see Figure 2.12) are pressure-rated containers in which materials are processed, treated, or stored.
- Heat exchangers (Figure 2.13) are used to transfer heat from one substance to another, usually without the two substances physically contacting each other.
- Pumps are used to move liquid materials through piping systems. The process industries use many different types of pumps. Each pump type has a unique symbol that appears on P&IDs. Table 2.1 shows some examples of pump symbols.

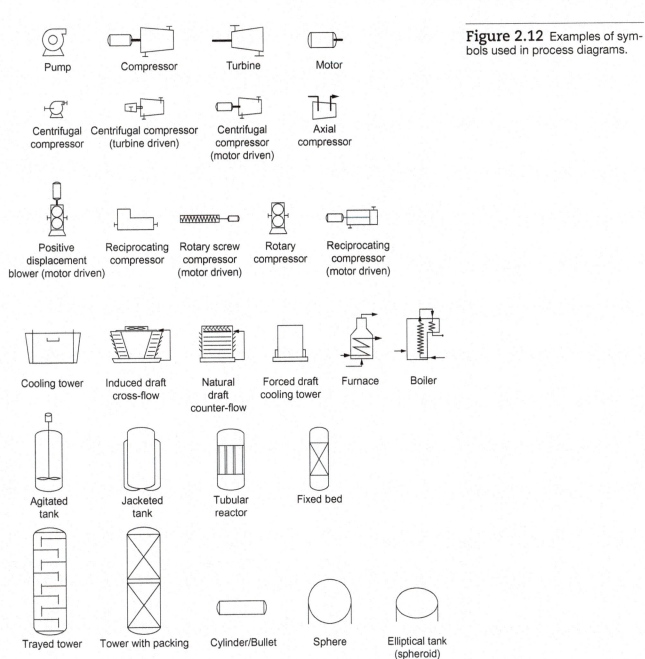

Figure 2.12 Examples of symbols used in process diagrams.

Piping Symbols

Figure 2.14 shows examples of some of the many piping symbols.

Actuator Symbols

Actuators are manual, pneumatic, electrical, hydraulic, or solenoid devices that move a control device, usually a valve.

- Manual actuators are operated by hand wheels or levers.
- Pneumatic actuators are operated by an air signal. A pneumatic actuator receives its signal from a transducer, which converts an electric signal to a corresponding air signal.

Figure 2.13 Heat exchanger symbols used in process diagrams.

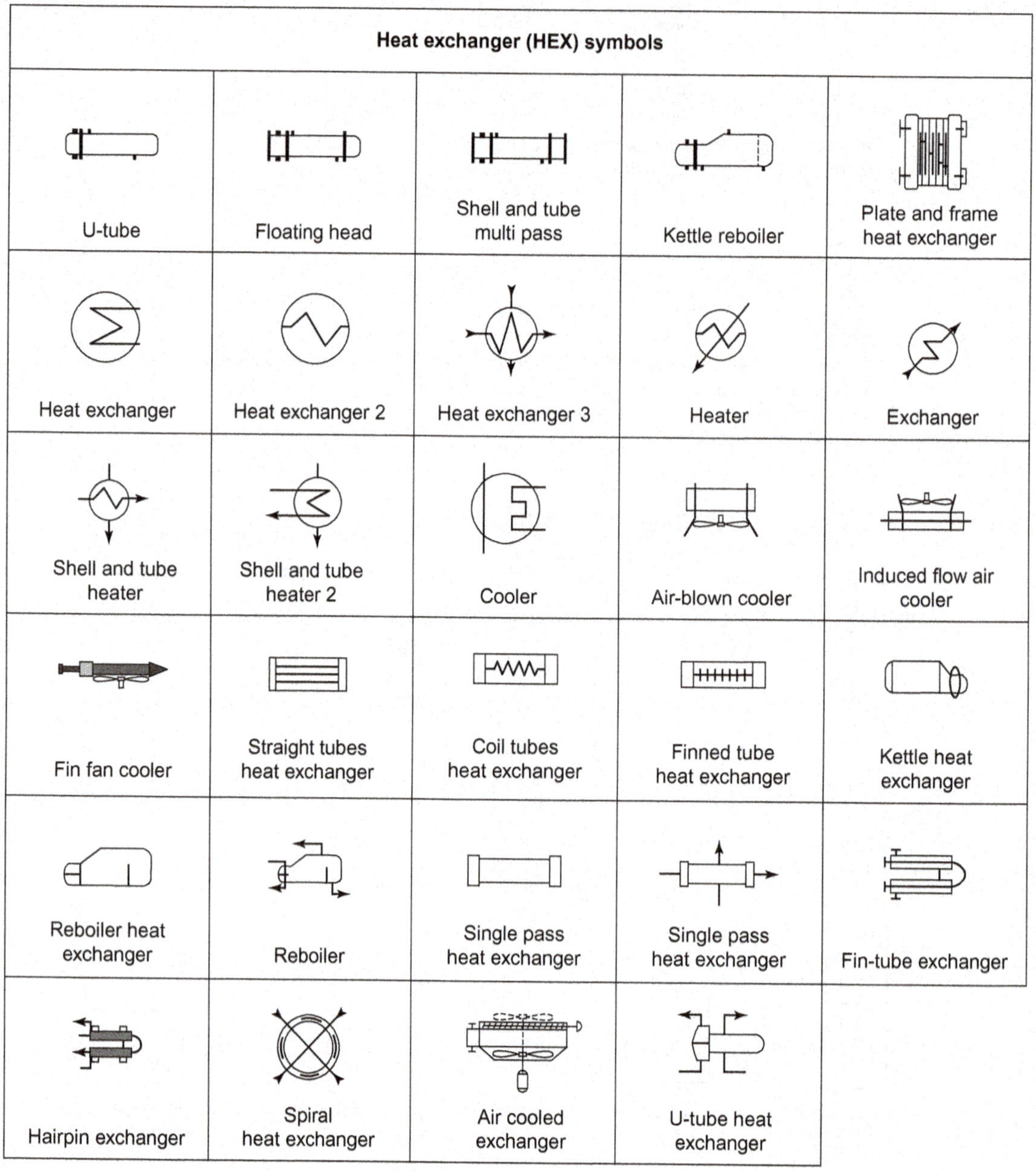

Heat exchanger (HEX) symbols				
U-tube	Floating head	Shell and tube multi pass	Kettle reboiler	Plate and frame heat exchanger
Heat exchanger	Heat exchanger 2	Heat exchanger 3	Heater	Exchanger
Shell and tube heater	Shell and tube heater 2	Cooler	Air-blown cooler	Induced flow air cooler
Fin fan cooler	Straight tubes heat exchanger	Coil tubes heat exchanger	Finned tube heat exchanger	Kettle heat exchanger
Reboiler heat exchanger	Reboiler	Single pass heat exchanger	Single pass heat exchanger	Fin-tube exchanger
Hairpin exchanger	Spiral heat exchanger	Air cooled exchanger	U-tube heat exchanger	

Table 2.1 Examples of Pump Symbols

Pump Type	Symbol	Pump Type	Symbol
Centrifugal pump	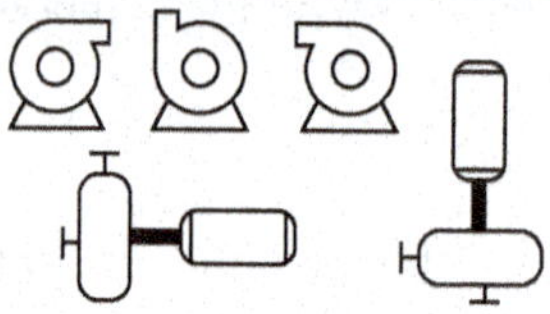 Horizontal Vertical	Rotary lobe pump	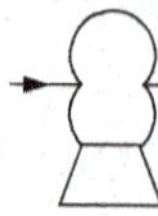
Positive displacement pump	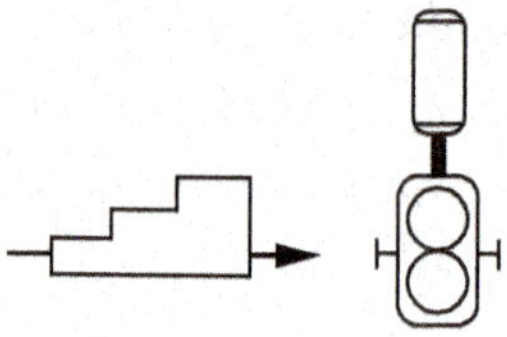	Turbine-driven equipment	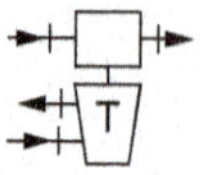

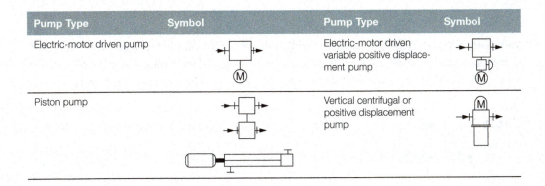

Pump Type	Symbol	Pump Type	Symbol
Electric-motor driven pump		Electric-motor driven variable positive displacement pump	
Piston pump		Vertical centrifugal or positive displacement pump	

Pipe fittings, types of connections		Tees	
Fitting	**Symbol**	**Fitting**	**Symbol**
Screwed ends		Tee	
Flanged ends		Tee, double sweep	
Bell and spigot ends		Tee, outlet down	
Welded and brazed ends		Tee, outlet up	
Soldered ends		Tee, single sweep, or plain t-y	
Elbows		**Other pipe fittings**	
Fitting	**Symbol**	**Fitting**	**Symbol**
Elbow , 90 degrees		Bushing	
Elbow , 45 degrees		Cap	
Elbow ,other than 90 or 45 degrees, specify angle	30°	Coupling	
		Plug	
Elbow ,long radius	LR	Reducer, concentric	
Elbow, reducing		Union, flanged	
Elbow, side outlet, outlet down		Union, screwed	
Elbow, side outlet, outlet up		Expansion joint, bellows	
Elbow, turned down		Expansion joint, sliding	
Elbow, turned up		Flexible hose	
Elbow, union		Flexible hose, flanged	

Figure 2.14 Examples of piping symbols used in process diagrams.

- Motor actuators are operated by an electrical motor.
- Hydraulic actuators are operated by fluid pressure and can use a pumping system or process fluid pressure. Hydraulic valves have fluid lines to one (single-acting) or both (dual-acting) sides of the actuator.
- Solenoid actuators are operated by energizing an on/off, open/shut electrical circuit.

Figure 2.15 shows examples of actuator symbols.

Actuator Types

Manual Pneumatic Motor Hydraulic Solenoid

Figure 2.15 Examples of actuator symbols.

Valve Symbols

Valves are used to control the flow of fluids through a pipe. Many types of valves are used in the process industries. Each of the many types of valves has a unique symbol to identify it on a process drawing. Table 2.2 shows some examples of different valve symbols.

Table 2.2 Examples of Valve Symbols Used in Process and Instrument Drawings (P&IDs)

Valve Type	Symbol	Valve Type	Symbol
Ball valve		Needle valve	
Butterfly valve		Nonreturn valve	NR
Check valve		Piston valve	
Actuator-operated valve		Plug valve	
Gate valve		Relief valve	
Globe valve		Stop-check valve	
Hand valve		Three-way valve	
Motor-operated valve			

Electrical Equipment and Motor Symbols

Electrical equipment can be used for a variety of functions. Each type has a unique symbol that is used to identify it on process drawings. Table 2.3 shows examples of electrical equipment symbols and their descriptions.

Table 2.3 Electrical Equipment and Motor Symbols

Symbol	Name	Description
	Transducer	A device that converts one type of energy to another, such as electrical to pneumatic
	Motor driven	A symbol that indicates a piece of equipment is motor driven (either AC or DC)
	Current transformer	A device that can provide circuit control and current measurement
	Transformer	A device that can either step up or step down the voltage of AC electricity
- - - - - -	Electrical signal	A signal that indicates voltage or current

Symbol	Name	Description
	Potential transformer	A device that monitors power line voltages for power metering
	Inductor	An electronic component consisting of a coil of wire
Motor	Motor	A motor (either AC or DC) that converts electrical energy into mechanical energy
	Outdoor meter device	A meter used to monitor electricity, such as a voltmeter (used to measure voltage) or ammeter (used to measure current)

Instrumentation Symbols

Instruments are devices used to measure and control process data (e.g. indicate flow, temperature, level, pressure, or analytical data). Instrumentation symbols, which are used to identify instrumentation throughout a facility, might or might not look like the physical devices they represent. In many process drawings, a $\frac{7}{16}$-inch (1.1-cm) diameter circle, called a balloon or bubble, is commonly used to represent any number of functionally different instruments. Figure 2.16 shows an example of an instrumentation balloon and what each letter in the balloon represents.

Figure 2.16 A. Instrumentation balloon. **B.** Instrument symbol interpretation key.

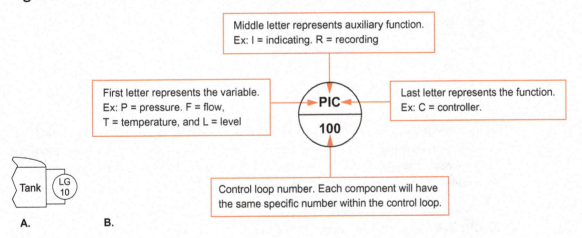

Considering the complexity of many control systems, this standardized approach works very well. The tag number is the primary key to defining the functionality of the instrument, and the lines (or absence of lines) within the balloon depict where the instrument is physically located.

A typical legend (see Figure 2.17) shows how various instruments are represented for a particular drawing.

Instrument tag numbers identify the measured variable, the function of the specific instrument, and the loop number. Instrument tag numbers give process technicians an indication of what that instrument is monitoring or controlling. An ISA instrument tag number (Figure 2.18) is described with both letters and numbers and is unique to that instrument.

The first letter identifies the measured or initiating variable, and the following letters describe the function of the instrument. For example, in Figure 2.18, F stands for "flow," I for "indicating," and C for "controller." Thus, this instrument is a flow-indicating controller,

Figure 2.17 Legend with instrument symbols.

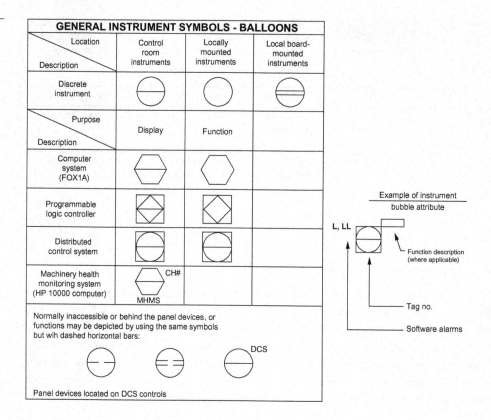

Figure 2.18 Sample ISA instrument tag number.

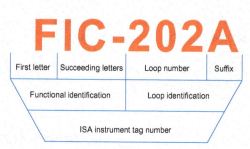

an instrument that controls flow and has an indicator on its faceplate. The ISA functional identification table (Table 2.4) lists examples of first letter identifiers with possible modifiers and succeeding letters. Table 2.5 shows examples of instrument tag letters and their interpretations based on the information provided in Table 2.4.

Table 2.4 Sample ISA Functional Identification Labels

Letter	First Letter (Process Variable)	Succeeding Letters (Type of Instrument)
A	Analysis	Alarm
B	Burner combustion	—
C	—	Controller
E	Electricity (voltage)	Element/sensor
F	Flow rate	—
G	—	Site or gauge glass/monitor
H	Hand	—
I	Current (electric)	Indicator
J	Power	—
K	Time	Control station
L	Level	Light

Letter	First Letter (Process Variable)	Succeeding Letters (Type of Instrument)
O	—	Orifice/restriction
P	Pressure/vacuum	Sample point
Q	Quantity	—
R	Radiation	Recorder
S	Speed/frequency	Switch
T	Temperature	Transmitter
V	Voltage/vibration	Valve/damper/louver
W	Weight/force	Well
Y	Event	Relay/compute/convert
Z	Position/dimension	Driver/actuator

The information found in this table was taken from the ISAS5.1 standard table, but it is subject to change. Thus, the most current ISA table should always be consulted for verification purposes.

Table 2.5 Sample Instrument Tag Letters with Functional Identifications

Letters	Functional Interpretation
FIC	Flow-Indicating Controller
FRC	Flow-Recording Controller
LI	Level Indicator
PC	Pressure Controller
PIC	Pressure-Indicating Controller
PT	Pressure Transmitter
PY	Pressure Relay
TE	Temperature Element
TT	Temperature Transmitter

2.7 Industry Standards

With the development of process drawings and industry standards, a system of symbols has been created to describe how both equipment and its associated instrumentation are interconnected. While various engineering and chemical companies have created symbol systems unique to their facilities or companies, many rely on the standards developed by industry organizations to address issues and standardize symbols across the various process industries.

- **ISA** (International Society of Automation) is a global, nonprofit technical society that develops standards for automation, instrumentation, control, and measurement. (ISA was founded as the Instrument Society of America.) For instrumentation, the ISA is the predominant source for instrumentation symbology under standard 5.1.

- The ISA standard 5.1 comprises both specific symbols and a coded system, built on the letters of the alphabet, that depicts functionality. Although many, if not most, large companies have moved toward adopting the ISA 5.1 standard in its entirety, other preferred symbols continue to be used. Therefore, process technicians should not assume that the ISA standard is used. All symbols, standard and nonstandard alike, should be identified in the legend of each drawing.

- The American National Standards Institute (**ANSI**) oversees and coordinates voluntary standards in the United States. ANSI accreditation is used as a baseline or "backbone" for standardization in various industries. ANSI develops and approves norms and

ISA The International Society of Automation (originally known as the Instrument Society of America), a global, nonprofit technical society that develops standards for automation, instrumentation, control, and measurement.

ANSI American National Standards Institute, an organization that oversees and coordinates voluntary standards in the United States.

API American Petroleum Institute, a trade association that represents the oil and gas industry in the areas of advocacy, research, standards, certification, and education.

ASME American Society of Mechanical Engineers, an organization that specifies requirements and standards for pressure vessels, piping, and their fabrication.

NEC National Electric Code; standard established by the National Fire Protection Agency (NFPA), which promotes the safe installation of electrical wiring and equipment.

OSHA Occupational Safety and Health Administration, a U.S. government agency created to establish and enforce workplace safety and health standards, conduct workplace inspections, provide worker training and education, and investigate serious workplace incidents.

guidelines that affect many business sectors while coordinating U.S. standards with international standards. This coordination allows American products to be used competitively worldwide.

- The American Petroleum Institute (**API**) is a trade association that represents the oil and gas industry. API, which speaks on behalf of the petroleum industry to the public and the various government branches, sponsors and researches economic analyses and provides statistical data to the public. API is a leader in the development of equipment and operating standards for the petroleum and petrochemical industries. API maintains more than 500 standards and recommendations and has a certification program for the inspection of industry equipment. API also conducts various education programs, including seminars, workshops, and conferences for the ongoing education of people involved in the petroleum industry.

- The American Society of Mechanical Engineers (**ASME**) specifies requirements and standards for pressure vessels, piping, and their fabrication.

- The National Electric Code (**NEC**) specifies practices that promote the safe installation of electrical wiring and equipment. NEC was established by the National Fire Protection Agency (NFPA).

- The Occupational Safety and Health Administration (**OSHA**), a US government agency, was created to establish and enforce workplace safety and health standards, conduct workplace inspections, provide worker training, and investigate serious workplace incidents.

Summary

Many different types of drawings are used within the process industries. Each drawing type represents different aspects of the process and various levels of detail. Studying combinations of these drawings provides a more complete picture of the processes at a facility.

Process drawings provide process technicians with visual descriptions and explanations of processes, their flows, equipment, and other important items in a facility. Process facilities use process drawings to assist with operations, modifications, training, and maintenance. The information contained within process drawings include a legend, title block, and application block.

Examples of process drawings include block flow diagrams (BFDs), process flow diagrams (PFDs), piping and instrumentation diagrams (P&IDs), and plot plans.

Block flow diagrams (BFDs) are the simplest drawings used in the process industry. While providing a general overview of the process, they contain few specifics. Block flow diagrams include the feed, product location, intermediate streams, recycle, and storage.

Process flow diagrams (PFDs) are basic drawings that use symbols and direction arrows to show the primary flow of a product through a process. PFDs describe the actual process including flow rate, temperature, pressure, pump capacities, heat exchangers, equipment symbols, equipment designations, reactor catalyst data, cooling water flows, and symbol charts.

Piping and instrument diagrams (P&IDs) are similar to PFDs but show more detailed process information, such as equipment numbers, piping specifications, instrumentation, and other detailed information.

Plot plans are scale drawings that show the layout of equipment, units, and buildings. They are drawn to scale so that everything is of the correct relative size and shows proper dimensions.

Symbols are figures used to represent types of equipment. Each specific piece of equipment has a unique symbol. Different industry organizations develop and publish symbology standards to improve consistency among process drawings. Although some industries have symbol systems for their own use, many rely on the standard symbols created by industry organizations.

Checking Your Knowledge

1. Define the following terms:
   a. Application block
   b. Legend
   c. Symbol
   d. Title block
   e. Block flow diagram (BFD)
   f. Electrical diagram
   g. Isometric drawing
   h. Piping and instrumentation diagram (P&ID)
   i. Plot plan
   j. Process flow diagram (PFD)
   k. Process schematic
   l. Utility flow diagram (UFD)

2. Which drawing provides a general overview of the process and contains few specifics?
   a. PFD
   b. BFD
   c. P&ID
   d. UFD

3. Process flow diagrams are typically drawn in a direction from _____.

4. Which of the following items are located on a process flow diagram? (Select all that apply.)
   a. Major process piping
   b. Equipment symbols
   c. Main pieces of equipment
   d. All instruments

5. (True or False) Utility flow diagrams provide process technicians with a PFD-type view of the instrumentation used for a process.

6. Which of the following are included in a process drawing? (Select all that apply.)
   a. Legend
   b. Electrical codes
   c. Application block
   d. Title block

7. (True or False) A balloon is an instrumentation symbol that is commonly used to represent the function of different instruments.

8. On an FIC ISA tag, what does the first letter "F" represent?
   a. Frequency
   b. Flow
   c. Force
   d. Function

9. Which society develops standards for automation, instrumentation, control, and measurement symbols?
   a. API
   b. NFPA
   c. ISA
   d. ASME

10. What are the two most common types of drawings that process technicians encounter?
   a. Process flow diagrams
   b. Block flow diagrams
   c. Utility flow diagrams
   d. Piping and instrumentation diagrams

11. (True or False) Flow arrows show the order and relationship of each component in a block flow diagram (BFD).

12. (True or False) On plot plans, major pieces of equipment are drawn twice as large as auxiliary equipment to indicate their relative importance.

13. What type of diagram shows all components and connections between instrumentation and the control room?
   a. Elevation diagram
   b. Process schematics
   c. Loop diagram
   d. Logic diagram

14. The National Fire Protection Agency established the ____ , which specifies practices that promote the safe installation of electrical wiring and equipment.

NOTE: Answers to Checking Your Knowledge questions are in the Appendix.

Student Activities

1. Write a paper about one or more standards organizations (e.g., ANSI, API, ASME, and ISA).

2. Work with a classmate to draw a flow sequence of a simple process and describe the basic steps associated with the process.

3. Identify all the symbols in Figure 2.2.

4. Complete the following chart by writing three to five sentences about each drawing and how it is used.

Drawing Type	Description and Use
Block flow diagram (BFD)	
Process flow diagram (PFD)	
Piping and instrument diagram (P&ID)	
Plot plan	

Chapter 3
Tools

Objectives

After completing this chapter, you will be able to:

3.1 List the types of tools used by technicians in the process industries. (NAPTA Tools and Equipment 1-7*) p. 41

3.2 Describe lifting equipment used in process industries. (NAPTA Tools and Equipment 4) p. 46

3.3 Describe basic safety guidelines for tools and devices described in this chapter. (NAPTA Tools and Equipment 4) p. 50

3.4 Describe the process technician's role in tool care and maintenance. (NAPTA Tools and Equipment 11, 12) p. 54

*North American Process Technology Alliance (NAPTA) developed curriculum to ensure that Process Technology courses will produce knowledgeable graduates to become entry level employees in process technology. Objectives from that curriculum are named here in abbreviated form. For example, "(NAPTA Tools and Equipment 4)" means that this chapter's objective relates to objective 4 of NAPTA's course content on tools and equipment.

Key Terms

Adjustable pliers—pliers that have a tongue and groove joint or slot that allows the jaws to widen and grip objects of different sizes; tongue and groove also are called Channellocks®; slotted design may be called slip joint pliers, **p. 42.**

Adjustable wrench—open-ended wrench with an adjustable jaw that contains no teeth or ridges (i.e., the jaw is smooth); also referred to as a Crescent® wrench, **p. 44.**

Crane—a mechanical lifting device used to lift and lower materials or move them horizontally, **p. 46.**

Dolly—a wheeled cart or hand truck used to transport heavy items such as barrels, drums, and boxes, **p. 49.**

Drill—a device that contains a rotating bit designed to bore holes into various types of materials, **p. 45.**

Electric tool—a tool operated by electrical power (either AC or DC), **p. 53.**

Flaring tool—a device used to create a cone-shaped enlargement at the end of a piece of tubing so the tubing can accept a flare fitting, **p. 43.**

Forklift—a motorized vehicle with pronged forks used to lift and transport equipment or items placed on a pallet, **p. 48.**

Fuel-operated tools—tools powered by the combustion of a fuel (e.g., gasoline), **p. 54.**

Gantry crane—a type of crane (similar to an overhead crane) that runs on elevated rails attached to a frame or set of legs, **p. 47.**

Grinder—a device that uses the rotating force of one material against another material (e.g., a stone grinding wheel against a metal blade) to reduce or modify the shape and size of an object, **p. 45.**

Hammer—a tool with a handle and a heavy head that is used for pounding or delivering blows to an object, **p. 41.**

Hand tool—a tool that is operated manually instead of being powered by electric, pneumatic, hydraulic, or other forms of power, **p. 41.**

Hoist—a device composed of a pulley system with a cable or chain used to lift and move heavy objects, **p. 46.**

Hydraulic tool—a tool that is powered using hydraulic (liquid) pressure, **p. 54.**

Impact wrench—a pneumatically or electrically powered wrench that uses repeated blows from small internal hammers that generate torque to tighten or loosen fasteners, **p. 44.**

Jib crane—a type of crane that contains a vertical rotating member and an arm that extends to carry the hoist trolley, **p. 47.**

Lineman's pliers—snub-nosed pliers with two flat gripping surfaces, two opposing cutting edges, and insulated handles that help reduce the risk of electrical shock, **p. 42.**

Locking pliers—pliers that can be adjusted with an adjustment screw and then locked into place when the handgrips are squeezed together; also referred to as Vise-Grips®, **p. 42.**

Monorail crane—a type of crane that travels on a single runway beam, **p. 47.**

Needle nose pliers—small pliers with long, thin jaws for fine work (e.g., holding small items in place, removing cotter pins, or bending wire), **p. 42.**

Nonsparking tools—tools that are manufactured from specific materials that will not produce sparks capable of igniting vapors; also referred to as *spark-resistant* or *spark-proof tools*, **p. 50.**

Overhead traveling bridge cranes—a type of crane that runs on elevated rails along the length of a factory and provides three axes of hook motion (up and down, sideways, and back and forth), **p. 47.**

Personnel lift—a lift that contains a pneumatic or electric arm with a personnel bucket attached at the end; also referred to as a manlift, cherry picker, Condor®, or JLG®, **p. 49.**

Pipe wrench—an adjustable wrench that contains two serrated jaws that are designed to grip and turn pipes or other items with a rounded surface; also referred to as a Stillson wrench, **p. 44.**

Pliers—a hand tool that contains two hinged arms and serrated jaws that are used for gripping, holding, or bending, **p. 41.**

Pneumatic tool—a tool that is powered using pneumatic (air or gas) pressure, **p. 53.**

Powder-actuated tool—a tool that uses a small explosive charge to drive fasteners into hard surfaces such as concrete, stone, and metal, **p. 54.**

Power tool—a tool that is powered by electric, pneumatic, or hydraulic power or is powder actuated, **p. 45.**

Reciprocating saw—a device that uses the back-and-forth motion of a saw blade to cut materials such as wood and metal pipe, **p. 45.**

Screwdriver—a device that engages with, and applies torque to, the head of a screw so that the screw can be tightened or loosened, **p. 42.**

Socket wrench—a tool that contains interchangeable socket heads of varying sizes that can be attached to a ratcheting wrench handle or other driver, **p. 44.**

Tool—a device designed to provide mechanical advantage and make a task easier, **p. 41.**

Torque wrench—a manual wrench that uses a gauge to indicate the amount of torque (rotational force) being applied to the nut or bolt, **p. 45.**

Valve wheel wrench—a hand tool used to provide mechanical advantage when opening and closing valves. These wrenches typically fit over the spoke of a valve wheel, **p. 45.**

Wrench—a hand tool that uses gripping jaws to turn bolts, nuts, or other hard-to-turn items, **p. 43.**

3.1 Introduction

Knowledge of hand and power tools is an integral part of the process technician's daily responsibilities. Knowing how to use tools safely and properly is critical. This chapter provides an overview of the types of tools most commonly used by process technicians, their applications, safety issues associated with the various tools, and proper tool care and maintenance.

Tools Used in Process Industries

A **tool** is a device designed to provide *mechanical advantage* to make a task easier. Mechanical advantage is the factor by which a device multiplies the force that is applied. The tools most often used by process technicians are hand tools.

Tool a device designed to provide mechanical advantage and make a task easier.

Hand Tools

Hand tools are operated manually instead of being driven by electricity, pneumatic power, hydraulic pressure, or other forms of power. Many process technicians carry an assortment of hand tools with them as they make their daily rounds. Some of the most common hand tools that a process technician might use include:

Hand tool a tool that is operated manually instead of being powered by electric, pneumatic, hydraulic, or other forms of power.

- Pliers
- Valve wheel wrench
- Pipe wrench
- Hammer
- Bolt cutter
- Knife
- Tubing cutter
- Wire stripper.

HAMMERS A **hammer** is a tool with a handle and a heavy head that is used for pounding or delivering blows to an object. The most common applications of hammers include driving nails, fitting components together, or breaking objects apart. While hammer shapes and sizes vary based on their applications, the most common features of a hammer are a long handle and a heavily weighted head. Figure 3.1 shows examples of some commonly used hammers.

Hammer a tool with a handle and a heavy head that is used for pounding or delivering blows to an object.

Claw hammer   Ball peen hammer   Sledge hammer   Rubber mallet

Figure 3.1 Examples of different types of hammers.

CREDIT: (left to right) Sergiy Kuzmin/ Shutterstock; Winai Tepsuttinun/ Shutterstock; Sashkin/Shutterstock; VectorOK/Shutterstock.

PLIERS **Pliers** are a hand tool that contains two hinged arms and serrated jaws used for gripping, holding, or bending. Pliers come in a variety of shapes and sizes.

When using pliers, it is important to remember that the gripping surfaces are serrated. These serrated edges will damage finished surfaces, so they should not be used on the flat

Pliers a hand tool that contains two hinged arms and serrated jaws that are used for gripping, holding, or bending.

Figure 3.2 Examples of different types of pliers.

Adjustable pliers   Locking pliers   Needle nose pliers   Lineman's pliers (Combination pliers)

Adjustable pliers   pliers that have a tongue and groove joint or slot that allows the jaws to widen and grip objects of different sizes.

Locking pliers   pliers that can be adjusted with an adjustment screw and then locked into place when the handgrips are squeezed together; also referred to as Vise-Grips®.

Needle nose pliers   small pliers with long, thin jaws for fine work (e.g., holding small items in place, removing cotter pins, or bending wire).

Lineman's pliers   snub-nosed pliers with two flat gripping surfaces, two opposing cutting edges, and insulated handles that help reduce the risk of electrical shock.

Screwdriver   a device that engages with, and applies torque to, the head of a screw so that the screw can be tightened or loosened.

sides of packing nuts, fittings, bolts, or fasteners. Figure 3.2 shows examples of different types of pliers.

Adjustable pliers may have different designs; they can be tongue and groove (also called Channellocks®) or slotted (slip joint) design. Adjustable pliers allow the jaws to widen and grip objects of different sizes. The jaw opening is typically adjusted by opening the pliers fully and selecting the desired groove or hinge pin position.

Locking pliers (sometimes referred to as Vise-Grips®) can be locked into place when the handgrips are squeezed together. Locking pliers contain an adjustment screw that allows the distance between the jaws to be adjusted. Because of their locking mechanism, locking pliers provide a firm grip and prevent slippage.

Needle nose pliers have long, thin jaws for fine work, for example, holding small items in place, removing cotter pins, or bending wire.

Lineman's pliers are snub-nosed pliers with two flat gripping surfaces, two opposing cutting edges, and insulated handles that help reduce the risk of electrical shock. Because of their versatility, lineman's pliers are most often used for cutting and bending wire, stripping wire insulation or cable jackets, and gripping pieces of wire so they can be twisted together.

SCREWDRIVERS   **Screwdrivers** are devices that engage with, and apply torque to, the head of a screw so the screw can be tightened or loosened. Screwdrivers, which consist of a handle and a shank, come in a variety of shapes and sizes. The tip of a screwdriver shank (the head) is shaped to match different screw head types. The most common types of screwdrivers used in the process industries are slotted (also referred to as flat head) and Phillips head. A slotted screwdriver has a flattened head, while the tip of a Phillips screwdriver has a cross pattern. Figure 3.3 shows examples of Phillips head and slotted screwdrivers.

Figure 3.3 Examples of different types of screwdrivers and screw heads.

Regular (Flat head) screwdriver   Phillips head screwdriver

When working with screwdrivers, process technicians should ensure that the tip is the proper size and shape for the screw head. Using an improperly sized screwdriver can damage (strip) the screw head or allow the screwdriver to slip, causing injury.

TUBING CUTTER   A tubing cutter is a tool used to cut copper, aluminium, or stainless steel tubing (Figure 3.4). Most devices include a tubing reamer to remove any burrs from the end of the tubing. Caution must be exercised when using this tool, or any type of cutter, since misuse can lead to serious injury.

Figure 3.4 Tubing cutter.

CREDIT: topimages/Shutterstock.

FLARING TOOL A **flaring tool** (shown in Figure 3.5) is a device used to create a cone-shaped enlargement at the end of a piece of tubing so the tubing can accept a flare fitting. As a general rule, the maximum flare diameter can be no more than 1.4 times the diameter of the tube to prevent the possibility of splitting the tubing material.

Flaring tool a device used to create a cone-shaped enlargement at the end of a piece of tubing so the tubing can accept a flare fitting.

Figure 3.5 Flaring tool.

CREDIT: Manuel Trinidad Mesa/ Shutterstock.

WRENCHES **Wrenches** are hand tools that use gripping jaws to turn bolts, nuts, or other hard-to-turn items. Wrenches come in many different shapes and sizes. Some examples of common wrenches include adjustable, impact, pipe, socket, torque, and valve wheel. Because the gripping surfaces on most wrenches are smooth, they can be used on finished surfaces and the flat sides of fittings, bolts, and fasteners. Figure 3.6 shows examples of different types of wrenches.

Wrench a hand tool that uses gripping jaws to turn bolts, nuts, or other hard-to-turn items.

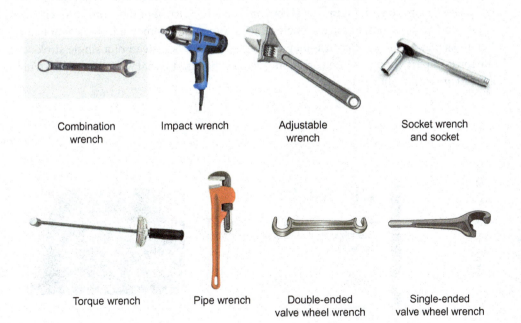

Combination wrench   Impact wrench   Adjustable wrench   Socket wrench and socket

Torque wrench   Pipe wrench   Double-ended valve wheel wrench   Single-ended valve wheel wrench

Figure 3.6 Examples of different types of wrenches.

CREDIT: (top row, left to right) balla family group/Shutterstock; ericlefrancais/ Shutterstock; Lotus_studio/Shutterstock; tawatchai lukmanee/Shutterstock; (bottom row, left to right) MARGRIT HIRSCH/ Shutterstock; Joao Virissimo/Shutterstock.

Adjustable wrench   open-ended wrench with an adjustable jaw that contains no teeth or ridges (i.e., the jaw is smooth); also referred to as a Crescent® wrench.

Socket wrench   a tool that contains interchangeable socket heads of varying sizes that can be attached to a ratcheting wrench handle or other driver.

Pipe wrench   an adjustable wrench that contains two serrated jaws that are designed to grip and turn pipes or other items with a rounded surface; also referred to as a Stillson wrench.

Adjustable wrenches are also referred to as Crescent® wrenches. They are open-ended wrenches with an adjustable jaw that contains no teeth or ridges (i.e., the jaw is smooth).

Socket wrenches contain interchangeable socket heads of varying sizes that can be attached to a ratcheting wrench handle or other driver. Most socket wrenches contain a ratcheting mechanism that allows a nut or bolt to be tightened or loosened in a continuous motion, as opposed to the socket or handle being removed and refitted after each turn.

Pipe wrenches are adjustable wrenches that contain two serrated jaws designed to grip and turn pipes or other items with a rounded surface. In this type of wrench, pressure on the handle causes the jaws to move together more tightly. Pipe wrenches come in many different sizes and shapes. Figure 3.7 shows an example of a pipe wrench with the components and direction of rotation labeled.

Figure 3.7 Pipe wrench with components and rotation direction labeled.

CREDIT: Bonezboyz/Shutterstock.

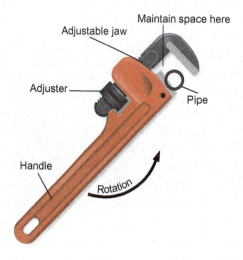

When using a pipe wrench, it is important to maintain space between the object being turned and the back of the jaws. It is also important to keep the pipe wrench in a suitable position to prevent injury or hand slippage.

An **impact wrench** is a pneumatically or electrically powered wrench that uses repeated blows from small internal hammers that generate torque to tighten or loosen fasteners. In some facilities, process technicians use impact wrenches to remove bolts from flanges or other pieces of equipment. Figure 3.8 shows an example of a pneumatic impact wrench.

Like socket wrenches, impact wrenches can accommodate removable sockets of many different shapes and sizes. Because they use repetitive blows instead of a single stroke of brute force, impact wrenches often are more effective than traditional wrenches at removing bolts that are extremely tight.

Impact wrench   a pneumatically or electrically powered wrench that uses repeated blows from small internal hammers that generate torque to tighten or loosen fasteners.

Figure 3.8 Pneumatic impact wrench.

CREDIT: Myibean/Shutterstock.

Did You Know?

Automotive mechanics use impact wrenches to remove the lug nuts from car wheels.

CREDIT: kurhan/Shutterstock.

Torque wrenches are manual wrenches that use a gauge to indicate the amount of torque (rotational force) being applied to the nut or bolt. Torque wrenches are used when precise tightening is crucial. Figure 3.9 shows an example of a torque wrench.

Torque wrench a manual wrench that uses a gauge to indicate the amount of torque (rotational force) being applied to the nut or bolt.

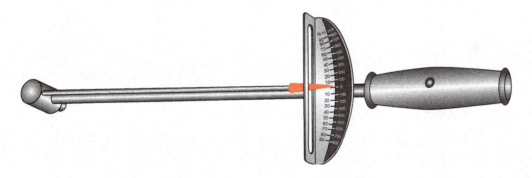

Figure 3.9 Torque wrench showing scale in newton-meters and foot-pounds.

CREDIT: MARGRIT HIRSCH/Shutterstock.

Valve wheel wrenches are used to provide mechanical advantage when opening and closing valves. Valve wheel wrenches (Figure 3.10) usually fit over the rim of a valve wheel, between spokes, to minimize the likelihood that the wrench will slip off the wheel when force is applied. Care must be taken when using this type of wrench to avoid applying so much force that it breaks the valve wheel or valve stem or causes the wrench to slip, which can result in injury.

Valve wheel wrench a hand tool used to provide mechanical advantage when opening and closing valves. These wrenches typically fit over the spoke of a valve wheel.

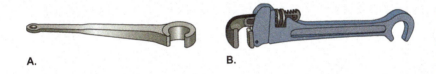

A.   **B.**

Figure 3.10 A. Single-ended valve wheel wrench. **B.** Combined pipe wrench and valve wheel wrench (also called an operator's wrench).

Power Tools

Power tools are tools driven by electricity, pneumatic power, hydraulic pressure, or powder charges (small charges filled with explosive materials such as gunpowder). Examples of common power tools (Figure 3.11) include drills, grinders, and saws.

A **drill** is a device that contains a rotating bit designed to bore holes into various types of materials.

A **grinder** is a device that uses the rotating force of one material against another material (e.g., a stone grinding wheel against a metal blade) to reduce or modify the shape and size of an object.

A **reciprocating saw** is a device that uses the back-and-forth motion of a saw blade to cut materials that range from wood to metal pipe.

When using power tools, process technicians must wear all required personal protective equipment and follow all company safety rules.

Power tool a tool that is powered by electric, pneumatic, or hydraulic power or is powder actuated.

Drill a device that contains a rotating bit designed to bore holes into various types of materials.

Grinder a device that uses the rotating force of one material against another material (e.g., a stone grinding wheel against a metal blade) to reduce or modify the shape and size of an object.

Reciprocating saw a device that uses the back-and-forth motion of a saw blade to cut materials such as wood and metal pipe.

Power drill   Electric grinder   Reciprocating saw

Figure 3.11 Examples of power tools.

CREDIT: (left to right) sbarabu/Shutterstock; vdimage/Shutterstock; Charles Brutlag/Shutterstock.

3.2 Lifting Equipment Used in Process Industries

The process industries use many different kinds of lifting equipment, including hoists, cranes, forklifts, personnel lifts, and dollies.

Hoists

Hoist a device composed of a pulley system with a cable or chain used to lift and move heavy objects.

A **hoist** is a device composed of a pulley or gear system with a cable or chain used to lift and move heavy objects.

HOIST SAFETY Basic safety requirements for hoists:

- Operation must be within specified weight limits. This information is usually stamped or stenciled on the hoist.
- Hoist systems should not be operated if the weight limit for the hoist is unknown or the weight of the object to be lifted is unknown or cannot be accurately estimated by experienced personnel.
- Chains and ropes should be free from kinks.
- Individuals must not work beneath equipment suspended by a hoist. Appropriate personnel must be present, and proper barricading must be in place.
- Chains should be inspected for bent or broken links and signs of corrosion.
- Ropes should be inspected for fraying and broken threads, and steel ropes should be inspected for corrosion.
- Loads should be properly secured and balanced before hoisting.

Figure 3.12 shows an example of manual lever and hand-chain (chain fall) operated hoists.

Figure 3.12 Examples of lever hoist and chain hoist.

CREDIT: A. Simon/Shutterstock.
B. Lesterman/Shutterstock.

A. Lever   B. Hand chain

Cranes

Crane a mechanical lifting device used to lift and lower materials or move them horizontally.

A **crane** is a mechanical lifting device that can be used to lift and lower materials or move them horizontally. There are four main types of overhead crane systems: overhead traveling bridge, gantry, jib, and monorail. Figure 3.13 shows examples of common types of cranes.

Overhead traveling bridge cranes run on an elevated rail system along the length of a warehouse and provide three axes of hook motion (up and down, sideways, and back and forth). Both single- and double-girder bridge designs have great flexibility, allowing the hook to be positioned very precisely. The bridge (which carries the hoist and trolley) is supported by a pair of rails called tracks that, in turn, carry a pair of wheels. Rails can be installed at either the floor or ceiling level. Figure 3.13A shows an example of an overhead traveling bridge crane.

Gantry cranes (similar to overhead cranes) run on elevated rails attached to a frame or set of legs. Gantry cranes provide the same performance characteristics as overhead bridge cranes. Figure 3.13B shows an example of a gantry crane.

Jib cranes consist of a pivoting head and boom assembly that carries a hoist and trolley unit. The pivoting head is supported either by a floor-mounted mast providing 360-degree boom rotation or by an existing building column that provides 180 degrees of boom rotation. Figure 3.13C shows an example of a jib crane.

Monorail cranes are cranes that travel on a single runway beam. These types of cranes are quite specialized and very effective when properly incorporated into a factory layout. Only two directions of hook travel are afforded by the monorail: up and down, and along the axis of the monorail beam. It is not recommended (and quite dangerous) to push the load out from under the centerline of the monorail beam. Monorail systems are most often

Overhead traveling bridge cranes a type of crane that runs on elevated rails along the length of a factory and provides three axes of hook motion (up and down, sideways, and back and forth).

Gantry crane a type of crane (similar to an overhead crane) that runs on elevated rails attached to a frame or set of legs.

Jib crane a type of crane that contains a vertical rotating member and an arm that extends to carry the hoist trolley.

Monorail crane a type of crane that travels on a single runway beam.

Figure 3.13 Common crane designs. **A.** Example of an overhead traveling bridge crane. **B.** Gantry crane. **C.** Jib crane. **D.** Monorail crane.

CREDIT: **A.** AlisLuch/Shutterstock. **B.** hanmon/Shutterstock. **C.** Tanasan Sungkaew/Shutterstock. **D.** PoohFotoz/Shutterstock.

A.

B.

C.

D.

integrated into continuous production systems for material transport (e.g., hot metal operations in foundries and material transport in paint booths). Figure 3.13D shows an example of a monorail crane.

CRANE SAFETY   Basic safety requirements for all cranes:

- Only personnel who have been trained and certified should be permitted to operate a crane.
- Proper load lift angle and balance must be maintained.
- Crane load limits must always be known and respected.
- Personnel must never work directly under cranes.
- Operational weight limits must not be exceeded.
- Chains and ropes must be free from kinks.
- Chains must be inspected for bent or broken links and signs of corrosion.
- Ropes must be inspected for fraying and broken threads, and steel ropes must be inspected for corrosion.
- Loads must be properly secured and balanced before hoisting.
- Correct parts must always be used when repairing a crane.

Forklifts

Forklift a motorized vehicle with pronged forks used to lift equipment or items placed on a pallet.

A **forklift** is a vehicle with elongated prongs (forks) used for lifting and transporting equipment or pallets of product. Forklifts are often used to lift items placed on a pallet to higher elevations for use in a unit or for storage. Occupational Safety and Health Administration (OSHA) regulations require that before operating a forklift, process technicians be trained and certified as competent to operate the equipment safely. Figure 3.14 shows an example of a commonly used forklift.

FORKLIFT SAFETY   Basic safety requirements for forklifts:

- Load limits must be known and followed.
- Only trained and certified personnel are permitted to operate a forklift.
- When carrying a load and moving the forklift, forks must not be overextended, raised, tilted, or misaligned. This could cause load imbalance or slippage.
- A forklift with a leaking hydraulic system must not be operated.
- The travel area must be kept clear of personnel.
- Workers must use caution when operating a forklift on a ramp or loading dock edge to avoid falling off.

Figure 3.14 Example of a forklift commonly used in the process industries.

- The forklift tires and the surface where the forklift will be used must match use requirements, and these usage rules must be obeyed. Many forklifts have small tires and can operate effectively only on smooth, solid surfaces.
- Fuel-powered forklifts must be operated only in properly ventilated spaces.

Forklifts are not intended to be used as personnel lifts. They are not equipped with the required fall protection systems and other safety features.

Personnel Lifts

Personnel lifts (also referred to as manlifts, cherry pickers, Condors®, or JLGs®) contain a pneumatic, hydraulic, or electric arm with a personnel bucket attached at the end. Many personnel lifts can be operated from both the ground and the personnel bucket. Figure 3.15 shows an example of a personnel lift.

Personnel lift a lift that contains a pneumatic or electric arm with a personnel bucket attached at the end; also referred to as a manlift, cherry picker, Condor®, or JLG®.

PERSONNEL LIFT SAFETY Basic safety requirements for personnel lifts:

- Personnel lifts must be operated only by trained and certified personnel.
- A second or third individual must be on the ground to monitor clearances and to warn other personnel when a personnel lift is in use in a busy or congested area.
- Power lines and overhead obstructions must be avoided.
- Proper safety procedures and personal protective equipment (PPE) (e.g., a secured harness) must be in place at all times.
- Tools must be secured so they do not drop onto personnel working below.
- The area must be barricaded when a personnel lift is in use.
- Outrigger stabilizer feet (if available) must be properly positioned on a solid surface.

Figure 3.15 Example of a personnel lift used in the process industries.

CREDIT: sarawuth wannasathit/Shutterstock.

Dollies

A **dolly** is a wheeled cart or hand truck that is used to transport heavy items such as barrels, drums, and boxes. Figure 3.16 shows examples of different types of dollies.

Dolly a wheeled cart or hand truck used to transport heavy items such as barrels, drums, and boxes.

DOLLY SAFETY Basic safety requirements for the operation of dollies:

- Dollies must be pushed rather than pulled so they cannot run over the driver.
- Dollies must be operated below recommended weight limitations.
- The load must be secured before moving it.
- The path needs to be clear and well-lighted.
- Slip-resistant footwear must be worn.
- A safe angle must be maintained to prevent tipping.

Figure 3.16 Examples of dollies.

CREDIT: (left to right) Iamnee/Shutterstock. gualtiero boffi/Shutterstock.

3.3 Basic Hand and Power Tool Safety

OSHA states that employers shall be responsible for the safe condition of tools and equipment used by employees, but it is up to the employee to ensure that tools are used properly and safely.

In some highly hazardous work areas, the use of **nonsparking** or spark-resistant hand tools and intrinsically safe power tools is required. These spark-resistant hand tools are manufactured with specific materials, such as brass, bronze, copper alloys, or plastic, which will not produce sparks capable of igniting any vapors that might be present in the area. Intrinsically safe power tools are tools in which the amount of electrical and thermal energy is limited to a point below that which is required for ignition of a hazardous atmosphere (produced by vapors or dust).

Nonsparking tools   tools that are manufactured from specific materials that will not produce sparks capable of igniting vapors; also referred to as *spark-resistant* or *spark-proof tools*.

Safety Tips For Tool Use

General tool usage safety tips:

- Never engage in horseplay in the workplace, especially when working with tools.
- Always obtain the required work permits before beginning any job (e.g., a hot work permit for grinding operations).
- Use the right tool for the job.
- Understand how to properly use and maintain the tools with which you are working.
- Review safety procedures and familiarize yourself with the hazards associated with specific tools.
- Inspect the condition of the tool before use.
- Check that edged tools are sharp.
- Inspect metal tools for slivers, cracks, or rough spots.
- Check for wear or damage on any plastic or rubber parts or coatings.
- Inspect wooden handles for splinters, chips, or weathering.
- Do not use damaged tools. Instead, make sure they are labeled "Do Not Use" and report the tool condition to your supervisor.
- Wear the appropriate personal protective equipment (e.g., always wear eye protection when using tools).

- Avoid wearing loose clothing or jewelry, as these can get tangled in tools or machinery.
- Make sure the work area is clear and obstacles are removed.
- Keep a firm grip on the tool and maintain proper balance and footing.
- Do not leave tools in walkways or high traffic areas.
- Make sure the floor is clean and dry to prevent slips or falls when you are working with or around tools.
- Make sure the work area is properly lighted.
- Maintain a safe distance from other workers when using tools.
- Do not distract others while they are using a tool.
- Clean the tool when done and return it to the proper storage location.
- Be aware of other work that may affect the usage of the tool in that area.

General Hazards

General hazards presented by tools include:

- Sparks (especially from grinding tools) in flammable or hazardous environments, which can result in a fire or explosion
- Impact from flying parts or fragments (from the tool or material being worked on)
- Burns from any material surface or tool, especially power tools, that becomes heated from use
- Lacerations, contusions, and muscle strain caused by improper body position, misuse, or repetitive use
- Harmful dusts, fumes, mists, vapors, or gases released by the material being worked on
- Impalement on sharp edges
- Injury from a tool that is dropped from a height
- Ergonomic hazards such as carpal tunnel syndrome or tendonitis from improper design or use, or repetitive motion
- Noise hazards, such as a hand tool striking a surface or a loud noise generated by power tools (e.g., a jackhammer)
- Falls while trying to use the tool at heights
- Forceful ejections of projectiles (e.g., nails from a nail gun or pressurized water from a water blaster).

Hand Tools

Hand tools are tools that are operated manually instead of being powered by electric, pneumatic, hydraulic, or other forms of power. Hand tools come in a wide range of types and designs.

The greatest hazards associated with use of hand tools are caused by misuse and improper maintenance. The following are some general tips for proper use and safety when using hand tools:

- Check wooden-handled tools for splinters, cracks, chips, or weathering. Make sure the handle of a hammer is securely attached to the head.
- Never modify a tool to do a task for which it was not designed.
- Do not use screwdrivers as chisels or gouges.
- Check impact tools (e.g., chisels and wedges) for heads that have been flattened with repeated use (called mushroom heads) because these can shatter.

- Make sure the jaws of wrenches and pliers are not sprung (loose) to the point that slippage occurs.
- Inspect spark-resistant tools for wear or damage because they are made of materials that are softer than other tools and can wear down more quickly.
- Maintain a proper grip, holding the handle firmly across the fleshy part of your hand.
- Do not overexert yourself. If you feel your hand or arm strength weakening, take a break and resume the task later.
- In moist environments, make sure your hands remain dry and your vision is not blurred by sweat.
- Do not overtighten fasteners.

WRENCHES   General safety suggestions for wrenches:

- Valve wheel wrenches should fit over the valve wheel rim completely. The wheel rim must be in contact with the top of the wrench opening when applying force.
- The body should be braced and the wrench should be pulled rather than pushed. If the wrench must be pushed, the flat of the hand should be used (rather than gripping around the wrench).
- The wrench should be kept in the plane of the valve wheel rim. Pulling at an angle increases the chance of slippage.
- A pipe extension or cheater bar should never be used to apply excessive leverage to a wrench.
- Wrenches should not be struck with hammers or other objects.
- A pipe wrench must not be used as a valve wheel wrench. Wrench teeth create sharp edges and burrs on a valve wheel that can cut an unprotected hand.

Power Tools

There are various types of power tools. The main types are electric, pneumatic, fuel-operated, hydraulic, or powder-actuated. While each of these tools can enhance performance, they also can pose a variety of hazards, including:

- Power-source related hazards such as electrocution (electrical) or line whip (pneumatic)
- High-speed impact from projectiles hurled by broken tools or materials
- Injuries from contact with moving parts or entanglement
- Fall or trip hazards from cords or hoses
- Vibration and noise hazards
- Slipping hazard from leaking fluids
- Hot surfaces caused by activity or prolonged operation (e.g., drill bits and motor housings).

In addition to the general safety procedures outlined previously, the following are some safety procedures for power tools:

- Follow all manufacturers' specifications for proper and safe use, and read all warning labels.
- Understand the capabilities, limitations, and hazards of the tool being used.
- Check that all safety features work properly.
- Never point the tool toward a person.

- Check that all guards and shields are in place.
- Do not use tools that are too heavy or too difficult to control.
- Inspect all cords or hoses for wear, fraying, or damage.
- Make sure the tool is kept clean and properly maintained, including lubrication, filter changes, and other manufacturer-recommended practices.
- Avoid using tools in dangerous environments (e.g., wet, flammable, extremely hot or cold).
- Use only intrinsically safe or explosion-proof tools in flammable environments.
- Do not carry the tool by its cord or hose.
- Do not yank the tool cord or hose to disconnect it.
- Keep cords and hoses away from heat, oil, and sharp edges.
- Do not let cords or hoses get knotted or tangled with other cords or hoses.
- Make sure cords or hoses are not lying across a walkway, presenting a trip hazard.
- Keep cords or hoses away from rotating or moving parts.
- Choose the correct accessories and use them properly.
- Check all handles to make sure they are secure and all grips are on tight.
- Keep your finger off the switch when carrying the tool or before it is in position to avoid accidental startup.
- Hold or brace the tool securely.
- Disconnect the tool before servicing it or when changing accessories (e.g., blades and bits).

ELECTRIC TOOLS **Electric tools** are tools operated by electrical power (either AC or DC). The following are some safety practices related to electric tools:

Electric tool a tool operated by electrical power (either AC or DC).

- Make sure all electric tools are properly grounded or double insulated.
- Always use a ground fault circuit interrupter (GFCI) when using electrical power tools.
- Know the hazards of electricity, including electrocution, shocks, and burns.
- Be aware of potential secondary hazards from electricity (e.g., a mild shock from an electric tool that startles a worker, causing a fall from a ladder).
- Never remove the grounding plug from a cord.
- Understand the mechanical hazards of the tool.
- Remember to avoid wet or damp environments.
- Do not use a tool if it becomes wet.

PNEUMATIC TOOLS **Pneumatic tools** are tools powered by pneumatic (air or gas) pressure. The following are safety practices related to pneumatic tools:

Pneumatic tool a tool that is powered using pneumatic (air or gas) pressure.

- Make sure attachments are properly secured using a tool retainer or safety clip.
- Check the hose and hose connection. Use only approved connectors.
- Do not exceed the manufacturer's safe operating pressure for hoses, pipes, valves, filters, and fittings.
- For hoses exceeding a one-half inch (1.25-cm) inside diameter, use a safety device at the source of supply or branch line to reduce pressure in the event of hose failure.
- Never kink the hose to cut off the air supply. Instead, turn off the air using the valve.
- Check for proper pressure before using the tool.
- Ensure that hose connections and overhead tools are secured with a lanyard.

Hydraulic tool a tool that is powered using hydraulic (liquid) pressure.

Fuel-operated tools tools powered by the combustion of a fuel (e.g., gasoline).

Powder-actuated tool a tool that uses a small explosive charge to drive fasteners into hard surfaces such as concrete, stone, and metal.

HYDRAULIC TOOLS **Hydraulic tools** are tools powered by hydraulic (liquid) pressure. Many of the safety precautions for hydraulic tools are the same as those for pneumatic tools. However, one added hazard is exposure to hot and/or harmful hydraulic fluids.

FUEL-OPERATED TOOLS **Fuel-operated tools** are tools powered by the combustion of a fuel (e.g., gasoline). Process technicians should always use caution when operating these tools and ensure that proper grounding practices are used when refuelling. Both filling and tool operation produce static electricity, which can result in a fire or explosion. Tools should be refueled and operated in open air environments to avoid build-up of harmful fuel vapors or equipment exhaust.

POWDER-ACTUATED TOOLS **Powder-actuated tools** are tools that use a small explosive charge to drive fasteners into hard surfaces such as concrete, stone, or metal. Figure 3.17 shows an example of a powder-actuated tool designed to be struck by a hammer.

Because powder-actuated tools use powder charges that are explosive, they should always be handled with caution. Most of these types of tools will not activate unless the tip is pressed against a work surface. However, accidents do happen. Technicians should only place the tip of a powder-actuated tool toward the intended work surface, and never point it at another person.

Figure 3.17 Example of a powder-actuated tool.

3.4 Process Technician's Role in Tool Care and Maintenance

Like pumps, compressors, or any other type of equipment, hand tools must be properly maintained for safe operation. Hand tools should be stored inside a properly sealed container or toolbox to prevent exposure to moisture and the potential for corrosion.

With power tools, routine checks and lubrication of various mechanical moving parts should always be performed before and after use. Electric tools should be inspected for frayed cords and bent or loose plugs.

With all hand tools, the handles should fit tightly. If the handle is cracked or otherwise damaged, replace or repair the tool before using. As a general rule, always inspect your tools for defects before using them.

Such housekeeping chores might seem tedious and unnecessary, but simple precautions can mean the difference between successful completion of a project and serious or fatal injury. Keep your equipment in top working condition to make the job easier and safer.

Summary

There are many tool and construction hazards in the process industries. Process technicians use hand tools and powered tools while doing their jobs and must do so safely. Additionally, while process technicians are not likely to be involved in construction-related tasks, they might encounter construction areas around their facility. Thus, technicians must be aware of the safety hazards that construction areas and tools present.

Hand and power tools used on a regular basis include hammers, pliers, cutters, and many types of wrenches. Wrenches are used to make adjustments and to tighten and remove fasteners as needed.

Lifting equipment is used in industry to assist the process technician in raising heavy objects to elevated positions. Hoists, cranes, and forklifts are all commonly used lifting equipment. Lifts that contain a pneumatic or electric arm with a bucket attached to the end can be used to lift personnel to elevated locations as needed. Safety is a primary concern with large devices such as hoists, cranes, and forklifts. Process technicians should use caution to prevent potentially fatal accidents when using this equipment.

A safe work environment requires the proper care, use, and maintenance of tools. Before working with tools, process technicians should select the proper tool for the job and inspect the tool for wear, damage, or other defects that could make it unsafe to use. Technicians must never modify tools or use them for tasks for which they were not designed. Proper techniques and safety practices must be maintained when using tools. This includes wearing proper PPE, clearing obstacles or hazards from the work site, and ensuring that other individuals are a safe distance away before work begins.

Checking Your Knowledge

1. Define the following terms:

   a. Tool

   b. Dolly

   c. Crane

   d. Hand tool

   e. Power tool

   f. Hoist

   g. Hydraulic tool

   h. Pneumatic tool

   i. Powder-actuated tool

   j. Nonsparking tool

2. It is the responsibility of every employee to ensure that tools are used _____ and _____.

   a. properly and safely

   b. quickly and safely

   c. properly and quickly

   d. with supervision and permission

3. _____ pliers have a tongue and groove joint that allows the distance between jaws to be altered.

   a. Needle nose

   b. Adjustable

   c. Locking

   d. Lineman's

4. A _____ wrench is a device used to apply a precise amount of force to a fastening device such as a nut or bolt.

   a. pipe

   b. socket

   c. torque

   d. valve

5. A(n) _____ crane runs on an elevated runway system along the length of a warehouse and provides three axes of hook motion (up and down, sideways, and back and forth).

   a. overhead traveling bridge

   b. gantry

   c. jib

   d. monorail

6. (True or False) Process technicians have to obtain certification to operate a forklift.

7. (True or False) A pipe wrench can be used as an alternative to a valve wheel wrench.

8. If a tool is found to be damaged upon inspection prior to use, the correct next steps are:

   a. To throw it away and find a replacement

   b. To see if it is still functional and report it at the end of shift

   c. To put in a requisition for a replacement and use the tool carefully

   d. To label it "Do Not Use" and report its condition to the supervisor

9. Why should forklifts never be used as personnel lifts? (Select all that apply.)

   a. The forks are slippery and a fall hazard.

   b. They can tip over easily.

   c. They are not stable for personnel to stand on.

   d. They are not equipped with the required fall protection systems.

10. (True or False) Keeping equipment in top working condition makes the job easier and safer.

NOTE: Answers to Checking Your Knowledge questions are in the Appendix.

Student Activities

1. Based on your personal experience, name some hazards you have encountered working with tools, along with any injuries you or others suffered, at work or at home. How could these hazards have been addressed and the injuries prevented? Discuss this with your fellow students.

2. Research tool safety. Write a two-page paper for submission and prepare a presentation, focusing on a specific tool, its usage and maintenance, safe operation, and potential hazards.

3. Given a set of assorted tools, identify each tool type and describe or demonstrate proper and improper use with a classmate. Be prepared to present this information to the class.

4. Consider the use of wrenches in the process industries. Work in a team and discuss the various types of wrenches, how they are used, and the safety-related effects of proper and improper usage.

5. Using sewing thread and weights, experiment with the physics of proper and improper cable loading.

6. Demonstrate the proper use of tools for tightening a bolt or a flange.

7. Research and present OSHA regulations for forklift, crane, and/or power tool use.

Chapter 4

Piping, Gaskets, Tubing, Hoses, and Fittings

Objectives

After completing this chapter, you will be able to:

4.1 Describe the purpose, types, and uses of piping, gaskets, tubing, hoses, blinds, and accessories. (NAPTA Piping 1-3, 14*) p. 58

4.2 Discuss the uses, advantages, and constraints of various construction materials. (NAPTA Piping 5) p. 64

4.3 Explain selection, sizing criteria, and schedules related to piping thickness and flange ratings. (NAPTA Piping 1, 6-8) p. 67

4.4 Identify different types of connections, fittings, and sealants for piping, tubing, and hoses. (NAPTA Piping 4, 9, 10, 13) p. 69

4.5 Discuss the potential hazards involved with piping, gaskets, tubing, hoses, and fittings. p. 76

4.6 Identify the process technician's responsibility with regard to potential hazards, operation, and maintenance of for piping, gaskets, tubing, and hoses. (NAPTA Piping 11-13, 15, 16) p. 77

*North American Process Technology Alliance (NAPTA) developed curriculum to ensure that Process Technology courses will produce knowledgeable graduates to become entry level employees in process technology. Objectives from that curriculum are named here in abbreviated form. For example, "(NAPTA Piping 5)" means that this chapter's objective relates to objective 5 of NAPTA's course content on piping.

Key Terms

Alloys—compounds composed of two or more metals or a metal and nonmetal (e.g., carbon steel) that are mixed together in a molten solution, **p. 60.**

Blinds (or blanks)—solid plates or covers that are installed between pipe flanges to prevent the flow of fluids and to isolate equipment or piping sections when repairs are being performed; typically made of metal, **p. 62.**

Corrosion—deterioration of a metal by a chemical reaction (e.g., iron rusting), **p. 62.**

Expansion loop—segment of pipe that allows for expansion and contraction during temperature changes, **p. 63.**

Gasket—flexible material used to seal components together so they are air- or watertight, **p. 59.**

Heat tracing—a coil of heated electric wire or steam tubing that is wrapped around a pipe to increase the temperature of the process fluid, reduce fluid viscosity, and facilitate flow, **p. 74.**

Insulation—any substance that prevents the passage of heat, light, electricity, or sound from one medium to another, **p. 73.**

Jacketed pipe—a pipe-within-a-pipe design that allows hot or cold fluids to be circulated around the process fluid without the two fluids coming into direct contact with each other, **p. 73.**

MAWP—maximum allowable working pressure; a safety limit for components in a piping system; the safe pressure level determined by the weakest element in the system, **p. 69.**

Nominal pipe size—a number that is used to represent pipe size, **p. 67.**

Pipe clamp—piping support that protects piping, tubing, and hoses from vibration and shock; also, a ring-type device used to stop a pipe leak temporarily, **p. 63.**

Pipe hanger—piping support that suspends pipes from the ceiling or other pipes, **p. 63.**

Pipe shoe—piping support that supports pipes from beneath, **p. 63.**

Schedule—piping reference number that pertains to the wall thickness, which affects the inside diameter (ID) and specific weight per foot of pipe, **p. 67.**

Steam trap—device used to remove condensate from the steam system or piping, **p. 75.**

Stress corrosion—type of corrosion that results in the formation of cracks (called stress cracks), **p. 65.**

Tensile strength—the pull stress, in force per unit area, required to break a given specimen, **p. 64.**

4.1 Introduction

Piping, gaskets, tubing, hoses, and fittings are the most prevalent pieces of equipment in the process industries and account for a significant portion of the initial investment when building a new facility.

In most plants, you will see large segments of pipe going from one location to another. These pipes carry chemicals and other materials into and out of various processes and equipment.

When building a process facility, it is important to select proper construction materials and connectors. Process technicians need to be aware that some materials and connectors are inappropriate for certain processes, pressures, or temperatures. Improper material selection and improper operation can lead to leaks, wasted product, or conditions that are harmful to people and/or the environment.

Piping, Gaskets and Flanges, Tubing, Hoses, Blinds, Supports, and Expansion Items

Piping

Pipes are long, hollow cylinders through which fluids are transmitted. In the process industries, piping is used to transport fluids throughout a facility and between different parts of a process. Piping also is used to carry materials or products outside a facility and across long distances. An example of this type of piping is the 800-mile-long Trans-Alaska Pipeline.

Piping comes in many different sizes and materials. It can be rigid (e.g., metal) or flexible (e.g., plastic). It can be rated for high pressure (e.g., steam) or low pressure (e.g., instrument air) use. It can be permanent (e.g., welded joints) or temporary (e.g., threaded ends), and it can carry fluids over long or short distances.

Did You Know?

The Trans-Alaska Pipeline was designed and constructed to move oil from the North Slope of Alaska to the northernmost ice-free port of Valdez, Alaska.

The pipeline itself is 800 miles (1287 kilometers) long and 48 inches (1.2 meters) in diameter. It crosses three mountain ranges and more than 800 rivers and streams

The construction of the pipeline, which began in March 1975 and took three years and $8 billion to complete, was the largest privately funded construction project at that time.

Since its construction, more than 17 billion barrels of oil have moved through the Trans-Alaska pipeline.

CREDIT: Saraporn / Shutterstock.

Gaskets and Flanges

Gaskets are components made from flexible material used to make a seal between two piping or equipment connection points. They can also be used to dampen equipment vibration, minimizing the damage that can result from excessive vibration. They are made from both metallic and nonmetallic materials, including compressed fiber, graphite, and rubber. Gaskets are shaped and sized to fit the connection point.

Gasket flexible material used to seal components together so they are air- or watertight.

Gaskets are used between all flange connections to seal the connection. As flange bolts are tightened to join two piping sections or pieces of equipment, the gasket is compressed between the two surfaces. It fills any imperfections in the sealing surfaces, providing a sealed connection.

Full-face gaskets are those that have holes cut in them for the flange bolts to pass through. They are usually used with flat-faced flanges. Ring-type gaskets are those without bolt holes; they are used with raised-face flanges. Another gasket type, called a ring type joint (RTJ) is a metal ring, which fits into a groove in the two flanges. In applications where the flange is an unusual shape or is extremely large, the gasket can be a segmented type in which two or more gaskets are arranged together and overlap to provide a continuous sealing surface.

Types of gaskets include spiral wound, jacketed, and corrugated. Gasket material and temperature/pressure rating must match the application.

Piping gasket material must be compatible with the substance inside the pipe. Otherwise, the gasket could corrode, fail, and leak process fluid. Figure 4.1 shows examples of gaskets.

Gaskets and flanges are rated by classes. Common classes are 150, 300, 400, 600, 900, 1500, and 2500. Temperature ratings are dependent on the flange material and gasket material used, and range from cryogenic temperatures to more than 2000 degrees Fahrenheit (1093 degrees Celsius).

Figure 4.1 Gaskets: **A.** Ring type, semi-metallic spiral wound. **B.** Full face, rubber. **C.** Metallic, ring type joint.

CREDIT: **A.** Oil and Gas Photographer/Shutterstock. **B.** Ph.wittaya/Shutterstock. **C.** robertonencini/Shutterstock.

A.   **B.**   **C.**

The proper bolt-tightening pattern is important whenever a new gasket is installed in a flanged connection. This pattern includes tightening opposite bolts first, then moving 90 degrees for the next set. So (using clock positions), an example of a correct pattern would be 12 o'clock, 6 o'clock, 3 o'clock, 9 o'clock, 1 o'clock, 7 o'clock, 10 o'clock, and 4 o'clock. Tightening flange bolts in the proper pattern maintains the correct placement of the gasket.

The gasket should be sized properly and should never be folded or crimped. Even, but not excessive, pressure on all of the flange bolts should be maintained to prevent leakage. If the bolts are tightened excessively, the gasket can be damaged and the seal will be compromised. Gaskets should never restrict the inside diameter of the pipe.

Tubing

Tubing is manufactured using an extrusion process to form it in the shape of small diameter piping, and it can be flexible or rigid. Tubing is used to transport small amounts of process fluids and usually has a diameter of less than 1 inch (2.5 cm).

Tubing can be made from a variety of materials, including copper, synthetics, plastics, and **alloys** (compounds composed of two or more metals or a metal and nonmetal mixed together in a molten solution), and it can be used in a variety of applications (e.g., instrument air lines, heat trace tubing, metered chemical additions, and sampling systems).

Metal tubing can be bent using a tubing bender (Figure 4.2). Attempting to bend metal tubing without a tubing bender can kink the tubing and cause it to leak. The connecting methods used on metal tubing are compression fittings (most common), flared fittings, and brazed or soldered joints. Table 4.8 on page 71 describes tubing connection methods for metal and plastic tubing.

Alloys compounds composed of two or more metals or a metal and nonmetal (e.g., carbon steel) that are mixed together in a molten solution.

Figure 4.2 Tubing bender.

CREDIT: curraheeshutter/Shutterstock.

Hoses

Hoses are flexible tubes that carry fluids. Hoses can be used in a variety of applications, including steam for cleaning or flushing, plant air for air-driven equipment, water for washing, and nitrogen or other gas services. Also, if a section of pipe is damaged or working improperly, a hose can sometimes be installed temporarily until a replacement pipe can be fabricated.

Because of the operating pressure and temperature as well as the corrosive nature of some fluids, the hose material selected must be able to withstand the service for which it is being used. For example, utility hoses are used to deliver air, water, nitrogen, and steam. These types of hoses are made from a variety of materials, including plastic, rubber, fiber, metal, or a combination of materials. Some of the specialized hoses and their uses are described below.

Care must be taken to prevent damage to hoses (for example, caused by the tires of heavy vehicles or by kinking). If hose lengths will be laid out across roads where vehicles will travel, they should be protected from crushing by boards laid on each side and parallel to the hose. When they are no longer in use, hoses should be drained, neatly coiled, and properly stored.

One unique feature of utility hoses is the style of their connectors. Each connector is designed to connect to a specific service or utility (e.g., air hoses can be connected only to air lines and water hoses can be connected only to water lines). This prevents cross-contamination by someone accidentally connecting a particular hose to an incorrect service line (e.g., connecting an air hose to a steam line).

Because of the heat and pressure associated with steam, steam hoses are typically constructed of a heat resistant rubber inner tube, reinforced with multiple layers of braided steel fiber. This durable construction allows for operation at high temperatures and pressures.

Composite hoses are flexible thermoplastic hoses reinforced with a woven metal inner and/or outer helix. They are usually used for connecting piping to a transport vehicle for loading or offloading. Many hoses of this type have quick-lock connectors that use a camlock coupling to lock the hose in place. The metal coil strengthens the hose and protects it from kinking and collapsing, while the cover provides durability for long-term use. These hoses have specific pressure, temperature, and chemical ratings. Figure 4.3 shows an example of a hose with a quicklock connector.

Figure 4.3 Hose with quick-lock camlock coupling connector.

CREDIT: yanin kongura / Shutterstock.

Color-coded flexible hoses (shown in Figure 4.4) are flexible hoses in different colors to indicate different uses. Each type of hose is also matched with a specific hose fitting (see Table 4.9 on page 72). Color coding and specific hose fittings help to ensure that the proper hose is used for a particular service (e.g., water hoses might be one color, while nitrogen hoses are another color). Color coding helps process technicians know which service the hose is intended for and reduces the risk of error or injury. The colors and fittings that are used for each type of hose vary from company to company.

Figure 4.4 Color-coded flexible hoses

CREDIT: **A.** PRILL/Shutterstock. **B.** Voyagerix/Shutterstock. **C.** Chromatic Studio/Shutterstock.

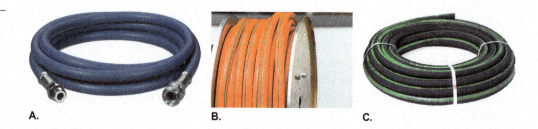

A.   B.   C.

LIMITATIONS OF HOSES AND FITTINGS   All hoses have a maximum pressure rating at a specific temperature. Pressure ratings usually are determined at ambient temperature 70°F (21°C), and maximum working pressure will decrease with increasing temperature. The maximum hose pressure must not be exceeded because of the danger of rupture, which could cause personal injury or environmental hazards. Before use, the hose must be inspected for any damage. The inspection date should be located to ensure that the hose is still within the current inspection period.

When there is a chance that the temperature, pressure, or chemical limitations of a hose will be reached or exceeded, the hose must be replaced with piping. The replacement piping material must be able to withstand the high temperature and pressure conditions and must be compatible with the process chemicals in the line.

Blinds

Blinds (or blanks)   solid plates or covers that are installed between pipe flanges to prevent the flow of fluids and to isolate equipment or piping sections when repairs are being performed; typically made of metal.

Corrosion   deterioration of a metal by a chemical reaction (e.g., iron rusting).

Blinds (or blanks) are solid plates or covers that are installed between pipe or equipment flanges to prevent the flow of fluids and to isolate equipment or piping sections when repairs are being performed. Blinds are normally made of metal and have properties that are compatible with the normal service of the piping in which they are installed. For example, if the normal service is highly corrosive, then the blind will be made of a material resistant to **corrosion** to prevent deterioration of the metal by a chemical reaction.

Two common types of blinds that process technicians can encounter are paddle blinds and spectacle blinds (Figure 4.5).

Figure 4.5 A. Paddle blind. **B.** Spectacle blind.

A.   B.

Both paddle and spectacle blinds are installed between two flanges. Paddle blinds have a handle that sticks out of the flange (see Figure 4.6A). This handle provides a quick visual reference that tells process technicians the pipe has been blocked with a blind. Paddle blinds are not designed to handle the system pressure and are installed to ensure all flow in a pipe has been blocked. On the other hand, spectacle blinds (or pressure blinds) such as the one in Figure 4.6B have a disk (either solid or with a hole in it) sticking out of the flange instead of a handle. Spectacle blinds are thick because they are designed to handle system pressure. The exposed side of the spectacle blind indicates to a process technician that the flange contains a blind and that the flow path is either open or blocked. If the side with the hole in it is visible from the outside, then the technician knows the pipe is blocked. If the solid side is visible, the technician knows the pipe is unblocked and ready for service. Figure 4.6C shows a cutaway of a paddle blind installed in a flange.

Because of the hazards associated with the blinding process, blank/blind checklists are frequently used to ensure complete isolation of the equipment or line during repair and to ensure that all blinds are removed before normal operations are resumed.

Figure 4.6 **A.** Paddle blind installed. **B.** Spectacle blind installed. **C.** Pipe flange paddle blind installed.

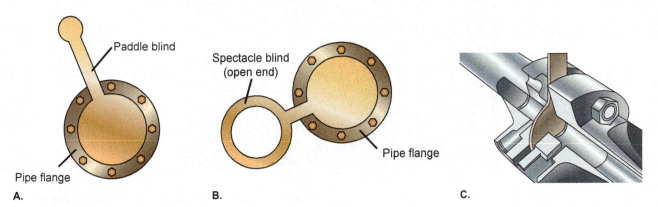

A.   B.   C.

Piping Supports

It is quite common to find extremely long runs of pipe in the process industries, and the long pipe runs must be supported properly in order to function well. Figure 4.7 shows a few of the piping supports that are available.

Pipe shoes are piping supports that hold pipes from underneath. **Pipe hangers** suspend pipes from above. **Pipe clamps** are piping supports that protect piping, tubing, and hoses from vibration and shock.

Piping supports hold piping in place to prevent movement or damage from the weight of the piping run. They also allow for thermal and mechanical expansion and contraction. Without proper support, the piping and the equipment associated with these supports can be easily damaged. For example, if a line is supported improperly, it could throw a pump out of alignment, causing seal, coupling, and impeller damage.

Pipe shoe   piping support that supports pipes from beneath.

Pipe hanger   piping support that suspends pipes from the ceiling or other pipes.

Pipe clamp   piping support that protects piping, tubing, and hoses from vibration and shock.

Figure 4.7 A sampling of piping supports. **A.** Pipe shoe. **B.** Pipe hanger. **C.** Pipe clamp.
CREDIT: B. Wichien Tepsuttinun / Shutterstock. **C.** Viktor Chursin / Shutterstock.

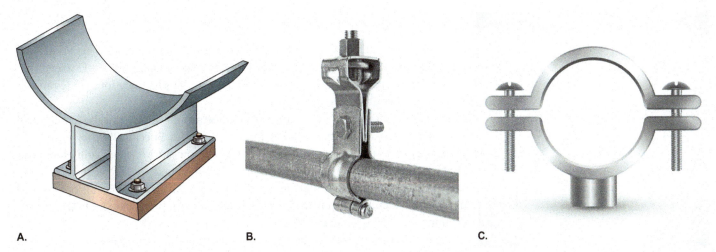

A.   B.   C.

Expansion Loops and Joints

An **expansion loop** is a segment of pipe that allows for expansion and contraction during temperature changes. Figure 4.8 shows an example of expansion loop in a pipeline.

Temperature inside a pipe can change dramatically during normal operations. For example, 650 PSIG superheated steam is approximately 725°F (385°C). As the steam cools, the temperature decreases to 450°F (232°C) and the pipe contracts. Expansion loops allow

Expansion loop   segment of pipe that allows for expansion and contraction during temperature changes.

Figure 4.8 Expansion loops allow for thermal expansion and contraction.

CREDIT: Jochen Tack / Alamy Stock Photo.

Figure 4.8 Expansion loops allow for thermal expansion and contraction.

CREDIT: Jochen Tack / Alamy Stock Photo.

for this type of thermal expansion and contraction and minimize its impact. Without an expansion loop, the pipe could rupture, damage the foundation, or break the supports. The pipe also could become misaligned because of expansion and contraction, thereby damaging the seals on pumps and turbines. This type of damage can be very costly and could cause serious physical and environmental hazards.

An *expansion joint* is a bellowslike component on piping that allows for thermal expansion and vibration (Figure 4.9).

Figure 4.9 Piping expansion joint.

CREDIT: Image Source / Alamy.

4.2 Materials of Construction

Industrial piping, tubing, hoses, and fittings can be made of many different materials, such as carbon steel, alloy steel, stainless steel, exotic metals, glass, plastic, or clay. However, carbon steel is the most common piping material because it is appropriate for a wide range of temperatures and is relatively economical.

When designing piping and valve systems, engineers or designers must be familiar with the process and the substances that will flow through the pipes. Specifically, they must know the temperature of the substance, its viscosity, how much pressure it exerts, and other properties (flammability, corrosiveness, and reactivity). Failure to take these factors into account could have serious effects on the safety and performance of the plant once it is operational. For example, some metals become brittle at extremely low temperatures. Table 4.1 contains a list of common construction materials and their characteristics.

Process technicians must know these characteristics and limitations for their equipment as well.

Carbon Steel and its Alloys

Carbon steel is the most common material used for construction of piping systems where corrosion is not a problem. The benefits of carbon steel, compared to stainless steel or zirconium, are its availability, affordability, and increased **tensile strength** (the pull stress, in force per unit area, required to break a given specimen). The greater the tensile strength, the thinner the pipe walls can be.

Tensile strength the pull stress, in force per unit area, required to break a given specimen.

Table 4.1 Common Construction Materials and Their Characteristics

Construction Material	Temperature Ranges	Description
Carbon steel	−20°F to 800°F (−29°C to 426°C)	The most commonly used construction material because of its wide temperature range, weldability, strength, and relatively low cost.
Stainless steel	−150°F to > 1400°F (−101°C to > 760°C)	Less brittle than carbon steel at extremely low temperatures; appropriate for use in high-temperature applications; more corrosion-resistant than carbon steel, especially in acid service; susceptible to stress corrosion cracking; costlier than carbon steel.
Brass, bronze, and copper	−50°F to 450°F (−45.6°C to 232°C)	Corrosion-resistant at low or moderate temperatures; excellent heat transfer properties; higher cost than steel
Alloys (e.g., hastelloy and inconel)	Varies with the alloy	Can be used in high-temperature applications (above 800°F/426°C) (e.g., furnace tubes) and highly corrosive service; high cost.
Plastics	Varies with the plastic	Used for low-pressure applications and corrosive services; easy to install and lightweight; low in cost.

While some construction materials are pure metals, others are alloys, which are compounds composed of two or more metals (or metal and nonmetal) mixed together in a molten solution. Bronze, for example, is an alloy of copper and tin. Carbon steel (also known simply as steel) is an alloy of iron and carbon. Alloys improve the properties of single-component metals and provide special characteristics that are not present in the original metals.

Carbon steel is used routinely in the construction of piping for organic chemicals, as well as neutral or basic aqueous solutions at moderate temperatures. In instances where corrosion is an issue, the walls of the pipe are sometimes lined with a protective coating (e.g., Teflon®, glass, or polymers such as butyl rubber); their wall thickness can be increased, or an alloying material can be added to change the properties of the carbon steel. The result is a more corrosion-resistant and higher-strength material. The most common alloying elements are nickel, manganese, chromium, and molybdenum.

Stainless Steel

Stainless steel is an iron-carbon alloy with chromium that is commonly used in the process industries because it resists heat and corrosion, such as rust. Because of these properties, stainless steel often is used in pump shafts, valves, gears, and vessels.

The use of stainless steel, however, does have some drawbacks. For example, stainless steel is significantly more expensive than carbon steel, and it is subject to **stress corrosion**, a type of corrosion that results in the formation of stress cracks. Stress corrosion can occur when stainless steel is exposed to chlorides (e.g., sea water) or caustics (e.g., sodium hydroxide).

Stainless steel comes in several grades, which are determined by the amount of chromium and other elements present (e.g., nickel, molybdenum, silicon, and manganese). Stainless steel alloys can be grouped into three categories: martensitic, ferritic, and austenitic. These categories are based on their composition, molecular structure, and physical properties.

Martensitic stainless steel (e.g., SS 410) is a small group of magnetic steels that typically contain 12% chromium, a moderate level of carbon, and a very low level of nickel. Martensitic steel can be hardened with heat treatment. The result is a very hard, high-strength alloy.

Ferritic stainless steel (e.g., SS 430) has low carbon content and contains chromium as the main alloying element. Ferritic steels are very resistant to oxidation and cannot be hardened by heat treatment.

Austenitic stainless steel (e.g., SS 304 and SS 316) describes a group of chromium-nickel steels that are nonmagnetic, are very corrosion resistant, and cannot be hardened by heat treatment.

Table 4.2 lists some of the more common grades of stainless steel, their types, compositions, properties, and uses.

Stress corrosion type of corrosion that results in the formation of cracks (called stress cracks).

Table 4.2 Common Stainless Steel Grades and Properties

Type	Grade	% Chromium	% Nickel	% Other	Properties/Uses
Martensitic	SS 410	11.5–13.5	—	—	Very hard; magnetic; used to create turbine blades and valve trim
Ferritic	SS 304	18–20	8–12	1% Silicon (Si)	Corrosion resistant; magnetic; used in process piping
Austenitic	SS 316	16–18	10–14	2–3% Molybdenum (Mo)	Nonmagnetic; improved corrosion resistance over SS 304; used in process piping

Other Metals

Medium and high alloys are proprietary groups of alloys that have better corrosion resistance than stainless steels. Table 4.3 lists examples of different types of alloys and nonferrous metals and their descriptions.

Table 4.3 Common Alloys and Other Metals

Medium Alloy Steels	High Alloy Steels	Nonferrous Metals
Description: Moderate corrosion and temperature resistance	**Description:** High corrosion and temperature resistance	**Description:** Extremely high corrosion and temperature resistance
Examples: ■ Durimet 20 ■ Carpenter 20 ■ Incoloy 825 ■ Hastelloy G-3	**Examples:** ■ Hastelloy B-2 ■ Chlorimet 2 and 3 ■ Hastelloy C-276, C-4 ■ Inconel 600	**Examples:** ■ Titanium ■ Zirconium ■ Tantalum (composed of nickel, aluminum, and copper)

Nonmetals

A variety of nonmetallic piping is used in corrosive service (e.g., acid or high pH service). These materials are used because they provide superior corrosion resistance. Table 4.4 provides examples of some nonmetallic piping.

Table 4.4 Examples of Common Nonmetallic Piping

Substance	Description
Clay	Different varieties are used to construct piping associated with settling basins and waste treatment ponds; an inexpensive, economical material
Elastomeric	Has the elastic properties of rubber and made from rubber, epoxy, or other resins; flexible material compensates for vibration and expansion/contraction
Glass	Resistant to most acids except hydrofluoric acid and hot, concentrated phosphoric acids; brittle and subject to thermal shock; impact, abrasion, and thermal shock resistance are added properties in a nucleated crystalline ceramic-metal form of glass that is similar to, or better than, the corrosion resistance of conventional glass
Plastics	Good for applications that do not have extreme high or low temperatures, high pressures, and are not corrosive; include polyvinyl chloride (PVC), chlorinated PVC, high-density polyethylene, polypropylene, and polybutylene
Porcelain and stoneware	Acid resistant, but brittle; brick-lined construction used for many severely corrosive conditions

4.3 Pipe and Flange Selection and Sizing Criteria

Pressure, temperature, corrosiveness, and flow of the process fluid determine the proper piping material and size. The amount of material that will flow through a piping system determines the line size. The corrosiveness of the fluid determines the proper material. The temperature and pressure determine the pipe schedule and flange class. Design engineers calculate the anticipated flow rates and determine the required material, piping size, and class when a unit is designed. Operating a pipe outside the design flows can cause pipe failure.

Flanges and their gaskets are rated by class. The higher the class, the thicker the flange and the higher the pressure rating for a given temperature. When working with corrosive fluids, it is important to use appropriate gasket materials. With some fluids, corrosion occurs much more quickly when the pressure and temperature of the fluid are high.

Pipe Thickness and Ratings

Piping comes in many different dimensions and ratings. To standardize these dimensions and ratings, the American National Standards Institute (ANSI) published a set of guidelines that industries follow. ANSI was sponsored by the American Society for Testing Materials (ASTM) and the American Society of Mechanical Engineers (ASME). The dimensions and characteristics specified in the ANSI guide are referred to as a pipe's schedule.

Piping is sized according to a standardized system which uses two non-dimensional numbers: **nominal pipe size** (NPS), which represents the pipe diameter, and pipe **schedule**, which represents the pipe wall thickness. In pipes up to 12 inches (30 cm), the NPS of standard thickness pipe (schedule 40) is equal to the outside diameter (OD) minus twice the wall thickness. In pipes larger than 12 inches (30 cm), the nominal pipe size is also the actual outside diameter (OD). For any given nominal pipe size, the outside diameter (OD) remains constant as the pipe schedule changes. As seen in Figure 4.10, what changes with schedule (wall thickness) is the inside diameter (ID).

Nominal pipe size a number that is used to represent pipe size.

Schedule piping reference number that pertains to the wall thickness, which affects the inside diameter (ID) and specific weight per foot of pipe.

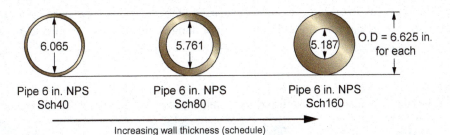

Figure 4.10 Pipe schedule reference chart (in inches).

The schedule of the pipe describes the pipe's design capacity for pressure and temperature. An archaic method of distinguishing different weights of pipe includes standard wall (STD) (commonly called sch 40), extra strong (XS) or extra heavy (XH) wall (sch 80), and double extra strong wall (XX), which is slightly thicker than schedule 160.

As the schedule increases, the wall thickness increases, the ID is reduced, and the allowable working pressure is increased (see Table 4.5).

Table 4.5 Carbon Steel Piping Properties (Nominal Pipe Sizes 3, 4, 6, 8, 10)

Nominal pipe size inch (mm)	Schedule No.	Weight of pipe lb./ft	O.D. (in.)	Wall thickness (in.)	I.D. (in.).	Allowable working pressure for temp. (in °F) not to exceed						
						−20 to 100	200	300	400	500	600	700
3 (80 mm)	S40	7.58	3.500	0.215	3.068	1640	1640	1640	1640	1550	1419	1353
	X80	10.25		0.300	2.900	2552	2552	2552	2552	2412	2207	2105
	160	14.33		0.438	2.624	4122	4122	4122	4122	3895	3566	3401
	XX	18.58		0.600	2.300	6089	6089	6089	6089	5754	5267	5024
4 (100 mm)	S40	10.79	4.5	0.237	4.026	1439	1439	1439	1439	1360	1244	1187
	X80	14.99		0.337	3.826	2275	2275	2275	2275	2150	1968	1877
	160	22.51		0.531	3.438	3978	3978	3978	3978	3760	3441	3282
	XX	27.54		0.674	3.152	5307	5307	5307	5307	5015	4590	4378
6 (150 mm)	S40	18.98	6.625	0.280	6.065	1205	1205	1205	1205	1139	1042	994
	X80	28.58		0.432	5.761	2062	2062	2062	2062	1948	1783	1701
	160	45.30		0.718	5.187	3753	3753	3753	3753	3546	3246	3097
	XX	53.17		0.864	4.897	4659	4659	4659	4659	4403	4030	3844
8 (200 mm)	S40	28.56	8.625	0.322	7.981	1098	1098	1098	1098	1037	950	906
	X80	43.40		0.500	7.625	1864	1864	1864	1864	1761	1612	1537
	XX	72.40		0.875	6.875	3554	3554	3554	3554	3359	3074	2932
	160	74.70		0.906	6.813	3699	3699	3699	3699	3496	3200	3052
10 (250 mm)	S40	40.50	10.75	0.365	10.02	1022	1022	1022	1022	966	884	843
	X80	64.40		0.594	9.75	1484	1484	1484	1484	1403	1284	1224
	160	115.7		1.125	8.50	3736	3736	3736	3736	3531	3232	3082

Flange Class and Ratings Service

In high-pressure systems, piping, fittings, valves, and flanges must be appropriate for the pressure rating of the piping system. Pipe flanges are rated by class. The classes of common flanges range from 150 to 2500. As the class is increased, the flange thickness increases, and the allowable working pressure is increased (Table 4.6). For pipe sizes up to 3" and class 150, a flange has four bolts. The number of flange bolts increases in increments as the pipe size increases.

Table 4.6 Flange Pressure and Temperature Rating for Carbon Steel Pipe Flanged Fittings

Not to exceed °F	(°C)	Class Pressure-Temperature Rating for Carbon Steel Pipe Flanged Fittings						
		150	300	400	600	900	1500	2500
−20 to 100	−29 to 73	285	740	990	1480	2220	3705	6170
200	93	260	675	900	1250	2025	3375	5625
300	149	230	655	875	1315	1970	3280	5470
400	204	200	635	845	1270	1900	3170	5270
500	260	170	600	800	1200	1795	2995	4990
600	316	140	550	730	1095	1640	2735	4560
650	343	125	535	715	1075	1610	2685	4475
700	371	110	535	710	1065	1600	2665	4440

CAUTION: Piping should never be used in a service that exceeds its pressure and temperature ratings!

Pipe sizes and schedules are assigned to piping to help piping system designers specify safe installations. Schedule numbers range from 5 (thinnest wall) to 160 (thickest wall). Refer to Table 4.5 to see the relative increase in the wall thickness of selected schedules of pipe.

Pressure ratings for all components in a piping system must meet or exceed the piping system's maximum allowable working pressure, or **MAWP**. The schedule number defines the wall thickness of the pipe and the MAWP at a specific temperature. For flanges, the class (e.g., Class 300 or Class 600) defines the MAWP at a specific temperature. The weakest component must meet or exceed the MAWP in order for it to be used safely in a piping system. The larger a flange class number, the higher that flange's MAWP for a given temperature.

Did You Know?

You can tell if a piece of pipe is intended for high- or low-pressure applications just by looking at the thickness of the walls.

Thicker-walled piping is designed for high-pressure applications.

Low pressure

High pressure

MAWP maximum allowable working pressure; a safety limit for components in a piping system; the safe pressure level determined by the weakest element in the system.

Did You Know?

You can determine the MAWP when you know the temperature, pipe size and schedule, and flange class.

Example: If a piping system consists of 8" Sch 80 carbon steel pipe and class 900 carbon steel flanges operating at a maximum of 500 degrees F, what would be the MAWP of this system?

On Table 4.5, see this pipe's MAWP at 500 degrees F = 1761 PSI.

On Table 4.6, see the flange's MAWP = 1795 PSI.

Since the weaker component determines the MAWP, in this example the system's MAWP would be 1761 PSI.

4.4 Connecting Methods

Pipe Connections

Pipes and valves can be connected together in a variety of ways. They can be threaded (screwed), flanged, or bonded (e.g., welded, glued, soldered, or brazed).

The intended use of the pipe determines which connection type is the most appropriate. For example, if the pipe is used in low-pressure water service, a threaded joint might be appropriate because it is cheaper and easier to install (and disassemble) than welded or flanged joints. However, if the pipe is to be used in high-pressure, flammable, or corrosive service, a welded joint would be a better option because it is less likely to leak than threaded joints. In general, connections of process piping moving hazardous material will be welded or flanged, or threaded if the pipe size is less than 2 inches (50 mm). Flange joints with gaskets are more expensive and are used in locations where equipment will need to be removed for maintenance or blinded.

THREADED Threaded (screwed) connections involve joining two pipes together through a series of tapered threads. In a threaded connection, the pipe is cut with "male" threads and the connector is cut with "female" threads so the two can be joined together. These threads are cut with precision to ensure the two pieces fit together tightly to avoid leaks, however, this tightness can make it difficult to connect the two pieces. Because of this, thread sealing compound or Teflon® tape often is employed to lubricate the joints, facilitate the connection, and seal against leaks.

Threaded connections are prone to leaks because the connections are weak (i.e., more easily broken because of vibration or external force). For this reason, threaded connections are typically used for low-pressure, nonflammable, and nontoxic service.

Sealant compounds and tapes are used to lubricate and seal the threads on a threaded pipe. Fresh sealant and tape should be used every time a threaded pipe is connected or reconnected. Process technicians also should inspect pipe connections routinely for leakage. Leakage could be an indication that the sealant compound or tape (Figure 4.11) was not applied correctly. Table 4.7 lists some examples of sealant materials.

Figure 4.11 Teflon® tape.

CREDIT: thodonal88 / Shutterstock.

Table 4.7 Types of Sealing Compounds

Sealant	Description
Sealant compound (pipe dope)	Paste applied to the threads of a pipe to seal the threaded connection.
Resin compound	Does not harden like pipe dope and is resistant to shrinkage
Teflon® tape	A thin, white tape made of Teflon®, applied to male pipe threads, to lubricate and seal the threaded connection (see Figure 4.11)

When dismantling a pipe section where sealant or tape is used, the old sealant or tape must be removed before any new sealant material is applied. Failure to remove the old sealant or tape completely can cause leaks. If tape is applied improperly, it can bunch up on the threading and prevent a proper seal or connection.

FLANGED Flanged connections are typically used where the piping might need to be disconnected from another pipe or a piece of equipment. Flanges come in different styles: weld neck, slip-on, socket weld, threaded, lap joint, and blind. The application will determine the correct style to use.

In a flanged connection, two mating surfaces are joined together with bolts. Between the two mating plates is a gasket (a flexible material used to seal components to prevent leakage). As the bolts are tightened, the gasket is compressed between the two surfaces, increasing the tightness of the seal and preventing leakage.

BONDED Bonded pipe joints can be glued, brazed, or soldered. Gluing involves fusing joints together with glue. Glued joints are typically found on plastic lines and pipes, such as lines made of polyvinyl chloride (PVC) or high-density polyethylene (HDPE).

Soldering or brazing involves fusing joints together using a molten filler metal heated to its melting point. Since the metal filler has a lower melting point than the metal being joined, it does not get hot enough to melt the base metal. Soldered joints are found on metal pipes like copper water pipes.

Welded joints are created when sufficient heat is applied to a joint to cause the base metal to melt and fuse the two pipes or fittings together. A molten filler material is used at a high temperature, which when cooled, forms a very strong bond. A butt weld is used to connect two pipes of the same diameter. If one pipe is small enough to fit snugly inside the other, a socket weld is used. Figure 4.12 shows examples of butt and socket welds.

A.   **B.**

Figure 4.12 **A.** Butt weld.
B. Socket weld.

CREDIT: A. shinobi / Shutterstock.
B. LightCooker / Fotolia.

BELL AND SPIGOT Bell and spigot pipe joints (Figure 4.13), are commonly used in clay and concrete piping. The bell is the larger diameter end of the pipe and the spigot is the end that goes into the bell. The joint is sealed using a sealing compound (caulk, cement, etc.) or a compressible rubber ring. The bell and spigot connection is generally used in low-pressure applications, such as sewer piping.

Figure 4.13 Bell and spigot connection.

CREDIT: JL-Pfeifer / Shutterstock.

Tubing Connections

Tubing is most commonly used for instrument air to control valves, although it also has a wide variety of other applications. As shown in Table 4.8, the three primary methods of connecting tubing are with compression fittings, flared fittings, and quick-connect fittings. Tubing, which can be metal or plastic, can be bent and therefore elbows are not usually needed. Generally tubing is used in applications where the ID is 1 inch (25.4 mm) or less.

Table 4.8 Tubing Connection Methods

Fitting Name	Description
Compression fitting	Tubing connection that is sealed by a slip-on tapered sleeve (ferrule) and that presses against the joining tubing when pressure is applied by a threaded connection that fits over the sleeve (also referred to as a *ferrule fitting*) (commonly used with both metal and plastic tubing)
Flared fitting	Fitting for ductile (soft) copper, or aluminum tubing where the ends are flared in a bell shape and sealed with a tapered fitting (used with metal tubing)
Quick connect	Special tubing and piping connections that are sealed by spring-loaded seals so they can be connected and disconnected easily (commonly used with plastic tubing or plastic piping)

Hose Connections

Flexible or flex hose connections (see Table 4.9) are usually used to provide temporary connections. Flex hose should always be thoroughly cleaned between uses. Otherwise, cross-contamination or mixing of chemicals could create conditions in which chemicals react with one another, potentially causing a spill, leak, fire, or explosion.

Table 4.9 Hose Connection Fittings

Fitting Name	Description
Chicago pneumatic (CP) coupling	Type of coupling with two prongs and a sealing gasket used primarily to connect air or water supplies to a hose; also called a crow's foot
Camlock	Fitting composed of a cam and a groove that is used on hoses for quick connect and disconnect
Quick connect	Special hose connections that are sealed by spring-loaded seals so they can be connected and disconnected easily
Boss fitting	Screwed fitting (as opposed to quick connect type), used to connect steam hoses

When using flex hose connections, it is essential to check the pressure, temperature, and corrosiveness of the material for which they will be used. By doing so, the process technician can make sure the flex hose connection is appropriate for the conditions in which it will be used.

Fitting Types

Fittings are piping system components used to connect together two or more pieces of pipe. There are many different types of fittings in the process industries. Figure 4.14 shows examples of some of the most common fittings. Table 4.10 contains a description of the pipe fittings shown in Figure 4.14.

Figure 4.14 Examples of common pipe fittings.

A. Wye (Y) fitting   B. Cap   C. Bushing   D. Flange   E. Plug   F. 45° Elbow   G. Cross

H. Coupling   I. Bell reducer   J. 90° Elbow   K. Union   L. Tee   M. Allthread (close nipple)   N. Nipple

Table 4.10 Descriptions of Pipe Fittings Shown in Figure 4.14

Fitting Name	Description
A. Wye (Y) fitting	A pipe fitting that can be used as a sampling point, a blowdown point, or a strainer to trap particles before entering process equipment
B. Cap	Used to cap (seal) the open end of a pipe; can be threaded, glued, or welded
C. Bushing	Threaded on both the internal and external surfaces; used to join two pipes of differing sizes
D. Flange	Consists of a mating plate which is connected to another mating plate to join two pieces of pipe or a valve to a piece of pipe
E. Plug	Used to close the end of a piping run; similar to caps, except that the threading is male and it is used with a female-threaded fitting; can be glued, threaded, or welded into place
F. and J. Elbows	Used to change the direction of flow; usually a 45- or 90-degree angle
G. Cross	Allows four pipes to be connected together; a cross shape with 90-degree angles

Fitting Name	Description
H. Coupling	Used to connect two threaded pipe sections or hoses; female threading on both ends
I. Bell reducer	Used to connect two pipes of different diameters (large gauge to smaller gauge); both ends are female, with one end being larger than the other
K. Union	Joins two sections of threaded pipe but allows them to be disconnected without cutting or disturbing the position of the pipe
L. Tee	Allows the splitting or joining of flows; a T-shape in which the openings normally join at a 90-degree angle, but they can have a lesser angle
M. Allthread (close nipple)	A short length of pipe (usually less than 3 inches [7.6 cm]) with threads throughout
N. Nipple	Can be used to extend the length of a pipe or make temporary connections for maintenance purposes; short length of pipe (usually less than 6 inches [15 cm]) with threads on both ends

Operation of Piping

All piping components and systems are designed to operate within three specified limits: temperature, pressure, and compatibility with the process material. ASME, ANSI, and other standards organizations rate all piping and fittings for these three conditions. The rating assigned to a given pipe is based on a pressure/temperature limit that, if exceeded, could result in a rupture.

Pressure safety valves or relief valves are commonly used to keep piping within the safe operating zone of these limits. The pressure safety valves and relief valves are inspected periodically. In some cases, automatic equipment shutdown systems are used to keep the pressures and temperatures within limits.

JACKETED PIPES **Jacketed pipes** have a pipe-within-a-pipe design so hot or cold fluids can be circulated around the process fluid without the two fluids coming into direct contact with each other. In other words, a jacketed pipe operates like a heat exchanger. Figure 4.15 shows an example of jacketed pipe. The operator monitors the flow and temperature of both the inner pipe and the pipe jacket.

Jacketed pipe a pipe-within-a-pipe design that allows hot or cold fluids to be circulated around the process fluid without the two fluids coming into direct contact with each other.

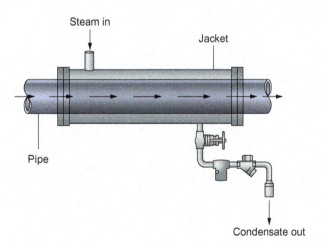

Figure 4.15 Jacketed pipe. Note the steam entering the piping jacket and the condensate draining so the jacket will remain full of steam.

INSULATION **Insulation** is any substance that prevents the passage of heat, light, electricity, or sound from one medium to another. In the process industries, pipe insulation is primarily used to maintain the temperature of process fluids and refrigeration systems and to protect workers from thermal burns that could occur if a worker were to come in contact with piping that is extremely hot or extremely cold. Protecting the process fluid from temperature changes from ambient conditions (i.e., temperatures outside the pipe) is desirable because it reduces energy costs for the process unit. Figure 4.16 shows examples of different size fiberglass pipe insulation.

Insulation any substance that prevents the passage of heat, light, electricity, or sound from one medium to another.

Insulation can be made from many materials, including:

- Rigid foam (cellular glass or polyethylene)
- Calsil (calcium silicate)
- Kaowool (ceramic fiber)
- Mineral wool
- Fiberglass
- Insulation brick

Figure 4.16 Examples of pipe insulation.

CREDIT: John Kasawa / Shuttershock.

When selecting insulation, it is important to consider the fluid type and temperature, the need for tracing, the type of piping, the type of covering, and the need for personal protection. The insulation must be able to handle the working temperatures of the fluid. The type of metal pipe will need to be considered, because the wrong insulation could accelerate pipe corrosion. Soft wrap insulation blankets are made to fit piping and equipment in order to prevent heat loss. This form of insulation is easily installed and removed.

In high temperature systems, insulation to prevent burns should be applied to any lines that people might contact. Insulation also can be applied to conserve heating or cooling energy. An insulation jacketing, made from aluminum or plastic, covers the insulation to protect it.

Operators should inspect the insulation on all piping and piping system components. If a portion of the insulation is missing, the piping could drop below working temperatures of the fluid, causing an upset.

Heat tracing   a coil of heated electric wire or steam tubing that is wrapped around a pipe to increase the temperature of the process fluid, reduce fluid viscosity, and facilitate flow.

TRACING   Heat tracing is a coil of heated electric wire or steam tubing that is attached to or wrapped around a pipe to increase the temperature of the process fluid, reduce fluid viscosity, and facilitate flow. Heat tracing can be very important in cold climates because frozen or extremely viscous process fluids can damage piping and equipment and can result in a unit upset condition. All tracing must be inspected to ensure that it is functioning correctly.

Steam heat tracing is a constant flow of steam through small diameter tubing that runs beside the pipe to keep the pipe warm, thus reducing viscosity and preventing the materials within the pipe from freezing.

Electrical heat tracing involves an electrical wire that is wrapped around a pipe and is controlled by a thermostat. The thermostat cycles on and off at preset temperatures to keep the fluid in the pipe warm.

Figure 4.17 shows an example of heat tracing.

Figure 4.17 Heat tracing.

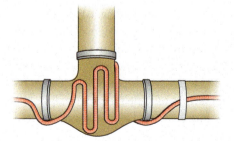

In extremely cold climates, heat tracing might not be sufficient to prevent freezing. In these instances, freezing is prevented by draining process fluids from the piping and equipment. Some systems, such as safety showers, eye baths, and fire protection, are protected by small bleed valves that allow a continuous water drip. The dripping maintains a small flow through the piping to prevent freezing.

STEAM TRAPS **Steam traps** are devices used to remove condensate from a steam system or piping and from steam tracing. Most steam traps will also remove air and noncondensable gases from steam systems. When steam cools, condensate forms, so steam traps were invented to catch and remove the condensate, but prevent live steam from escaping. Removing condensate from steam lines helps to maintain the temperature of piping and prevent freezing. It also prevents water hammer, which can cause steam piping leaks or failure from occurring. Removing air and noncondensable gases helps to maximize the heat transfer rate in steam system equipment.

Steam trap device used to remove condensate from the steam system or piping.

Steam traps can vary in design, but they operate by using one of three basic principles: density difference between steam and condensate, temperature difference between steam and condensate, or velocity difference between steam and condensate. The choice of design depends on the operating system and the application. Examples of various steam trap types are shown in Figure 4.18.

Figure 4.18 Cutaways of different steam trap types: **A.** Mechanical trap (bucket). **B.** Thermostatic trap (bellows). **C.** Thermodynamic trap (impulse).

CREDIT: Courtesy of Gayle Cannon.

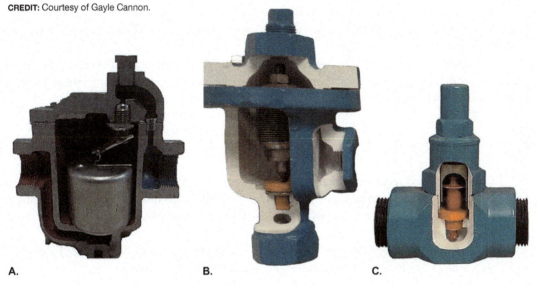

A.   B.   C.

The three types of steam traps utilizing these operating principles are mechanical, thermostatic, and thermodynamic. Mechanical traps use the principle of differing density between steam and water. Two common designs of mechanical traps are the inverted bucket and the float. The inverted bucket steam trap employs an upside down bucket shaped component, into which steam enters. The entering steam causes it to rise, which keeps the discharge valve shut. Condensate causes the bucket to fall, which opens the discharge valve to release the condensate.

Thermostatic traps operate using the difference in temperature between steam and condensate. Both bimetallic and liquid filled designs are common types of thermostatic steam traps. As steam enters a bellows style trap, it heats the liquid contained in a flexible bellows. This causes the bellows to expand and close the discharge valve when steam passes through the trap. The bellows contracts and opens the discharge valve when cooler condensate enters it.

The third type of steam trap is a thermodynamic design, which operates by sensing the difference in velocity between steam and condensate. Two designs of thermodynamic traps are the disc and impulse traps. In an impulse trap, the velocity of steam is much greater than that of condensate. While the slower moving condensate discharges from the trap, the higher speed of the steam creates a pressure drop which causes the discharge valve to close. Figure 4.19 shows a thermodynamic type trap and a mechanical type trap installed on steam piping.

Figure 4.19
A. Thermodynamic disc type steam trap. **B.** Mechanical inverted bucket type steam trap.

CREDIT: **A.** Denis Karelin/Shutterstock. **B.** Courtesy of Gayle Cannon.

A.

B.

If the steam trap fails closed, condensate will accumulate in the steam piping, reducing its temperature and the temperature of the process line. If the steam trap fails open, there will be excessive steam use and, in closed systems, steam will be blown into the condensate header.

Testing methods used to determine a steam trap's operating condition include visual observation of its condition and operation, temperature measurement using infrared testing devices, and ultrasonic detection.

4.5 Potential Hazards

In some process facilities, process technicians are responsible for the care and inspection of piping, gaskets, tubing, hoses, and fittings. If a process technician has any doubt about whether any of these elements can be used for a particular job, he or she should consult with a supervisor.

Many hazards are associated with piping, gaskets, tubing, hoses, and fittings. Figure 4.20 shows an example of corroded pipes. Table 4.11 lists some of these hazards.

Figure 4.20 Corroded pipes.

CREDIT: A Lesik / Shutterstock.

Table 4.11 Hazards and Possible Impacts

Improper Operation	Possible Impacts			
	Individual	Equipment	Production	Environment
Gasket incompatible with material in pipe	■ Exposure to chemicals	■ Leaking connection point	■ Loss of production ■ Process shutdown	■ Atmospheric release ■ Spill to the environment
Using the wrong hose type	■ Exposure to chemicals ■ Burn or injury	■ Weakened/leaking hose		■ Atmospheric release ■ Spill to the environment
Piping corrosion	■ Exposure to chemicals	■ Leaking piping	■ Loss of production ■ Process shutdown	■ Atmospheric release ■ Spill to the environment
Cross connection of hoses	■ Exposure to chemicals ■ Serious injury/death	■ Equipment damage	■ Loss of production ■ Process shutdown	■ Atmospheric release ■ Spill to the environment
Incorrect piping material selected	■ Injury	■ Stress cracking ■ Piping failure	■ Loss of production ■ Process shutdown	■ Atmospheric release ■ Spill to the environment
Steam trap fails closed		■ Upset conditions	■ Loss of production ■ Process shutdown	■ Atmospheric release ■ Spill to the environment
Steam trap fails open	■ Burn or injury		■ Wasted steam	

4.6 Process Technician's Role in Operation and Maintenance

When working with piping, gaskets, tubing, hoses, and fittings, process technicians conduct monitoring and maintenance activities to ensure that the piping and equipment are functioning properly. Process technicians should always look at, listen to, and check for the items listed in Table 4.12.

Table 4.12 Process Technician's Role in Operation and Maintenance

Look	Listen	Check
■ Look for leaks in abnormal locations ■ Inspect fittings ■ Check gauges	■ Listen for hissing noises that are not normally present.	■ Monitor for temperatures within acceptable ranges (using care not to touch hot pipes) ■ Be alert for vibrations in piping, signifying problems with connected equipment

Equipment Inspection and Leak Testing for Piping and Hoses

PIPING INSPECTION Initially upon installation and from time to time, it is necessary to perform a *pressure tightness test* to verify that the piping system can still safely handle the system pressure. The most common types of tests are hydrostatic and pneumatic.

Hydrostatic testing, also known as a *hydro or tightness test*, uses a hydraulic pump to increase the pressure in a water-filled system in order to test for leaks. Once the pressure has been increased (to a multiplier greater than the MAWP), the pump is shut off and the pressure is monitored for a period of time. Because water is a noncompressible fluid, a decrease in pressure caused by a leak can be detected quickly. After the leak is identified, the test pressure is released, repairs are made, and a retest is performed.

One disadvantage of hydrostatic testing is that the ambient temperature must be within a certain range (i.e., cannot be too low or water will freeze). Before conducting a test of this type, technicians should always check ANSI standards to determine temperature limitations for all of the materials used in the system test.

In cases where water or other liquids cannot be used, pneumatic testing is employed. In *pneumatic testing*, the system is isolated and then air is introduced and compressed to a desired pressure. The piping is then monitored for leaks.

A disadvantage of pneumatic testing is that it takes longer to depressurize and repressurize the system if the test is unsuccessful. Also, potential for a rupture exists in high-pressure, large-volume systems.

When conducting pressure tightness tests, it is important to realize that pressure greater than MAWP is being applied to the piping structure during the test and that overpressurization risks exist. Thus, it is necessary to refer to the ANSI Code for Pressure Piping for various pressure tolerances of the piping prior to conducting the test.

Other types of piping inspection include a variety of nondestructive testing (NDT). This can include radiography (X-ray) (RT), magnetic particle inspection (MT), dye penetrant testing (PT), ultrasonic inspection (UT), and visual inspection (VT). Carbon steel pipelines are inspected regularly with smart pigs that use magnetic particle inspection (MT).

HOSE INSPECTION   All hoses should be inspected visually prior to each use and thoroughly inspected at regular intervals. The hoses should be hydrostatically tested periodically above the MAWP of the hose. Companies might have different policies on the frequency, method of inspection, and recertification labeling. Using a hose that is out of inspection can result in hose rupture and injury.

Did You Know?

There are pigs in the pipeline! Not real pigs, mind you, but pipe pigs.

Pipe pigs are mechanical devices that clean and inspect pipelines. Some pigs, called smart pigs, inspect pipelines using sensors, cameras, and isotopes to detect corrosion.

CREDIT: fahim abdelmajid / Alamy

Piping Symbols

Piping and instrumentation diagrams (P&IDs) show the equipment, piping, and instrumentation contained in a process in a facility. Contained on a P&ID are many different symbols that process technicians must be able to identify and interpret. Figure 4.21 shows some of the more commonly used symbols for piping.

It is important to note that, although there are standards, some symbols can vary slightly from plant to plant.

Figure 4.21 Common piping symbols.

Piping Symbols

....	Future equipment
▬▬▬▬▬▬	Major process
─────────	Minor process
─/─/─/─/─	Pneumatic
─∟─∟─∟─	Hydraulic
─×─×─×─×	Capillary tubing
●─●─●─●	Mechanical link
─ ─ ─ ─ ─	Electric
◁▭▷	Jacketed or double containment
─o─o─	Software or data link

Summary

Piping, gaskets, tubing, hoses, and fittings are some of the most prevalent pieces of equipment in the process industries. In any facility, it is not unusual to see large segments of pipe going from one location to another. Pipes are connected using gaskets and fittings.

Proper operation of piping, tubing, and hoses is extremely important. Each of these items has a maximum allowable working pressure that cannot be exceeded. The incorrect flow (which has a direct impact on the fluid velocity) can also result in pipe failure. Pipes often move high temperature fluids. Failure to operate the piping, tubing, and hoses correctly can result in personnel injuries, production losses, and environmental releases.

Industrial piping, tubing, hoses, and fittings can be made of many different materials, including carbon steel, stainless steel, alloy steel, iron, exotic metals, elastomeric material, and plastics.

The selection and sizing criteria of piping material is critical, and such decisions must consider pressure, temperature, flow rate, and corrosiveness of the process fluid. Corrosion occurs much more quickly as the pressure and temperature of a process increases. All gasket materials must be appropriate for the chemical properties of the fluid, and operating temperature will dictate if insulation is required.

Piping connections can be welded, flanged, threaded, or bonded. The intended use of the pipe determines which connection type is the most appropriate. Sealant compounds are used to lubricate and seal the connections on threaded pipe. Gaskets are used to seal the connections on flanged pipe. Flared fittings and compression fittings are commonly used for tubing connections. These are mechanical seals and do not normally require separate sealing materials.

All of the materials used within the process industries have pressure, temperature, and chemical property limitations. These limitations are important to the safety and performance of the unit.

Equipment testing is used to check pipes for structural problems. The most common equipment tests are hydrotesting, pneumatic testing, and nondestructive testing.

Process technicians must constantly be aware of their surroundings and must analyze the entire system in order to completely understand the functionality of a unit. A part of this understanding includes using the proper types of piping, gaskets, tubing, hoses, and fittings for the job.

Identifying the need for maintenance and repair is a vital role of the process technician. While the process technician might not perform the repairs, in-depth familiarity with the equipment is necessary to ensure safe operation and maintenance.

Checking Your Knowledge

1. Define the following terms:

 a. Alloys

 b. Schedule

 c. Gasket

 d. Expansion loop

 e. Insulation

 f. Corrosion

 g. Tensile strength

 h. Jacketed pipe

2. Which of the following is the primary material for piping?

 a. Glass

 b. Carbon steel

 c. Plastic-lined material

 d. Solid plastic

3. A type of support that suspends pipe from the ceiling is a(n):

 a. Pipe clamp

 b. Pipe hanger

 c. Pipe shoe

 d. Expansion loop

4. (True or False) Caps can be threaded, glued, or welded.

5. Carbon steel can be used in which of the following Fahrenheit temperature ranges?

 a. −50°F to 450°F

 b. −150°F to > 1400°F

 c. −20°F to 800°F

 d. −60°F to 600°F

6. Which of the following is NOT a material used for piping insulation?

a. Plastic

b. Fiberglass

c. Calsil

d. Rigid foam

7. _____ might be an indication that the sealant compound or tape was not applied correctly.

8. Which of the following is a white tapelike substance used to lubricate the thread connections of a male pipe thread?

a. Pipe dope

b. Oil

c. Sealant compound

d. Teflon® tape

9. Which of the following tests are used to perform a pressure tightness test on a piping system? (Select all that apply.)

a. Hydrostatic testing

b. Pneumatic testing

c. Temperature testing

d. Hypertesting

10. To prevent the liquid contents in a pipe from freezing, engineers often apply coils of heated electric wire or steam tubing called:

a. Insulation

b. Heat shielding

c. Heat tracing

d. Steam traps

11. For a carbon steel piping and flange system that is 6-inch schedule 40 with class 600 flanges operating at a max of 400°F, what is the MAWP?

a. 1200

b. 1205

c. 1270

d. 2062

12. Match the type of pipe fitting with its description:

Type	Description
I. Plug	a. Used to close (seal) the open end of a pipe; can be threaded, glued, or welded.
II. Cap	b. Threaded on both the internal and external surfaces; used to join two pipes of differing sizes.
III. Bushing	c. Consists of a mating plate used to join two pieces of pipe or a valve to a piece of pipe.
IV. Flange	d. Used to close the end of a piping run; the threading is male and fits into a female-threaded fitting; can be glued, threaded, or welded into place.

13. (True or False) If there is any doubt about whether a component can be used for a particular job, a process technician should check with a supervisor.

14. List three potential hazards of using the wrong hose for the task.

NOTE: Answers to Checking Your Knowledge questions are in the Appendix.

Student Activities

1. Make a sketch of the various types of piping in and around your home (e.g., water, sewage, and natural gas). Identify the different piping materials and the characteristics of the service (e.g., high pressure, low temperature, corrosive).

2. Research a type of piping and connection assigned by your instructor and prepare a two-page report.

3. Given various piping, gaskets, tubing, hoses, and fittings in a lab setting, identify the different types of connections and practice connecting them (e.g., connect a steam hose to a boss fitting or connect a water hose to a Chicago pneumatic coupling/crow's foot).

Chapter 5
Valves

Objectives

After completing this chapter, you will be able to:

5.1 Describe the purpose, types, and components of valves in the process industry. (NAPTA Valves 1-5*) p. 82

5.2 Identify typical problems associated with valves. (NAPTA Valves 10) p. 94

5.3 Describe safety and environmental concerns associated with valves. (NAPTA Valves 7) p. 95

5.4 Describe the process technician's role in typical procedures and safe valve operation and maintenance. (NAPTA Valves 8 and 9) p. 98

*North American Process Technology Alliance (NAPTA) developed curriculum to ensure that Process Technology courses will produce knowledgeable graduates to become entry level employees in process technology. Objectives from that curriculum are named here in abbreviated form. For example, "(NAPTA Valves 1-5)" means that this chapter's objective relates to objectives 1 to 5 of NAPTA's course content on valves.

Key Terms

Ball check valve—a valve used to limit the flow of fluids to one direction; available in both horizontal and vertical designs, **p. 86.**

Ball valve—a type of valve that uses a flow control element shaped like a hollowed-out ball attached to an external handle to control flow, **p. 86.**

Block valve—a valve used to block flow to and from equipment and piping systems (e.g., during outage or maintenance). Block valves differ from some other types of valves, in that they should not be used to throttle flow, **p. 92.**

Bonnet—a bell-shaped dome mounted on the body of a valve where the valve disc is held when the valve is opened, **p. 89.**

Butterfly valve—a type of valve that uses a disc-shaped flow control element to control flow, **p. 88.**

Check valve—a type of valve that allows flow in only one direction and is used to prevent reversal of flow in a pipe, **p. 85.**

Control valve—a valve through which flow is controlled by a signal from a controller, **p. 92.**

Damper—a movable plate that regulates the flow of air or flue gases, **p. 94.**

Diaphragm valve—a type of valve that uses a flexible, chemical-resistant, rubber-type membrane (diaphragm) to control flow, **p. 88.**

Gate valve—a positive shutoff valve that uses a gate or guillotine that, when moved between two seats, causes a tight shutoff; it should never be used to throttle flow, **p. 83.**

Globe valve—a type of valve that uses a disc and seat to regulate the flow of fluid through the valve body, **p. 84.**

Handwheel—the mechanism that raises and lowers a valve stem to control the flow of fluid through a valve, **p. 89.**

Lift check valve—a valve, with a design similar to that of a globe valve, which allows flow in only one direction; it can be installed in both horizontal and vertical lines, **p. 85.**

Louvers—movable, slanted slats that are used to adjust the flow of air, **p. 94.**

Manual valve—a hand-operated valve that is opened or closed using a handwheel or lever, **p. 93.**

Multiport valve—a type of valve used to split or redirect a single flow into multiple directions or direct flows from multiple sources to the same destination, **p. 92.**

Needle valve—a variation of a globe valve that controls small flows using a long, tapered plug, **p. 85.**

Nonrising stem valve—a valve in which the stem does not rise through the handwheel when the valve is opened or closed, **p. 89.**

Packing—a substance such as Teflon® or graphite-coated material that is used inside the stuffing box to prevent leakage from occurring around the valve stem, **p. 89.**

Plug valve—a type of valve that uses a flow control element shaped like a hollowed-out plug attached to an external handle to control flow, **p. 87.**

Relief valve—a safety device designed to open slowly if the pressure of a liquid in a closed space, such as a vessel or a pipe, exceeds a preset level, **p. 90.**

Rising stem valve—a valve in which the stem rises out of the valve handwheel when the valve is opened, **p. 89.**

Safety valve—a safety device designed to open quickly if the pressure of a gas in a closed vessel exceeds a preset level, **p. 91.**

Straight-through diaphragm valve—a valve that contains a flexible membrane (diaphragm) that extends across the valve opening; long diaphragm movements are required to operate it; better for viscous liquids than the weir diaphragm valve, **p. 88.**

Swing check valve—a valve used to control the direction of flow and prevent contamination or damage to equipment caused by backflow, **p. 85.**

Throttling—partially opening or closing a valve to restrict or regulate fluid flow rates, **p. 83.**

Valve body—the lower portion of the valve; contains the fluids flowing through the valve, **p. 89.**

Valve disc—the section of a valve that attaches to the stem and that can fully or partially block the fluid flowing through the valve, **p. 90.**

Valve knocker—a device usually attached to a chain operator, used to facilitate the movement (opening or closing) of an overhead valve, **p. 93.**

Valve seat—a section of a valve where the flow control element rests when the valve is completely closed; it is designed to maintain a leak-tight seal when the valve is shut, **p. 90.**

Valve stem—a long, slender shaft that attaches to the flow control element in a valve, **p. 89.**

Weir diaphragm valve—a valve that contains a dam-like device, sometimes called a saddle, that functions as a seat for the diaphragm, **p. 88.**

5.1 Introduction

Valves are piping system components used to control or stop the flow of fluid through a pipe. Valves allow process technicians to achieve a desired flow rate or pressure which can, in turn, be used to achieve a desired variable, such as tank level. Valve classifications are based on the design of the internal components, how the valve is operated, or the function of the valve.

When determining the type of valve to be used in a process, engineers must consider the function the valve will perform (open/closed vs. throttling), the type of fluid that will be passing through the system (gas, liquid, slurry, etc.), and the operating pressure and temperature of the fluid.

Valves are manufactured using many different materials, including cast iron, brass, carbon steel, steel alloys, bronze, fiberglass, plastic, and glass. The selection of materials used in the construction of a valve is based on the intended use, particularly the fluid that will be passing through the system. For example, valves made from elastomers or plastics are not appropriate for use with hydrocarbon-based materials, since those materials will break down the plastic/rubber and lead to valve failure. Certain chemicals such as hydrochloric acid are not compatible with most metals. Chlorides such as those found in salt water can result in severe corrosion of many metals.

Valve rating classes are based on the temperature and pressure of the intended service of the valve and the materials used in its construction. A valve's pressure rating decreases as temperature increases. For example, a carbon steel valve in the 150 class has a maximum pressure rating of 285 PSIG at 100 degrees Fahrenheit (37.8 degrees Celsius), but that maximum pressure rating decreases to 140 PSIG at 600 degrees Fahrenheit (316 degrees Celsius).

Types of Valves

Many different types of valves are used in the process industries. The most common valve types include:

- Gate valves
- Globe valves
- Check valves
- Ball valves
- Plug valves
- Butterfly valves
- Diaphragm valves
- Relief and safety valves
- Multiport valve

Process technicians must be familiar with each of the different valve types and their maintenance and operating characteristics.

Gate Valves

Gate valves are positive shutoff valves that use a gate or guillotine that, when moved between two seats, causes a tight shutoff. A wedge-shaped disc or gate is raised or lowered into the body of the valve by operating a handwheel. Gate valves are the most common type of valve in the process industries. They are designed for on/off service and are not intended for **throttling** (a condition in which a valve is partially opened or closed to restrict or regulate the fluid flow rates). Figure 5.1 shows two views of a gate valve with the gate open and closed.

In a rising stem gate valve, the valve stem does not enter the valve body and does not contact the process fluid. Valve position is indicated by the position of the stem above or below the handwheel. In a nonrising stem-type gate valve, the stem is threaded to the disc and the disc moves within the valve body; the stem does not rise out of the valve. The stem is exposed to the process material, so this type of valve is usually used with noncorrosive fluids. Because the stem does not indicate the position of the valve (i.e., whether it is opened or closed), many nonrising stem valves have a pointer or indicator threaded onto the upper end of the stem to indicate the position of the gate.

Gate valve a positive shutoff valve that uses a gate or guillotine that, when moved between two seats, causes a tight shutoff; should never be used to throttle flow.

Throttling partially opening or closing a valve to restrict or regulate fluid flow rates.

Figure 5.1 **A**. Open gate valve showing flow (notice the gate is in the upper valve body). **B**. Closed gate valve showing flow stopped by the gate.

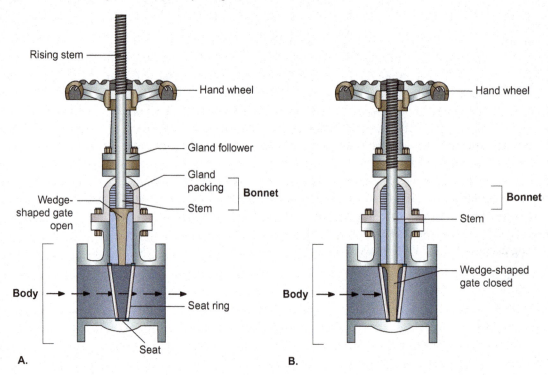

A.

B.

The body of the valve contains two seat rings that seal around each side of the gate. When the gate is fully closed, flow is blocked completely. Because of their design, gate valves should not be used for throttling, since throttling can cause disc chatter, metal erosion, and seat damage that can prevent the valve from sealing properly.

Globe Valves

Globe valve   a type of valve that uses a disc and seat to regulate the flow of fluid through the valve body.

Globe valves use a disc and seat to regulate the flow of fluid through the valve body. Flow passes through a globe valve in an S-shaped pattern, and flow rate is determined by how much the disc is lifted off the valve seat. These types of valves are designed to regulate flow in one direction and usually are used in throttling service. Globe valves are very common in the process industries. Figure 5.2 shows two types of globe valves with the valve in the open position.

In a globe valve, fluid flow is controlled by raising or lowering the disc or flow-control element. These flow-control elements come in a variety of shapes, including ball, cylindrical,

Figure 5.2 A. Opened globe valve with ball-shaped disc. **B**. Cutaway of an open globe valve with a cylindrical-shaped disc.

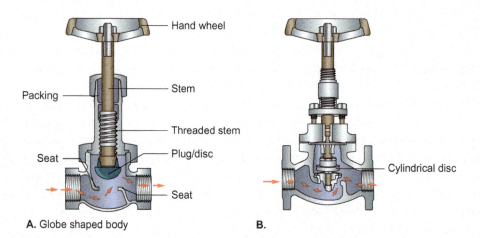

A. Globe shaped body

B.

and needle. The seats in these valves are designed to accommodate the shape of the disc. Figure 5.3 shows an example of these different disc shapes.

A **needle valve** is a variation of a globe valve used for very precise control of small flows using a long, tapered plug attached to the valve stem, raised and lowered onto the seat to control flow. Some needle valves have a three-way design that allows them to be used directly in the process piping line. These valves are used for metering or sampling flows (for example, metering a specific additive flow to a tank or other vessel, or sampling material from a process product line).

Needle valve a variation of a globe valve that controls small flows using a long, tapered plug.

Ball-shaped   Cylindrical-shaped   Needle-shaped

Figure 5.3 Globe valve disc designs.

Check Valves

Check valves are used to prevent backflow (reverse flow in piping when the flow stops). By eliminating backflow, these valves prevent equipment damage and contamination of the process. All check valves are directional and must be placed in the line correctly.

Check valves, sometimes called *nonreturn valves*, are installed in pump discharge lines and other process piping (Figure 5.4). The most common types of check valves are swing check, lift check, and ball check.

Check valve a type of valve that allows flow in only one direction and is used to prevent reversal of flow in a pipe.

Figure 5.4 A. Swing check valve. **B**. Swing check valve with flow through the valve. **C**. Closed swing check valve preventing backflow.

CREDIT: A. Surasak_Photo/Shutterstock.

 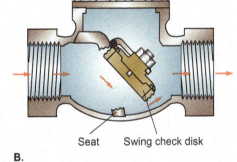 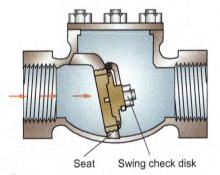

Seat   Swing check disk   Seat   Swing check disk

A.   **B.**   **C.**

Swing check valves are used to control the direction of flow and prevent process fluid contamination or damage to equipment caused by backflow. These valves usually have an arrow stamped on the body that signifies the direction of flow through them.

Swing check valves contain discs and seats that are set at an angle. As the fluid moves forward through the valve, the valve disc is lifted (forward flow). If the flow stops or changes direction (backward flow), gravity or back pressure forces the valve closed, preventing backflow.

Lift check valves have a disc-shaped flow control element that lifts up off the seat as fluid flows through it (forward flow) and drops back onto the valve seat if the fluid flow stops or attempts to flow in a reverse direction (backward flow). Lift check valves are available in horizontal and vertical designs.

Figure 5.5 shows examples of a lift check valve in the open and closed position, respectively.

Swing check valve a valve used to control the direction of flow and prevent contamination or damage to equipment caused by backflow.

Lift check valve a valve with a design similar to that of a globe valve, which allows flow in only one direction; it can be installed in both horizontal and vertical lines.

Figure 5.5 A. Open lift check valve showing flow through valve. **B**. Closed lift check valve showing flow stopped.

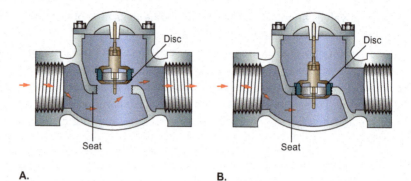

Disc

Seat

A.

Disc

Seat

B.

Ball check valve   a valve used to limit the flow of fluids to one direction; available in both horizontal and vertical designs.

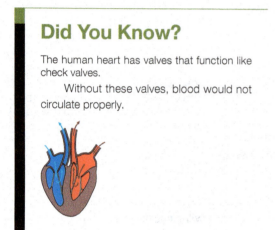

Ball check valves are especially well-suited for use with fluids containing solids or viscous materials and are available in horizontal, vertical, and angle designs. Ball check valves have a ball, or sphere-shaped, flow control element that responds to gravity or back pressure by moving back onto the seat when flow stops, preventing backflow. Figures 5.6 A and B show examples of a ball check valve in the open and closed positions, respectively.

Figure 5.6 A. Open ball check valve showing flow through valve. **B**. Closed ball check valve showing flow stopped.

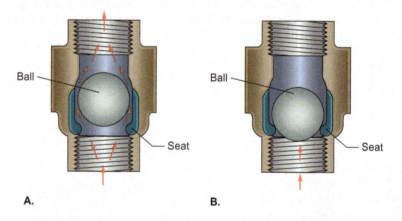

Ball

Seat

A.

Ball

Seat

B.

Ball Valves

Ball valve   a type of valve that uses a flow control element shaped like a hollowed-out ball attached to an external handle to control flow.

A **ball valve** is a device that uses a flow control element shaped like a hollowed-out ball, attached to an external handle, to control flow. Ball valves are quick opening and are used where a low pressure drop is required. Ball valves are referred to as quarter-turn valves. In other words, turning the valve's stem a quarter of a turn (90 degrees) brings it to a fully open or fully closed position (in comparison to other valves that require multiple turns to open or close fully).

Ball valves use a sphere held against a cup-shaped seat to control flow. The seat has a circular opening with a smaller diameter than the ball. Stops are provided on the valve handle to indicate open and closed positions. Figures 5.7 A, B, and C show examples of ball valves in different operating positions.

Ball valves are typically used for on/off service. When a ball valve is opened, the hollowed-out portion of the ball (sometimes referred to as the port) lines up perfectly with the inner diameter of the pipe and the valve handle is parallel to the piping. When a ball valve is closed, the port aligns with the wall of the pipe, and the valve handle is perpendicular to the piping.

The hole in most ball valves is smaller in diameter than the internal diameter of the pipe. Some ball valves, however, are designed to have an opening that is the same diameter as the internal diameter of the pipe. These are called full port ball valves.

Figure 5.7 A. Cutaway of a ball valve in the open position, **B**. Partially opened ball valve showing flow through the valve. **C**. Closed ball valve showing flow stopped.

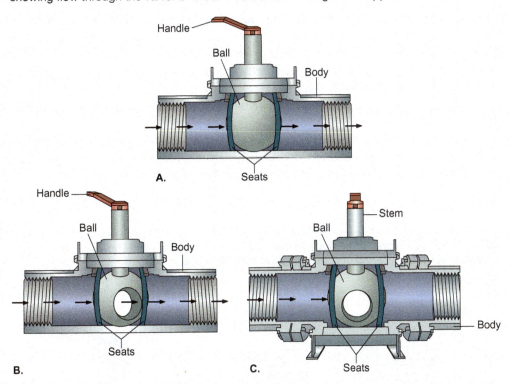

Plug Valves

A **plug valve** uses a flow control element shaped like a hollowed-out plug, attached to an external handle, to control flow. Plug valves are quick-opening, quarter-turn valves that are similar to ball valves.

Like ball valves, plug valves are quarter-turn valves that use a hollowed-out flow control element to control flow. In plug valves, the opening for the flow is machined through a tapered cylindrical plug. The opening is rotated perpendicular (at a 90-degree angle) to the direction of flow to close.

Plug valves are designed for on/off service and are well suited for applications such as slurry, sludge, wastewater, and fuel gas. Figure 5.8 shows a plug valve in different operating positions.

Plug valve a type of valve that uses a flow control element shaped like a hollowed-out plug attached to an external handle to control flow.

Figure 5.8 A. Partially open plug valve showing the handwheel flow indicator, and flow through the valve. **B**. Closed plug valve showing the handwheel flow indicator, and flow stopped at the plug.

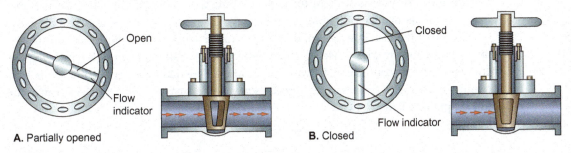

Butterfly Valves

Butterfly valve a type of valve that uses a disc-shaped flow control element to control flow.

A **butterfly valve** uses a disc-shaped element to control flow. Like ball valves, butterfly valves can be fully opened or closed by turning the valve handle one-quarter of a turn.

Butterfly valves are designed much like dampers, except the interior body of the butterfly valve is shaped to fit the disc. This results in a tight seal when the valve is closed. Because of the way they are designed, butterfly valves are best suited for low-temperature, low-pressure applications such as cooling water systems.

Figure 5.9 shows a butterfly valve in both the open and closed positions and a partially opened butterfly valve.

Unlike many other valves, butterfly valves can be used for throttling and on/off service. However, the throttling capabilities of a butterfly valve are not linear (for example, opening the valve halfway might provide a flow that is near maximum).

Figure 5.9 A. Partially open butterfly valve. **B**. Closed butterfly valve showing flow stopped. **C**. Cutaway photo of a partially open butterfly valve.

CREDIT: C. freeman98589/Fotolia.

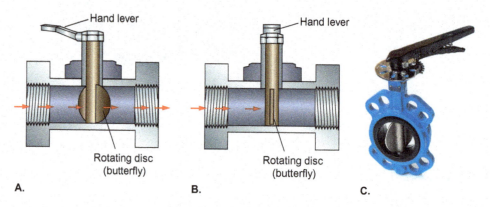

Diaphragm Valves

Diaphragm valve a type of valve that uses a flexible, chemical-resistant, rubber-type membrane (diaphragm) to control flow.

Weir diaphragm valve a valve that contains a dam-like device, sometimes called a saddle, that functions as a seat for the diaphragm.

Straight-through diaphragm valve a valve that contains a flexible membrane (diaphragm) that extends across the valve opening; long diaphragm movements are required to operate it; better for viscous liquids than the weir diaphragm valve.

Diaphragm valves use flexible, chemical resistant, rubber-type diaphragms to control flow. Their flexibility and material composition make them different from other flow control elements. In this type of valve, the diaphragm seals the parts above it (e.g., the plunger) from the process fluid.

Weir diaphragm valves contain a dam-like device, sometimes called a saddle, which functions as a seat for the diaphragm. The flow must go over the top of the weir and under the diaphragm to exit.

Straight-through diaphragm valves contain a flexible diaphragm that extends across the valve opening. Long diaphragm movements are required for operation of this type of diaphragm valve. There is very little pressure drop across this type of valve when open.

Figure 5.10 A shows a straight-through (or straightway) diaphragm valve; parts B and C show examples of a weir diaphragm valve in the open and closed position, respectively.

Because of their unique design, diaphragm valves work well with process substances that are exceptionally sticky, viscous, or corrosive, with the straight-through type being the better choice for viscous fluids. However, they are not adequate for applications with high pressures or excessive temperatures. Diaphragm valves are easy to clean, but leaks can occur if the diaphragm fails.

Figure 5.10 Diaphragm valves. **A**. Closed straight-through (straightway) valve. **B**. Open weir diaphragm valve displaying flow through valve. **C**. Closed weir diaphragm valve showing flow stopped.

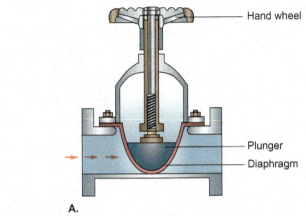

Hand wheel

Plunger

Diaphragm

A.

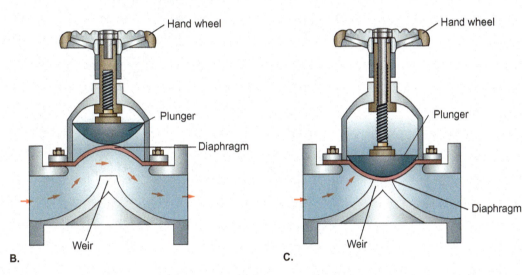

Hand wheel

Plunger

Diaphragm

Weir

B.

Hand wheel

Plunger

Diaphragm

Weir

C.

Valve Components

No matter the mechanism used to control flow, most valves have similar components (see Figure 5.1). All valves contain a closure device attached to a stem that passes through the valve body. A **handwheel** is a mechanism that is turned in a counterclockwise direction to raise a valve stem and in a clockwise direction to lower a valve stem, to allow or restrict the flow of fluid through the valve.

Packing is a substance such as Teflon® or graphite-coated material that prevents leakage from around the valve stem. The packing material is located within a stuffing box inside the valve bonnet (the top of the valve). This packing is held tightly in place by a packing gland and packing gland nuts.

The **valve body** is the lower portion of the valve that contains the fluids flowing through the valve and the valve seat. The **bonnet** is the part of a valve through which the stem leaves the body; it contains the valve packing. The bonnet is a bell-shaped dome mounted on the body of a valve.

The **valve stem** is the long slender shaft that attaches to the flow control element in a valve. Gate (and some globe) valves are classified as either rising stem, or nonrising stem. In a **rising stem valve**, the stem rises out of the valve when the valve is opened. Rising stem valves are used when it is important to know if the valve is opened or closed. In a **nonrising stem valve**, the valve stem remains in place when the valve is opened or closed. Because the stem is threaded into the gate, the gate travels up and down along the stem as the handwheel

Handwheel   the mechanism that raises and lowers a valve stem to control the flow of fluid through a valve.

Packing   a substance such as Teflon® or graphite-coated material that is used inside the stuffing box to prevent leakage from occurring around the valve stem.

Valve body   the lower portion of the valve; contains the fluids flowing through the valve.

Bonnet   a bell-shaped dome mounted on the body of a valve where the valve disc is held when the valve is opened.

Valve stem   a long, slender shaft that attaches to the flow control element in a valve.

Rising stem valve   a valve in which the stem rises out of the valve handwheel when the valve is opened.

Nonrising stem valve   a valve in which the stem does not rise through the handwheel when the valve is opened or closed.

Did You Know?

Bridge wall markings, located on the body of valves, provide information such as the size and rating class of the valve, materials of construction, or direction of flow.

CREDIT: Winai Tepsuttinun/Shutterstock.

Valve disc the section of a valve that attaches to the stem and that can fully or partially block the fluid flowing through the valve.

Valve seat a section of a valve where the flow control element rests when the valve is completely closed; it is designed to maintain a leak-tight seal when the valve is shut.

Relief valve a safety device designed to open slowly if the pressure of a liquid in a closed space, such as a vessel or a pipe, exceeds a preset level.

Did You Know?

Process technicians often use the slang term *pop valve* to refer to safety valves because they open quickly, or pop, once the pressure threshold has been exceeded.

CREDIT: Dale Stagg/Shutterstock.

is rotated. Because the stem does not indicate the position of the valve (i.e., whether it is opened or closed), many nonrising stem valves have a pointer or indicator threaded onto the upper end of the stem to indicate the position of the gate. Nonrising stem valves are useful when tight space will not allow for the stem to rise above the handwheel. Figure 5.11 shows examples of rising and nonrising stem valves.

The **valve disc** is the section of a valve that attaches to the stem and can fully or partially block the fluid flowing through the valve. Disc design can vary depending on the type of valve and the type of blockage required. The **valve seat** is a section of a valve designed to maintain a leak-tight seal when the valve is shut. The seat is located in the interior and near the bottom of the valve.

Figure 5.11 A. Rising stem valve. **B**. Nonrising stem valve. Both valves are in the open position.

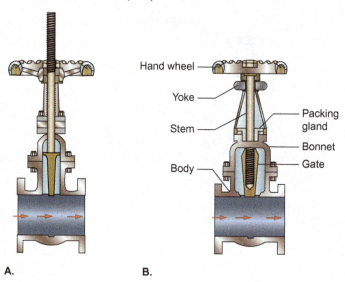

Relief and Safety Valves

Relief and safety valves are used to protect equipment and personnel from the hazards of overpressurization. They are used only after all other methods of pressure control prove to be ineffective. Both of these valves are designed to open and discharge when the pressure in a line, vessel, or other equipment exceeds a preset threshold. The differences between them are the speed at which they open, the size of their outlet port, and the type of service for which they are intended (liquid versus gas).

RELIEF VALVES Relief valves (Figure 5.12) are safety devices designed to open slowly if the pressure of a liquid in a closed space, such as a vessel or a pipe, exceeds a preset level. These valves open more slowly and allow less volume to escape than safety valves because liquids are virtually noncompressible, which means a small release, requiring a smaller outlet port, is all that is required to correct overpressurization.

Because they open slowly, relief valves are not intended for use in gas service. These valves can be found in liquid drain lines in which the contents are vaporized and sent to a flare system, on the discharge side of positive displacement pumps, or on the top of tanks that are

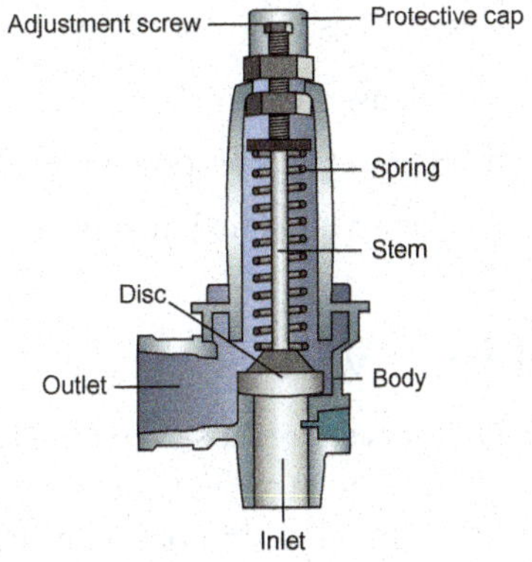

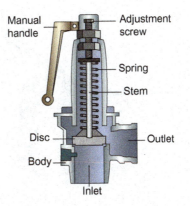

Figure 5.12 Relief valve components.

intended to operate liquid full. For example, a relief valve is found on the top of most residential hot water heaters.

In a relief valve, a flow-control disc is held in place by a spring. When the pressure in the system exceeds the threshold of the spring, the valve is forced open (proportional to the increase in pressure) and liquid is allowed to escape into a containment receptacle or other safety system. As the pressure drops below the threshold, the spring forces the flow control element back into the seat, resetting the valve.

SAFETY VALVES   **Safety valves** are designed to open quickly if the pressure of a gas in a closed vessel exceeds a preset level. A few examples of safety valve applications include steam lines, compressor discharge lines, and overhead tower vapor lines. Safety valves are designed to open and vent excess pressure to a flare header or through a large exhaust port into the atmosphere (depending on the substance being vented).

Conventional safety valves contain a spring that is used to set the opening pressure. When the opening pressure is reached, the disc begins to lift off the seat and system pressure enters an area called the huddling chamber. In the huddling chamber, a larger area is exposed to the increased pressure causing the valve to "pop" open quickly. Figure 5.13 shows components of a safety valve.

Safety valve   a safety device designed to open quickly if the pressure of a gas in a closed vessel exceeds a preset level.

Figure 5.13 Safety valve components.

To protect equipment and piping, safety valves do not close until the system pressure falls beneath the design specification. Some safety valves will reseat themselves after activation; others must be manually reset by a qualified technician.

According to ASME Code, safety valves on boilers must be equipped with a lifting device (or handle) to allow for manual activation for in-service testing and clearing of any deposit that might interfere with the operation of the valve. There are also certain instances when other safety/relief valves—those in air, steam, or hot water (exceeding 140 degrees Fahrenheit/ 60 Celsius) service—are required to have a lifting lever or handle.

Safety valves differ from relief valves because they:

- Are used with gas service, not liquid
- Have a faster response time
- Have a larger exhaust port.

Block Valves

Block value   a valve used to block flow to and from equipment and piping systems (e.g., during outage or maintenance). Block valves differ from some other types of valves, in that they should not be used to throttle flow.

Block valves are those valves used to block flow to and from equipment and piping systems (e.g., during an outage or maintenance). Block valves differ from some other types of valves in that they should not be used to throttle flow. The term is usually associated with gate valves but can also be used to describe ball valves or plug valves.

DOUBLE BLOCK AND BLEED VALVES   Certain specialized valves, such as double block and bleed valves, are sometimes used in process industries. These valves are able to perform the work of two isolation valves and one drain valve (between them). They are used in combustible gas trains in many industrial applications. They may be used to prevent product contamination, to test seals, or to allow repair of essential equipment while the unit continues in operation.

Multiport Valves

Multiport valve   a type of valve used to split or redirect a single flow into multiple directions or direct flows from multiple sources to the same destination.

A **multiport valve** is used to split or redirect a single flow into multiple flows or direct flows from multiple sources to the same destination. Multiport valves can contain three-, four-, or six-way openings. These types of valves usually have globe valve type closures that allow the flow to be directed two ways in the three-way valve, three ways in the four-way valve, and so forth. An example of a multiport valve is a three-way valve used to redirect the process flow from a vessel to a flare system.

In Figure 5.14, flow enters the three-way valve though the inlet. The normal direction has the flow going through the left outlet (to the vessel). When the flow is redirected through the outlet on the right, as shown, it exits to the flare. Since the flow from multiport valves can be routed to more than one location, another common use for these valves is on the inlet and outlet of heat exchangers to facilitate backwashing (reversing flow in order to clean the exchanger of debris or other unwanted material).

Figure 5.14 Multiport valves are used to split or redirect a single flow into multiple directions.

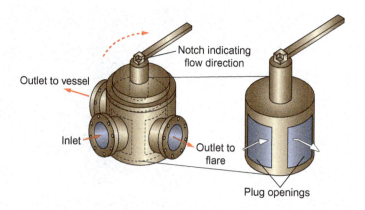

Control Valves

Control valve   a valve through which flow is controlled by a signal from a controller.

Control valves (shown in Figure 5.15) are valves through which the flow of fluid is controlled by a signal, usually received from a controller. An actuator is the control system component that receives a signal from a controller and moves the control valve. Actuators can operate using pneumatic (air), electronic, or hydraulic power.

Control valves are mechanically controlled and can be operated automatically or manually. In the automatic mode, a control valve is adjusted by a controller. In the manual mode, the output of a control valve is adjusted by a process technician.

Figure 5.15 A. Internal elements of a control valve. **B.** Control valve.

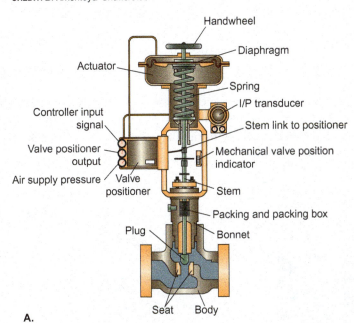

Handwheel
Diaphragm
Actuator
Spring
I/P transducer
Controller input signal
Stem link to positioner
Valve positioner output
Mechanical valve position indicator
Air supply pressure
Valve positioner
Stem
Packing and packing box
Plug
Bonnet
Seat Body

A.

B.

Manual Valves

A **manual valve** is a hand-operated valve that is opened or closed using a handwheel or lever. A valve wheel wrench is often required to operate manual valves. Chain operators and valve knockers may be installed on valves located in piping that is above grade and inaccessible.

VALVE KNOCKERS   A **valve knocker** (also called a *hammer blow actuator*) is used to facilitate the movement (opening or closing) of a hard-to-operate valve (Figure 5.16). The purpose of valve knockers is to help operate a valve handwheel when valves are in hard-to-reach places or are being used for high-pressure service. Valve knockers are constructed of a heavy anvil and a mechanism that delivers a hammerlike striking action through the anvil to the valve handwheel. The striking action causes small movements of the valve that can either open or close the valve.

Manual valve   a hand-operated valve that is opened or closed using a handwheel or lever.

Valve knocker   a device usually attached to a chain operator, used to facilitate the movement (opening or closing) of an overhead valve.

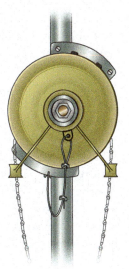

Figure 5.16 Valve knocker.

Louvers and Dampers

Louvers and dampers are adjustable plates used to regulate the air flow (draft) or flue gas flow in furnaces, boilers, cooling towers, and air fin heat exchangers.

Louvers are movable, slanted slats that are used to adjust the flow of air. Axles and gears attach to each plate, allowing these plates to open or close together. Louvers are commonly used to control air supply to furnaces, boilers, cooling towers, and fin-fan heat exchangers. Figure 5.17 shows an example of a louver.

Dampers are movable plates that regulate the flow of air or flue gases. Dampers consist of a metal plate attached to an axle that allows the plate to turn perpendicular (at right angles) to the flow to open or close. They primarily control the flow of stack gases in furnaces. Figure 5.18 shows an example of a damper.

Figure 5.17 Open louvers on an industrial cooler.

CREDIT: GTS Productions/Shutterstock.

Figure 5.18 Damper in a furnace duct.

Louvers and dampers can be operated by hand, pneumatically, or by an electric motor. They also might be part of a control loop and manipulated as part of a system designed to control the amount of air, or draft, through a furnace or boiler.

It is important to know that louvers and dampers cannot provide a total shutoff of flow; some leakage can occur. Additionally, care must be taken when opening and closing louvers and dampers. They can be damaged easily and might stick in an open or closed position, which can cause damage to the equipment associated with the damper system.

5.2 Potential Valve Problems

Process technicians are often required to troubleshoot valve problems and determine the best solution to return the valve to normal operating conditions. The most common valve conditions that a process technician might need to investigate are pluggage and leakage.

Valves can become plugged when the flow in the process is reduced and solids settle in the valve. If the valve is a drain valve, solids can collect because the valve is usually the lowest point in the system. Over time, these solids can build up and plug the valve. Pluggage can prevent a valve from closing tightly to allow equipment or line isolation.

Valve blockage can be addressed by flushing the system. Pluggage of small bleed, drain, vent, and instrument tap valves can be cleared using an approved rod-out tool (Figure 5.19). In cases of extreme pluggage, it might be necessary to remove the valve and its associated piping from service to clear the obstruction.

Valve leakage can occur either across the valve (within the pipe) or into the atmosphere. Atmospheric leakage usually occurs through the valve stem packing or flanges.

Figure 5.19 Rod-out device.

Leakage from the valve stem packing is addressed by tightening down on the packing nuts to compress the packing to tighten it around the stem. Flange leaks are addressed by tightening the flange bolts. The appropriate flange bolting pattern should be followed to ensure equal pressure is applied to the flange gasket, and care must be taken to not over torque the flange bolts. If the gasket is fully compressed and leakage has not stopped, it might be necessary to remove the line from service to replace the gasket.

Causes of leakage include corrosion (internal or external), erosion (usually internal), vibration, wear (external), and damage from an external source (e.g., impact from something striking the pipe). Because valve leakage can result in fire, explosion, personnel exposure, and environmental impact, it is important to resolve leaks as quickly as possible.

5.3 Hazards Associated with Improper Valve Operation

Many hazards are associated with piping and valves. Table 5.1 lists some of those hazards and their impacts.

Table 5.1 Safety and Environmental Concerns Associated with Valves

Improper Operation	Possible Impacts			
	Individual	Equipment	Production	Environment
Throttling a valve that is not designed for throttling		Valve damage to the point that it will not seat and stop flow, even when closed	Off-specification product because of improper flows	Valve leakage causing vessel overflow into the atmosphere
Use of excessive force when opening or closing a valve	Slipping and falling, as well as muscle and back strain	Damage to the valve seat, the packing, or the valve stem, causing leakage and making the valve difficult to open or close	Off-spec product because of improper flows	Valve leakage
Failure to clean and lubricate valve stems	Possible injuries as a result of a valve wrench slipping off the valve handle	Valve stem seizure or thread damage, making the valve difficult to open and close. The valve handle or wheel could break or the stem could shear		
Improperly closing a valve on a process line	Possible injury because of equipment overpressurization	Equipment damage (e.g., over-pressurizing a pump)		Possible leak to the environment
Failure to wear proper protective equipment when operating valves in high temperature, high-pressure, hazardous, or corrosive service	Burns or chemical exposure and other serious injuries			

Proper Body Position

An operator should never lean over the valve handle when operating a valve. If the valve bonnet were to fail under high pressure, the bonnet could fly up and strike the process technician in the face. Instead, the technician should operate the handwheel with his or her body positioned as shown in Figure 5.20 to prevent potential injury and exposure.

When opening a bleed or vent valve, the process technician should never stand in front of or directly under the valve to avoid the possibility of injury or chemical exposure.

Figure 5.20 Proper body position when opening valves. Note that the process technician's body is upright and that no part of the body is directly over the valve.

CREDIT: Xmentoys/Shutterstock.

To prevent bruises, abrasions, and "knuckle busting" from slipping handwheels or levers, the process technician should always wear gloves and use proper tools and body positioning when working on a valve that is stuck. Improper body positioning can cause slipping, falling, or back strain because of overexertion or hyperextension.

When working with valves, the process technician should always wear appropriate PPE, use a valve wrench to reduce the effort required, and maintain proper body position to prevent injury.

Operating Overhead Valves

Valves are sometimes located in positions that can create hazards. Whenever possible, valves should be operated from a point above or level with the valve. This helps prevent injury from dripping fluid that might be present as a result of valve leakage or failure. Fluids flowing through valves might be hazardous and could cause serious injuries if released.

Elevated valves (e.g., valves in pipe racks) might have a chain-operator extension that allows the technician to operate the valve from ground level. Even if chain valve wheels have an attached safety cable to prevent the wheel from falling should the attachment mechanism fail, caution must be taken when operating this type of valve. Care should be taken when manipulating a chain-operator to ensure the chain does not become dislodged from its cradle around the handwheel or get tangled in clothing. Figure 5.21 shows an example of a chain-operated valve.

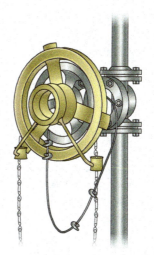

Figure 5.21 Chain-operated valve with safety cable.

Safe Footing

When adjusting a valve, the optimum operating position is usually at ground level. If that is not possible, the process technician should obtain solid footing by firmly planting both feet on the surface being used for elevation (e.g., ladder, scaffold, or secured platform). To prevent slipping or falling, ensure the platform is stable and has adequate traction (i.e., it is dry and clear of any oily residue or debris).

When adjusting a valve, use a pushing motion instead of a pulling motion. A pushing motion allows the process technician to react more effectively if he or she begins to slip, especially when working from an elevated work surface.

Use of Valve Wrenches and Levers

The process technician should always use the proper valve wrench to adjust a valve. See Figure 5.22 for examples of valve wrenches. Make sure that the valve wrench is the proper size and in good condition. Also, be sure to check for pinch points before using the valve wrench or lever. Ensure that hands, or any other parts of the body, do not enter those pinch points.

The process technician should always use the correct tool for the job and never improvise with the closest tool at hand (e.g., never use a pair of pliers or pipe wrench instead of a valve wrench). Using the wrong tool can cause injury to the process technician, or damage to the valve and handwheel.

When using a valve wrench or lever, the process technician should establish a firm grip on the valve wrench handle, secure his or her footing, and then position the body to prevent a strain or to avoid striking an object. To utilize greater mechanical advantage, the valve

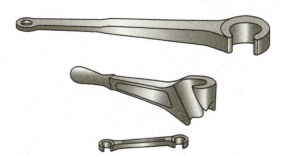

Figure 5.22 Examples of valve wrenches.

wrench should be grasped closer to the end of the handle than the hook of the wrench. Use body weight instead of back muscles when adjusting valves that are difficult to open or close. Seek help when operating particularly stubborn valves.

5.4 Process Technician's Role in Operation and Maintenance

Depending on company policy, a process technician might be responsible for incidental (minor) maintenance and repair of valves. Minor maintenance can include tightening valve packing or replacing a small valve. The most critical repairs are performed by maintenance personnel with specialized expertise.

When monitoring valves, process technicians are generally responsible for inspecting for leaks, cleaning, greasing and lubricating stems, and monitoring the operation and position of valves.

Inspecting for Leaks and Corrosion

Inspections are performed to check for leaks and signs of external corrosion. The most obvious places to look for leaks and corrosion include areas around the packing, bonnets, gaskets, flanges, or threads.

Greasing and Lubricating

Valves must be greased and lubricated regularly for proper operation. When greasing and lubricating, process technicians typically should grease the valve stems, lubricate additional components, adjust the packing, and inspect the packing for leaks.

Monitoring

When monitoring valves, technicians must always remember to look at, listen to, and check for the items listed in Table 5.2.

Failure to perform proper maintenance and monitoring could have an impact on the process and result in personal injury and equipment damage.

Table 5.2 Process Technician's Role in Operation and Maintenance

Look	Listen	Check
■ Inspect valves to make sure there are no leaks or signs of external corrosion ■ Monitor valve for excessive wear ■ Check to make sure the valve stems are properly lubricated	■ Listen for abnormal noises (e.g., check for valve "chatter")	■ Make sure the valve is not overly tightened ■ Make sure the handwheels of wide-open valves move freely

Summary

Valves regulate and direct the flow of materials through piping, and are some of the most prevalent and important pieces of equipment in the process industries. Many different types of valves are used in the process industries, and they all share the ability to control flow by placing an obstruction in the flow path of a fluid. The most widely used types of valves in the process industry are ball, butterfly, check, diaphragm, gate, globe, plug, safety, and relief.

Valves can be categorized in a number of ways: by their intended purpose (starting/stopping or throttling flow), by their flow control element (gate, plug, ball, etc.), by their method of connection (welded, flanged, etc.), by the characteristics of the flow through them (high/low pressure or temperature, viscosity, corrosive or erosive nature, etc.), by their material of construction, or by their method of control (linear or rotational movement).

The operation of some valves is controlled—either manually or automatically—while the operation of other valve types requires no operator interface (safety/relief and check valves).

Valves must never be installed or operated under conditions that exceed their safe pressure and temperature operating limits. System valves should never be closed in such a way that a line or piece of equipment is blocked in or "deadheaded," which could result in overpressurization and catastrophic failure.

Excessive wear and damage to valves can be caused by throttling valves that are not designed for throttling service, using excessive force to open or close a valve, or failing to clean and lubricate valve stems. This damage could lead to valve leakage or failure.

When making rounds, process technicians should always inspect valves for leaks and perform proper maintenance and lubrication procedures to prevent damage or excessive wear.

Checking Your Knowledge

1. Define the following terms:
   a. Ball valve
   b. Block valve
   c. Butterfly valve
   d. Check valve
   e. Control valve
   f. Diaphragm valve
   g. Gate valve
   h. Globe valve
   i. Multiport valve
   j. Manual valve
   k. Needle valve
   l. Plug valve
   m. Relief valve
   n. Safety valve
   o. Bonnet
   p. Damper
   q. Handwheel
   r. Louvers
   s. Packing
   t. Throttling
   u. Valve seat
   v. Valve knocker

2. (True or False) Gate valves are the most common type of valve designed for on/off service and NOT for throttling a process.

3. Globe valves use a _____ to regulate the flow.
   a. Disc and seat
   b. Gate
   c. Diaphragm
   d. Plate

4. Check valves are used to prevent _____ in piping when the flow stops.
   a. overflow
   b. continuous flow
   c. trickle flow
   d. backflow

5. Which of the following are types of check valves? (Select all that apply.)
   a. Swing
   b. Lift
   c. Butterfly
   d. Gate

6. (True or False) A ball valve opens completely with a quarter turn of the stem using the valve handle.

7. (True or False) The two types of diaphragm valves are membrane diaphragm valves and straight-through diaphragm valves.

8. A _____ valve would be selected for use with extremely viscous or corrosive materials.

   a. globe

   b. plug

   c. diaphragm

   d. ball

9. (True or False) Safety valves operate quickly to prevent overpressurization by gases, which can cause equipment damage or injury.

10. A _____ can be used to address pluggage in small valves.

    a. valve wrench

    b. rod-out tool

    c. chain operator

    d. welding rod

11. Name three causes of leakage in valves.

12. Which of the following improper operations can lead to damage to the valve stem? (Select all that apply.)

    a. Use of excessive force when opening or closing a valve

    b. Failure to clean and lubricate valve stems

    c. Closing a valve under high pressure

    d. Closing a valve at low temperatures

13. (True or False) Use the tool closest at hand when adjusting valves.

14. Match the valve type with its description.

Type	Description
I. Ball	a. Designed to open if the pressure of a gas exceeds a preset threshold
II. Butterfly	b. Uses a disc-shaped flow control element to control flow
III. Check	c. Uses a hollowed-out plug to control flow
IV. Diaphragm	d. Uses a rubber-type bendable component to control flow
V. Gate	e. Uses a guillotine to block the flow of fluids
VI. Globe	f. Uses a hollowed-out ball to control flow
VII. Plug	g. Designed to open if the pressure of a liquid exceeds a preset level
VIII. Relief	h. Uses a ball, cylindrical, or needle-shaped disc to block fluid flow
IX. Safety	i. Used to prevent accidental backflow

15. List two common places to look for leaks and corrosion in valves.

NOTE: Answers to Checking Your Knowledge questions are in the Appendix.

Student Activities

1. Given a hand valve (globe or gate), dismantle the valve and label all the separate components.

2. Using a simulator or process skid unit, identify all the valve types in use.

3. Given a P&ID, label the valves and identify their types and the direction of flow.

4. In your home, identify at least five valves (e.g., the valve under your kitchen sink). Tell where each valve is located and identify what type of valve you think it might be (e.g., ball, gate, or globe valve) and its purpose.

5. Given a drawing of a valve, identify all of the components (e.g., stem, seat, and flow control element).

Chapter 6
Pumps

 ## Objectives

After completing this chapter, you will be able to:

6.1   Explain the purpose and selection of pumps in the process industry. (NAPTA Pumps 1*) p. 103

6.2   Identify common pump types and their components. (NAPTA Pumps 2, 6, 10) p. 104

6.3   Identify associated components and their purpose. (NAPTA Pumps 4, 5, 8, 9) p. 120

6.4   Identify operating principles of pumps. (NAPTA Pumps 6) p. 125

6.5   Explain the purpose of a pump curve and demonstrate its use. (NAPTA Pumps 3) p. 127

6.6   Identify potential problems associated with pumps. (NAPTA Pumps 7, 11) p. 129

6.7   Describe safety and environmental hazards associated with pumps. (NAPTA Pumps 12) p. 131

6.8   Describe the process technician's role in pump operation and maintenance. (NAPTA Pumps 14) p. 133

6.9   Identify typical procedures associated with pumps. (NAPTA Pumps 13) p. 133

*North American Process Technology Alliance (NAPTA) developed curriculum to ensure that Process Technology courses will produce knowledgeable graduates to become entry level employees in process technology. Objectives from that curriculum are named here in abbreviated form. For example, "(NAPTA Pumps 1)" means that this chapter's objective relates to objective 1 of NAPTA's course content on pumps.

Key Terms

Axial pump—a dynamic pump that uses a propeller or row of blades to propel liquids along the shaft, **p. 111.**

Bearings—mechanical components that keep the pump shaft in alignment with the casing and absorb axial and radial forces, **p. 124.**

Canned pump—a sealless pump that ensures zero emissions. It is often used on EPA-regulated liquids, **p. 108.**

Cavitation—a condition inside a pump in which the liquid being pumped partially vaporizes because of variables such as temperature and pressure drop, and the resulting vapor bubbles then implode, **p. 130.**

Centrifugal force—the energy that causes something to move outward from the center of rotation, **p. 104.**

Centrifugal pump—a type of dynamic pump that uses an impeller on a rotating shaft to generate pressure and move liquids, **p. 104.**

Cutwater—a thick plate in the discharge nozzle of a pump that breaks the vortex, **p. 126.**

Deadhead—also called shut-off pressure; it is the maximum pressure (head) that occurs when a pump is operating with zero flow (discharge valve shut), **p. 106.**

Diaphragm pump—a mechanically or air-driven positive displacement reciprocating pump that uses a flexible membrane to move liquid, **p. 119.**

Discharge static head—the vertical distance between the centerline of a pump and the surface of the liquid on the discharge side of the pump (if the discharge line is submerged) or the pipe end (if the discharge line is open to the atmosphere), **p. 125.**

Double-acting piston pump—a type of piston pump that takes suction and discharges by reciprocating motion on every stroke, **p. 119.**

Dynamic pump—a type of pump that converts the spinning motion of a blade or impeller into dynamic pressure to move liquids, **p. 104.**

External gear pump—a type of positive displacement rotary pump in which two gears rotate in opposing directions, allowing the liquid to enter the space between the teeth of each gear in order to move the liquid around the casing to the discharge, **p. 113.**

Head—measurement of pressure caused by the weight of a liquid, measured in feet or meters, determined by the height of the liquid above the centerline of a pump, **p. 125.**

Impeller—a device with vanes that spins a liquid rapidly in order to generate centrifugal force, **p. 105.**

Internal gear pump—a type of positive displacement rotary pump, called a "gear within a gear" pump, in which two gears rotate in the same direction, one inside the other, and trap liquid between the teeth of the gears to move liquid from suction to discharge, **p. 113.**

Liquid head—pressure developed from the pumped liquid passing through the volute, **p. 106.**

Lobe pump—a type of positive displacement rotary pump consisting of a single or multiple lobes; liquid is trapped between the rotating lobes and is subsequently moved through the pump, **p. 114.**

Magnetic (mag) drive pump—a type of pump that uses magnetic fields to transmit torque to an impeller, **p. 109.**

Mechanical seals—seals that typically contain two flat faces (one that rotates and one that is stationary), that are in close contact with one another in order to prevent leaks, **p. 121.**

Multistage centrifugal pump—a type of pump that uses two or more impellers on a single shaft (generally used in high-volume, high-pressure applications, such as boiler feed water pumps), **p. 108.**

Net positive suction head (NPSH)—the liquid head (pressure), minus the vapor pressure, that exists at the suction end of a pump, **p. 104.**

Net positive suction head available (NPSH$_a$)—NPSH available from the pump system; it must always be equal to or greater than the NPSH$_r$, **p. 104.**

Net positive suction head required (NPSH$_r$)—pump characteristic established by manufacturer testing that indicates the amount of NPSH necessary for the pump to function properly, **p. 104.**

Piston pump—a type of positive displacement reciprocating pump that uses a piston inside a cylinder to move fluids, **p. 118.**

Plunger pump—a type of positive displacement reciprocating pump that displaces liquid using a plunger and maintains a constant speed and torque, **p. 119.**

Positive displacement pump—a type of pump that uses pistons, diaphragms, gears, or screws to deliver a constant volume with each stroke, **p. 104.**

Priming—the process of filling the suction line and casing of a pump with liquid to remove vapors and eliminate the tendency for it to become vapor-bound or to lose suction, **p. 107.**

Pump—a mechanical device that transfers energy to move liquids through piping systems, **p. 103.**

Pump performance curve—a specification that describes the capacity, speed, horsepower, and head needed for correct pump operations, **p. 127.**

Reciprocating pump—a type of positive displacement pump that uses the inward stroke of a piston or diaphragm to draw liquid into a chamber (intake) and then positively displace the liquid using an outward stroke (discharge), **p. 117.**

Rotary pump—a type of positive displacement pump that moves in a circular motion to move liquids by trapping them in a specific area of a screw or a set of lobes, gears, or vanes, **p. 113.**

Screw pump—a type of positive displacement rotary pump that displaces liquid with a screw. The pump is designed for use with a variety of liquids and viscosities and a wide range of pressures and flows, **p. 116.**

Seal—a device that holds lubricants and process fluids in place while keeping out foreign materials where a rotating shaft passes through a pump casing, **p. 120.**

Seal flush—a small flow (slip stream) of pump discharge or externally supplied liquid that is routed to the pump's mechanical seal. This acts as a barrier liquid between the two faces of the seal to reduce friction and remove heat, **p. 121.**

Single-acting piston pump—a type of piston pump that pumps by alternating suction and discharge actions on each piston stroke, **p. 118.**

Stuffing box— the area in a pump's casing that contains the packing material, **p. 120.**

Suction static head—the vertical distance between the centerline of a pump and the surface of the liquid on the suction side of the pump, **p. 125.**

Total head—a measure of a pump's ability to move liquid through a pumping system, **p. 125.**

Vane pump—a type of positive displacement rotary pump having either flexible or rigid vanes designed to displace liquid, **p. 114.**

Vanes—raised ribs on the impeller of a centrifugal pump designed to accelerate a liquid during impeller rotation, **p. 105.**

Venturi—a device consisting of a converging section, a throat, and a diverging section; its purpose is to create a constriction in a pipe, **p. 111.**

Viscosity—the degree to which a liquid resists flow under applied force (e.g., molasses has a higher viscosity than water at the same temperature), **p. 104.**

Volute—a widened spiral casing in the discharge section of a centrifugal pump designed to convert liquid speed to pressure, **p. 105.**

Wear rings—close-running, noncontacting replaceable metal rings located between the impeller and casing of a centrifugal pump; wear rings allow for a small clearance between the two components, **p. 124.**

6.1 Introduction

Pumps are mechanical devices that transfer energy to move liquids through piping systems. Pumps are used in many applications, including filling or emptying tanks, wells, pits, and trenches; supplying water to boilers; providing circulation for systems; moving material between process vessels, supplying fire control water; lubricating equipment; and drawing process samples.

The two main categories of pumps are dynamic and positive displacement. Dynamic pumps use impellers to generate centrifugal force, which is then converted to dynamic

Pump a mechanical device that transfers energy to move liquids through piping systems.

pressure to move liquids, or they use propellers to move liquids axially (in a straight line). Positive displacement pumps use pistons, diaphragms, screws, lobes, gears, or vanes to move or push liquids. Dynamic pumps tend to be used more often than positive displacement pumps because they are less expensive, easier to operate, and require less space and maintenance.

When working with pumps, process technicians should conduct monitoring and maintenance activities to ensure that pumps are properly lubricated and that there is no leakage from any of the pump components. Process technicians also should monitor pumps for damaging conditions such as overpressurization, overheating, and cavitation.

Selection of Pumps

To select the appropriate pump for a job, several factors must be considered. Consideration first must be given to the properties of the process liquid for which the pump is intended. One of those properties is the viscosity of the liquid. **Viscosity** is the degree to which a liquid resists flow under applied force. Liquids that are less viscous flow more easily than others. The pump selected must be appropriate for the liquid's viscosity.

Other properties affecting pump selection are density and specific gravity, which affect the horsepower required by the pump. The denser the fluid, the more horsepower will be needed.

The vapor pressure of the liquid affects the **net positive suction head (NPSH)**, or the pressure that exists at the suction end of a pump. The **net positive suction head available (NPSH$_a$)** must always be equal to or greater than **net positive suction head required (NPSH$_r$)** to ensure reliable operation and prevent cavitation.

A selected pump must be able to meet the required suction head, discharge head, and flow rate. These design factors include how much suction head is provided to the pump, how much discharge head the pump is required to supply, and how much flow the pump is required to provide.

Other factors that determine pump selection include the required output of the pump, the speed (which affects capacity), the altitude (which affects suction head), and the temperature (which determines the construction material of the pump). If a pump is required for slurry service, it also must be able to handle suspended solids.

6.2 Types of Pumps

The two main categories of pumps are dynamic and positive displacement. Within each of these categories are subcategories. Figure 6.1 shows a diagram of the different types of pumps and their subcategories.

Dynamic pumps raise a liquid's pressure by first using the spinning motion of a blade or impeller (centrifugal force) to produce velocity energy, and then converting that velocity energy into pressure energy to move liquids. Dynamic pumps are classified as either centrifugal or axial.

Positive displacement pumps use pistons, diaphragms, gears, vanes, lobes, or screws to deliver a constant volume with each stroke. Positive displacement pumps are classified as either reciprocating or rotary. Unlike dynamic pumps, positive displacement pumps deliver the same amount of liquid, regardless of the discharge pressure.

Dynamic Pumps

CENTRIFUGAL PUMPS **Centrifugal force** is the force that causes something to move outward from the center of rotation. **Centrifugal pumps** use an impeller on a rotating shaft to

---

Viscosity   the degree to which a liquid resists flow under applied force (e.g., molasses has a higher viscosity than water at the same temperature).

Net positive suction head (NPSH)   the liquid head (pressure), minus the vapor pressure, that exists at the suction end of a pump.

Net positive suction head available (NPSH$_a$)   NPSH available from the pump system, it must always be equal to or greater than the NPSH$_r$.

Net positive suction head required (NPSH$_r$)   pump characteristic established by manufacturer testing that indicates the amount of NPSH necessary for the pump to function properly.

Dynamic pump   a type of pump that converts the spinning motion of a blade or impeller into dynamic pressure to move liquids.

Positive displacement pump   type of pump that uses pistons, diaphragms, gears, or screws to deliver a constant volume with each stroke.

Centrifugal force   the energy that causes something to move outward from the center of rotation.

Centrifugal pump   a type of dynamic pump that uses an impeller on a rotating shaft to generate pressure and move liquids.

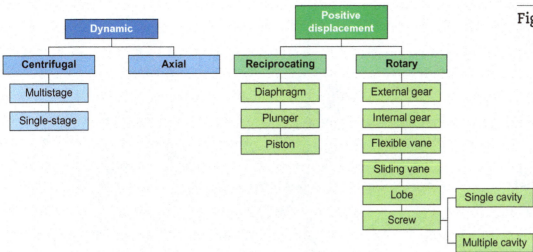

Figure 6.1 Pump family tree.

generate pressure and move liquids to the outer casing within the pump and then out through the discharge nozzle. Pressure is created in the liquid as it passes through a widening spiral casing in the discharge section known as a **volute**. Figure 6.2 shows the outside and cutaway views of a centrifugal pump.

Centrifugal pumps are used to move large volumes of low to medium viscosity liquid. Centrifugal pumps differ from positive displacement pumps because the amount of liquid they deliver depends on the discharge pressure, not on the size of the chamber.

Volute a widened spiral casing in the discharge section of a centrifugal pump designed to convert liquid speed to pressure.

A.   B.

Figure 6.2 A. Outside of a centrifugal pump showing complete system. **B.** Cutaway of a centrifugal pump.

CREDIT: A. Photo smile/Shutterstock. **B.** Design Assistance Corp

Centrifugal Pump Components The main components of a centrifugal pump include the suction, inlet (suction eye), outlet (discharge), impeller, bearings and seals, shaft, and casing or housing. Figure 6.3 illustrates and identifies components of a centrifugal pump.

In a centrifugal pump, liquid enters the suction and flows through the suction eye to a spinning impeller. The impeller is attached to a shaft that is coupled to the driver (motor or steam turbine). An **impeller** is a vaned device that spins a liquid rapidly in order to generate the centrifugal force necessary to move the liquid. The **vanes** on an impeller are raised ribs designed to accelerate a liquid during impeller rotation. This rotation results in centrifugal force that causes the liquid to spin toward the outer edge of the impeller. Figure 6.4 shows an example of an open impeller, a semi-open impeller, and a cutaway of closed impellers.

Impeller a device with vanes that spins a liquid rapidly in order to generate centrifugal force.

Vanes raised ribs on the impeller of a centrifugal pump designed to accelerate a liquid during impeller rotation.

Figure 6.3 **A.** Centrifugal pump components. **B.** Cutaway of a centrifugal pump.

CREDIT: B. smspsy/Shutterstock.

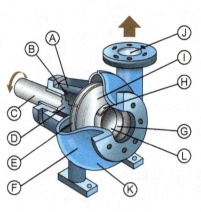

A Stuffing box
B Packing/seal
C Shaft
D Shaft sleeve
E Vane
F Casing/housing
G Eye of impeller
H Casing wear ring
I Impeller
J Discharge/outlet
K Volute
L Suction intake/inlet

A.

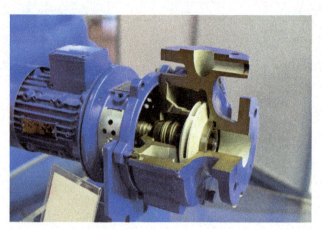

B.

Figure 6.4 **A.** Open impeller.
B. Semi-open impeller.
C. Cutaway of a pump with closed impellers.

CREDIT: A. fantasy/Fotolia. **C.** yotanan chankheaw/Shutterstock.

A. Open impeller   **B.** Semi-open impeller

C. Cutaway of a pump with closed impellers.

Liquid head   pressure developed from the pumped liquid passing through the volute.

Deadhead   also called shut-off pressure; it is the maximum pressure (head) that occurs when a pump is operating with zero flow (discharge valve shut).

As the liquid is forced through the pump, the velocity energy is converted to pressure energy through a widening of the casing known as a volute. The pressure that is developed is called **liquid head**. The amount of liquid head developed is a function of the tip speed of the impeller (impeller revolutions per minute and diameter). The higher the tip speed, the higher the liquid head.

The maximum head occurs with the pump in operation at zero flow (with the discharge valve shut) and is called **deadhead** or shutoff pressure.

IMPORTANT NOTE:

Deadheading a pump causes friction in the liquid, which is converted to heat. This heat can cause vapor lock or cavitation that can damage the pump. Always follow proper procedures and make sure the discharge valve is opened slightly before starting a centrifugal pump.

Figure 6.5 shows the path of the liquid as it enters a pump, leaves the impeller, and approaches the discharge nozzle. The following explains what is happening at each step in Figure 6.5:

Step 1: The liquid enters into the suction and is picked up by the motion of the impeller.

Step 2: As the liquid moves at high speed from the close clearance area (impeller to pump case) to the wider clearance area of the volute, the velocity energy in the liquid converts to pressure energy as the liquid slows down.

Step 3: The high-pressure liquid approaches the discharge nozzle. In the discharge nozzle, the vortex (the cone formed by the swirling liquid) is broken.

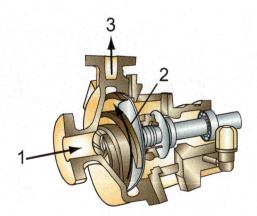

Figure 6.5 Horizontal centrifugal pump showing flow through the pump.

Some centrifugal pumps, including most multistage centrifugal pumps, use stationary diffuser vanes installed within the casing (instead of a volute) to slow the liquid's velocity and convert the velocity energy to pressure.

While the majority of centrifugal pumps in the process industry are horizontally situated, a vertical centrifugal pump is used with some applications. The operating principles are the same. The most notable differences are the vertical orientation of the shaft and the horizontal mounting of the impeller. Advantages of a vertical centrifugal pump are that it takes less space and usually can withstand higher operating temperatures and pressures than a horizontal pump. The greatest disadvantage of vertical centrifugal pumps is that the motor is situated on top of the pump and must be removed to work on the pump.

When working with centrifugal pumps, it is important to know that pumps cannot remove gas (air) from the suction line, so a pump must be primed before it is started. **Priming** a pump involves filling the suction of the pump with liquid to remove any vapors that might be present. This reduces the likelihood of the pump cavitating, becoming vapor-bound, or losing suction. Pump priming is discussed in more detail later in this chapter.

Priming the process of filling the suction line and casing of a pump with liquid to remove vapors and eliminate the tendency for it to become vapor-bound or to lose suction.

Did You Know?

When a pump cavitates, it sounds as if rocks are being poured into the suction line.

These bursts of liquid and gas create an effect like a "liquid hammer" that can seriously damage the inside of the pump and the impeller. Pump impellers that have been damaged by cavitation exhibit signs of gradual erosion and pitting on the impeller vanes.

CREDIT: Naronta / Shutterstock.

Single-Stage Versus Multistage Pumps   Centrifugal pumps can be single-stage or multi-stage. (Note, a stage is one pressure level increase). If the pressure differential between pump suction and discharge is greater than 150–200 pounds (68–91 kg), it would be impractical and costly to use a single-stage pump. In such cases, a multistage pump is used.

In a single-stage pump, the liquid enters the pump, the pressure is increased one time, and then the liquid exits the pump. In a multistage pump, the liquid enters the pump and the pressure is increased multiple times (e.g., two, three, or four times) before it exits the pump.

Single-stage centrifugal pumps consist of a disk-shaped impeller mounted on a shaft and fitted into a case. The impeller's role is to impart kinetic energy (the energy associated with mass in motion) and velocity (speed) to the pumped liquid. An everyday example of a single-stage centrifugal pump is the cooling water pump in an automobile engine.

A **multistage centrifugal pump** (shown in Figure 6.6) is used in high-volume, high-pressure applications such as boiler-feed water pumps. Multistage pumps contain two or more impellers on a single shaft. The discharge of one impeller enters into the suction of the next to increase the pressure of the liquid in sequential steps, so a multistage pump is like having several pumps in one.

Multistage centrifugal pump   a type of pump that uses two or more impellers on a single shaft (generally used in high-volume, high-pressure applications, such as boiler feed water pumps).

Figure 6.6 Multistage centrifugal pump.

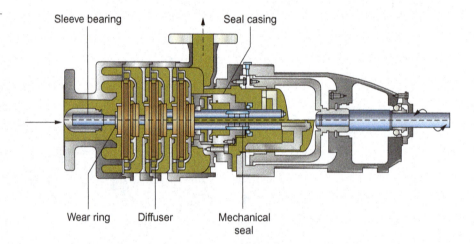

Sleeve bearing   Seal casing

Wear ring   Diffuser   Mechanical seal

The advantages and disadvantages of using a multistage centrifugal pump are listed in Table 6.1.

Table 6.1 Advantages and Disadvantages of Using a Multistage Centrifugal Pump

Advantages	Disadvantages
■ Effective	■ High initial cost
■ Requires a smaller amount of space than two single-stage pumps	■ Parts can be expensive
■ Minimal maintenance	■ Proper installation and operation are critical
	■ More complex bearing coolers, seal flush systems, and operating guidelines
	■ Increase in operating problems because of the complexity

Specialty Centrifugal Pumps   In addition to the centrifugal pumps already mentioned, other types of specialty pumps include canned, magnetic drive, high-speed centrifugal, and jet pumps.

Canned pumps are sealless pumps that ensure zero emissions. These pumps are used with liquids that are difficult to seal, such as liquefied gases, and liquids regulated by the Environmental Protection Agency (EPA). Some states also require the use of sealless pumps for certain hazardous processes. Canned pumps can provide zero emissions because there are no seals to leak to the atmosphere. Figure 6.7 shows an example of a canned pump.

Canned pump   a sealless pump that ensures zero emissions. It is often used on EPA-regulated liquids.

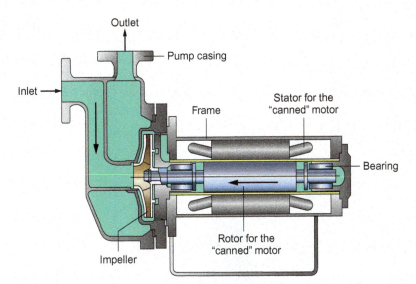

Figure 6.7 Canned pump.

In a canned pump, the rotating part of the electric motor attaches to the pump's shaft in an enclosure similar to a can. Windings surround the can and induce a magnetic field that causes the entire rotor assembly to rotate when electrical current is applied.

The advantages and disadvantages of using a canned pump are listed in Table 6.2.

Table 6.2 Advantages and Disadvantages of Using a Canned Pump

Advantages	Disadvantages
▪ Seals are not required because everything is contained within the pump enclosure ▪ No leakage of contents to the environment ▪ No lubrication required	▪ High cost ▪ Pumped liquid must be compatible with motor components ▪ Typically cheaper to replace the pump than to repair it

Magnetic (mag) drive pumps use magnetic fields to transmit torque to an impeller. Like canned pumps, magnetic drive pumps are used with liquids that are difficult to seal, such as liquefied gases and liquids regulated by the EPA. Figure 6.8 shows an example of a magnetic (mag) drive pump.

Magnetic (mag) drive pump a type of pump that uses magnetic fields to transmit torque to an impeller.

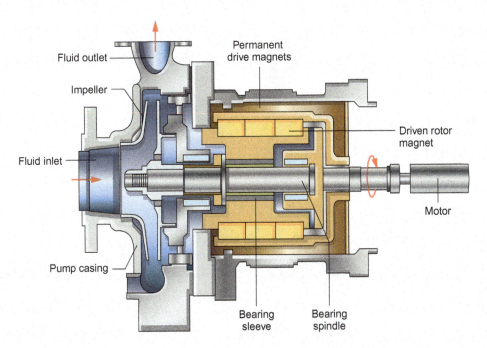

Figure 6.8 Magnetic (mag) drive pump.

In a magnetic drive pump, a magnet couples the pump to the driver to eliminate shaft leakage and the need for mechanical seals. A magnetic force passing through a stainless-steel canister drives the inner coupling (which includes the magnets on the drive shaft and the magnets on the pump shaft). These magnets are separated by a shroud that keeps the process contained inside the pump. Figure 6.9 shows examples of magnetic couplings.

Figure 6.9 A. Magnetic drive coupling used to connect shafts for power transmission. **B.** Operating principles of the magnetic coupling.

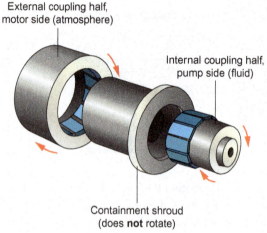

A. Magnetic drive coupling

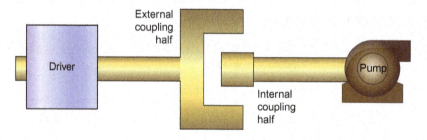

No physical / mechanical connection

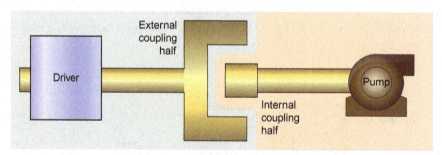

Hermetically separates two zones in order to prevent fluid or gas leakage from one area to another

B.

The advantages and disadvantages of using a magnetic drive pump are listed in Table 6.3.

Table 6.3 Advantages and Disadvantages of Using a Magnetic Drive Pump

Advantages	Disadvantages
▪ Elimination of costs associated with mechanical seal repair ▪ Elimination of expensive monitoring equipment for EPA-regulated liquids	▪ Expensive to purchase ▪ Subject to slippage (magnets failing to rotate at the same speed) ▪ Must be "bump-started," i.e., started and stopped quickly to ensure the magnets on the drive shaft line up with the magnets on the pump shaft

High-Speed Centrifugal Pumps   High-speed centrifugal pumps are used to generate high pressures. The pressure generated by a centrifugal pump is proportional to the *square* of the impeller rotations per minute (rpm). So, a tenfold increase in impeller rpm results in a 100-fold increase in discharge pressure. Impeller speeds of 20,000 to 30,000 rpm can be achieved using a gearbox coupled to the drive shaft. This results in relatively high discharge pressures in proportion to the size of the pump.

Jet pumps are used to lift liquids from wells. A jet pump consists of a centrifugal pump that recycles up to three-fourths of its discharge into a jet ejector, which consists of a nozzle and a **Venturi** (a device consisting of a converging section, a "throat," and a diverging section), installed in its suction line. Figure 6.10 shows an example of a jet pump.

The recycled liquid enters the jet ejector nozzle at a high pressure. As it goes into the jet, through the Venturi, velocity is increased, while pressure decreases (Bernoulli's Principle). This area of low pressure allows additional fluid from the well to enter the pump suction line and be drawn up into the pump.

Venturi   a device consisting of a converging section, a throat, and a diverging section; its purpose is to create a constriction in a pipe.

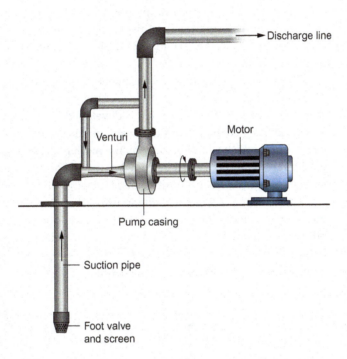

Figure 6.10 Jet pump with Venturi.

AXIAL PUMPS   Axial pumps are dynamic pumps that use a propeller or row of blades to propel liquids in a horizontal direction, parallel to the pump's shaft (as opposed to centrifugal pumps, which use an impeller to force liquids radially, to the outer wall of the pump casing.). An everyday example of the operating principle of an axial pump is the operation of a boat's motor, in which a propeller forces water out behind it. Figure 6.11 shows the main components of an axial pump.

Axial pump   a dynamic pump that uses a propeller or row of blades to propel liquids along the shaft.

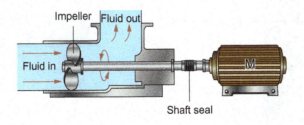

Figure 6.11 Axial pump showing components.

Axial pumps are not capable of generating much head, but they can move very large volumes of liquid. They are widely used in applications where a large amount of liquid is circulated without much pressure. Examples of industrial applications are:

- Commercial and municipal sewage
- Power plant operation (providing water from lakes or reservoirs for cooling)
- Chemical plants (moving liquids used in evaporators and crystallizers)
- Agricultural use (irrigation and drainage).

Positive Displacement Pumps

Positive displacement pumps (shown in Figure 6.12) use pistons, diaphragms, vanes, gears, lobes, or screws to deliver a constant volume with each stroke. Unlike dynamic pumps, positive displacement pumps deliver the same amount of liquid, regardless of the discharge pressure.

Positive displacement pumps move liquid by trapping it in the space between the pumping elements and forcing (displacing) it into the discharge nozzle. In some pumps, the size of the space between the pumping elements decreases between the suction and discharge sides.

The meaning of positive displacement is that when the pumping element (gear, piston, screw, lobe, vane, or diaphragm) moves, fluid moves into the pump and displaces the material in front of it. The pressure that is developed is not a function of pump speed. There is essentially no theoretical limit to the amount of pressure developed by a positive displacement pump, which is limited only by the mechanical strength of the pump and piping and the amount of power available from the driver. For this reason, process technicians should *never* deadhead or block a positive displacement pump while it is running.

The flow rate of a reciprocating positive displacement pump is determined by the volume of the cylinder per pump stroke times the number of strokes per minute. The flow rate can be controlled precisely by varying the stroke rate. Because of this, these pumps are frequently used as metering pumps (precision pumps that deliver an exact amount of liquid with each stroke or revolution of the handle).

Figure 6.12 Positive displacement reciprocating pump showing flow in and out of the cylinder. **A.** Vertical orientation. **B.** Horizontal orientation.

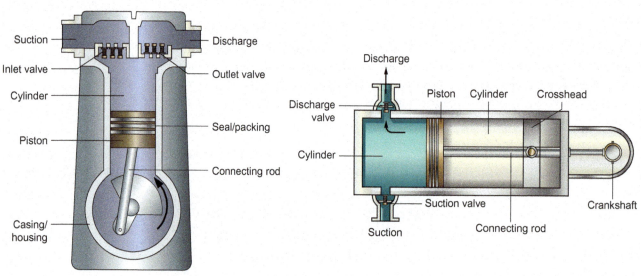

The two categories of positive displacement pumps are rotary and reciprocating. Rotary pumps displace the liquid at a constant rate by using a rotary motion, while reciprocating pumps use a piston or plunger that moves back and forth to displace the liquid.

Because positive displacement pumps move a certain volume of material to create pressure, they are self-priming (i.e., they can remove gas or air from the suction line). This makes them useful in many applications where a centrifugal pump would not work (e.g., removing waste from a drum).

ROTARY PUMPS **Rotary pumps** are positive displacement pumps that move liquids by rotating a screw or a set of gears, lobes, or vanes. As these screws, lobes, gears, or vanes rotate, the liquid is drawn into the pump (intake) by lower pressure on one side and forced out of the pump (discharged) by higher pressure on the other side.

External Gear Pumps **External gear pumps** contain two rotating gears, one of which is driven by the motor. The gears rotate in opposite directions, allowing the liquid to enter the space between the teeth of each of the gears (Figure 6.13). The second gear is driven by the first gear, as they mesh and unmesh in their rotation. The pumped liquid does not enter the space between the two gears, but rather is trapped between the pump casing and the spaces between the gear teeth as it is moved around the casing to the outlet.

Rotary pump a type of positive displacement pump that moves in a circular motion to move liquids by trapping them in a specific area of a screw or a set of lobes, gears, or vanes.

External gear pump a type of positive displacement rotary pump in which two gears rotate in opposing directions, allowing the liquid to enter the space between the teeth of each gear in order to move the liquid around the casing to the discharge.

Figure 6.13 A. Components of an external gear pump. **B.** External gear pump in casing. **C.** Water flowing through external gear pump.

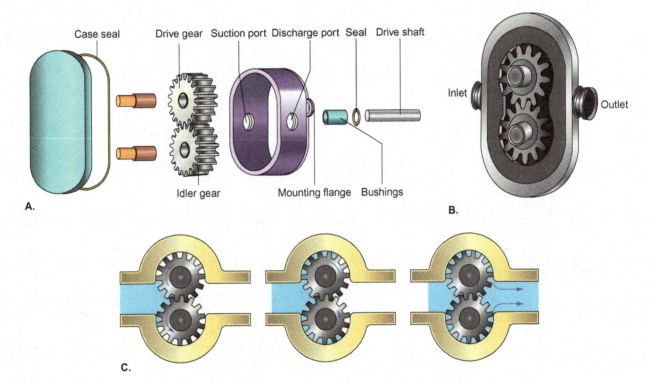

Internal Gear Pumps An **internal gear pump** is referred to as "gear-within-a-gear" pump because of its design (Figure 6.14). As with the external gear pump, one of the gears is driven by the motor. This outer gear is the larger of the two and is called the rotor or driven gear. It drives the smaller, internal gear (called the idler gear), which is off center in the pump cavity. In the internal gear pump, the gears are not side by side, but instead the idler gear is inside the driven gear. The gears become unmeshed at the suction port, creating a low-pressure area where the liquid enters the area between the gear teeth. A stationary crescent-shaped

Internal gear pump a type of positive displacement rotary pump, called a "gear within a gear" pump, in which two gears rotate in the same direction, one inside the other, and trap liquid between the teeth of the gears to move liquid from suction to discharge.

Figure 6.14 A. Components of an internal gear pump. **B.** Water flowing through an internal gear pump.

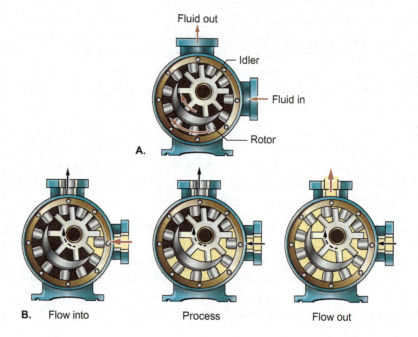

Figure 6.14 A. Components of an internal gear pump. **B.** Water flowing through an internal gear pump.

divider, located at the point where the gears are out of mesh, prevents liquid from returning to the suction side by creating a seal between the suction and discharge sides of the pump. As the liquid approaches the discharge port, the gears mesh, forcing the liquid out of the pump.

The advantages and disadvantages of using a gear pump are listed in Table 6.4.

Table 6.4 Advantages and Disadvantages of Using a Gear Pump

Advantages	Disadvantages
■ Suitable for medium pressures	■ Not suitable for solids
■ Quiet operation	
■ Accommodates a variety of materials	
■ Used at high speeds	

Lobe pump a type of positive displacement rotary pump consisting of a single or multiple lobes; liquid is trapped between the rotating lobes and is subsequently moved through the pump.

Lobe Pumps **Lobe pumps** are similar in operation to external gear pumps, in that two rotors rotate in opposite directions and trap liquid between a pumping element (a lobe in this case) and the pump casing to move it from suction to discharge side. Lobe pumps differ from external gear pumps in that both rotors are driven with their movements synchronized by timing gears and the lobes on the rotors don't touch each other as they rotate. Rotors can have between one and six lobes, which have larger spaces between them than the teeth in gear pumps, so they are ideal for pumping very viscous materials and are capable of handling solids that can be present in a liquid. Figure 6.15 shows a drawing of a lobe pump with the components labeled and the flow path through the pump.

The advantages and disadvantages of using a lobe pump are listed in Table 6.5.

Vane pump a type of positive displacement rotary pump having either flexible or rigid vanes designed to displace liquid.

Vane Pumps **Vane pumps** are rotary pumps that have flexible or rigid vanes that are designed to displace liquids. Vane pumps are used for low-viscosity liquids, and they can be run dry for short periods of time without damage. Vane pumps can develop a good vacuum, and the vanes can be replaced or reversed when they become worn. An everyday

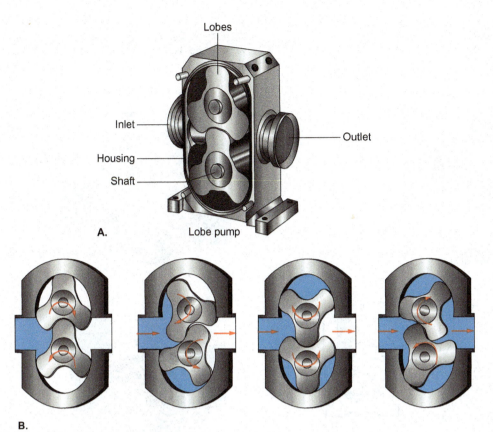

Lobes

Inlet

Housing

Shaft

Outlet

A.   Lobe pump

B.

Figure 6.15 **A.** Components of a lobe pump. **B.** Internal flow in a lobe pump.

Table 6.5 Advantages and Disadvantages of Using a Lobe Pump

Advantages	Disadvantages
■ Can handle slurries with large particle size	■ Not suitable for low-viscosity liquids
	■ Inferior loading characteristics
	■ Low suction ability
	■ Might require factory service for repair

example of a vane pump is the power-steering pump found in many automobiles. Figure 6.16 shows a sliding vane pump with the components labeled and the direction of flow through the pump.

Sliding vane pumps consist of a cylindrical rotor that is eccentrically mounted (i.e., it has an axis or point of support that is not centrally placed) in a cylindrical case. Because the rotor and case are eccentric, there is a crescent-shaped void between the rotor and the casing. Small plates, or vanes, are set into slots in the rotor. Centrifugal force and/or springs, hydraulic pressure, or other forces cause the vanes to slide in and out of the rotor as it turns, forming pumping chambers for the liquid and maintaining a tight seal. As the rotor turns, liquid enters the case at one narrow end of the crescent-shaped void and is swept along by the vanes through the widest part of the crescent and out the discharge at the other narrow end of the crescent.

The design of the flexible vane pump is similar to the sliding vane, except that the vanes bend (or flex) as they go through the narrow part of the crescent. At the suction port of the pump, the vanes are fully extended to create large pumping chambers for the liquid to enter. The size of the chambers decreases as the discharge port is reached, forcing the liquid out. Figure 6.17 shows a flexible vane pump.

Figure 6.16 A. Sliding vane pump. **B.** Flow through a sliding vane pump.

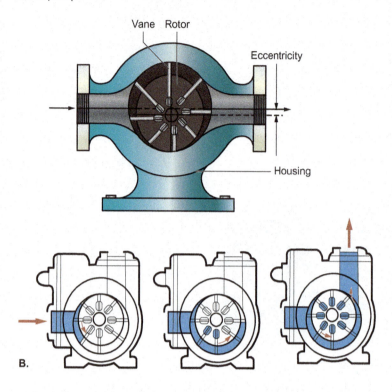

Figure 6.17 A. Components of a flexible vane pump. **B.** Flow through the flexible vane pump.

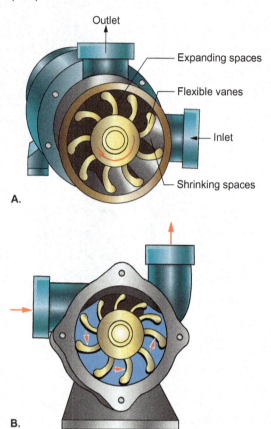

Screw pump a type of positive displacement rotary pump that displaces liquid with a screw. The pump is designed for use with a variety of liquids and viscosities and a wide range of pressures and flows.

Screw Pumps **Screw pumps** are rotary pumps that displace liquid with a screw. This screw design accommodates a wide range of liquids with varying viscosities, pressures, and flows. Screw pumps are available in different configurations, with the main differences being the number of screws and the pitch of the blades on the screw. Most modern day screw pumps have multiple screws. A variation of the single screw pump, called a progressive cavity pump, closely resembles the Archimedes screw. Screw pumps, such as those illustrated in Figure 6.18, consist of a casing with a rotating screw impeller containing helical threads. The liquid enters the pump, is trapped between the case and the helical threads, and then is moved along the screw until it is pushed out through the discharge. In multiple screw designs, one of the screws is connected to the driver and that screw powers the other screws. Some multiple screw pumps have two separate inlets, which join together at the discharge.

The advantages and disadvantages of using a screw pump are listed in Table 6.6.

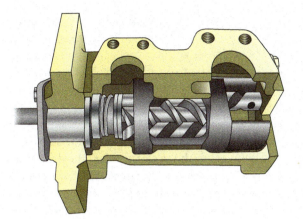

Figure 6.18 Screw pump.

Table 6.6 Advantages and Disadvantages of Using a Screw Pump

Advantages	Disadvantages
■ Used at high speeds ■ Good suction characteristics ■ Compact design that creates low vibration	■ Expensive ■ Low level of efficiency

RECIPROCATING PUMPS   Reciprocating pumps are positive displacement pumps that use first an inward stroke of a piston or diaphragm to draw liquid into a chamber, then a subsequent outward stroke to positively displace that liquid. Figure 6.19 shows how a reciprocating pump operates and how liquid is drawn in and discharged.

Most reciprocating pumps consist of a liquid end and a drive end. The liquid end contains a device that displaces a fixed liquid volume to the pump suction for each stroke from the drive end. The suction and discharge flows are typically directed by the positioning of check valves.

Inlet and outlet valves open and close to prevent a backflow of liquid. As the stroke of the pump draws liquid from the suction (suction stroke), the inlet valve allows the suction to fill while the discharge valve closes to prevent backflow. As the stroke transfers liquid to the discharge (discharge stroke), the discharge valve opens to allow the liquid to flow through, and the suction valve closes.

To better illustrate the actions of a reciprocating pump, think of a syringe. As a syringe plunger is pulled out of its housing, liquid is drawn in (intake). As the plunger is pushed back into the housing, liquid is forced out (discharge).

Reciprocating pump   a type of positive displacement pump that uses the inward stroke of a piston or diaphragm to draw liquid into a chamber (intake) and then positively displace the liquid using an outward stroke (discharge).

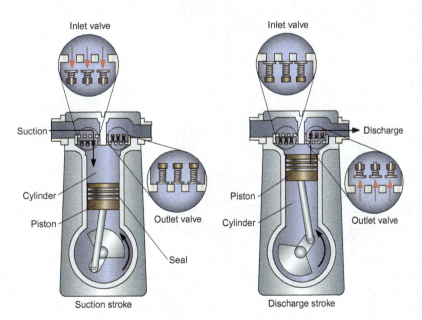

Figure 6.19 Piston-type reciprocating pump showing suction and discharge strokes.

Piston pump a type of positive displacement reciprocating pump that uses a piston inside a cylinder to move fluids.

Piston Pumps **Piston pumps**, like the one shown in Figure 6.19, are reciprocating pumps that are driven by a motor, air, hydraulics, or direct-acting steam. Piston pumps use a piston inside a cylinder within a casing to move liquids. Motor-driven pumps drive the piston by a crankshaft to convert the rotary motion of the driver to the back-and-forth motion of the piston. Steam-driven pumps drive the piston directly (i.e., the steam driver's piston moves the pump's piston). Discharge volume is equal to the volume displaced by the piston, less slippage past the piston packing rings. Both simplex and multiplex designs are available. Direct-acting steam pumps allow for variable pressure and flow by throttling the steam inlet (the point at which the steam enters the driver).

A simplex design contains a single piston for the suction and discharge. A multiplex design has multiple pistons for suction and discharge (e.g., duplex pumps have two pistons, triplex have three, and quadraplex have four). Figure 6.20 shows examples of simplex and multiplex pumps.

Figure 6.20 Simplex pump and multiplex pump designs.

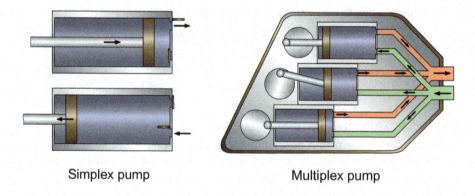

Simplex pump   Multiplex pump

Single-acting piston pump a type of piston pump that pumps by alternating suction and discharge actions on each piston stroke.

Piston pumps also can be single- or double-acting. Figure 6.21 shows examples of single- and double-acting piston pumps. A **single-acting piston pump** pumps by reciprocating motion on every other stroke. In other words, it fills the cylinder when the piston moves in

Figure 6.21 Comparison of single- and double-acting piston pumps.

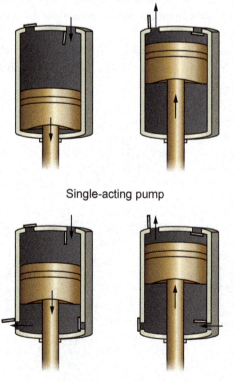

Single-acting pump

Double-acting pump

one direction (called the suction stroke), and forces the liquid out of the cylinder as the piston moves in the other direction (called the discharge stroke).

A **double-acting piston pump** takes suction and discharges by reciprocating motion on every stroke (suction on one side of the piston, and discharge on the other side). In other words, as one end is filling, the other end is emptying, or discharging.

Plunger Pumps **Plunger pumps** displace liquid using a plunger and are similar to piston pumps. Plunger pumps maintain a constant speed and torque. There is no change in the capacity of the pump when constant stroke rate is maintained. Plunger pumps are capable of operating at higher pressure than piston pumps. The most significant design difference between the two, besides the shape of the pumping element, is the way the cylinders are sealed. In a piston pump, the packing is attached to the piston and moves with it, while a plunger pump's sealing system is stationary, with the plunger moving through it on its stroke (see Figure 6.22). The plunger is driven by a crankshaft and can be powered by an electric motor or by steam, pneumatic, or hydraulic power.

Double-acting piston pump a type of piston pump that takes suction and discharges by reciprocating motion on every stroke.

Plunger pump a type of positive displacement reciprocating pump that displaces liquid using a plunger and maintains a constant speed and torque.

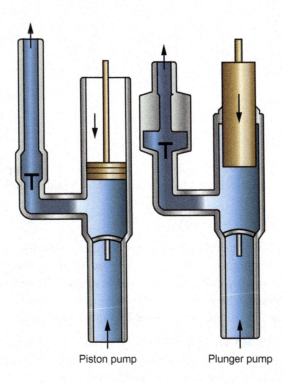

Piston pump   Plunger pump

Figure 6.22 Comparison of piston and plunger pumps.

Diaphragm Pumps **Diaphragm pumps** are mechanical or air-driven pumps that use a flexible membrane to move liquid. Basic designs of diaphragm pumps include those consisting of a single sealed diaphragm, with the fluid to be pumped on one side and air or a pumping medium on the other side. As pressure is increased and decreased, the diaphragm flexes and liquid enters or exits the pumping chamber. As with the piston and plunger pumps, inlet and outlet valves that open with the corresponding suction or discharge stroke prevent backflow of liquid.

The other type of diaphragm pump consists of two diaphragms connected by a common shaft. The fluid that is being pumped is on the outside of both diaphragms. The diaphragms, which are forced by a mechanical linkage, compressed air, or some other liquid, move in a reciprocating (back-and-forth) motion. The pump discharges with one diaphragm, while the other diaphragm exhausts air and allows liquid to enter the suction side of the chamber. Adjusting the compressed air inlet valve and pressure allows a variable capacity and head. Figure 6.23 shows an example of a diaphragm pump.

The advantages and disadvantages of using a diaphragm pump are listed in Table 6.7.

Diaphragm pump a mechanically or air-driven positive displacement reciprocating pump that uses a flexible membrane to move liquid.

Figure 6.23 Diaphragm pump suction and discharge strokes.

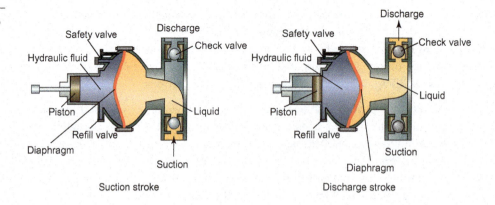

Suction stroke   Discharge stroke

Table 6.7 Advantages and Disadvantages of Using a Diaphragm Pump

Advantages	Disadvantages
■ No contact between liquid and the reciprocating piston, which eliminates the possibility of leaks ■ Can pump highly viscous liquids and slurries ■ Self-priming	■ Limited in head and capacity range ■ Require check valves in the suction and discharge nozzles

6.3 Associated Components

Pumps have associated components that are important to their operation. These include pump seals (mechanical and packing), seal flushes, seal pots, bearings, and wear rings.

Pump Seals

Seal a device that holds lubricants and process fluids in place while keeping out foreign materials where a rotating shaft passes through a pump casing.

Seals contain process fluids within a piece of equipment (pump, mixer, etc.), while keeping out foreign materials. They are necessary for applications where a rotating shaft passes through a pump's casing. Seals are especially critical in preventing environmental incidents because they can be used to prevent leakage of hazardous materials. The two main types of sealing arrangements used in pumps are mechanical packing and mechanical seal.

MECHANICAL PACKING Packing is used in nontoxic, nonflammable, and nonpolluting applications, such as process water and cooling water pumps. Packing material is manufactured using materials such as graphite or carbon fibers, which are treated with a lubricant such as Teflon®. Strands of the fibers are braided together into a rope-like product, which is cut into rings and inserted into the stuffing box, around the shaft. (A **stuffing box** is the area in a pump's casing that contains the mechanical packing material.)

Stuffing box the area in a pump's casing that contains the packing material.

The *packing gland* holds the packing rings in place. When the gland bolts are tightened, the packing is compressed and forced against the shaft, forming a seal. The friction created by the packing rubbing against the rotating shaft generates heat, which is addressed by the lubricant embedded in the packing and a quench stream. If the packing gland bolts are over-tightened, the shaft can overheat, causing the packing to become brittle or the shaft to break. Since the packing rings require lubrication and cooling, some amount of leakage of the cooling fluid (steam or water) from the stuffing box is expected. Because of this, mechanical packing sealing systems are not used in pumps that process toxic or flammable materials. Figure 6.24 displays an example of a stuffing box that uses mechanical packing.

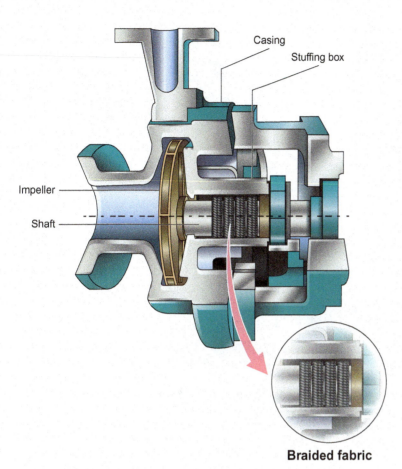

Casing

Stuffing box

Impeller

Shaft

Braided fabric

Figure 6.24 Stuffing box using mechanical packing.

MECHANICAL SEALS   A **mechanical seal** is composed of two flat surfaces (faces), one stationary and one rotating with the shaft, that are in close contact with (nearly touching) each other to create a seal and prevent leakage. The faces are pushed toward each other by the hydraulic force provided by the liquid being pumped and by the use of springs that provide tension between a hard surface face and a softer surface face. In order for the two seal faces to provide a tight seal, they must be machined to an extremely high degree of flatness—usually within 0.00003 inches or 0.762 microns. The faces are lubricated with a thin boundary layer of liquid or sometimes vapor (if the liquid has evaporated) between the faces. This film layer separating the two seal faces is about 20 millionths of an inch (a half-micron). Because they operate so close to each other, some amount of contact does occur, generating heat and causing wear on the seal. Figure 6.25 shows examples of mechanical seals used to prevent leaks.

Mechanical seals   seals that typically contain two flat faces (one that rotates, and one that is stationary), that are in close contact with one another in order to prevent leaks.

Seal Flushes and Seal Pots

Many pumps use some type of seal flush to keep the seal lubricated and free of debris. A **seal flush** is a small flow (slip stream) of pump discharge or externally supplied liquid that is routed to the pump's mechanical seal. This acts as a barrier liquid between the two faces of the seal to reduce friction and remove heat. A flush can use the liquid being pumped through the process (self-contained) or external liquids. Typically, the self-contained flush is used when the liquid being pumped is clean, noncorrosive, and nontoxic. Figure 6.26 shows an example of a seal flush.

Seal flush   a small flow (slip stream) of pump discharge or externally supplied liquid that is routed to the pump's mechanical seal. This acts as a barrier liquid between the two faces of the seal to reduce friction and remove heat.

Figure 6.25 Mechanical seals used to prevent leaks. **A.** and **B.** Mechanical seals. **C.** Photo of mechanical seal installed in pump.

CREDIT: **A & B.** nayladen/Shutterstock. **C.** Surasak_Photo/Shutterstock.

A.

B.

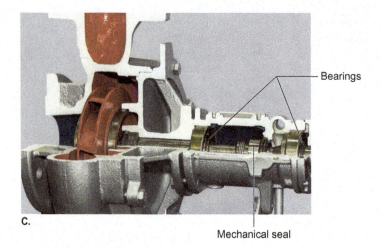

C.

Bearings

Mechanical seal

Figure 6.26 Seal flush.

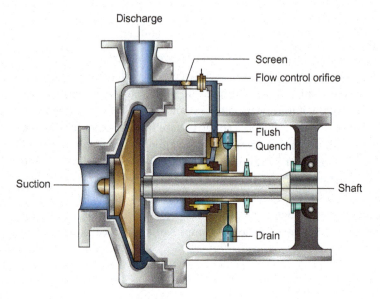

Discharge

Screen

Flow control orifice

Flush

Quench

Suction

Shaft

Drain

Many pumps use a reservoir called a *seal pot* to contain a secondary seal liquid. The seal pot pressure is slightly higher than the pump seal pressure. A change in pressure or a seal pot that empties too quickly can indicate a pump seal leak. If a seal leak occurs, the leaking material is the material from the seal pot (usually nonflammable and nontoxic). Figure 6.27 shows an example of a condensate seal pot.

Two types of seal systems employ a seal pot: tandem seal and double seal. A *tandem seal* has a primary seal and a backup seal. If the primary seal leaks, then the pressure in the seal

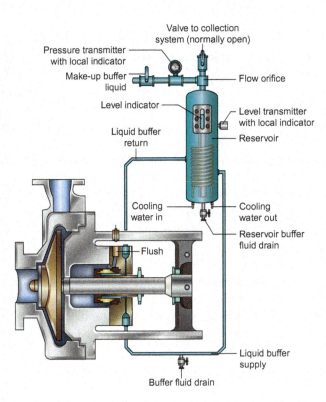

Figure 6.27 Seal pot.

pot increases dramatically. If the backup seal fails, then the seal pot is drained of its contents. In both instances, maintenance needs to be performed.

A *double seal* uses pressurized liquid from the seal pot to pressurize the seal in two directions: internally to the process, and externally to the atmosphere. If the primary seal fails, then the neutral liquid inside the seal pot will leak into the process, indicating there is a seal failure. Because of this, a double seal is highly effective on flammable, toxic, and corrosive materials and those with solids and abrasive material. Figure 6.28 shows an example of a double seal.

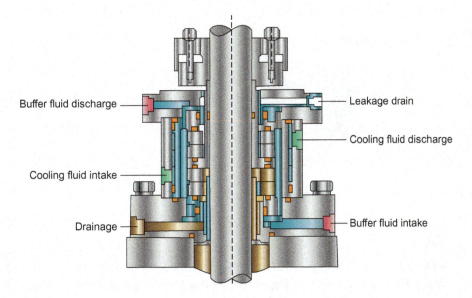

Figure 6.28 Double seal.

Bearings

Bearings mechanical components that keep the pump shaft in alignment with the casing and absorb axial and radial forces.

Bearings (Figure 6.29) are mechanical components that keep the pump shaft in alignment with the casing and absorb axial and radial forces. In pumps, bearings provide support for the shaft or rotor, minimize friction between moving parts and maintain the shaft's alignment by restricting its motion inside the pump casing while allowing the shaft to turn freely. When a pump is in operation, both radial and axial forces (or *loads*) act upon the shaft (Figure 6.30). Radial loads are those that are perpendicular to the shaft, while axial (or thrust) loads are those that are parallel to the shaft. The purpose of bearings is to absorb the loads created by the rotation of the shaft and the hydraulic pressure.

A rolling contact bearing is a set of two metal rings, called *races*, with a number of metal balls or rollers between them. The individual balls or rollers are evenly spaced and separated from each other by a bearing cage, and all of the components are contained in a bearing housing.

The inner race of a bearing assembly is attached to the shaft, and rotates with it, as do the balls or rollers, while the outer race is stationary, attached to the pump.

Bearing failure is the second leading cause of pump shutdown/repair. The life of bearings is affected by operating conditions such as adequacy and quality of lubrication, presence of contaminants, shaft loads, shaft vibration, and shaft speed and temperature.

Figure 6.29 Examples of bearings.

CREDIT: **A.** pelfophoto/Shutterstock. **B.** noomcm/Shutterstock. **C.** Billion Photos/Shutterstock.

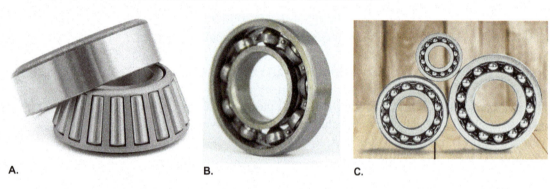

A.   **B.**   **C.**

Figure 6.30 Radial/axial forces on shaft.

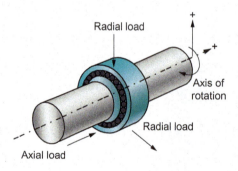

Radial load

Axis of rotation

Radial load

Axial load

Wear Rings

Wear rings close-running, noncontacting replaceable metal rings located between the impeller and casing of a centrifugal pump; wear rings allow for a small clearance between the two components.

Wear rings (shown in Figure 6.31) are rings that allow the impeller and pump casing to seal tightly together without wearing each other out. Wear rings are close-running, noncontacting, replaceable, pressure breakdown devices located between the impeller and casing of a centrifugal pump. One ring, called the impeller wear ring, rotates with the impeller. The other, called the casing wear ring, is stationary. Wear rings increase efficiency by minimizing discharge-to-suction recirculation. They control axial thrust (back-and-forth motion of the shaft) by reducing the discharge pressure acting on the impeller. They also minimize seal and chamber pressure and can be replaced to restore clearances between the moving and stationary rings.

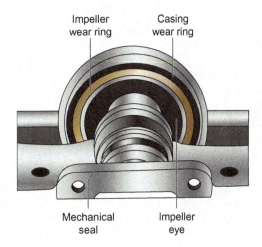

Figure 6.31 Wear ring.

6.4 Operating Principles

It is important to understand how pumps operate. Specifically, process technicians need to be familiar with inlet flows, outlet flows, head, pressure, and the internal workings of the pump.

Head

Head is the pressure of a liquid caused solely by the weight of the liquid; the amount of head is determined by the height of the liquid. Static head is measured when liquid is not moving through the pump. **Suction static head** is the vertical distance between the centerline of a pump and the surface of the liquid on the suction side of the pump. **Discharge static head** is the vertical distance between the centerline of a pump and the surface of the liquid on the discharge side of the pump (if the discharge line is submerged) or the pipe end (if the discharge line is open to the atmosphere).

Total head in a pumping system is a measure of a pump's ability to move liquid through the system. The amount of suction head is deducted from the discharge head, since the measurement is only of the pump's capability, without any assistance from pressure on the suction side. If the suction is lower than the pump's centerline, then that number is added to the discharge head, since some of the pump's energy must provide the lift required to raise the liquid into the pump's suction. If the pump is idle, the head measured is called total static head. Figure 6.32 shows calculation methods for suction, discharge, and total static head.

When liquid flows through the system, resistance (friction) caused by the length/diameter of the piping and fittings/valves in the line also must be considered in the head calculation, because they will consume some of the pump's energy. The total head (or total dynamic head) is the difference between the pressure on the discharge side and suction side of the pump, when the pump is in operation and factors such as friction and velocity are considered.

To convert the head to pressure (PSI), you must multiply by the density of the liquid. For water, the conversion factor is 0.433 PSI/ft (62.4/144). (Pure water at 39.2 Fahrenheit [4 degrees Celsius] has a specific gravity of 1.) If the liquid has a specific gravity that is less than 1.0, the pump develops less pressure than when pumping water and consumes less horsepower.

Head measurement of pressure caused by the weight of a liquid, measured in feet or meters, determined by the height of the liquid above the centerline of a pump.

Suction static head the vertical distance between the centerline of a pump and the surface of the liquid on the suction side of the pump.

Discharge static head the vertical distance between the centerline of a pump and the surface of the liquid on the discharge side of the pump (if the discharge line is submerged) or the pipe end (if the discharge line is open to the atmosphere).

Total head a measure of a pump's ability to move liquid through a pumping system.

Figure 6.32 Calculated static head.

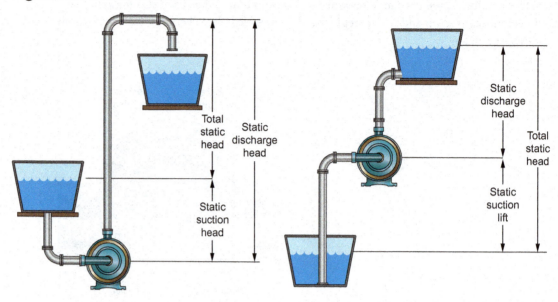

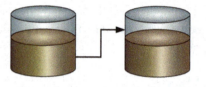

Did You Know?

Some students have asked why technicians can't just pump liquid into the top of the tank so they don't have to try to overcome the head at the bottom of the tank.

While this sounds like a reasonable solution, it actually is not. If the liquid were pumped up, it still would be necessary to overcome discharge head in the piping so that the liquid could reach the top of the tank. This pressure could be equal to or greater than the pressure encountered simply by pumping it into the bottom of the tank.

In addition, liquids that are free-falling from the top of a tank produce static electricity. If a volatile liquid (e.g., gasoline) is being pumped, this electricity could ignite the vapors.

Inlet (Suction)

The inlet (suction) is the point at which the liquid enters the pump.

Some pumps require a suction strainer or screen to filter out debris particles and prevent damage to the pump. Pumps are monitored for indications of plugging, such as a decrease in the flow or an increase in the differential pressure across the suction strainer. When indicated, the pump is taken out of service and the screens are cleaned. Suction screens also are used when starting up a new facility, to capture items that might have been left inadvertently in the piping systems during construction or maintenance activities (e.g., bolts, tools, or rags).

What Happens Inside the Pump

CENTRIFUGAL PUMPS   In centrifugal pumps, pressure is added to the liquid by first increasing its velocity through centrifugal force. Liquid enters the suction pipe and flows into the impeller eye at the center of the pump. Liquid is guided from the impeller eye to the impeller outlet by *impeller vanes* (raised ribs on the impeller used to catch and sling the liquid) and is accelerated in the direction of impeller rotation. As the liquid leaves the impeller, it enters the volute, a widening area of the casing, where the velocity energy in the liquid is converted to pressure energy. As the high-pressure liquid approaches the discharge nozzle, it is directed through the nozzle by a **cutwater** (a thick plate in the discharge nozzle of the pump that breaks the vortex).

Cutwater   a thick plate in the discharge nozzle of a pump that breaks the vortex.

RECIPROCATING PUMPS   Reciprocating pumps use the inward stroke of a piston, plunger, or diaphragm to draw (intake) liquid into a chamber, followed by a subsequent outward stroke to positively displace (discharge) the liquid. Reciprocating pumps consist of a liquid end and a drive end. The liquid end contains a device that displaces a fixed liquid volume for each stroke from the drive end. The suction and discharge flows are typically directed by the positioning of inlet and outlet valves. As the stroke of the pump draws liquid from the suction, the inlet valve allows the suction to fill while the discharge valve closes to prevent backflow. As the stroke transfers the liquid to the discharge, the discharge valve opens to allow flow through the discharge, while the suction valve closes.

Did You Know?

Some pumps can pump with such an intense force that they actually make a "hole" (vortex) in the liquid material.

   The lack of liquid created by the vortex can cause pump cavitation if steps are not taken to prevent it.

Outlet (Discharge)

Several variables affect the discharge flow rate, including piping size, downstream vessel size, discharge head, pressure differential, and pump design. Causes of no or low discharge flow include reduced head, closed valves, line blockage, system redesigns, and blocked suction strainers/screens.

6.5 Pump/System Curves

It is important for a process technician to understand the basic concept of pump performance and system curves. A **pump performance curve** is a specification that describes the relationship between flow rate and head needed for correct operation of the pump (Figure 6.33). The manufacturer determines the performance curve based on the type of pump and the related process.

Pump performance curve   a specification that describes the capacity, speed, horsepower, and head needed for correct pump operations.

Figure 6.33 Typical centrifugal pump curve.

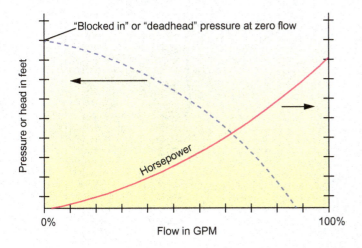

A system curve is a representation of the system conditions—head, friction, piping size, etc. By comparing the system curve and the pump performance curve, a manufacturer can determine which pump is the best one for the process (i.e., which one has the proper operating capacity).

Each centrifugal pump has a characteristic pump curve. Liquid head is the pressure developed from the pumped liquid passing through the volute. As the flow rate is increased, the liquid head decreases. This continues until the maximum pumping rate of the pump is reached.

The maximum liquid head occurs with the pump in operation at zero flow (with the discharge valve closed) and is called the deadhead or shutoff pressure. Some centrifugal pumps (especially very high speed and multistage pumps) can have excess vibration or undesired overheating if they are deadheaded, so deadheading should be avoided.

Centrifugal pump manufacturers supply performance pump curves, which usually are based on evaluation with water and contain the information about:

- Head
- Flow rate
- Efficiency
- Horsepower
- NPSH required.

Figure 6.34 provides a pump curve for a pump operating at 3,500 rpm with various impeller diameters. If you look at the pump curve data, you will notice that the deadhead or shut-off pressure on the 7¼″ impeller is 230 feet (70.104 m). At a flow rate of 400 GPM, the head is 215 feet (65.532 m), the horsepower is 33, the required NPSH is 13 feet (3.96 m), and the efficiency is 68 percent.

Figure 6.34 Sample pump curve.

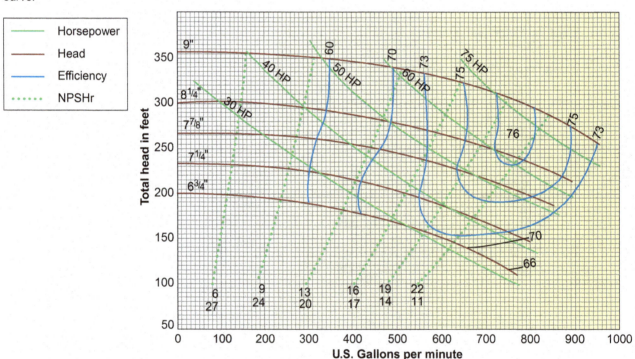

Total Head

The pump's total head is usually on the y axis and is stated in feet or centimeters. The head developed by the pump does not depend on the material being pumped (e.g., water or organic).

Flow Rate

The flow rate is usually on the x axis and is stated in gallons per minute (GPM) or liters per minute. The maximum flow rate that can be achieved increases as the impeller diameter is increased. The maximum head occurs at zero flow. As the flow rate is increased, the head decreases until the maximum flow rate of the pump is achieved.

Efficiency

The efficiency is the ratio of the liquid horsepower to the brake horsepower. Pumps are designed so the best efficiency point (BEP) is at *design operating condition* (the condition at which the pump was designed to run). Operations at other points on the curve result in reduced efficiency (e.g., more brake horsepower for the same liquid horsepower).

Horsepower

The horsepower required is usually based on evaluation with water. The horsepower increases as the impeller diameter increases or as the flow rate is increased. If the specific gravity of the material is less than 1.0, less horsepower is required. If the specific gravity of the material is greater than 1.0, more horsepower is required.

The viscosity of the liquid also affects the horsepower required. Water has a viscosity of one *centipoise* (a common measure of the viscosity of a liquid) at room temperature. If the liquid is more viscous than water, the required horsepower increases. If the liquid is less viscous than water, the required horsepower decreases.

NPSH Required

NPSH is the net positive suction head. The NPSH given on the pump curve is the required NPSH for the pump. The NPSH available from the process must be calculated and must be equal to or greater than the NPSH required, or pump cavitation can occur. NPSH is usually stated in feet or centimeters and is calculated using:

- Atmospheric pressure above the liquid
- Vapor pressure of the liquid
- Liquid height above pump centerline
- Pressure drop in the suction line.

   The formula for NPSH is:

$$\text{NPSH available} = \text{atmospheric pressure above liquid} +/- \text{liquid height}$$
$$- \text{vapor pressure} - \text{pressure drop}$$

If the NPSH available is too low, the technician has several options to increase it. One option would be to increase the pressure above the liquid. Another option would be to raise the liquid level. The technician also can reduce liquid temperature, which lowers the vapor pressure. Finally, the technician can reduce the flow rate, which lowers the pressure drop and decreases the NPSH required.

6.6 Potential Problems

When working with pumps, process technicians should always be aware of potential hazards such as overpressurization, overheating, cavitation, and leakage.

Overpressurization

Pump overpressurization can occur if the valves beyond the pump are incorrectly closed or blocked. Consider the example in Figure 6.35. In the figure, valves A and B are both closed,

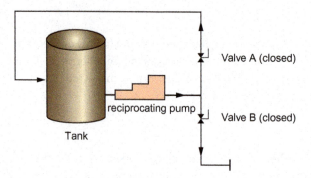

Figure 6.35 Improper valve alignment ("deadheading").

giving the liquid no place to go. The result is excessive back pressure or deadheading. Both can cause serious personal injury, pump seal failures, piping failure, or pump damage.

Overheating

Pump overheating is often caused by improper lubrication. Without lubrication, bearings fail and equipment surfaces rub together. As these surfaces rub against one another, friction is produced and heat is generated. This can cause mechanical failures, leakage, and decomposition of the process liquid.

If the pump is deadheaded, the liquid in the pump is heated by the mechanical energy of the motor, which can cause vapor lock and cavitation. To prevent either condition, recycle loops (minimum flow lines) might be added to allow flow through the pump, even if the valves downstream are closed. Recycle loops help prevent overheating of the liquid.

Process technicians should always monitor rotating equipment for excessive heat. Operating equipment under these conditions can lead to permanent equipment damage or personal injury (e.g., burns).

Leakage

Process technicians should check pumps for leaks because leaks can introduce slipping hazards, exposure to harmful or hazardous substances, process problems (e.g., inferior product produced as a result of improper feed supply), and releases to the atmosphere. Leaks most frequently occur where the pump shaft exits the pump casing. Packing or mechanical seals are usually used to prevent leakage from occurring. Pump seal leaks are usually corrected by tightening or replacing the packing or by replacing the mechanical seal.

Cavitation

Cavitation a condition inside a pump in which the liquid being pumped partially vaporizes because of variables such as temperature and pressure drop, and the resulting vapor bubbles then implode.

Cavitation is a condition inside a pump wherein the liquid being pumped partly vaporizes because of variables such as temperature increase and pressure drop. These variable changes cause vapor pockets (bubbles) to form and collapse (implode) inside a pump.

Cavitation occurs when the pressure in the eye of a pump impeller falls below the vapor pressure of the liquid being pumped. This is a very serious problem in dynamic pumps, especially centrifugal pumps. Cavitation can also be a problem in vacuum operations because low-pressure liquids vaporize at lower temperatures.

Key characteristics of cavitation include large pressure fluctuations, inconsistent flow rate, and severe vibration. Cavitation has been described as sounding like the pump is pumping rocks. Technicians who identify cavitation should always try to eliminate it as quickly as possible because it can cause excessive wear on the pump seal, impeller, bearings, and casing.

To prevent cavitation, a pump should always be primed before it is started. As mentioned earlier, priming fills the suction of a pump with liquid to remove any vapors that might be present. This reduces the likelihood that the pump will become vapor-bound or lose suction.

Cavitation can still occur, however, even if a pump is properly primed. For example, if the liquid becomes too hot, vapor bubbles can be created. Certain processes (ethane and ethylene liquids) require the pump to be chilled as well as primed prior to startup. Another way to stop cavitation is to raise the level of the liquid in the suction line to increase the suction pressure on the pump. Table 6.8 displays causes of and solutions for cavitation.

Table 6.8 Causes of and Solutions for Cavitation

Cause	Solution
Suction pressure reduction	Increase suction pressure, slow down pump, check the level of the suction vessel, and check the design. Check for restrictions
Liquid temperature increase	Lower liquid temperature
Flow rate too high	Reduce flow rate
Separation and contraction of flow because of a change in viscosity	Check design, and find cause of viscosity change
Suction strainer/screen pluggage	Clean strainer/screen
Undesirable flow conditions because of obstructions or sharp turns	Locate obstructions or sharp turns and correct
Pump is not suitable for the system curve	Check design
Low liquid level	Increase liquid level

Excessive Vibration

Excessive vibration can cause mechanical damage and pump shutdown, especially on a pump equipped with vibration sensors. Because of this, a vibration analysis should be conducted when excessive vibrations are detected. When misalignment occurs because of vibration, a realignment of the pump is necessary.

Pump Shutdown

Pumps can shut down if a deviation from normal conditions occurs. This shutdown can cause a loss of process flow. Table 6.9 shows the causes and solutions for the most common causes of pump shutdown.

Table 6.9 Causes of and Solutions for Pump Shutdown

Cause	Solution
Overspeed trip (steam-driven)	Check turbine overspeed trip for proper operation; check suction for proper flow; check discharge line for large failure
Power failure	Determine cause of power failure (e.g., breaker, fuse, loss of commercial power). Repair and restore power as appropriate
Flow, pressure, or vibration (on a system with sensors)	Determine cause (e.g., leak, valve closed, instrumentation malfunction elsewhere in the process); correct problem to return pump to normal operation
Motor trip	Pump is using too much horsepower, flow rate is too high, or mechanical problem exists Correct problem to return pump to normal operation

Vapor Lock

When a pump suffers from vapor lock, the pump loses its liquid prime (the liquid being pumped). This causes vibrations and abnormal noises. The solution (especially if the pump is improperly vented) is to shut down the pump, bleed off the vapor, and restart the pump.

6.7 Safety and Environmental Hazards

A number of safety and environmental concerns are associated with pumps, including chemical hazards and hazards caused by the equipment itself. Hazards with normal and abnormal pump operations can affect personal safety, equipment operation, plant operations, and the environment.

Personal Safety Hazards

Pumps can cause personal injury if attention is not paid to safety regulations. Possible hazards involving pumps include exposure to hazardous liquids and materials, slipping or tripping because of leaks, overpressurization, and the physical hazard of rotating equipment. When hazardous chemicals such as acids, caustics, reactive materials, and high temperature petroleum liquids are pumped at high rates and pressures, a small problem can escalate quickly to a serious hazard.

During normal pump operations, process technicians must always wear personal protective equipment (PPE), including safety glasses or goggles, work gloves, a hard hat, and hearing protection. Additional PPE may also be required based on the liquid being pumped and other factors. During routine and preventive maintenance, additional measures should be taken. These measures might include housekeeping duties such as cleaning up any material that might have leaked from the pump, ensuring coupling guards and motor guards are in place, and removing trash and debris from around pumps.

Lockout/tagout and isolation procedures must also be followed, and the pump should be placed in a zero energy state before maintenance is performed. *Zero energy state* means that all forms of energy have been removed from the pump (e.g., pressure, rotation, chemical hazards). This protects maintenance personnel from hazards while working on the pump.

Equipment Operation Hazards

There are several equipment operation hazards associated with pumps during normal and abnormal operations. Table 6.10 lists some of these hazards and their consequences.

Table 6.10 Equipment Operation Hazards During Normal and Abnormal Operations

Situation	Consequences
Insufficient NPSH (cavitation)	Causes seal damage, bearing damage, and impeller damage. Cavitation is caused by inadequate suction pressure
Abnormal speed	The pump is designed to operate at a predetermined speed. Excess speed results in overheating and increasing the temperature of the product, thereby increasing the vapor pressure. This can lead to cavitation and damage to the pump internals. Low speed affects the pump output. (Note: Pumps that are motor-driven are constant speed devices)
Vibration	Causes mechanical failure of pump components if operation continues. Vibration is caused by cavitation, worn bearings, or misalignment of the pump
Abnormal temperature	Abnormal operating conditions could result in damage to the pump and hazards from the material being too hot. Exposure to the process technician (high or low temperature) also can occur, causing severe burns, and creating a personal safety hazard
Overpressure	Caused by starting a positive displacement or multistage pump with the discharge closed. Actual catastrophic failure in such a case would be an unlikely, yet possible, outcome; more probably, motor overload and/or internal equipment damage would result
Lack of lubrication	Causes overheating or scoring and failure of rotating parts

Facility Operation Hazards

Reliable pump operation is vital to the operation of the facility. In continuous operations (e.g., refineries), process equipment depends on continuous liquid flow for proper operation. Loss of liquid flow can cause process equipment failure, which can result in equipment downtime, fires, or explosions. Loss of production can cause facility shutdowns, pump replacement costs because of damage, and loss of product.

Standby pumps are used in many cases to maintain flow if the primary pumps fail. Steam-driven standby pumps are useful if the pump failure results from power loss. Many standby pumps start automatically when a given process variable (e.g., flow rate or pressure) reaches a preset low point.

Environmental Hazards

Environmental impact depends on the pumped liquid. Impact can range from excess water usage (because of leaks) to facility or community evacuations if the liquid is extremely hazardous (e.g., toxic or flammable). The Occupational Safety and Health Administration (OSHA) and the EPA have set limits for the discharge of specific toxic and flammable liquids. Violations of these limits can result in heavy fines and penalties.

Potential problems associated with the environment include discharge onto the ground or into the sewer, discharge into the air, or violation of government regulations. Additional problems are seal leaks where environmental regulations limit the amount of fugitive emissions into the environment. If these limits are exceeded, the leak must be reported to the regulatory agencies involved. If there is a massive leak, it could affect the area and community outside the facility.

6.8 Process Technician's Role in Operation and Maintenance

The process technician plays a vital role in the operation and maintenance of the pumps within a process industry. Process technicians must have a basic understanding of the pumps within their process area and be aware of the potential problems that can occur with pumps at any given time.

Process technicians should be trained in operational and emergency procedures associated with pumps, including startup and shutdown. In addition, monitoring must be conducted on a regular basis to eliminate potential problems or hazards before they occur or to deal with them as quickly as possible if they do occur.

When working with pumps, process technicians should look, listen, and check for the items listed in Table 6.11. Failure to perform proper maintenance and monitoring could affect the process and result in equipment damage.

Table 6.11 Process Technician's Role in Operation and Maintenance

Look	Listen	Check
■ Monitor oil levels to make sure they are satisfactory ■ Make sure water is not collecting in the lubricating oil (water is not a lubricant, so it can cause bearing failure) ■ Examine seals and flanges to make sure there are no leaks ■ Observe suction and discharge pressures ■ Monitor differential pressure across strainers/screens	■ Listen for abnormal noises	■ Check for excessive vibration ■ Check for excessive heat

6.9 Typical Procedures

Specific equipment procedures must be followed when working with pumps. For example, pump operation and maintenance tasks performed by the process technician can include monitoring, lockout/tagout, routine and preventive maintenance, startup, and shutdown.

Monitoring

When monitoring pump operation, the process technician typically is directed to perform several steps.

1. Check and record suction/discharge pressures.
2. Check lubricant levels.
3. Check seal flush system (if applicable).

4. Check for abnormal noise and vibration, using senses (sight, hearing, touch, etc.) to ensure proper operation.

5. Check for proper temperatures.

6. Check pump rates and instrument outputs to the flow controllers of the pump.

7. Check for missing coupling guards.

8. Check that suction and discharge valves are lubricated.

Startup

Steps can vary from equipment to equipment. The process technician should always follow the proper startup procedure for a given pump. A general startup procedure has seven steps.

1. Ensure all downstream equipment is lined up and ready.

2. Ensure the pump is ready for startup.

3. Ensure all auxiliary systems (seal flush, cooling water, etc.) are lined up.

4. Warm up or chill all equipment as needed.

5. Prime and then start pump.

6. Ensure the flow rate is as expected.

7. Conduct routine inspections to check suction/discharge pressures, to check the machine operation, and to check for leaks.

Lockout/Tagout

The control of hazardous energy sources and electrical hazards is mandated by OSHA and covered in these standards:

- Control of Hazardous Energy (Lockout/Tagout) - 29 CFR 1910.147
- Selection and Use of Work Practices - 29 CFR 1910.333.

The purpose of these standards is to ensure that, before any worker attempts to conduct service or maintenance activities on any piece of equipment that could unexpectedly become energized and cause injury, the equipment is placed in a safe condition by performing a lockout/tagout (LOTO) procedure.

The following is a generic lockout/tagout procedure:

1. De-energize all equipment drivers per an energy isolation plan. This could mean closing a run switch and/or de-energizing a breaker in the motor control center or electrical room. If possible, physically disconnect the driver from the machine.

2. Close all block valves to isolate the equipment following lockout/tagout (LOTO) and standard operating procedure (SOP). Any valve that could provide a source of energy into the machine should be closed and secured by a cable or chain using a specific color-coded lock.

3. Depressurize and drain the equipment. All process pressure should be removed through vent valves to the flare or vent system. Any source of pressure such as steam at the driver should be removed through the vent valves to the atmosphere. These valves should be tagged open using the appropriate LOTO tags. Energy sources of any type, including chemical, thermal, pneumatic, hydraulic, mechanical, gravity, and electrical must be isolated.

Again, the procedures may differ from facility to facility, so process technicians must always follow the procedures specific to their unit. Figure 6.36 shows an example of a lockout/tagout tag.

Figure 6.36 Lockout/tagout equipment.

CREDIT: Paul Jantz/123RF.

The pump-specific LOTO process would include the following steps:

1. Close the pump suction and discharge so liquid does not enter the pump.
2. De-energize the driver by opening the breaker to the motor or blocking the inlet and exhaust steam valves on the turbine.
3. Drain the pump to the designated collection system.
4. Shut down any auxiliary systems, such as lube oil and seal flush systems.
5. Lock and/or tag the valves and breaker for the motor or valves on the turbine (or the driver used). Lock and/or tag the suction and discharge valves on the process pump. (Any bleed valves to the atmosphere should be tagged open.)
6. Flush/purge the pump to clear the pump of the chemical contained in the pump.

Shutdown

Shutdown steps can vary from equipment to equipment. The process technician should always follow the proper shutdown procedure. A general shutdown procedure follows.

1. Ensure the pump is ready to be shut down (the standby pump is running or the process is being shut down).
2. Shut off the pump.
3. If the pump is being shut down following the startup of a standby pump, immediately check the standby pump's flow rate. Loss of flow from the standby pump can occur if the discharge check valve on the pump which was shut down sticks open. If this occurs, closing the discharge block valve on the shut down pump should restore the flow.
4. Close the suction and discharge valves if the pump is being removed from service for maintenance. The valves can be left open if the pump is to remain in standby service.
5. Prepare the pump for maintenance as needed.

Emergency Procedures

Emergency procedures include additional actions by a technician beyond those usually required during a shutdown. Process technicians should always follow the specific emergency operating procedures for their unit and company in order to protect personnel and the environment.

Summary

Pumps play a major role in process operations. They move products through the piping systems, serving as the "heart" of a given process because they transfer liquids from one place to another and provide flow through process equipment. In continuous operations, such as refineries, process equipment depends on continuous liquid flow for proper operation. In process-critical flows, standby pumps are commonly used and can automatically start up in the event of a loss of flow or pressure in the pumping system.

Two main categories of pumps are dynamic and positive displacement pumps. Dynamic pumps convert centrifugal force to pressure to move liquids. They are classified as either centrifugal or axial. Positive displacement pumps are piston, plunger, lobe, diaphragm, gear, or screw pumps that deliver a constant volume with each stroke. Unlike dynamic pumps, positive displacement pumps deliver the same amount of liquid, regardless of the discharge pressure.

The internal operation of a pump depends on the pump type. Dynamic pumps operate differently from positive displacement pumps. The selection of pumps for use with a particular process is based on a number of factors, including liquid viscosity, line pressures and flows, equipment speed, altitude, and temperature, and whether the material is a slurry or liquid. Variables that affect pump discharge are calculated into the performance of the pump. Pump manufacturers supply optimum pump curves that are used in system design to size the pump or driver.

Process technicians play an important role in ensuring that pump operations remain at normal conditions. Close monitoring of conditions is required to ensure problems can be resolved before they escalate to hazardous situations. Pumps can cause personal injury if attention is not paid to the required safety regulations. Environmental impact can range from excess water usage (because of leaks) to facility or community evacuations if the liquid is extremely hazardous (toxic or flammable).

Problems typically associated with pumps in a process include overheating, cavitation, leakage, unexpected pump shutdown, and excessive vibration.

Checking Your Knowledge

1. Define the following terms:
 a. Axial pump
 b. Centrifugal pump
 c. Diaphragm pump
 d. Dynamic pump
 e. Positive displacement pump
 f. Rotary pump
 g. Screw pump
 h. Vane pump
 i. Cavitation
 j. Impeller
 k. Lobe pump
 l. Piston pump
 m. Priming
 n. Pump performance curve
 o. Seal
 p. Vanes
 q. Viscosity
 r. Volute

2. Pump selection is based on (select all that apply):
 a. Viscosity
 b. Required discharge pressure
 c. Available suction pressure
 d. Flow rate
 e. Amount of gas in the suction line

3. (True or False) A gear pump is a type of centrifugal pump.

4. Which of the following are rotary-type positive displacement pumps? (Select all that apply.)
 a. Gear
 b. Lobe
 c. Vane
 d. Axial
 e. Diaphragm

5. _____ pumps are sealless pumps that ensure zero emissions.

6. (True or False) A mechanical packing type sealing system should be used only in pumps handling hazardous materials.

7. (True or False) Suction head refers to pumping from a point higher than the liquid (for example, from a sump).

8. Match the potential cause of pump shutdown with the potential troubleshooting solution:

i. Power failure	a. Check turbine overspeed trip for proper operation; check suction for proper flow; check discharge line for large failure.
ii. Flow, pressure, or vibration	b. Determine cause (for example, breaker, fuse, or loss of commercial power). Repair and restore power as appropriate.
iii. Overspeed trip	c. Pump is using too much horsepower, flow rate is too high, or mechanical problem exists. Correct problem to return pump to normal operation.
iv. Motor trip-out	d. Determine cause (for example, leak, valve closed, or instrumentation malfunction elsewhere in the process). Correct problem to return pump to normal operation.

9. When a pump suffers from _____ lock, the pump loses liquid prime, vibrates, and makes abnormal noises.

a. vibration

b. vapor

c. cavitation

d. bearing

10. On the diagram below, identify the following parts of a centrifugal pump:

a. Casing

b. Discharge nozzle

c. Eye of impeller

d. Shaft

e. Volute

f. Packing

g. Suction intake

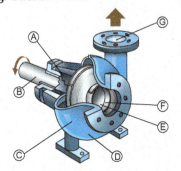

11. Two types of seal systems that employ a seal pot are _____ seals and _____° seals.

12. _____ in a pumping system is a measure of a pump's ability to move liquid through the system.

13. What option would a technician have to increase NPSH available when it is too low?

a. Decrease the pressure above the liquid

b. Decrease the liquid level

c. Increase the pressure above the liquid

d. Increase the liquid temperature

14. (True or False) During normal pump operations, process technicians must always wear personal protective equipment (PPE).

15. (True or False) Zero energy state means that all forms of energy have been removed from the pump.

16. (True or False) When working with pumps, technicians should make sure water is collecting in the lubricating oil.

17. Identify the correct sequence for the Lockout/Tagout procedure for a pump:

____ Drain the pump to the designated collection system.

____ Lock and/or tag the valves and breaker for the motor or valves on the turbine.

____ Close the pump suction and discharge so liquid does not enter the pump.

____ Flush/purge the pump to clear the pump of the chemical contained in the pump.

____ Shut down any auxiliary systems, such as lube oil and seal flush systems.

____ De-energize the driver by opening the breaker to the motor or blocking the inlet and exhaust steam valves on the turbine.

NOTE: Answers to Checking Your Knowledge questions are in the Appendix.

Student Activities

1. Using a cutaway model (or drawing) of a reciprocating piston pump, locate the following:

   a. Inlet check valve

   b. Outlet check valve

   c. Piston/plunger

   d. Packing

   e. Crankshaft

   f. Connecting rod

   g. Cylinder

   h. Seal

2. Using the pump curve in Figure 6.34, determine the head, horsepower, and NPSH required for a 9-inch (23-cm) impeller at 600 GPM flow rate.

3. Using a pump model, demonstrate a task associated with typical pump operations, such as lockout/tagout, startup, or shutdown.

4. Demonstrate how you would perform the following maintenance and monitoring activities on a typical pump:

   a. Inspect for abnormal noise.

   b. Inspect for excessive heat.

   c. Check oil levels.

   d. Check for leaks around seals and flanges.

   e. Check for excessive vibration.

5. Write a short paper comparing and contrasting the two sealing systems used by pumps: mechanical packing and mechanical seals.

Chapter 7
Compressors

*North American Process Technology Alliance (NAPTA) developed curriculum to ensure that Process Technology courses will produce knowledgeable graduates to become entry level employees in process technology. Objectives from that curriculum are named here in abbreviated form. For example, "(NAPTA Compressors 7)" means that this chapter's objective relates to objective 7 of the NAPTA curriculum about compressors).

Key Terms

Antisurge—recycle flow returned from the discharge of a compressor stage to a lower pressure suction; serves to prevent surging, **p. 153.**

Antisurge protection—control instrumentation designed to prevent damage to the compressor by preventing it from operating at or near an undesirable pressure and by preventing flow conditions that result in surge, **p. 153.**

Axial compressor—a dynamic compressor that contains a rotor with contoured blades followed by a set of stationary blades (stator). In this type of compressor, the flow of gas is *axial* or parallel to the shaft, **p. 143.**

Centrifugal compressor—a dynamic compressor in which the gas flows from the inlet, located near the suction eye, to the outer tip of the impeller blades, **p. 142.**

Compressor—a mechanical device used to increase the pressure of a gas or vapor, **p. 140.**

Cylinder—a cylindrical chamber in a positive displacement compressor in which a piston compresses and then expels the gas, **p. 146.**

Deadheading—creating a condition in which all outlet paths from a compressor's discharge line are closed, **p. 155.**

Demister—a device that promotes separation of liquid from gas, **p. 154.**

Dry carbon rings—an easy-to-replace, low-leakage type of seal consisting of a series of carbon rings that can be arranged with a buffer gas to prevent process gases from escaping, **p. 152.**

Dynamic compressor—a compressor that uses centrifugal or axial force to accelerate a gas and then to convert the velocity to pressure, **p. 142.**

Inlet guide vane—a stator, located in front of the first stage of a compressor, which directs the gas into the compressor at the correct angle; also called *intake guide vane*, **p. 143.**

Interlock—a type of hardware or software that does not allow an action to occur unless certain conditions are met, **p. 156.**

Labyrinth seal—a shaft seal designed to restrict flow by requiring the fluid to pass through a series of ridges in an intricate path, **p. 152.**

Liquid buffered seal—a close-fitting bushing in which oil or water is injected in order to stop the process fluid from reaching the atmosphere, **p. 152.**

Liquid ring compressor—a rotary compressor that uses an impeller with vanes to transmit centrifugal force into a sealing fluid (e.g., water), driving it against the wall of a cylindrical casing to form a compression area, **p. 148.**

Lubrication system—a system that circulates and cools sealing and lubricating oils, **p. 151.**

Multistage compressor—a device designed to compress gas multiple times by delivering the discharge from one stage to the suction of another stage, **p. 149.**

Positive displacement compressor—a device that uses screws, sliding vanes, lobes, gears, or pistons to deliver a set volume of gas with each stroke or rotation, **p. 144.**

Reciprocating compressor—a positive displacement compressor that uses the inward stroke of a piston to draw (intake) gas into a chamber and then uses an outward stroke to positively displace (discharge) the gas, **p. 145.**

Rotary compressor—a positive displacement compressor that uses a rotating motion to pressurize and move gas through the device, **p. 147.**

Seal system—system designed to prevent process gas from leaking out of the compressor shaft, **p. 152.**

Separator—device used to physically separate two or more components in a mixture, **p. 154.**

Single-stage compressor—a device designed to compress gas a single time before discharging it, **p. 149.**

Surging—the intermittent flow of gas through a compressor that occurs when the discharge pressure is fluctuating, resulting in flow reversal and instability within a compressor, **p. 150.**

7.1 Introduction

Compressor a mechanical device used to increase the pressure of a gas or vapor.

Compressors are an important part of the process industry. A **compressor** is a device that transports gases and vapors from one place to another and increases their pressure for use in applications that require higher pressures. For example, compressors can be used to compress gases such as carbon dioxide, nitrogen, and light hydrocarbons. They can provide the compressed air required to operate instruments or equipment. They also can be used for cooling (refrigeration, air conditioning).

The difference between the operation of compressors and the operation of pumps has to do with the physical properties of gases and liquids. While they operate similarly, compressors cannot move liquids, and pumps cannot move gases. Compressors move gases, and pumps move liquids.

The type of compressor that is used for a particular application depends on several factors. These include the type of gas being compressed, the flow rate, and the discharge pressure. Flow rate is expressed as cubic feet per minute (cfm) or meters cubed per second (m^3/s). Discharge pressure is expressed as pounds per square inch (PSI) or kilopascals (kPa).

Types of Compressors

All compressors require a drive mechanism such as an electric motor or turbine to operate, and all are rated according to their discharge capacity and flow rate. Compressors can be single- or multistage, depending on the compression ratio required. Most compressors require auxiliary components for cooling, lubrication, filtering, instrumentation, and control. Some compressors require a gearbox between the driver and compressor to increase the speed of the compressor.

Depending on their operating principle, compressors are classified as either dynamic or positive displacement type. Dynamic compressors use impellers or blades to accelerate a gas and then convert that velocity into pressure. Dynamic compressors are more widely used than positive displacement compressors because they are less expensive, more efficient, have a larger capacity, and usually require less maintenance. Positive displacement compressors use components such as pistons, lobes, screws, or vanes to compress a fixed amount of gas with each stroke or rotation and deliver a constant volume.

Each of the two classes of compressors have many different types (see Figure 7.1). The most common type of compressor used in the process industries, however, is the centrifugal compressor.

Figure 7.1 Major compressor types.

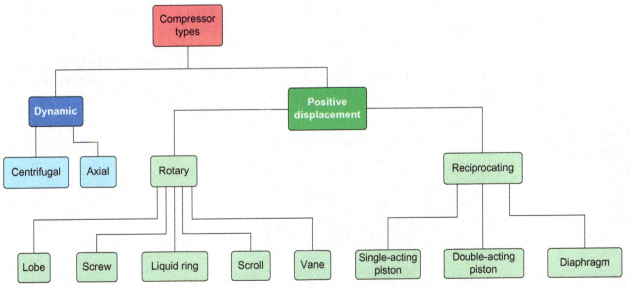

Dynamic compressor   a compressor that uses centrifugal or axial force to accelerate a gas and then to convert the velocity to pressure.

Centrifugal compressor   a dynamic compressor in which the gas flows from the inlet, located near the suction eye, to the outer tip of the impeller blades.

Dynamic Compressors

Dynamic compressors use centrifugal or axial force to accelerate the gas and then convert the velocity of the gas into pressure. The two major types of dynamic compressors are centrifugal and axial.

CENTRIFUGAL COMPRESSORS   **Centrifugal compressors** are dynamic compressors in which the gas flows from the inlet, located near the suction eye, to the outer tips of the impeller blades. In a centrifugal compressor, the gas enters at the low-pressure end and is forced through the impeller by the rotation of the shaft. The velocity is increased greatly as the gas moves from the center of the impeller toward the outer tips of the blades. When the gas leaves the impeller and enters the volute, the velocity is converted into pressure because of the decreased velocity of the molecules, which is caused by a narrowing of the casing.

Centrifugal compressors are used throughout industry because they have few moving parts, are energy efficient, and provide higher flow rates than similarly sized reciprocating compressors. Centrifugal compressors are also popular because their seals allow them to operate nearly oil-free, and they have a high degree of reliability. They are effective in toxic gas service when the proper seals are used, and they can compress high volumes at low pressures. The primary drawback of centrifugal compressors is that they cannot achieve the high compression ratio of reciprocating compressors without multiple stages. The main components of a centrifugal compressor include bearings, a housing (casing), an impeller, an inlet and outlet, a shaft, and seals. Figure 7.2 shows an example of a centrifugal compressor.

Figure 7.2 Centrifugal compressor with driver.

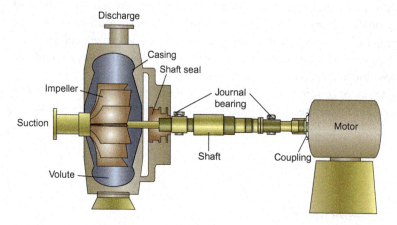

Centrifugal compressors are best suited for continuous-duty applications, operating at full capacity such as refrigeration and cooling systems, compressed air systems, and other uses that require high volumes but relatively low pressures.

In a centrifugal compressor, there is a direct relationship between impeller speed, velocity, pressure, and flow (Figure 7.3). As the impeller speed increases, velocity increases. As velocity increases, pressure increases. As pressure increases, flow increases.

Figure 7.3 Process variables in a centrifugal compressor.

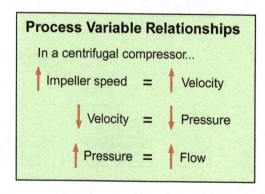

Centrifugal compressors are single-stage or multistage, and the stages might be contained in one casing or several different casings. (Note: Multistage compressors are discussed in more detail later in this chapter.) Figure 7.4 shows a single-stage centrifugal compressor.

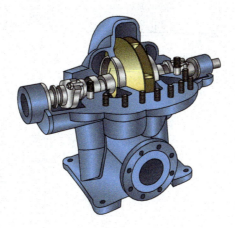

Figure 7.4 A single-stage centrifugal compressor.

> ## Did You Know?
>
> Physics teaches us that as heat is added to a substance, molecular motion and vapor pressure increase.
> In other words, heat causes gases to expand. As the temperature of a gas increases in a closed container, so does its pressure. Conversely, if you want to compress more gas into the same space, you must cool it.

AXIAL COMPRESSORS   **Axial compressors** are dynamic compressors in which the flow of gas is *axial* (parallel to the shaft). A typical axial compressor has a rotor that looks like a fan with contoured rotor blades followed by a set of stationary blades, called *stator* blades.

Rotor blades attached to the shaft spin and send the gas over stator blades, which are attached to the internal walls of the compressor casing. These blades decrease in size as the casing size decreases. Rotation of the shaft and its attached rotor blades causes flow to be directed axially along the shaft, building higher pressure toward the discharge of the unit.

Each pair of rotors and stators is referred to as a *stage*. Most axial compressors have a number of such stages placed in a row along a common power shaft in the center. Figure 7.5 shows an example of an axial compressor with rotor and stator blades.

The stator blades are required to ensure efficiency. Without stator blades, the gas would rotate with the rotor blades. Each stage is smaller than the last because the volume of gas is reduced by the compression of the preceding stage. This is the reason axial compressors generally have a conical shape, widest at the inlet and narrowest at the outlet. This compressor type typically has between nine and fifteen stages. The main components of an axial compressor are the housing (casing), inlet and outlet, rotor and stator blades, shaft, and **inlet guide vanes**.

Axial compressors are efficient compressors, but they are not used as frequently in industry as reciprocating and centrifugal compressors because of their high purchase and maintenance costs. The most common use for axial compressors is in jet engines that power aircraft and some ships. In process industries, axial compressors are also used in gas turbines, to compress air required for combustion, or with generators to generate electric power.

Axial compressor   a dynamic compressor that contains a rotor with contoured blades followed by a set of stationary blades (stator). In this type of compressor, the flow of gas is *axial* or parallel to the shaft.

Inlet guide vane   a stator, located in front of the first stage of a compressor, which directs the gas into the compressor at the correct angle; also called *intake guide vane*.

Figure 7.5 Axial compressor showing rotor and stator blades.

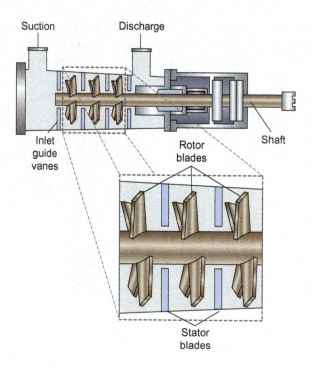

Compression Ratio

Compression ratio, sometimes called *pressure ratio,* is an important consideration in the selection of a compressor. Compression ratio determines the maximum pressure increase that can be expected to be achieved in each stage of compression. Compression ratio is calculated by dividing the absolute discharge pressure by the absolute suction pressure.

Calculating Compression Ratio

Compression ratio = Discharge pressure ÷ Suction pressure

Example: A compressor with a suction pressure of 400 PSIA and a discharge pressure of 1400 PSIA has a compression ratio of 3.5.

1400 PSIA ÷ 400 PSIA = 3.5 compression ratio

Compression ratio is often limited by the acceptable discharge temperature of the gas. Discharge temperature increases along with compression ratio. As compression occurs, gas molecules are forced closer together, which increases their rate of movement and contact with each other. This, in turn, results in a temperature increase, which sometimes can exceed acceptable levels. In cases where the discharge temperature will be too high, a solution might be to increase the number of compression stages, add intercoolers, or decrease the compression ratio that occurs in each stage.

Positive Displacement Compressors

Positive displacement compressor a device that uses screws, sliding vanes, lobes, gears, or pistons to deliver a set volume of gas with each stroke or rotation.

Positive displacement compressors (see Figure 7.6) are devices that use pistons or diaphragms, screws, sliding vanes, lobes, or gears to deliver a set volume of gas with each stroke. Most positive displacement compressors work by trapping a specific quantity of gas and forcing it into a smaller volume. The two categories of positive displacement compressors are reciprocating and rotary, with reciprocating being the most commonly used.

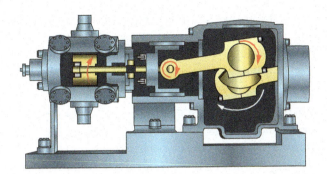

Figure 7.6 Positive displacement compressor showing motion.

RECIPROCATING COMPRESSORS   The term *reciprocating* refers to the back-and-forth movement of a compression device (a piston or other device). **Reciprocating compressors** use a back-and-forth or up-and-down motion to pressurize and move gas through the device (Figure 7.7). A common application for the reciprocating compressor is an instrument air system. They also are used for off-gas compression (flammable gases).

Reciprocating compressor
a positive displacement compressor that uses the inward stroke of a piston to draw (intake) gas into a chamber and then uses an outward stroke to positively displace (discharge) the gas.

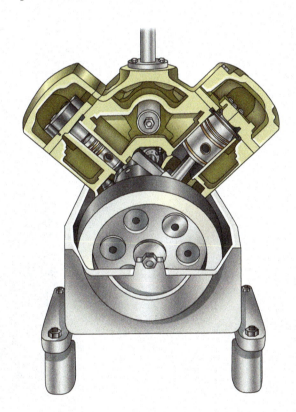

Figure 7.7 Cutaway of a reciprocating compressor.

Did You Know?

A hand-operated bicycle pump is an example of a reciprocating compressor.

In this type of compressor, air is drawn into the chamber when the handle (which is attached to a piston) is pulled up, and it is compressed and forced out when the handle is pushed down. The laws of physics state that an increase in pressure creates an increase in temperature, so the barrel of the bicycle pump will become warm to the touch.

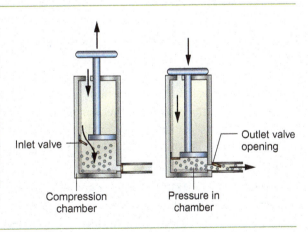

In theory, reciprocating compressors are more efficient and less expensive to purchase and install than centrifugal compressors. However, problems with pulsation and mechanical reliability, along with limited flow capability, cause these compressors to be less desirable than centrifugal compressors for most industrial applications. More potential problems associated with compressors are discussed later in this chapter.

PISTON COMPRESSORS In a piston-type compressor, a piston receives force from a driver and transfers that force to the gas being compressed. Piston-type compressors can be single or double-acting. In a single-acting compressor, compression occurs on only one side of the piston, and the compressor alternates between suction and discharge strokes. Double-acting compressors use both sides of the piston. Both suction and discharge actions occur with each piston stroke. During the suction stroke, gas is trapped on one side of the piston, while it is compressed and discharged on the other side.

The **cylinder** is the chamber in which a piston compresses gas and from which gas is expelled. Figure 7.8A displays a cutaway picture of a piston type reciprocating compressor.

Cylinder a cylindrical chamber in a positive displacement compressor in which a piston compresses and then expels the gas.

Figure 7.8 A. Piston-type compressor. **B.** Gas flow in a piston-type reciprocating compressor, with labeled parts.

CREDIT: A. SergeV/Shutterstock.

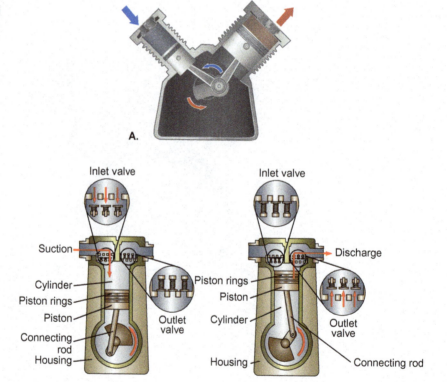

The pistons are attached to a connecting rod, which is attached to the crankshaft. This rod converts the rotational motion of a driver to the reciprocating motion of the piston. The piston's motion pulls gas into a cylinder from the suction line and then displaces it from the cylinder through the discharge line. Valves (check valves) on the suction and discharge allow the flow of the gas in one direction only. Figure 7.8B displays the major components of the compressor, which include a cylinder, inlet, outlet, valves, housing, piston, piston rings, and connecting rod. Figure 7.9 shows an example of single-acting and double-acting compressors.

DIAPHRAGM COMPRESSORS Another type of reciprocating compressor is the diaphragm compressor, which can be used for a wide range of pressures and flows. In a diaphragm compressor, hydraulic fluid is forced against one side of the diaphragm, which

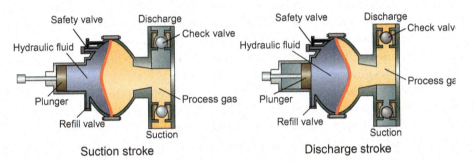

Single-acting compressor   Double-acting compressor

Figure 7.9 Examples of single-
and double-acting compressors.

flexes it and pushes it into the process gas side of the space, compressing and pressurizing
the process gas. Figure 7.10 displays the basic components of a diaphragm compressor.

Because the process gas in a diaphragm compressor does not come in contact with the
fluid, product purity is assured. This is useful in laboratory or medical applications and in
toxic gas service.

Suction stroke   Discharge stroke

Figure 7.10 Diaphragm compressor with labeled parts.

ROTARY COMPRESSORS   Rotary compressors, such as the ones shown in Figure 7.11,
move gases by turning a set of screws, vanes, or lobes. As these screws, vanes, or lobes rotate,
gas is drawn into the spaces between these components and moves through the compressor
until being forced out of the compressor through the discharge. Rotary compressors can
operate effectively with as little as 10 percent of their maximum flow capacity. This makes
them especially useful in applications where the operating conditions (flow/pressure) vary
greatly. Rotary compressors have the advantage that, from a constant suction pressure, they
maintain a more constant discharge pressure than piston compressors do. The maintenance
costs associated with most rotary compressors also are usually less than those of reciprocating compressors, because of their design.

Rotary compressor   a positive
displacement compressor that uses
a rotating motion to pressurize and
move gas through the device.

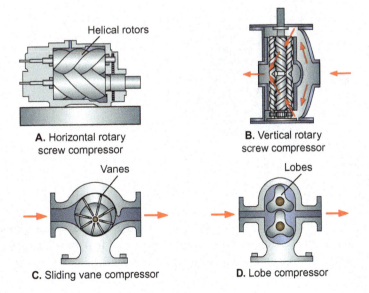

A. Horizontal rotary
screw compressor

B. Vertical rotary
screw compressor

C. Sliding vane compressor

D. Lobe compressor

Figure 7.11 Examples
of rotary compressors.
A. Horizontally oriented rotary
screw compressor. **B.** Vertically
oriented rotary screw compressor.
C. Sliding vane compressor.
D. Lobe compressor.

In a *rotary screw compressor*, two helical (spiral) rotors (sometimes called screws) mesh together as they turn and move a gas or vapor from the suction to the discharge end. Although the two rotors do not contact each other, the space between them is very small, typically 0.003–0.006 inches (0.08–0.15 mm). Two designs of rotary screw compressors are oil flooded and oil free. In the oil flooded screw compressor, one of the rotors is connected to a mechanical driver and is called the "drive" rotor. The other rotor (called the "driven" rotor) is driven in the opposite direction through the oil film on the rotor, pushed by the motion of the driven rotor. The oil-free design requires gears to keep both rotors turning at the proper rate of speed to keep them synchronized. Gas is drawn into the void between the two helical rotors at the compressor's inlet. As these rotors mesh together, the space between them becomes smaller, which reduces the volume of the gas and raises its pressure. The gas proceeds axially along the length of the rotor until it reaches the discharge end.

A *sliding vane compressor* employs an *eccentric* (off-center) rotor with vanes that move freely along longitudinal slots in the rotor. The vanes are forced out against the casing wall by centrifugal force, creating individual chambers of gas that are compressed as the eccentrically mounted rotor turns. The size of each chamber decreases as it approaches the discharge port, compressing the gas into a smaller volume for discharge at a pressure higher than when it entered the compressor. Refer back to Figure 7.11 for examples of rotary compressors.

A *lobe compressor* is unique in that it doesn't compress the gas as it moves through the machine. This type of compressor is often called a *Roots blower*, named after the Roots brothers who invented it in 1854. Gas is compressed when it gets to the discharge port of the compressor and is exposed to back pressure from the discharge line. As the pressure increases to overcome the system pressure, the gas is compressed. Gears are used to rotate each of the rotors and maintain clearances between them. There are usually two or three lobes on each of the rotors.

Another kind of rotary compressor is a liquid ring compressor. A **liquid ring compressor** uses an eccentric (off-center) impeller with vanes to transmit centrifugal force to a sealing fluid (e.g., water), driving it against the wall of a cylindrical casing. Gas is drawn into the vane cavities at the inlet (suction) port, located near the rotor. As the rotor turns, the distance between the casing wall and the vanes decreases, forcing the water inward and compressing the gas in each of the vane cavities. The liquid is used in place of a piston; it compresses the gas without friction. Figure 7.12 shows an example of a liquid ring compressor.

Liquid ring compressor a rotary compressor that uses an impeller with vanes to transmit centrifugal force into a sealing fluid (e.g., water), driving it against the wall of a cylindrical casing, to form a compression area.

Figure 7.12 Liquid ring compressor.

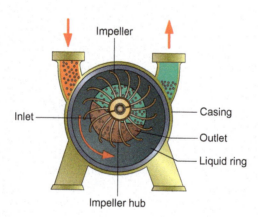

Inlet — Impeller — Casing — Outlet — Liquid ring — Impeller hub

In a recycling system, the sealing fluid is replenished and cooled in an external reservoir and then recirculated back to the compressor. In a once-through cooling system, the sealing fluid is removed and replaced with fresh fluid.

Single-Stage Versus Multistage Compressors

Single-stage compressors, which compress the entering gas one time, are generally designed for high gas flow rates and low discharge pressures.

Multistage compressors compress the gas multiple times by delivering the discharge from one stage to the suction of another stage in order to achieve higher pressures. Figure 7.13 shows a functional diagram of a multistage compressor. Figure 7.14 shows an example of a multistage compressor.

Multistage compressors raise the gas to the desired pressure in steps or stages (chambers). Between stages, an intercooler may be used to remove the heat developed during compression. If condensation occurs as a result of that cooling, any liquid must be removed from the gas stream and not allowed to enter the compressor. Liquids are not compressible and could cause severe damage to the compressor.

In all compressors, the temperature of a gas increases as it is compressed. The amount of temperature increase is a function of the gas's specific heat and the compression ratio (the ratio of the discharge pressure to inlet pressure in absolute pressure units). To avoid extremely high discharge temperatures, the compression ratio in compressors usually is limited to around 3.75 for each stage.

Frequently in the process industry, the desired discharge pressure is more than 10 times that of the inlet pressure. In such cases, a single-stage compressor cannot be used because of the high temperature generated. Instead, a multistage compressor is required. Cooling takes place when gas exits a compressor casing either to the next casing or to the discharge.

Multistage compressors can be centrifugal, axial, or reciprocating (piston) compressors. Large multistage compressors can be extremely complex, with many subsystems, including bearing oil systems, seal oil systems, and extensive vibration detection systems.

Single-stage compressor
a device designed to compress gas a single time before discharging it.

Multistage compressor   a device designed to compress gas multiple times by delivering the discharge from one stage to the suction of another stage.

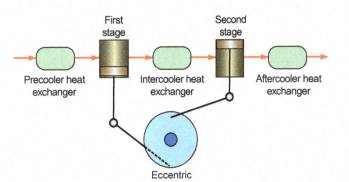

Figure 7.13 Functional diagram of a multistage compressor.

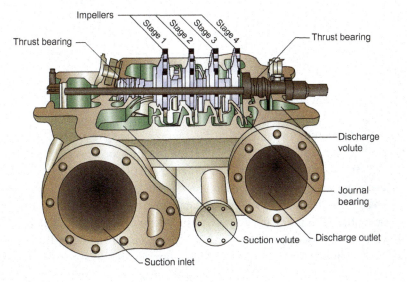

Figure 7.14 Multistage compressor with components labeled.

7.2 Centrifugal Compressor Performance Curve

Design and selection of compressors involve several factors, including physical property data and a description of the process gas. When a compressor is selected, a performance curve is provided with the compressor, which allows engineers to predict its operating efficiency.

In order for centrifugal compressors to perform adequately and operate on their performance curve, they must be selected to match the existing operating conditions. (This is also true for centrifugal pumps.) A typical compressor performance curve is shown in Figure 7.15. The vertical axis is feet of head (H_p), which means feet of head for a gas being compressed. The horizontal axis is actual cubic feet per minute (ACFM) of gas compressed. (In metric measurement, centimeters and cubic centimeters would be used.) The compressor operates on its specified curve unless a problem occurs (e.g., it malfunctions, is mechanically defective, or there is a change in the composition of the gas.

Figure 7.15 Performance curve used by engineers to predict operating efficiency.

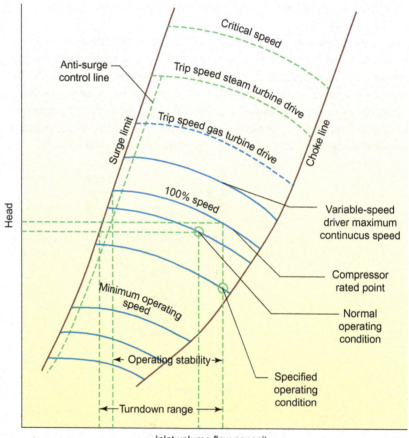

Inlet volume flow capacity

In compressor performance curves, the H_p increases and the volume of gas compressed (ACFM) decreases as the compressor discharge pressure increases. In dynamic compressors, when the volume of gas entering the compressor drops below a critical flow rate, the compressor is backed up to its surge point on the compressor's performance curve and it could stall.

Surging is a condition in which gas flow decreases to the point where the compressor becomes unstable. When compressor discharge pressure falls below the pressure in the discharge line, forward flow stops and gas from the discharge line flows backward toward the suction. At this point, the downstream line pressure drops until it falls below the compressor discharge pressure, and forward flow begins again. The condition causes severe vibration in the compressor and its associated equipment. If the issue that caused the surge condition is not corrected, the cycle will repeat. Severe damage can occur to compressor components if a surge condition is not corrected immediately.

Surging the intermittent flow of gas through a compressor that occurs when the discharge pressure is fluctuating, resulting in flow reversal and instability within a compressor.

Data such as flow rate, temperature, and pressure conditions are collected on compressors. This data is provided to engineers who analyze it to ensure that the compressor is operating at the designed output. If the data show abnormalities, the compressor might be assigned to maintenance for repair. It is important to keep compressors properly maintained to prevent compressor failure, reduce maintenance costs, increase equipment life, and prevent process interruption.

7.3 Associated Utilities and Auxiliary Equipment

Compressors are part of process systems. A wide variety of auxiliary equipment is needed to enhance equipment productivity. Auxiliary equipment associated with compressors includes lubrication systems, seal systems, antisurge protection, intercoolers, aftercoolers (heat exchangers), and separators.

Lubrication System

When gases are compressed, the temperature of the compressor bearings and seals usually increases. As a result, the temperature of the oil used to lubricate the bearing and seals also increases. This heat is removed by cooling the lubricants.

A **lubrication system** circulates and cools the sealing and lubricating oils. In some applications, the sealing and lubricating oils are the same. Figure 7.16 shows an example of a lubricant cooling system.

Lubrication system   a system that circulates and cools sealing and lubricating oils.

Figure 7.16 Diagram of lubrication system components.

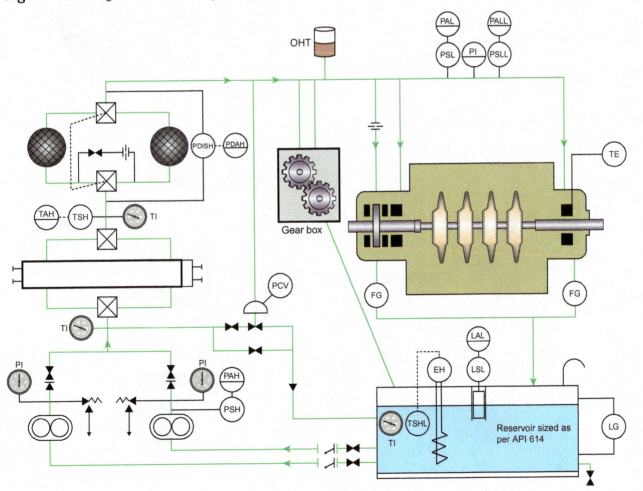

Seal system system designed to prevent process gas from leaking out of the compressor shaft.

Labyrinth seal a shaft seal designed to restrict flow by requiring the fluid to pass through a series of ridges in an intricate path.

Liquid buffered seal a close-fitting bushing in which oil or water is injected in order to stop the process fluid from reaching the atmosphere.

Dry carbon rings an easy-to-replace, low-leakage type of seal consisting of a series of carbon rings that can be arranged with a buffer gas to prevent process gases from escaping.

Seal System

A **seal system** (shown in Figure 7.17) prevents process gases from leaking to the atmosphere where the compressor shaft exits the casing. Compressor shaft seal types include labyrinth seals, liquid buffered seals, and dry carbon rings.

A **labyrinth seal** (shown in Figure 7.17B) is designed to restrict flow by requiring the fluid to pass through a series of ridges in an intricate path. A purge of an inert gas (barrier gas) is provided is an intermediate injection point on the seal to prevent external leakage of process gas.

A **liquid buffered seal** is a close-fitting bushing in which a liquid is injected in order to seal the process from the atmosphere.

Dry carbon rings (see Figure 7.17C) are low-leakage types of seals that can be arranged with a buffer gas that separates the process gas from the atmosphere.

Figure 7.17 A. Typical seal oil system. **B.** Labyrinth seal used to minimize leakage around a compressor shaft. **C.** Example of dry carbon rings with buffer gas.

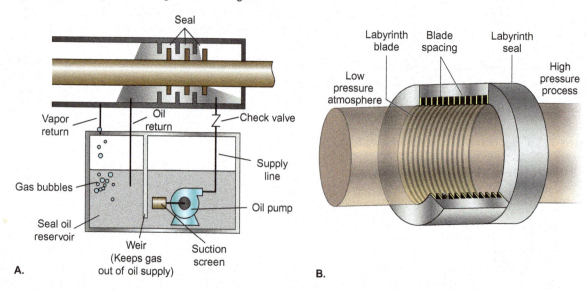

A.

B.

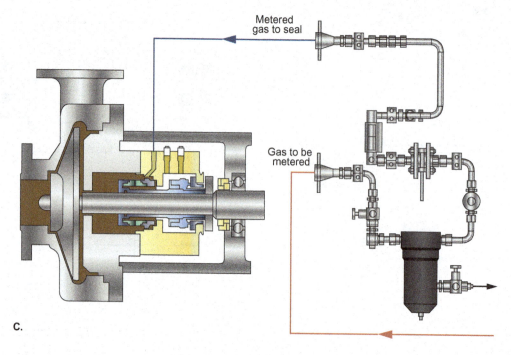

C.

Packing is used in reciprocating compressors to prevent leakage where the piston rod passes through the crank end of the cylinder. Typical packing consists of rings of fibrous material fitted around the piston rod (contained in a stuffing box) and compressed by an adjustable gland. Friction between the packing and piston rod is reduced by the injection of lubricating oil into the stuffing box.

Antisurge Devices

Because surge is so detrimental to equipment, centrifugal compressors are equipped with **antisurge protection**. **Antisurge** systems (Figure 7.18) are important because surging can cause bearing and shaft failures and destruction of machinery. The antisurge protection is designed by engineers to ensure proper and safe compressor operation.

Antisurge protection control instrumentation designed to prevent damage to the compressor by preventing it from operating at or near an undesirable pressure and by preventing flow conditions that result in a surge.

Antisurge recycle flow returned from the discharge of a compressor stage to a lower pressure suction; serves to prevent surging.

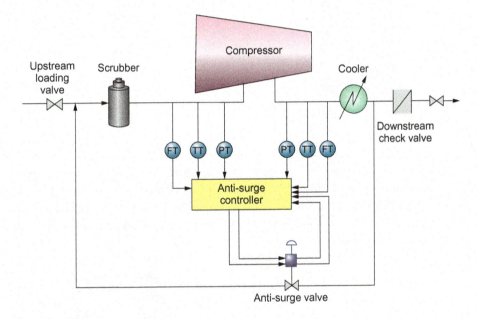

Figure 7.18 Diagram of an antisurge control system.

One form of antisurge device is a minimum flow recycle device. In this type of device, the compressor discharge flow is measured. When the flow rate falls below the rate set on the recycle instrument (a rate above the minimum critical flow rate), a recycle valve located downstream of the compressor discharge opens very rapidly. This returns some of the flow back to the compressor suction line to avoid damage to the equipment. The recycled gas allows the compressor to maintain a high enough suction flow rate to prevent surging.

Another form of antisurge protection is a variable speed driver, which controls the discharge pressure. When steam turbine drives are used, steam flow to the turbine is used to control the compressor speed.

Compressors also are designed with automatic shutdown systems to prevent damage to the compressor during abnormal conditions.

Coolers

Coolers are used to cool gases at various stages in the process.

Precoolers are used to cool the gas before it enters a compressor. Intercoolers are used in multistage compressors to remove some of the heat generated by each stage of compression. In other words, intercoolers cool the discharge of the first stage before it enters the suction of the second stage. Interstage coolers (intercoolers) prevent the process gas and equipment from overheating and also cool antisurge recycle flows. Aftercoolers cool the gas downstream from the last stage when the compression cycle is complete.

Separators

Separators　devices used to physically separate two or more components in a mixture.

Demister　a device that promotes separation of liquid from gas.

Separators physically separate two or more components in a mixture. Examples of separators include demisters and desiccant dryers.

A **demister** in a compressor system (shown in Figure 7.19) is a device used to promote separation of liquid from gas. Demisters contain a storage area designed to collect the liquid that is separated from the gas, so it can be removed from the system to prevent damage.

Figure 7.19 Demister.

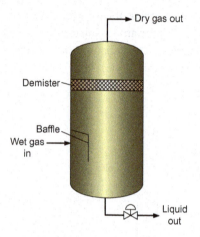

Desiccant dryers (shown in Figure 7.20) use chemical drying agents (desiccants) to remove or adsorb moisture. Desiccant dryers usually are used in tandem (one in service and one out of service). When the desiccant in one dryer becomes saturated, the dryer is taken out of service and hot gas is sent through the dryer bed to dry it out. Because they are more efficient than demisters, desiccant dryers are often used for sensitive applications such as instrument air.

Figure 7.20 A. Desiccant dryer. **B.** Tandem desiccant dryer.

CREDIT: Motionblur Studios/ Shutterstock.

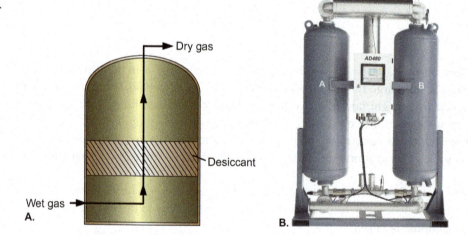

7.4 Potential Problems

The role of the process technician is vital to the production and safety of any process unit. It is the responsibility of process technicians to have a clear understanding of the process in which they are working and always maintain a keen awareness of their surroundings. Process technicians must have a basic understanding of troubleshooting techniques to prevent damage to a compressor if it malfunctions.

To prevent potential problems, compressors are equipped with various protection devices. Problems that can arise include overpressurization, overheating, surging, leaks, loss of lubrication, vibration, malfunction of interlock systems, loss of capacity, and motor overload. Process technicians must be able to recognize these conditions and respond appropriately.

Overpressurization

Compressor overpressurization can occur if the valves associated with the compressor are incorrectly closed or blocked. Consider the example in Figure 7.21A. In this example, valve A is open and valve B is closed. This means the gas is flowing through valve A and then back into the tank. If a technician were to redirect the flow of the gas through valve B, he or she must open valve B before closing valve A. If both valves are closed at the same time, as shown in Figure 7.21B, the gas has no place to go. This is referred to as **deadheading**. The result is increased pressure and temperature that can cause surging, damage the compressor, or cause serious personal injury. Positive displacement compressors carry a higher inherent risk of overpressurizing the system than do dynamic compressors.

Deadheading creating a condition in which all outlet paths from a compressor's discharge line are closed.

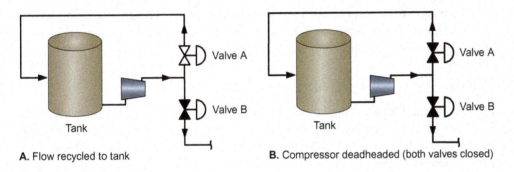

A. Flow recycled to tank   **B.** Compressor deadheaded (both valves closed)

Figure 7.21 A. Proper valve line-up with flow recycled to tank. **B.** Deadheaded (both valves closed).

Overheating

Excessive heat generation can be very detrimental to processes and equipment. Compressor overheating can be caused by loss of cooling, improper lubrication, or valve malfunction. Without lubrication, bearings fail and equipment surfaces rub together. As surfaces rub against one another, friction is produced and heat is generated. Process technicians should always monitor compressors for unusual sounds and excessive heat. Operating compressors under these conditions can lead to permanent equipment damage and/or personal injury (e.g., burns).

Surging

Another operational hazard is surging (explained earlier in the chapter). Surging is caused by a temporary flow reversal condition in one or more stages of the compressor. This flow reversal causes the compressor speed to fluctuate wildly and vibration to increase dramatically. Surging is associated with dynamic type compressors. As discharge flow drops below acceptable levels, antisurge protection activates to ensure there is no reverse flow through the compressor.

Surging can result when the compressor throughput and head fall into an unstable region of the head/capacity curve. This can cause many problems, including bearing failures, shaft failures, and destruction of the machinery.

A compressor's surge point is a variable function of gas properties and composition, system pressure, machine speed, compressor differential pressure, and compressor rate. Surging can occur at a low compressor rate or at a high speed and high differential pressure.

Surge prevention is critical to prevent damage, so process technicians should adjust the variables listed above (to the extent they have control of them) in a direction away from conditions that could lead to surging.

Seal Oil Problems

Seal oil problems can cause loss of process gases to the atmosphere. Examples of seal oil problems include dirty seal oil (which causes plugging of passages), vibration or cracking of internal oil tubing, and internal seal failure.

Leaks

Process technicians should always check compressor cylinders or housing for leaks because leaks can introduce harmful or hazardous substances into the atmosphere and create process problems. Excessive flows of barrier gas purges are a sign of seal leakage. Monitoring for the presence of process gas leakage to the atmosphere is one way technicians can determine the condition of the atmospheric seal.

Compressors can be damaged if liquid, which is noncompressible, is inadvertently introduced into compression chambers or between compressor components. Because of this, compressors employ a liquid removal step for gases that could condense inside the compressor. In addition, interstage liquid removal points are provided to prevent liquid from causing compressor damage.

When equipment is operating under vacuum conditions, it is necessary to monitor process fluid for oxygen to determine whether air is leaking into the compressor. The presence of oxygen in the process fluid indicates leakage.

Lubrication Fluid Contamination

Lubrication fluids can be contaminated with materials such as water, process fluids, and metal particulates. As the level of contamination increases, the cooling and lubricating capacity of the fluid decreases. Because of this, frequent monitoring of the seal and bearing lubricants is necessary. Monitoring includes frequent visual inspection of bearing areas, as well as temperature probes to identify any unusual temperature increase. It should also include a check for low flow or pressure of lubricant going to the bearings.

Vibration

Excessive vibration can cause serious problems in compressors. Common causes of vibration include surging, impeller imbalance or damage, worn bearings, piping pulsation, shaft misalignment, damaged couplings, and bearing or seal damage. In most cases, vibration monitoring devices are attached to equipment to activate an alarm when vibration becomes excessive. Some sites, however, depend on process technicians to detect elevated or unusual vibration conditions.

Interlock System

Interlock a type of hardware or software that does not allow an action to occur unless certain conditions are met.

An **interlock** is a type of hardware or software that does not allow an action to occur if certain conditions are not met. For example, modern automobiles contain interlocks that prevent you from taking the key out of the ignition unless the transmission is in park. Interlocks are used to ensure that a proper sequence is followed. They also can be used to monitor vital aspects of a process and, if necessary, to shut down a process. To prevent major damage, compressors are typically equipped with interlock systems. These interlock systems are designed to prevent startup of the compressor before certain conditions are met (e.g., suction valve open, lubricating oil circulating) and will shut down the compressor if all of the conditions necessary for safe operation cease to be met.

Examples of interlock protection in compressors include indicators and shutdown systems for low lubrication, low oil pressure or flow, excessive vibration, high liquid level in a suction vessel, high discharge temperature, high differential pressure, or high power consumption. Many compressors are designed with emergency shutdown (ESD) instrumentation that engages if an abnormal condition exists. Most of these systems are designed to *fail safe* (i.e., shut down the system in a manner that causes little or no equipment damage).

Voting sensor systems are used to reduce the frequency of false shutdowns. In a voting system, multiple temperature transmitters monitor the discharge temperature. The voting sensor system uses two-out-of-three logic to determine an action. In other words, if two out of three transmitters indicate that the temperature is too high, then the compressor will shut down.

This logic prevents the failure of one transmitter from shutting down the compressor unnecessarily. It also allows faulty transmitters to be repaired without shutting down the compressor.

Loss of Capacity

Most compressors are designed with safety factors that allow continuous operation up to the driver's rated capacity. Exceeding the design rating can cause damage to the compressor because of the possibility of liquid carryover from the suction drum into the machine. Unless the liquid carried into the suction is in the form of a *slug* (a large amount all at once), the protective instrumentation is designed to protect against significant damage. However, liquid droplets in small quantities carried into suction lines can damage the machine's components over time.

Loss of capacity in a reciprocating compressor can be caused by leaking compression valves and/or leaking piston rings. A leaking compression valve can be indicated by higher-than-normal temperature on the valve's cover plate located on the exterior of the cylinder. Leaking piston rings also can be identified by a decline in rate with normal or lower-than-normal valve cover temperatures.

Motor Overload

Compressor motor overload and motor interlock shutdown can occur for a variety of reasons, including higher than designed discharge pressure, high motor amperage, or bearing failure. In some cases, the motor overload occurs if the compressor is used to compress a gas with a molecular weight above what the compressor is designed to handle. Compressors with variable speed electric drivers can suffer motor overload situations if the torque demand is high when the revolutions per minute (rpm) are low. This situation is most likely to occur with rotary compressor types.

High/Low Flow

Lubricant flow that is too high or too low can indicate imminent bearing or seal failure. Lubricant flow monitoring systems are designed to activate an alarm if flow that is too high or too low is detected.

Gas flow that is too high through the compressor could be caused by a compressor operating at excessive rpm, which could result in compressor damage. Gas flow that is too low through the compressor could indicate suction-side restrictions or the onset of compressor surge conditions.

7.5 Safety and Environmental Hazards

Hazards associated with normal and abnormal compressor operations can affect personal safety, equipment, and the environment. Table 7.1 lists some of these hazards and their effects.

Table 7.1 Hazards Associated with Compressors

Personal Safety	Equipment	Environment
■ Hazardous fluids/materials	■ Rotating equipment	■ Discharge on ground or in sewer sump (seal leaks)
■ Excessive noise levels	■ Flange gasket or seal leakage or failure	■ Discharge of harmful gas or fluid into the air
■ Slipping/tripping because of leaks	■ Fires resulting from seal blowout	
■ Chemical exposure	■ Hot surfaces or cryogenic material releases	
■ Skin burns	■ Excessive vibration	
■ Rotating equipment hazards	■ Inadequate lubrication	
■ Leaks	■ Overpressure causing equipment failure	
■ High or very low temperatures	■ Blown cylinder heads	
■ High pressure	■ Shearing off of axial compression blade	
■ Electric shock	■ High voltage electricity	

7.6 Process Technician's Role in Operations and Maintenance

When working with compressors, process technicians must conduct monitoring and control activities to ensure the compressor is not overheating or leaking. Process technicians should check vibration, oil flow, oil level, temperature, and pressure. They also should ensure that connectors, hoses, and piping are in good condition. They should examine safety systems and instrumentation and oil levels, and check for leaks at seals, packing, and flanges. All these activities contribute to the safety of the people in the area as well as reliability of the equipment.

Process technicians are responsible for knowing and understanding the startup, shutdown, and emergency procedures associated with compressors in their assigned work area.

Process technicians work with shift supervisors to ensure that other plant personnel are aware of any sudden change in gases being introduced to the equipment or gases being discharged into the plant header. This helps to ensure that material balance is maintained and that other process units downstream of the equipment are not affected.

Failure to perform proper monitoring and control could affect the process and result in equipment damage. All process technicians should monitor equipment and guard against hazards, and they must follow appropriate procedures, including:

- Wearing the appropriate personal protective equipment and observing all safety rules
- Observing compressor operation for signs of compressor failure or maintenance needs during normal operation
- Checking and recording pressures, temperatures, and flow rates periodically to identify a problematic trend
- Checking rotating equipment frequently for mechanical problems, such as lack of lubrication, seal leakage, overheating of compressors and drivers, and excessive vibration
- Checking lubrication systems to ensure proper temperatures and lubricant levels.

When working with compressors, process technicians should look, listen, and check for the items listed in Table 7.2. In addition to operational maintenance, process technicians might perform other routine and preventive maintenance tasks.

Table 7.2 Process Technician's Role in Operation and Maintenance

Look	Listen	Check
■ Note oil levels to make sure they are satisfactory ■ Look at seals and flanges to make sure there are no leaks ■ Observe vibration monitors to ensure they are within operating range ■ Verify that the liquid level in the suction drum is within limits ■ Observe flow rates ■ Monitor suction and discharge pressures and temperatures	■ Listen for abnormal noises	■ Check for excessive vibration ■ Check for excessive heat

7.7 Typical Procedures

Process technicians should be aware of the various compressor procedures such as startup, shutdown, lockout/tagout, and emergency procedures to ensure safety.

Startup and Shutdown

Startup and shutdown procedures vary greatly, depending on the type of compressor driver and the type of compressor. While some machines are started under no-load or low-load conditions by opening the recycle valves and establishing minimum flow, others are started with both suction and discharge valves wide open.

The following is one example of the steps necessary for starting a compressor:

1. Ensure that the machine is completely assembled and ready for startup and that the process is ready for the compressor to come on line.

2. Place ancillary lubricating and seal oil systems, housing vent lines, and other auxiliary components into operation.

3. Use an inert gas, such as carbon dioxide or nitrogen, to purge the compressor case or cylinders of air, if required.

4. With pressure on the machine, drain the compressor's case or cylinders to remove any liquid, then check for leaks.

5. Adjust the controls (instrumentation) for minimum driver load at startup (e.g., unload reciprocating compressor cylinders).

6. Put the driver into condition to be started (e.g., prepare steam systems and superheaters and/or remove tagout devices from electrical switches).

7. Start the machine at low load. As applicable for variable speed drivers, bring the machine up to operating speed. Avoid running the equipment in its critical speed ranges.

8. For reciprocating compressors, adjust the cylinder loadings for efficient operation after confirming that the machine is running smoothly.

9. Closely monitor operation as the machine reaches normal operating conditions.

Lockout/Tagout

The control of hazardous energy sources and electrical hazards is mandated by OSHA and covered in these standards:

- Control of Hazardous Energy (Lockout/Tagout)—29 CFR 1910.147
- Lockout/Tagout Electrical Safe Workpractice Standard—29 CFR 1910.333.

The purpose of these standards is to ensure that equipment is placed in a safe condition (lockout/tagout) before any worker attempts to conduct service or maintenance tasks on any piece of equipment that could unexpectedly become energized and cause injury. See Chapter 6 for a generic LOTO procedure that can be adapted for compressors. Note: Although rules vary depending on the industry, there are limits to the voltages that process operators have the authority to work with in the electrical room or motor control center. Beyond these limits, an electrician is required.

Emergency Procedures

Emergency procedures include actions to be taken by a technician different from those required during a normal shutdown of equipment and units. Emergency procedures differ by process unit depending on the specific hazards and equipment present on each unit. Procedures exist at the site location level as well as the individual process units within the plant. Process technicians should always follow the specific emergency operating/shutdown procedures for their unit and follow company guidelines to ensure safety of all personnel as well as the equipment.

Emergency procedures include, but are not limited to:

- Loss of instrument air supply
- Loss of nitrogen supply
- Loss of cooling water supply
- Loss of steam supply
- Loss of plant or unit electrical power.

Summary

Compressors are used in the process industries to compress gases and vapors so that they can be used in a system that requires a higher pressure. Compressors are similar to pumps in the way that they operate. The primary difference is the service in which they are used. Pumps are used to move liquids and slurries; compressors are used in gas or vapor service.

Depending on their operating principles, compressors are classified as dynamic or positive displacement.

Dynamic compressors use centrifugal or rotational force to move gases. Types of dynamic compressors are centrifugal and axial compressors. In a centrifugal compressor, gas enters through the suction inlet and goes into the casing where a rotating impeller spins, creating centrifugal force. Axial compressors use a series of rotor and stator blades to move gas parallel to the shaft.

Positive displacement compressors use screws, sliding vanes, lobes, gears, or pistons. They include reciprocating and rotary compressor types. Rotary compressors move gas by rotating a screw, a set of lobes, or a set of vanes. Liquid ring compressors use an eccentric impeller with vanes to transmit centrifugal force into a sealing fluid and drive it against the wall of a cylindrical casing, to form a compression area for the gas.

Compressors can be single-stage or multistage. In a single-stage compressor, the gas enters the compressor and is compressed one time. In a multistage compressor, the gas is compressed multiple times, thereby delivering higher pressures.

Engineers use a compressor curve supplied by the manufacturer to design a compressor system.

Auxiliary equipment associated with compressors includes lubrication systems, seal systems, antisurge devices, intercoolers, aftercoolers, and separators.

Potential problems associated with compressors include overpressurization, overheating, surging, seal oil problems, leaks, lubrication fluid contamination, vibration, malfunction of interlock systems, loss of capacity, motor overload, and high/low flow.

Process technicians are required to be aware of safety and environmental hazards associated with compressors and to follow the standard operating procedures for the unit to ensure the safety of all personnel as well as the equipment. Along with their coworkers and supervisors, process technicians are responsible for preventing equipment failures and injuries.

Checking Your Knowledge

1. Define the following terms:
   a. Axial compressor
   b. Centrifugal compressor
   c. Dynamic compressor
   d. Liquid ring compressor
   e. Multistage compressor
   f. Positive displacement compressor
   g. Reciprocating compressor
   h. Rotary compressor
   i. Single-stage compressor
   j. Cylinder
   k. Interlock
   l. Surging

2. _____ ratio determines the maximum pressure increase that can be expected to be achieved in each stage of compression.

3. Name three reasons why centrifugal compressors are so widely used in the process industries.

4. The type of compressor that uses sliding vanes to force gases out of a chamber is classified as what kind of compressor?
   a. Rotary
   b. Centrifugal
   c. Axial
   d. Dynamic

5. Because of its design, a diaphragm compressor is especially well-suited for use in what types of service? (Select all that apply.)
   a. Toxic gas service
   b. Laboratory applications
   c. Wastewater treatment
   d. Food and beverage industry

6. List three factors that need to be considered when selecting a compressor for a specific service.

7. A lubrication system in a compressor:
a. provides circulation and cooling of sealing and lubrication oils
b. minimizes total leakage
c. increases the pressure on process gases
d. transfers heat to the seals

8. Surging can cause (select all that apply):
a. seal oil contamination
b. vibration
c. destruction of the machinery
d. deadheading conditions

9. (True or False) Interlock systems are used to bypass emergency shutdown systems in order to keep production running so a particular unit can meet its production goals.

10. (True or False) Compressors move liquids, and pumps move gases.

11. (True or False) A dynamic compressor might stall when it is backed up to its surge point on the compressor's performance curve.

12. What type of seal system is designed to restrict flow by requiring the fluid to pass through a series of ridges in an intricate path?
a. Lubrication
b. Labyrinth seal
c. Liquid buffered seal
d. Dry carbon rings

13. List three personal safety hazards associated with compressors.

14. Which of the following is an *equipment* hazard associated with compressors? (Select all that apply.)
a. Discharge of harmful gas into the air
b. Chemical exposure
c. High voltage electricity
d. Inadequate lubrication

15. List four things a process technician can do to monitor compressor equipment and guard against hazards.

NOTE: Answers to Checking Your Knowledge questions are in the Appendix.

Student Activities

1. Given a cutaway or diagram of a reciprocating compressor, identify the following components:
a. Casing
b. Connecting rods
c. Suction valve
d. Discharge valve
e. Piston

2. Using a P&ID provided by your instructor, locate and identify the compressors. Prepare a written report discussing potential problems that can arise from a compressor malfunction and how the system would be affected.

3. Using a P&ID provided by your instructor, locate the compressors. Prepare a presentation about the auxiliary equipment associated with each compressor and the role each plays in the compressor system.

4. In teams, use a P&ID or drawing of a centrifugal compressor provided by your instructor. Identify the antisurge line(s) and flow controllers. Have each team present its design to the class and explain its benefits.

Chapter 8
Turbines

 Objectives

After completing this chapter, you will be able to:

8.1 Identify the purposes, common types, and applications of turbines. (NAPTA Turbines 1-4*) p. 163

8.2 Describe the operating principles of turbines. (NAPTA Turbines 5) p. 172

8.3 Identify potential problems associated with turbines. (NAPTA Turbines 9) p. 174

8.4 Describe safety and environmental hazards associated with turbines. (NAPTA Turbines 6) p. 177

8.5 Describe the process technician's role in turbine operation and maintenance. (NAPTA Turbines 8) p. 177

8.6 Identify typical procedures associated with turbines. (NAPTA Turbines 7) p. 178

*North American Process Technology Alliance (NAPTA) developed curriculum to ensure that Process Technology courses will produce knowledgeable graduates to become entry-level employees in process technology. Objectives from that curriculum are named here in abbreviated form. For example, "(NAPTA Turbines 5)" means that this chapter's objective relates to objective 5 of the NAPTA curriculum about turbines).

Key Terms

Carbon ring—a seal component located around the shaft of the turbine that controls the leakage of motive fluid (typically steam) along the shaft or the entrance of air into the exhaust, **p. 169.**

Casing—a housing component of a turbine that holds all moving parts (including the rotor, bearings, and seals) and is stationary, **p. 167.**

Combustion chamber—a chamber between the compressor and the turbine where the compressed air is mixed with fuel. The fuel is burned, increasing the temperature and pressure of the combustion gases, **p. 170.**

Condensing steam turbine—a device in which exhaust steam is condensed in a surface condenser. The condensate is then recycled to the boiler, **p. 165.**

Fixed blade—a blade inside a turbine, attached to the casing, that determines the direction of the power fluid, **p. 168.**

Gas turbine—a device that uses the combustion of natural gas to spin the turbine rotor, **p. 170.**

Governor—a device used to control the speed of a piece of equipment, such as a turbine, **p. 168.**

Hydraulic turbine—a device that is operated or affected by a liquid (e.g., a water wheel), **p. 171.**

Impulse movement—motion that occurs when fluid is directed against turbine blades, causing the rotor to turn, **p. 166.**

Impulse turbine—a device that uses a directed flow of high velocity fluid against rotor blades to move the rotor; an example of Newton's second law of motion, **p. 165.**

Mechanical energy—energy of motion that is used to perform work, **p. 165.**

Motive fluid—a fluid whose pressure is used to drive a piece of equipment, **p. 163.**

Multistage turbine—a device that contains two or more stages (sets of blades), used as a driver for high-differential pressure, high-horsepower applications, and extreme rotational requirements, **p. 167.**

Noncondensing steam turbine—a turbine that exhausts steam at or above atmospheric pressure, allowing it to be further used in other equipment; also called a *back pressure turbine*, **p. 166.**

Nozzle—component used to drive the blades or rotor of the turbine; orients and converts the steam or motive fluid from pressure to velocity, **p. 168.**

Overspeed trip mechanism—an automatic safety device designed to remove power from a rotating machine if it reaches a preset trip speed, **p. 169.**

Radial bearing—a type of bearing designed to support and hold the rotor in place while offering minimum resistance to free rotation; prevents and offsets horizontal and vertical radial movement. Also called a *journal bearing*, **p. 169.**

Reaction turbine—a device that uses a reactive force of pressure or mass of the motive fluid to turn a rotor; this is an example of Newton's third law of motion, **p. 164.**

Rotating blades—components attached to the shaft of the turbine. The motive fluid causes the turbine to spin by impinging on the rotating blades, **p. 166.**

Rotor—a rotating assembly of a turbine comprised of the shaft and the rotating blades, **p. 168.**

Sentinel valve—a spring-loaded, small relief valve that provides an audible warning (high-pitched whistle) when the turbine nears its maximum operating pressure conditions, **p. 174.**

Shaft—a metal rotating component (spindle) that holds the blades and all rotating equipment in place, **p. 168.**

Single-stage turbine—a device that contains a set of nozzles or stationary blades and moving blades, called a *stage*, **p. 166.**

Steam chest—area where steam enters the turbine, **p. 168.**

Steam strainer—a mechanical device that removes impurities from the steam, **p. 168.**

Steam turbine—a device that is driven by the pressure of high-velocity steam, **p. 165.**

Thrust (axial) bearing—a type of bearing designed to support and hold the rotor in place while offering minimum resistance to free rotation; prevents and offsets back-and-forth axial movement, **p. 169.**

Trip and throttle valve—a valve designed to constrict (throttle) the inlet steam to control the turbine speed and to shut down the turbine in the event of excess rotational speed or vibration, **p. 168.**

Wind turbine—a device that converts wind energy into mechanical or electrical energy, **p. 171.**

8.1 Introduction

Turbines are used in the process industries to convert the kinetic and potential energy of a **motive fluid** (fluid used to drive a piece of equipment) into the mechanical energy necessary to drive a piece of equipment such as a pump, compressor, or generator.

Motive fluid a fluid whose pressure is used to drive a piece of equipment.

While the different turbine types vary in how they work, all of them use the same basic principles. That is, they use a fluid force to turn a rotor. The rotor, in turn, powers process devices that are connected to the turbine shaft.

Common Types and Applications of Turbines

The four main types of turbines are steam, gas, hydraulic, and wind. Turbines are classified according to how they operate and the fluid that turns them (i.e., steam, gas, liquid, or air). The two most common turbine types used in industry are steam turbines and gas turbines, which are the primary focus of this chapter.

All turbines fall into one of two classes: reaction turbines or impulse turbines. In modern day turbines, it is not unusual for a turbine to utilize both the impulse and reactive principles in the same machine. The primary differences between them are the type of blading and the way energy is extracted as the motive fluid moves through the turbine blades. Steam and hydraulic turbines can be either the reaction or impulse type. In both turbine designs, the force causes the rotor blades to turn, thereby rotating the shaft and any equipment coupled to the shaft.

Reaction Turbines

Reaction turbine a device that uses a reactive force of pressure or mass of the motive fluid to turn a rotor; this is an example of Newton's third law of motion.

The operating principle of a **reaction turbine** is that the force generated to turn the rotor is a result of a reactive force of pressure or mass of the motive fluid, not of direct force on the rotor blades. Fixed blades attached to the casing direct the motive fluid onto the moving blades and create the reactive force required to turn the rotor. A reaction turbine is an example of Newton's third law of motion, which states that for every action there is an equal and opposite reaction. Typically, the reaction design is used on lower pressure fluids.

An example of a simple reaction turbine is a Kaplan turbine, which would be used when the motive fluid flow volume is high but the suction head is low. Figure 8.1 shows an example of a reaction turbine.

Figure 8.1 A Kaplan turbine, which is a reaction type hydraulic turbine.
CREDIT: Fouad Saad / Shutterstock.

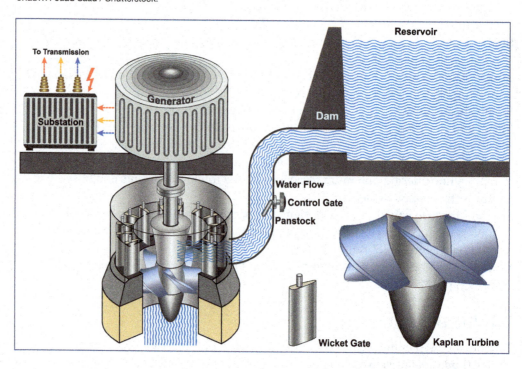

Impulse Turbines

Like a reaction turbine, an **impulse turbine** uses the motive fluid to move a rotor. However, in an impulse turbine, the rotor is acted upon by a focused source, not a reactionary force. Typically, the impulse design is used on higher pressure fluids. Figure 8.2 shows an example of an impulse turbine, called a Pelton wheel turbine.

A Pelton wheel is an impulse turbine that could be used in a hydroelectric dam. The Pelton wheel is used when the flow volume is low but the hydraulic head is high.

Impulse turbine a device that uses directed flow of high velocity fluid against rotor blades to move the rotor; an example of Newton's second law of motion.

Figure 8.2 A Pelton turbine, which is an impulse turbine.

CREDIT: M. Fuksa / Shutterstock.

Steam Turbines

A **steam turbine** is driven by the pressure of high-velocity steam discharged against the turbine's rotor. **Mechanical energy** is the energy of motion that is used to perform work. Steam turbines use the temperature and pressure energy of steam to turn a rotor and produce mechanical energy.

Steam turbines are the drivers for a variety of equipment such as pumps, generators, and compressors. Steam turbines have many advantages over electrical equipment. They are free from spark hazards and therefore are useful in areas where volatile substances are produced. They do not require electricity to run, so they can maintain processes during power outages. They also are suitable for damp environments that might cause electrical equipment to fail.

The rotating blades in a turbine are attached to a wheel. The combination of blades, wheel, and shaft is called the *rotor assembly*. As the steam progresses through the rotor assembly, the kinetic and potential energy of the steam is transformed to mechanical energy.

As stated earlier, steam turbines can be classified as impulse, reaction, or a combination of both types. It is common for larger multistage steam turbines to use a combination of impulse and reaction design. The high-pressure section is typically an impulse type, and the lower-pressure section is a reaction type. Steam turbines are also classified as condensing or noncondensing and as single-stage or multistage.

CONDENSING VERSUS NONCONDENSING TURBINES   A **condensing steam turbine** is a device in which exhaust steam is condensed in a surface condenser, and the condensate is then recycled to the boiler. The surface condenser operates at a vacuum because of the collapsing of steam to condensate and the removal of noncondensables via steam ejectors. A condensing turbine extracts the maximum amount of energy from the steam and is the type most often seen in electric power generation plants.

Steam turbine a device that is driven by the pressure of high-velocity steam.

Mechanical energy energy of motion that is used to perform work.

Condensing steam turbine a device in which exhaust steam is condensed in a surface condenser. The condensate is then recycled to the boiler.

Noncondensing steam turbine a turbine that exhausts steam at or above atmospheric pressure, allowing it to be further used in other equipment; also called a *back pressure turbine*.

A **noncondensing steam turbine**, also called a *back-pressure turbine*, functions like a pressure-reducing valve. Energy from the high-pressure steam is used to turn the turbine's rotor. Steam exits the turbine at a lower pressure, but maintains much of its heat energy and is used in other processes. The noncondensing turbine is the type most often seen in refineries and other plants where there is a need for large volumes of low pressure process steam.

Figure 8.3 illustrates an impulse noncondensing steam turbine.

Figure 8.3 Cutaway of an impulse noncondensing steam turbine.

CREDIT: Courtesy of Bayport Technical.

As steam enters an impulse turbine, it passes through a nozzle (see Figure 8.4). The nozzle orients the flow of the steam to strike the blades, causing rotation.

Figure 8.4 Impulse steam turbine nozzles.

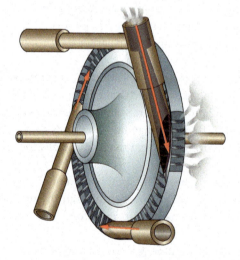

Single-stage turbine a device that contains a set of nozzles or stationary blades and moving blades, called a *stage*.

Rotating blades components attached to the shaft of the turbine. The motive fluid causes the turbine to spin by impinging on the rotating blades.

Impulse movement motion that occurs when fluid is directed against turbine blades, causing the rotor to turn.

SINGLE-STAGE VERSUS MULTISTAGE TURBINES A **single-stage turbine** is a device that contains a set of nozzles or stationary blades and moving blades, called a *stage*. In a reaction turbine, each set of stationary and rotating blades is a stage. **Rotating blades** attached to the shaft of the turbine are struck by the motive fluid, causing the turbine to spin. The motive fluid slows and is redirected as it transfers energy to the rotating blades.

In an impulse turbine, the nozzles direct the steam to the V-shaped rotor blades (called buckets or chevrons). In the Curtis wheel turbine, a stage has two rows of moving blades alternating with a row of fixed blades. Nozzles direct the steam to the first row of rotor blades, then the fixed blades, and finally to the second row of rotor blades. This causes **impulse movement** (movement that occurs when the steam first hits a rotor and the rotor begins to move). Figure 8.3 showed an example of a single-stage impulse turbine.

A **multistage turbine** is a device that contains multiple sets of blades, or stages. This type of turbine is typically used as a driver for high-differential pressure, high-horsepower applications, and extreme rotational requirements. Each stage in a multistage turbine provides additional energy to the rotor. Figure 8.5 provides an example of a multistage turbine.

Multistage turbine a device that contains two or more stages (sets of blades), used as a driver for high-differential pressure, high-horsepower applications, and extreme rotational requirements.

Figure 8.5 Multistage turbine.
CREDIT: arogant / Shutterstock.

In a multistage turbine, when one turbine stage turns, they all turn, because they are all fixed on the same shaft. As the steam moves down the line in a multistage turbine, the amount of steam energy decreases. This pressure differential is what helps provide motive force. (Note: The greater the pressure differential, the more pressure or force exists.)

STEAM TURBINE COMPONENTS   Turbine components vary depending on the type of turbine and its application. The most common components are turbine wheels, seals, bearings, steam inlets and outlets, a throttle valve, and a governor. Figure 8.6 shows steam turbine components.

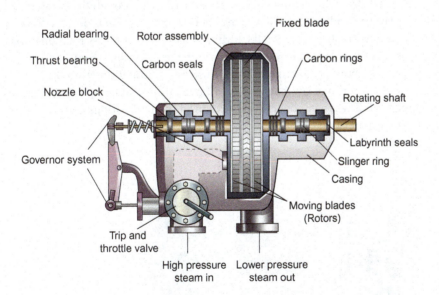

Figure 8.6 Steam turbine components.

The **casing** is the housing around the internal components of a turbine. The casing is stationary and holds all moving parts, including the rotor, bearings, and seals. Some turbine designs use split casings (top and bottom sections are bolted together) for ease of maintenance of the rotor, seal, and bearings. Figure 8.7 shows a turbine with a split casing design.

Casing a housing component of a turbine that holds all moving parts (including the rotor, bearings, and seals) and is stationary.

Figure 8.7 Turbine split casing covering the rotors.

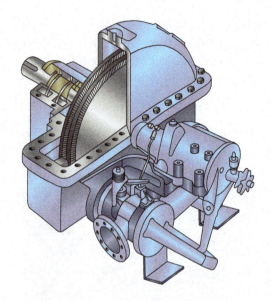

Steam chest   area where steam enters the turbine.

Steam strainer   a mechanical device that removes impurities from the steam.

Shaft   a metal rotating component (spindle) that holds the blades and all rotating equipment in place.

Fixed blade   a blade inside a turbine, attached to the casing, that determines the direction of the power fluid.

Rotor   a rotating assembly of a turbine that is comprised of the shaft and rotating blades.

Governor   a device used to control the speed of a piece of equipment, such as a turbine.

Nozzle   the component used to drive the blades or rotor of the turbine; orients and converts the steam or motive fluid from pressure to velocity.

Trip and throttle valve   a valve designed to constrict (throttle) the inlet steam to control the turbine speed and to shut down the turbine in the event of excess rotational speed or vibration.

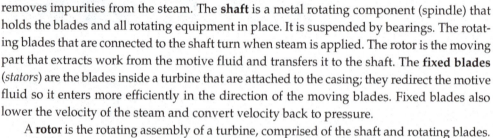

The **steam chest** is the area in a turbine where steam enters, and the **steam strainer** removes impurities from the steam. The **shaft** is a metal rotating component (spindle) that holds the blades and all rotating equipment in place. It is suspended by bearings. The rotating blades that are connected to the shaft turn when steam is applied. The rotor is the moving part that extracts work from the motive fluid and transfers it to the shaft. The **fixed blades** (*stators*) are the blades inside a turbine that are attached to the casing; they redirect the motive fluid so it enters more efficiently in the direction of the moving blades. Fixed blades also lower the velocity of the steam and convert velocity back to pressure.

A **rotor** is the rotating assembly of a turbine, comprised of the shaft and rotating blades. The **governor** is a device used to control the speed of the turbine. It can also be used to restrict steam flow as directed by the overspeed trip mechanism. A **nozzle** is a device that is designed to increase the velocity of steam. This restrictive component is used to drive the rotor of a turbine. The nozzle converts the steam or motive fluid from pressure to velocity. Steam enters the nozzle from the steam inlet and exits the turbine through the steam outlet.

The **trip and throttle valve** is the inlet steam valve that is designed to be used to throttle the turbine and to shut down the turbine in the event of excess rotational speed or vibration. This type of valve works in conjunction with the governor to control the speed of the turbine. Figure 8.8 shows a governor assembly used to control turbine speed.

Steam turbines are complex and have additional components that are not listed in Figure 8.6. These components can include additional bearings and seals, oil rings, and overspeed trip devices.

Figure 8.8 Governor assembly used to control turbine speed.

CREDIT: Courtesy of Eastman Chemical.

Thrust bearings, also called axial bearings, are designed to support and hold the rotor in place while offering minimum resistance to free rotation. These bearings prevent and offset axial (back and forth) movement of the rotor. **Radial bearings**, also called journal bearings, serve the same purpose, but protect against horizontal and vertical radial movement.

Carbon rings are components of seals that are located around the shaft of the turbine and that control the leakage of motive fluid (typically steam) along the shaft or the entrance of air into the exhaust. With *nonrotating carbon rings*, sealing occurs when the rings are forced against the shaft, minimizing leakage. Nonrotating carbon rings are kept from rotating by a stop on each ring.

Seals are mechanical devices that hold lubricants and process fluids in place while keeping out foreign materials. In a turbine, seals keep fluids from leaking out past the rotor shaft at the point where it exits the case. Figure 8.9 shows the top view of the carbon seals that are used to prevent steam from leaking out of the turbine casings.

The shaft sealing system keeps the high-pressure steam in the proper passages, preventing excessive leakage along the rotating and nonrotating boundaries. Typical seal designs are labyrinth and nonrotating carbon rings.

Labyrinth seals (shown in Figure 8.10) are close-fitting discs that can be rotating (mounted on the shaft), nonrotating (mounted on the stator), or an alternating pattern of rotating and nonrotating. Labyrinths consist of multiple rings that create a long pathway through which steam does not escape. The labyrinth seals are located along the turbine shaft to minimize leaks. Condensing steam turbines have gland sealing steam applied to the labyrinth seals. This steam seals the drive shaft as it exits the turbine case. The gland sealing steam is then condensed in a gland sealing steam condenser. Because the low pressure stage outlet of a condensing turbine operates under vacuum pressure, the gland sealing steam will stop air (oxygen) from leaking into the turbine. Gland sealing steam systems are not needed on noncondensing turbines because they do not operate under vacuum pressure.

Thrust (axial) bearing a type of bearing designed to support and hold the rotor in place while offering minimum resistance to free rotation; prevents and offsets back-and-forth axial movement.

Radial (journal) bearing a type of bearing designed to support and hold the rotor in place while offering minimum resistance to free rotation; prevents and offsets horizontal and vertical radial movement.

Carbon ring a seal component located around the shaft of the turbine that controls the leakage of motive fluid (typically steam) along the shaft or the entrance of air into the exhaust.

Figure 8.9 View of carbon seals.

Carbon seals

Figure 8.10 Example of a labyrinth seal used to keep fluids inside the turbine.

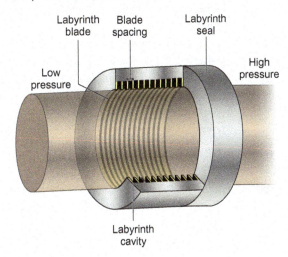

Labyrinth blade
Blade spacing
Labyrinth seal
Low pressure
High pressure
Labyrinth cavity

Sometimes buffer fluids, such as nitrogen gas, are used to regulate temperature at the seal and to exclude motive fluids from oil systems. Some leakage of the buffer gas is allowed in order to prevent leakage of the motive fluid into the system.

An **overspeed trip mechanism** is an automatic safety device designed to remove power from a rotating machine if it reaches a preset speed. The overspeed trip is the last-resort stop

Overspeed trip mechanism an automatic safety device designed to remove power from a rotating machine if it reaches a preset trip speed.

valve for inlet fluid to prevent damage to the turbine caused by excessive rotating speed. The overspeed trip protects the turbine when the governor fails to maintain the correct speed.

Gas Turbines

A **gas turbine** uses the combustion of natural gas to spin the turbine rotor. Figure 8.11 shows a drawing of a gas turbine with labels showing the internal components.

Gas turbine a device that uses the combustion of natural gas to spin the turbine rotor.

Figure 8.11 Gas turbine with internal components labeled.

CREDIT: Fouad A. Saad/Shutterstock.

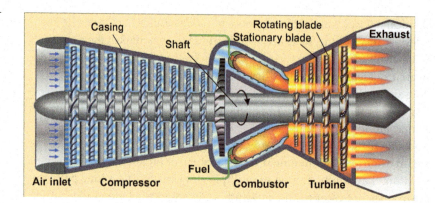

Gas turbines are more efficient, but typically more complex, than steam turbines. In a gas turbine, an electric motor is used to get the air compressor up to speed and start the machine. Once the turbine is started, it provides sufficient power to drive the integral air compressor and the connected process equipment (e.g., pumps, compressors, and generators). To conserve energy, the hot exhaust gases produced by the gas turbine can be used as a heat source for other parts of the process.

The axial compressor in a gas turbine pressurizes air for entry into the combustion chamber. All gas turbines have an axial compressor. In some cases, as the air is compressed through several stages, it is cooled to reduce the volume to be compressed in later stages. Heat exchangers called *intercoolers* are used between stages to cool gases before they reach the next stage.

Combustion chamber a chamber between the compressor and the turbine where the compressed air is mixed with fuel. The fuel is burned, increasing the temperature and pressure of the combustion gases.

The **combustion chamber**, located between the compressor and the turbine, is the area where compressed air is mixed with fuel. This fuel is burned, increasing the temperature and pressure of the air/fuel gas mixture. These hot gases flow into the turbine, where energy is extracted. This energy is used to drive the air compressor. Energy not required to drive the compressor can be used for powering equipment like ships or trains, or pumps and compressors in a plant. Gas turbines are also used to provide power for offshore oil and gas platforms. When connected to a generator, gas turbines are used in power plants to produce electricity.

MAJOR COMPONENTS OF COMBUSTION GAS TURBINES The major components of a combustion gas turbine are displayed in Figure 8.11. A combustion gas turbine has many components:

- An air inlet
- An axial air compressor to increase the pressure of the combustion air
- A combustion chamber, located between the compressor and the turbine, where compressed air is mixed with fuel and the fuel is burned, increasing the temperature and pressure of the combustion gases
- A turbine which powers the axial compressor and other connected equipment
- Exhaust flue ducting

Hydraulic Turbines

A **hydraulic turbine** is a turbine that is moved, operated, or affected by a liquid. In a hydraulic turbine, liquid flows across the rotor blades, forcing them to move. The faster the liquid flows, the faster the blades turn. An example is a conventional hydroelectric dam and hydraulic turbine, as shown in Figure 8.12.

The Francis turbine is a type of water turbine that combines radial and axial flow principles. It is the most common water turbine in use today and is primarily used for electrical power production. The Hoover Dam across the Colorado River in the southwestern United States, for example, was constructed using Francis turbines.

Hydraulic turbines also can be used to capture the energy of a high-pressure liquid whose pressure is being reduced. The energy recovered is used to do work, such as powering a pump.

Hydraulic turbine   a device that is operated or affected by a liquid (e.g., a water wheel).

Figure 8.12 A. Diagram of hydroelectric dam and hydraulic turbine. **B.** Francis hydraulic turbines at Hoover Dam.
CREDIT: **B.** CrackerClips/fotolia.

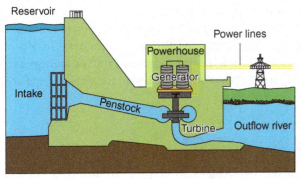

A.

B.

Wind Turbines

A **wind turbine** is a mechanical device that converts wind energy into mechanical or electrical energy. Windmills, such as those shown in Figure 8.13, are examples of wind turbines.

Wind turbine   a device that converts wind energy into mechanical or electrical energy.

Figure 8.13 A. Wind turbines. **B.** Wind turbine open nacelle showing generator.
CREDIT: **A.** Mimadeo / Shutterstock. **B.** Teun van den Dries / Shutterstock.

A.

B.

In a wind turbine, the air current moves across fanlike blades, causing them to turn. As they turn, a shaft is rotated. The rotation of the shaft turns a generator, which produces electrical power.

Auxiliary Equipment Associated With Turbines

Utilities and auxiliary equipment associated with turbine operation include lubrication systems and seals (described above).

LUBRICATION SYSTEM The lubrication system maintains lubrication and the temperature of turbine bearings. The lubrication system usually operates independently of the turbine, but in some systems, the oil pump is driven by the turbine's shaft.

The lubrication system consists of a pump, oil filters, oil coolers, supply and return lines, and an oil reservoir. The lubrication system also contains an emergency pressure tank so that oil pressure can be maintained temporarily if the oil pump stops.

To prevent damage in larger machines, oil flow must be established before shaft rotation begins and continue until after the shaft rotation stops. Because lubrication is so critical, many large machines are equipped with interlock devices to prevent turbine startup until the lubrication system is in service. Figure 8.14 shows a lube oil skid for a gas turbine and gas compressor on an offshore oil and gas platform.

Figure 8.14 Lube oil system skid for a gas compressor and gas turbine on an offshore oil and gas platform.

CREDIT: Gas and Oil Photographer/ Shutterstock.

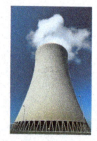
8.2 Operating Principles
Steam Turbines

Most steam turbines are designed to use only dry, superheated steam. Any water droplets in the incoming steam can result in severe vibration problems and can damage the turbine. Before starting up a steam turbine, the inlet steam line and turbine casing must be heated and free of any water. The steam must also be free of contaminants, and the supply pressure should be reasonably constant so that governor operation is stable.

After the steam has flowed through the turbine, it moves toward the outlet. The outlet flow is the point at which the steam or fluid pressure leaves the turbine after it is used and its energy has been converted into mechanical energy.

WHAT HAPPENS INSIDE A TURBINE?   A turbine operates because steam, gas, liquid, or air turns blades that turn the shaft. Kinetic (velocity) energy is gained when the fluid goes through the turbine's nozzles. As the fluid's kinetic energy moves through the turbine, it is transferred to the blades of the turbine, and it loses both pressure and temperature. The turbine converts that energy into the mechanical energy required to do work (e.g., power a generator or other process equipment).

CONTROL SYSTEMS   The principal control systems inside a turbine include the governor, the extraction/induction system, the oil system, and the safety/trip system. The safety/trip system consists of all sensors (e.g., high vibration, low oil pressure, overspeed, etc.) that initiate turbine shutdown, plus a logic controller that controls oil pressure at the trip and throttle valve. Removing oil pressure from this valve closes off the steam supply to the machine.

RADIAL AND AXIAL SHAFT MOVEMENT   Radial movement of the shaft is the vertical or horizontal movement that occurs when the rotor turns. The use of radial bearings helps minimize the vibration that radial movement can cause. Thrust is the axial, back-and-forth movement that is parallel to the shaft. The use of thrust bearings helps minimize the vibration that thrust can cause. Figure 8.15 shows examples of radial and axial movements.

Shaft misalignment can cause damage to shafts, bearings, turbine wheels, diaphragms, or casings. In large turbines, sensors are usually installed to continuously monitor vibration and shaft displacement so shutdown occurs before the machine is damaged. Shaft misalignment occurs when the shaft is not in the proper position, radially, axially, or angularly.

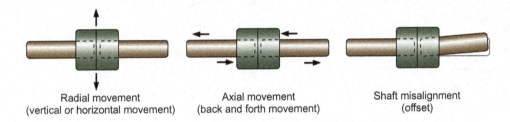

Radial movement
(vertical or horizontal movement)

Axial movement
(back and forth movement)

Shaft misalignment
(offset)

Figure 8.15 Illustration showing shaft movement and angular shaft misalignment.

GOVERNOR/SPEED CONTROL   Turbine operation is maintained by the governor system, which is a closed-loop controller. A closed-loop controller measures variables, compares them to the setpoint, and adjusts performance. In this case, the setpoint is the revolutions per minute (rpm) of the machine. The input sensor of the governor is typically an electronic shaft revolution counter that produces a millivolt or digital output signal proportional to the rpm. The governor system can also be a mechanical fly ball device (shown in Figure 8.16).

Figure 8.16 Turbine showing the fly ball governor.

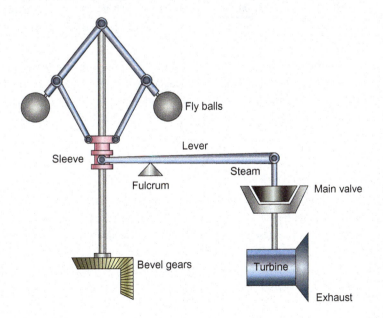

Fly balls

Lever

Sleeve

Fulcrum

Steam

Main valve

Bevel gears

Turbine

Exhaust

The output of the governor system is important to the position of the control rod (the link between the governor and the throttle valve). When the governor receives a signal to open or close the throttle valve, it adjusts the control rod to change the speed. The speed control in a car functions in a similar way. The accelerator (gas pedal) is linked to the throttle on the carburetor or fuel injector. The more the accelerator is applied, the more fuel is fed to the carburetor or fuel injector, and the faster the car runs.

The speed setpoint can be selected either by the process technician or by a master controller. As the amount of power demanded by the driven unit changes, the governor opens or closes the steam valve to maintain the required speed.

The controller portion of the governor system is either a mechanical link or an electronic, analog, or digital controller. The mechanical link arrangement operates the shaft of the governor valve. The **sentinel valve** is a spring-loaded, small relief valve that opens to provide an audible warning (high-pitched whistle) when the turbine nears its maximum operating pressure conditions.

Sentinel valve a spring-loaded, small relief valve that provides an audible warning (high-pitched whistle) when the turbine nears its maximum operating pressure conditions.

Combustion Gas Turbines

During the initial startup of a gas turbine, the compressor is started by an external starting mechanism, typically hydraulic, pneumatic, or electric. When the compressor starts, it delivers pressurized air to the combustion chamber where fuel and air are ignited. This ignition produces expanding combustion gases that drive the turbine. After the turbine is operational, the external starting mechanism can be disconnected from the compressor. This thermodynamic operating principle is called the Brayton cycle. Figure 8.11 showed the gas turbine cycle.

In the first stage of the Brayton cycle (see Table 8.1), ambient air is drawn into the compressor and pressurized. In the second stage, the compressed air enters the combustion chamber, where fuel is burned, and the compressed air is heated further. In the third stage, the high-pressure, high-temperature gas applies force to the turbine blades, which does work by turning the shaft.

Table 8.1 The Stages of a Combustion Gas Turbine (see Figure 8.11)

Cycle Name	Brayton Cycle; Open Cycle
Compression	Ambient air is drawn into the compressor, where it is pressurized
Heat addition	The compressed air then runs through a combustion chamber, where fuel is burned, heating that air; it is a constant-pressure process because the chamber is open to flow in and out
Expansion	The heated, pressurized air then gives up its energy, expanding through a turbine (or series of turbines). Some of the work extracted by the turbine is used to drive the compressor
Heat rejection	Heat rejection (in the atmosphere); heat recovery can take place at this point

In the final stage, gas exits at a reduced pressure and temperature, through the exhaust and heat recovery often takes place. An example of a heat recovery system is a waste heat boiler. A waste heat boiler is a heat exchanger that uses turbine exhaust gases to generate steam. This is called a combined cycle and it greatly improves efficiency.

8.3 Potential Problems

When working with turbines, process technicians should always be aware of potential problems such as turbine failure, loss of lubrication, bearing damage, excessive vibration, impingement, issues with the gland sealing steam, fouling, corrosion, erosion, pittting, assembly errors, turbine overspeeding hunting, and overpressurization. People are safest when equipment is running properly, so identifying and repairing these potential problems as soon as possible is important for everyone's safety.

Turbine Failure

If turbines are the prime movers of large and/or single-train machinery, turbine failure will cause a shutdown of a processing unit. This can have catastrophic consequences.

In the case of gas turbines and cogeneration facilities, the loss of a turbine and generator set does not cause widespread process unit outage. When electrical loss occurs, the backup utility power compensates for the loss of power.

TURBINE TRIPS   A turbine trip occurs when a turbine shuts down because of abnormal conditions. Large drivers typically have several shutdown activation sensors. The reasons for shutdown can include high speed, loss of oil pressure, excessive vibration, or other conditions. Some of these sensors reset to a "permit" state after re-establishing safe conditions. Others, however, might require a manual reset before the "run permit" state is activated. Documentation and training on the operation of shutdown interlocks must be thoroughly understood and followed to avoid delays in restarting the equipment. For example, process technicians should understand what interlock protection trips the driver, the reasons for this protection, and the recovery procedures from the tripped state. Recovery procedures are specific to pieces of equipment and the process facility. Process technicians should review the standard operating procedures for the piece of equipment before attempting any restart.

Loss of Lubrication

Loss of lubrication can result from oil pump failure or plugging of the oil filter. Larger steam turbine trains will have a low pressure lube oil trip to shut down the equipment when there is a loss of lubrication. Turbine bearing failure can result from dirt in the oil or a blockage in the oil supply line, so filtration must be capable of removing small contaminant particles.

When maintaining and lubricating turbines, process technicians should always use the correct oil as specified by the manufacturer or the standard operating procedures. (Note: Many facilities attach a label to the machine designating the correct lubricant. This reduces the chance of using an incorrect lubricant.) Using an incorrect lubricant (e.g., one that is too thick or too thin) can damage the equipment and create potentially hazardous situations such as reactivity with bearing and seal materials or process fluids.

Bearing Damage

Bearing damage can result from excessive temperatures and vibration caused by shaft bowing or other shaft misalignment problems. Process technicians should pay close attention and report excessive vibrations immediately so the condition can be eliminated before major damage or injuries occur. Furthermore, process technicians should not depend entirely on vibration interlocks and detectors because these devices sometimes fail.

Impingement

Impingement occurs when the steam quality decreases to the point that steam condensate exists in the turbine. Water droplets impinging on the turbine blades contribute to wear and erosion of the blades. To prevent impingement, the steam entering the turbine must be adequately superheated to avoid condensation.

Gland Sealing Steam

If the gland sealing steam and gland sealing steam condenser are not properly aligned in condensing steam turbines, a loss of vacuum could result in poor operation of the turbine.

Fouling, Corrosion, Erosion, and Pitting

Turbine *fouling* is caused by mineral deposits from the steam. The solution for fouling is a wheel wash or hot water wash to break down and carry off mineral deposits. Mineral deposits resulting from poor boiler feedwater quality can cause both the governor stem and the trip valve stem to stick when there is a loss of load. To prevent this, process technicians should always follow proper preventive maintenance procedures.

Many corrosion and erosion problems occur as a result of damage that takes place when the turbine is not running. To help prevent this, standby turbines should be kept free of moisture and other contaminants.

Pitting is a form of localized corrosion that creates small holes in metal components. It is often seen in out-of-service equipment. When steam leaks into and condenses inside the turbine, salts from the boiler water collect on surfaces and cause pitting. The solution for pitting is to ensure that the steam supply has double-block and bleed valves (see Chapter 5) to prevent steam leaks into the turbine.

Assembly Errors

Excessive wear, turbine failure, or turbine loss also can be caused by assembly error. Process technicians should pay close attention to the assembly and maintenance of the turbine and its lubrication system.

Turbine Overspeeding

Overspeeding of a turbine can lead to catastrophic failure of the machine. Destructive overspeeding can result from coupling failure, gearbox failure, or loss of load when the downstream equipment powered by the turbine fails. In such cases, the overspeed trip should shut down the turbine. If trip protection does not respond quickly enough or if it fails to respond at all, the equipment could be damaged.

Hunting

Hunting is a condition in which a control loop is improperly designed, installed, or calibrated; it may also occur when the controller itself is not aligned properly. Process technicians will observe this condition as a random behavior or cycling above and/or below the speed *setpoint* (the desired turbine rpm speed control point). Figure 8.17 shows an example of hunting.

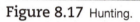

Figure 8.17 Hunting.

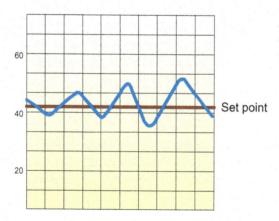

Overpressurization

Sentinel valves are safety valves used to warn process technicians of equipment overpressurization. These types of valves alert the process technician with a loud whistle to take corrective actions and thus prevent turbine shutdown or relief-valve activation.

8.4 Safety and Environmental Hazards

Hazards associated with normal and abnormal turbine operations can affect personal safety, process safety, equipment, production, and the environment. Table 8.2 lists several potential hazards and their effects.

Table 8.2 Hazards Associated with Improper Steam Turbine Operation

Improper Operation	Possible Effects			
	Individual	Equipment	Production	Environment
Uninsulated equipment and piping	Burns caused by exposure to heat or steam			Heat loss to the environment
Allowing hot steam to enter the turbine without following the proper warm-up procedure		Equipment damage from thermal shock	Downtime because of repair	
Running a turbine at a speed above or below the normal operating conditions	Personnel injuries from turbine failure	Hunting Turbine shutdown	Downtime, off-spec product	Releases to the environment
Failure to lubricate the linkage between the governor and the governor valve		Hunting Turbine shutdown	Loss of production	
Failure to maintain sufficient inlet steam pressure		Hunting Turbine shutdown	Reduced production from insufficient power	
Allowing "wet steam" or condensate into the turbine	Personnel injuries from turbine failure	Turbine shutdown	Loss of production, repair costs	Releases to the environment
High-discharge steam pressure		Turbine slowdown Relief valve popping	Loss of production	Noise hazard
Steam leaks	Burn injuries; wet, slippery surfaces	Turbine slowdown	Loss of production	Noise hazard
Improper lubrication	Personnel injuries from turbine failure	Severe damage or machine failure	Loss of production	Releases to environment Fire

8.5 Process Technician's Role in Operation and Maintenance

The process technician plays an important role in the safe operation and maintenance of turbines. Monitoring the lube oil levels and temperatures is important. Process technicians need to be alert to abnormal sounds or vibrations that can signal steam leaks or bearing problems, and they must listen to the turbine to ensure that there is no rubbing or dragging inside the housing.

Process technicians also should monitor for thermal stresses, which can cause turbine parts to warp. Warped parts can wear against each other, leading to friction and localized temperature increases. Warping can also result in excessive vibration, ruining the bearings and forcing the turbine to shut down.

Additional responsibilities of the process technician include watching for and reporting excessive vibration, surging, hunting, oil leaks, steam venting from the oil reservoir, or any other abnormal steam emissions.

It is important to check for adequate cooling of bearings and sudden changes in pressure, which could indicate that steam nozzles are plugged. Checking the lubrication system for corrosion is also important. Process technicians also should monitor:

- Inlet and exhaust steam temperatures (which could indicate dirty turbine blades) and pressures
- Oil levels, temperatures, flows, discoloration, foaming, filter pressure drops

- Speed governor lubrication
- Safety interlocks and devices.

Monitoring

Process technicians routinely monitor turbines to ensure they are operating properly. When monitoring and maintaining steam turbines, process technicians must always remember to look, listen, and check for the items listed in Table 8.3.

Process technicians must also monitor oil pressure and quality. High bearing temperatures and high turbine operating speeds, when combined with the steam used in the process, can elevate temperatures even higher and cause the oil to break down. High-quality oil reduces the risk of parts fusing together, which can cause the turbine to fail.

Vibration must also be monitored on a regular basis. Vibrations are indicative of an internal turbine problem or a problem with the turbine shaft. Vibration sensors are installed on many large turbines to detect vibrations. Process technicians, however, should still visually check the sensor readout and the turbine itself to ensure that there is no vibration.

Process technicians must also monitor speed, auxiliaries, and guards to ensure proper operation. Failure to perform proper maintenance and monitoring could affect the process and result in equipment damage or injury.

Table 8.3 Process Technician's Role in Operation and Maintenance

Look	Listen	Check
■ Look at oil levels to make sure they are satisfactory and that oil color is normal ■ Monitor cooling water or air flow to oil coolers, bearings, and seals ■ Check cooling water flow, pressure, and temperature to and from intercoolers ■ Look at seals and flanges to make sure there are no leaks ■ Look for water or foaming in the oil reservoir ■ Check for proper flow rates, steam pressure, and temperature throughout ■ Observe governor for proper operation	■ Listen for abnormal noises	■ Check for excessive vibration ■ Check for excessive heat ■ Check lubricants for grit, and smell them for breakdown odor

Routine and Preventive Maintenance

Process technicians should be familiar with routine and preventive maintenance procedures to ensure continued operation. For example, process technicians might need to check over-speed trip devices to ensure that the safety system is operating correctly.

Startup and shutdown are required for routine and preventive maintenance as needed within the facility. Turbines should be started and shut down slowly to reduce physical stress (thermal shock and warping of the shafts and bearings). Most facilities have written startup and shutdown procedures. If the turbines are driving complex or critical equipment, the procedures are specific for each piece of equipment.

8.6 Typical Procedures

Process technicians should be familiar with the appropriate procedures for monitoring, maintaining, and operating turbines.

Noncondensing Steam Turbine Startup Procedures

To ensure safety during turbine startup, process technicians must be familiar with proper startup procedures. Before attempting to energize a turbine, the process technician should always refer to the startup procedure for the steam turbines on each unit. Steps in a typical steam turbine startup procedure are:

1. Communicate with affected personnel about the impending turbine startup.
2. Ensure that the driven equipment connected to the steam turbine is ready for startup.
3. Verify the lube oil and cooling water are in service.
4. Open drains from the turbine casing, steam lines, and steam chest.
5. Open the turbine exhaust steam valve.
6. Open the steam inlet valve slowly, to begin slow rolling the turbine.
7. Observe the steam coming out of the drains and, when it becomes dry, close the drain valves.
8. Open the steam inlet valve to adjust turbine speed as needed.

Shutdown Procedures

An example of a steam turbine shutdown procedure is:

1. Switch to the spare turbine.
2. Close the inlet valve to the steam turbine. (Note: The inlet valve is usually closed first to prevent overpressurization of the turbine.)
3. Close the outlet steam valve from the turbine (if required) and open bleed valves on the turbine casing (if required).

Emergency Procedures

Process technicians must be knowledgeable about the emergency procedures relating to turbine operations for their units. Emergency procedures might include steps to address loss of steam, loss of cooling water, loss of lube oil, or mechanical breakdown of the turbine. Process safety incidents or personal injuries could occur if proper emergency procedures are not followed.

Lockout/Tagout

Depending on the type of maintenance required, lockout/tagout of the pump and turbine might be necessary (see Chapter 6). If lockout/tagout is required, make sure all energy-supplying devices are properly locked out and tagged out (includes inlet and outlet steam, oil supply and return, cooling water supply and return, and any other energy sources).

Summary

The basic purpose of a turbine is to convert the kinetic or potential energy of steam, gas, liquid, or wind into the mechanical energy required to drive equipment. Pumps, generators, and compressors are examples of equipment driven by turbines. High-velocity steam, gas, liquid, or air turns the blades that turn the shaft.

Steam turbines are driven by steam discharged at high pressures and temperatures against a set of turbine vanes. Reaction turbines are an example of Newton's third law of motion, which states that for every action, there is an equal but opposite reaction. Impulse turbines use steam to move a rotor directly and demonstrate Newton's second law of motion, which states that the rate of acceleration of an object is proportional to the force acting upon it. In condensing turbines, exhaust steam is condensed in a surface condenser and the condensate returns to the boiler. In a noncondensing turbine, only a portion of the steam energy is used in the turbine and the lower pressure exhaust steam can be used for other process equipment requirements.

Gas turbines produce hot gases in the combustion chamber and direct the gases toward turbine blades, causing the rotor to move.

Hydraulic turbines allow a liquid to flow across the rotor blades, forcing them to move. The faster the liquid flows, the faster the rotor turns.

Wind turbines convert wind energy using air pressure to move a rotor.

Turbine components consist of multiple parts that are similar in many types of turbines. A casing is the housing around the internal parts of the turbine. The shaft is the metal rotating component that holds all the rotating equipment in place. Some turbines have both moving blades and fixed blades. The governor is used to control speed, and the trip & throttle valve is used to regulate steam flow and shut down the turbine at excess rotational speed or vibration.

Most turbines used in the process industries are powered by either steam or gas. Steam turbines are used most often because of their convenient and efficient fit within the facility's steam distribution system.

Utilities and auxiliary equipment associated with turbine operation include lubrication systems. A turbine lubrication system requires the correct weight of high-quality oil to properly maintain turbine bearings, lubrication, and temperatures.

Improper operating conditions can result in severe bodily injury, total mechanical failure, lost product, lost revenue because of downtime, and environmental issues. When turbines are the prime movers of large and/or single-train machinery, turbine failure usually causes a shutdown of a process unit. Loss of lubrication, impingement, fouling, pitting, corrosion, assembly errors, and overspeed can trip or shut down turbine systems.

Bearing damage can result from excessive temperatures and vibration caused by shaft misalignment problems or improper lubrication. Vibration sensors are installed on many large turbines to alert the process technician to the excessive vibration condition so he or she can take appropriate corrective actions.

Process technicians must monitor the unit frequently to detect unusual conditions. Personal protective equipment (PPE) must be worn at all times while on the process units. Routine monitoring includes listening for unusual sounds; watching for excessive vibration; monitoring turbine speed; looking for oil or steam leaks; and maintaining proper oil levels, pressures, and temperatures. Turbines must be started and shut down slowly using the manufacturer's recommended procedures to reduce stress, thermal shock, and warping of shafts and bearings. During startup, the steam lines and equipment must be purged of accumulated condensate and warmed up to prevent water damage, and the machine should be placed on a slow roll to prevent thermal shock.

Process technicians must be familiar with the causes of and solutions for problems such as loss of lubrication, bearing damage, impingement, fouling, pitting or corrosion, assembly errors, and loss of overspeed shutdown systems in order to ensure process safety and personal safety. Performing proper preventive maintenance procedures can prevent or minimize most of these problems.

Checking Your Knowledge

1. Define the following terms:
   a. Gas turbine
   b. Hydraulic turbine
   c. Reaction turbine
   d. Steam turbine
   e. Wind turbine
   f. Impulse turbine
   g. Casing
   h. Governor
   i. Fixed blade
   j. Moving blade
   k. Nozzle
   l. Overspeed trip mechanism
   m. Rotor

2. (True or False) Gas and steam turbines are activated by the expansion of fluid on a series of curved rotor vanes attached to a central shaft.

3. Steam turbines are driven by the pressure of high-velocity _____ discharged at a high pressure and temperature against the turbine vanes.
   a. air
   b. water
   c. gas
   d. steam

4. (True or False) Newton's third law of motion states that for every action, there is a lesser but similar reaction.

5. A(n) _____ turbine is used as a driver for high-differential pressure and high horsepower applications.
   a. condensing
   b. noncondensing
   c. multistage
   d. impulse

6. A(n) _____ is used to control the speed of the turbine as steam is channeled through the restrictive component called the nozzle.
   a. governor
   b. trip and throttle valve
   c. thrust bearing
   d. overspeed trip

7. Inside a gas turbine, hot gases are produced in the _____ and are directed toward the turbine blades, causing the rotor to move.
   a. fuel line
   b. combustion chamber
   c. stage
   d. compressor

8. Turbines can be powered by which of the following? (Select all that apply.)
   a. Steam
   b. Gas
   c. Liquid
   d. Wind

9. Which of the following is true of a noncondensing steam turbine? (Select all that apply.)
   a. It is also called a back-pressure turbine.
   b. The condensate is recycled to the boiler.
   c. It functions like a pressure-reducing valve.
   d. Steam exits the turbine at a lower pressure.

10. (True or False) Labyrinth seals are single-layer seals used to allow small amounts of leakage inside the turbine.

11. _____ valves are safety valves used to warn process technicians about equipment overpressurization.

12. During what stage of a combustion gas turbine can heat recovery take place?
    a. Compression
    b. Heat addition
    c. Expansion
    d. Heat rejection

13. Turbine _____ is caused by mineral deposits from steam.

14. What equipment hazards are associated with a failure to maintain sufficient inlet steam pressure? (Select all that apply.)
    a. Hunting
    b. Machine failure
    c. Relief valve popping
    d. Turbine shutdown

15. List three things a process technician should monitor with regard to turbines.

16. Place the following noncondensing steam turbine startup procedure steps in the correct order.

___ Verify the lube oil and cooling water are in service.

___ Communicate with affected personnel about the impending turbine start up.

___ Open drains from the turbine casing, steam lines, and steam chest.

___ Open the steam inlet valve slowly, to begin slow rolling the turbine.

___ Observe the steam coming out of the drains and when it becomes dry, close the drain valves.

___ Ensure that the driven equipment connected to the steam turbine is ready for startup.

___ Open the steam inlet valve to adjust turbine speed as necessary.

___ Open the turbine exhaust steam valve.

NOTE: Answers to Checking Your Knowledge questions are in the Appendix.

Student Activities

1. Given a cutaway or a diagram of a steam turbine, identify the following components and explain the function of each.

  a. Casing

  b. Shaft

  c. Moving and fixed blades

  d. Governor

  e. Nozzle block

  f. Inlet

  g. Outlet

  h. Trip and throttle valve

2. Describe the theory of operation for a steam and a gas turbine.

3. Prepare a short presentation about how the governor inside a turbine helps control the speed and how it is related to the performance of the equipment.

4. Research and write a report about potential problems associated with turbines. In your report, list possible solutions and how the problems can affect other systems within a process facility.

5. Work with a classmate to write a startup, shutdown, emergency, or lockout/tagout procedure for a turbine.

Chapter 9
Motors and Engines

*North American Process Technology Alliance (NAPTA) developed curriculum to ensure that Process Technology courses will produce knowledgeable graduates to become entry level employees in process technology. Objectives from that curriculum are named here in abbreviated form. For example "(NAPTA Motors & Engines 1)" means that this chapter's objective relates to objective 1 of the NAPTA curriculum about motors and engines).

Key Terms

AC power source—a device that supplies alternating current, **p. 189.**

Bearing—a machine component that rotates, slides, or oscillates. Bearings reduce friction between the motor's rotating and stationary parts, **p. 189.**

Camshaft—a driven shaft fitted with rotating wheels of irregular shape (cams) that open and close the valves in an engine, **p. 190.**

Compression ratio—the ratio of the volume of the cylinder at the start of a compression stroke compared to the smaller (compressed) volume of the cylinder at the end of the stroke, **p. 193.**

Connecting rod—a component that connects a piston to a crankshaft, **p. 190.**

Coolant—a fluid that circulates around or through an engine to remove the heat of combustion. The fluid can be a liquid (e.g., water or antifreeze) or a gas (e.g., air or freon), **p. 190.**

Crankshaft—a component that converts the piston's up-and-down or forward-and-backward motion into rotational motion, **p. 190.**

Electromagnetism—magnetism produced by an electric current, **p. 191.**

Engine—a machine that converts chemical (fuel) energy into mechanical force, **p. 185.**

Engine block—the casing of an engine that houses the pistons and cylinders, **p. 190.**

Exhaust port—a chamber or cavity in an engine that collects exhaust gases and directs them out of the engine, **p. 190.**

Exhaust valve—a valve in the head at the end of each cylinder. It opens and directs the exhaust from the cylinder to the exhaust port, **p. 190.**

Fan—an assembly holding rotating blades inside a motor housing or casing that cools the motor by pulling air in through the shroud, **p. 189.**

Frame—a structure that holds the internal components of a motor and motor mounts, **p. 189.**

Head—the component of an engine on the top of the piston cylinders that contains the intake and exhaust valves, **p. 190.**

Induction motor—a motor that turns more slowly than the supplied frequency and can vary in speed based on the amount of load, **p. 186.**

Intake port—an air channel that directs fuel to an intake valve, **p. 190.**

Intake valve—a valve located in the head, at the top of each cylinder, that opens, allowing air–fuel mixture to enter the cylinder, **p. 190.**

Load—the amount of torque necessary for a motor to overcome the resistance of the driven machine, **p. 192.**

Motor—a mechanical driver that converts electrical energy into useful mechanical work and provides power for rotating equipment; sometimes called a *driver*, **p. 185.**

Oil pan / sump—a component that serves as a reservoir for the oil used to lubricate internal combustion engine parts, **p. 191.**

Piston—a component that moves up and down or backward and forward inside a cylinder, **p. 191.**

Rod bearing—a flat steel ring or sleeve coated with soft metal and placed between the connecting rod and the crankshaft, **p. 191.**

Rotor—the rotating member of a motor that is connected to the shaft, **p. 189.**

Shaft—a metal rod on which the rotor resides; suspended on each end by bearings, it connects the driver to the shaft of the driven end or mechanical device, **p. 189.**

Shroud—a casing over the motor that allows air to flow into and around the motor, **p. 189.**

Spark plug—a component in an internal combustion engine that supplies the spark to ignite the air–fuel mixture, **p. 191.**

Stator—a stationary part of the motor where the alternating current supplied to the motor flows, creating a magnetic field using magnets and coiled wire, **p. 189.**

Synchronous motor—a motor that runs at a fixed speed synchronized with the supply of electricity, **p. 185.**

Torque—a force that produces rotation, **p. 185.**

Universal motor—a motor that can be driven by either AC or DC power, **p. 186.**

Valve cover—a cover over the cylinder head that keeps the valves and camshaft clean and free of dust or debris; it also keeps lubricating oil contained, **p. 191.**

9.1 Introduction

In the process industries, motors provide power for rotating equipment such as pumps, compressors, fans, blowers, and conveyor drivers. Motors vary in size and the amount of power they provide. Because motors and other equipment require electricity in order to operate, electrical power distribution is an important concept in the process industries and is covered briefly in the next chapter to support the content on motors and engines.

The electricity used by motors and other equipment can be either alternating current (AC) or direct current (DC). Alternating current (AC) oscillates back and forth like a sine wave, while direct current (DC) flows in a single direction. The power from a typical electrical receptacle is AC, while batteries supply DC power. AC power is the more common type of power used in the process industries.

Although electricity is a common utility in process industries, significant hazards are associated with it. Electricity can be present without being visible, and serious injury or death by electrocution can occur, even at household voltages. When working around electrically powered equipment, process technicians must always follow safety procedures, be aware of potential hazards, and never perform any duties involving exposed live wires or connections. Tasks involving electrical equipment are conducted by specially trained technicians. The scope of work with electrical equipment that is performed by process technicians is usually very limited.

Purpose of Motors and Engines

A **motor** is a mechanical driver that converts electrical energy into useful mechanical work, providing power for rotating equipment. In the process industries, motors provide power for pumps, compressors, fans, blowers, conveyor drivers, valves, and other equipment.

An **engine** is a machine that converts chemical (fuel) energy into mechanical force. Originally, engines were considered any type of mechanical device that produced work. Most of the applications discussed here generate rotational force, or **torque**. Torque is used to operate other machinery. Generally, the most common types of engines used in process facilities are internal combustion engines (e.g., gasoline and diesel engines). The notable exception would be power plants and large gas and liquid natural gas facilities, where gas turbines are the most common type of internal combustion engines. In the process industries, engines might be referred to as *drivers*.

Engines are used for many of the same tasks as electric motors, but they might be selected instead of electric motors for a variety of reasons.

Motor a mechanical driver that converts electrical energy into useful mechanical work and provides power for rotating equipment; sometimes called a *driver*.

Engine a machine that converts chemical (fuel) energy into mechanical force.

Torque a force that produces rotation.

9.2 Types of Motors and Engines

Types of Motors

Electric motors use electricity for motive energy. Electric motors can use either AC or DC electricity and can be single speed or variable speed. Table 9.1 lists examples of some of the various types of motors.

Table 9.1 Selected Types of Motors

Motor Type	Description
Alternating current (AC)	The most common type of motor used in the process industries, primarily because of the simplicity of its construction
Direct current (DC)	Most commonly used in situations where varying amounts of speed and torque are needed
Single or fixed speed	Motors that run at a single rotator speed; generally, AC motors
Variable speed	Allows for the rotation speed of the motor to be adjusted; can be either an AC or a DC motor

There are two basic types of AC motors: synchronous and induction. **Synchronous motors** run at a fixed speed that is synchronized with the supply electricity. Synchronous motors rotate at exactly the supply frequency or at a fixed fraction of the supply frequency (e.g., one-half or one-third).

Synchronous motor a motor that runs at a fixed speed that is synchronized with the supply of electricity.

Induction motor a motor that turns slightly more slowly than the supplied frequency and can vary in speed based on the amount of load.

Induction motors, on the other hand, turn slightly slower than the supplied frequency and can vary in speed based on the amount of load. Most induction motors are of the "squirrel cage" design, although the wound rotor design might be used where variable speed is necessary. Figure 9.1 shows an example of a squirrel cage rotor and a wound rotor.

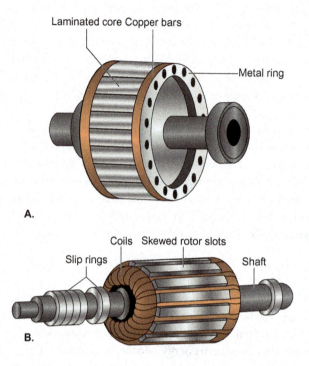

Figure 9.1 A. Illustration of a squirrel cage induction rotor. **B**. Wound rotor design.

Three-phase, single-speed squirrel cage AC induction motors are by far the most common type in the process industry. They are smaller, less expensive, and they require significantly less maintenance than wound rotor motors.

DC motors are useful for variable speed applications because adjusting the voltage can vary their speed easily. They are useful in applications where speed and load requirements vary, such as in the case of a crane loading or offloading a ship. When a load is attached, maximum torque is required to move it. Once the load has been placed, the quicker the crane can return to lift another load, the sooner the task will be completed, so speed is more important during this phase of the task. The amount of current that DC motors draw is roughly proportional to the torque load they drive. DC motors are also useful in applications that require very high torque at startup because series wound DC motors develop their greatest torque at slow speeds. This makes them particularly good as starting motors for small diesel engines and similar applications.

Another motor design, the **universal motor**, can be driven by either AC or DC power. Universal motors are typically used in small, high-speed applications, such as angle grinders and other small power tools. They are suitable only for intermittent use, however, because they require frequent maintenance.

Universal motor a motor that can be driven by either AC or DC power.

SELECTION OF ELECTRIC MOTORS VERSUS OTHER DRIVERS Various factors affect the selection of electric motors over other devices (e.g., internal combustion engines, steam turbines, hydraulic motors, and pneumatic motors). For example, internal combustion engines present toxic emission hazards and cannot be used in confined spaces. Likewise, pneumatic drivers that use nitrogen or other hazardous gases must not be used in confined spaces because of the risk of asphyxiation.

Some areas might prohibit the use of a gas or diesel engine because of the flammability of the fuel and the potential for fire. In these situations, an electric motor might be the motor of choice. An explosive environment might require a special classification of motor (e.g., intrinsically safe motors that do not produce sparks or thermal effects that will ignite flammable vapors) or an alternate type of driver.

In some applications, steam turbines (see Chapter 8) might be more suitable than electric motors. In other applications, hydraulic motors might be the motor of choice because they are smaller than electric motors. The availability of the energy source is also considered when deciding which type of driver to use. Steam or hydraulic power might not be practical because of the piping required. Other important factors are the physical location (e.g., elevation from the ground) and the space available to place a motor safely.

Motors must be located where they are protected from external factors, yet remain accessible for repairs. Because electric motors need power, fire pumps, cooling tower pumps and fans, and other critical devices that are electrically powered are often backed up with diesel-driven or gasoline-driven engines. In the event of an electrical power outage, they will still function.

Long-term costs often determine the choice of a driver. Electric motors are often preferred because of their reliability, ease of use, low maintenance costs, and ease of repair.

Long-term costs of operation must be evaluated carefully against reliability and convenience. In many instances, the initial cost involves more than just the purchase price of the driver. For example, a motor that exceeds the capacity of the facility's current electrical distribution system would require a costly electrical distribution upgrade.

Types of Engines

Engines are typically one of two types: internal combustion or external combustion. Within these two types are several subtypes, which include gasoline and diesel engines and steam turbines (discussed in Chapter 8). The piston steam engine was popular in the 18th and 19th centuries. Steam turbine technology, however, gradually has replaced the reciprocating (piston) steam engine in virtually all commercial usage.

INTERNAL COMBUSTION ENGINES   Internal combustion engines are engines in which the combustion (burning) of fuel and an oxidizer (usually air) occurs in a confined space called a *combustion chamber*. As the fuel is combusted, an *exothermic* (heat-producing) reaction occurs that creates high temperatures. High-pressure gases expand and act on various parts of the engine (e.g., pistons or rotors) and cause them to move.

Gasoline and diesel engines are both internal combustion engines. They are similar in their operation, but they differ in how the fuel is ignited.

In both diesel and gasoline engines, fuel and air are introduced into the cylinders above the pistons. This fuel is then burned, producing heat and pressure. Gasoline engines use a spark (e.g., from a spark plug) to ignite the fuel. Diesel engines use the heat which results from compression of air in the cylinder (instead of a spark) to ignite the fuel. Compressed natural gas (CNG) and liquefied petroleum gas (LPG) are also popular fuels for internal combustion engines. Their operation is similar to that of a gasoline engine.

In gas turbines (see Table 8.1 in Chapter 8 Turbines), natural gas or other fuel is combined with compressed air to create high-energy (high-pressure and high-temperature) combustion gases.

EXTERNAL COMBUSTION ENGINES   External combustion engines are engines in which an internal working fluid is heated (usually by an external source such as a heat exchanger) until it expands. The expanding fluid then acts on an engine mechanism to provide motion and usable work.

External combustion engines use a combustion chamber outside the engine to heat a separate working fluid, which in turn does work (e.g., moving pistons or turbines). An example of an external combustion engine is a steam turbine. The output of a steam turbine is similar to other engines, but the method for producing rotational power is different. In a steam turbine, the source of the rotational force is high-pressure steam that is produced in a boiler.

Did You Know?

Gasoline engines were invented in 1876 by Nicolaus Otto.

Rudolf Diesel developed the idea for the diesel engine and obtained the German patent in 1892.

SELECTION OF DIFFERENT TYPES OF ENGINES   Steam turbines convert energy efficiently. The steam required to drive these engines can be produced through combustion of waste products. This combustion, which occurs in a furnace or other device, produces heat, which is used to generate the steam necessary to power the turbine. Steam turbines might also be selected when a powerful engine is needed to drive large equipment (e.g., pumps, compressors, or generators) because they are smaller and more compact than electric motors providing the same amount of power. It is also easier to vary the speed and energy output of steam turbines than motors.

A disadvantage of steam turbines is that they typically require a boiler and all its auxiliary equipment to produce steam.

One benefit of a gas turbine is that it can be used as a backup electrical generating source because it is powered by natural gas. Gas turbines can also be configured as combined-cycle engines, in which energy is extracted from the combustion exhaust gas to improve efficiency. That is, although the turbine extracts most of the energy, the gases exiting the turbine are still very hot and can be used to boil water and produce steam. Table 9.2 provides a comparison of different engine types and their uses.

Table 9.2 Comparison of Engine Types Used in Process Industries

Engine Characteristic	Gasoline	Diesel	Gas Turbine	Steam Turbine
Remote use	✕	✕		
Mobile use	✕	✕		
Stationary use		✕	✕	✕
Intermittent/standby use	✕	✕	✕	✕
Continuous use			✕	✕
Emergency electric generation		✕	✕	✕

NOTE: Although diesel engines and gas engines are capable of continuous use, this is not common in most process facilities. Diesel engines are used continuously in applications on drilling rigs, ships, and remote electric generators.

Did You Know?

The original locomotives of the 1800s were steam-driven, piston engines. Many modern locomotives are diesel-driven.

CREDIT: Neil roy Johnson/Shutterstock.

USES OF ENGINES IN THE PROCESS INDUSTRIES Engines are critical for daily operations in the process industries. They are used to drive pumps, fans, compressors, and electric generators. Engines are used for many of the same tasks as electric motors, but they might be selected instead of electric motors for a variety of reasons.

Engines are sometimes used in place of motors as a backup source of power. Diesel or gasoline-powered backup generators are often used during power outages to continue operations or to allow for the safe shutdown of the facility. They can also be used to power a backup firewater pump if a fire occurs during a power outage.

Diesel and gasoline engines can also be used to power portable equipment. For example, if a service pump is needed infrequently and at different locations around a facility, it might be preferable for the pump to be driven by an engine. In remote locations where it is impractical to run large power cables, a machine with integral engines and generators (e.g., a large welding machine) can be used. Such portable devices are frequently required during turnarounds when the normal electrical power might be unavailable.

Gasoline and diesel engines can perform many of the same tasks. Diesel engines tend to be better suited for large tasks or tasks that require continuous operations. Gasoline engines are better suited for smaller tasks and intermittent operation. Diesel engines are also more appropriate in areas where gasoline (which is more flammable than diesel) poses an undue risk.

Turbines can also be used in place of engines for some tasks (e.g., driving pumps, compressors, and electric generators). Like gasoline and diesel engines, turbines provide a power source other than electricity. Turbines are almost always stationary and typically are used where a large engine is required. Although reciprocating engines are less expensive to install, more efficient at idle speeds, quicker to start, and more responsive to speed changes, steam or gas turbines have many advantages and are the better choice for many applications.

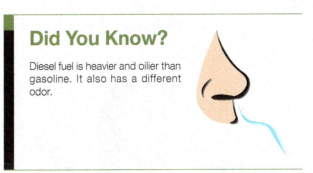

Did You Know?

Diesel fuel is heavier and oilier than gasoline. It also has a different odor.

9.3 Components of Motors and Engines

Motors

AC motors are composed of a frame, a shroud, a stator, a rotor, a fan, and bearings. Figure 9.2 displays an AC motor with these components identified.

- The **frame** is a structure that holds the internal components of a motor and motor mounts.
- The **shroud** is a casing over the motor that allows air to flow into and around the motor. The air keeps the temperature of the motor cool.
- The **rotor** is a rotating member of a motor or turbine that is connected to the shaft; it usually consists of an iron core with copper bars attached to it. When the stator creates a rotating magnetic field, it creates a second magnetic field in the rotor. The magnetic fields from the stator and rotor interact, causing the rotor to turn.
- The **shaft** is a metal rotating component (spindle) that holds the rotor and all rotating equipment in place.
- The **stator** is a stationary part of the motor to which the alternating current supplied to the motor flows, creating a magnetic field using magnets and coiled wire.
- The **fan** is an assembly holding rotating blades attached to the shaft inside a motor housing or casing that cools the motor by pulling air in through the shroud.
- The **bearing** is a machine component that rotates, slides, or oscillates. Bearings reduce friction between the motor's rotating and stationary parts.
- The **AC power source** supplies alternating current to the stator. Figure 9.3 shows a fan, a frame, and a rotor and motor.

Frame a structure that holds the internal components of a motor and motor mounts.

Shroud a casing over the motor that allows air to flow into and around the motor.

Rotor the rotating member of a motor that is connected to the shaft.

Shaft a metal rod on which the rotor resides; suspended on each end by bearings, it connects the driver to the shaft of the driven end or mechanical device.

Stator a stationary part of the motor where the alternating current supplied to the motor flows, creating a magnetic field using magnets and coiled wire.

Fan an assembly holding rotating blades inside a motor housing or casing that cools the motor by pulling air in through the shroud.

Bearing a machine component that rotates, slides, or oscillates. Bearings reduce friction between the motor's rotating and stationary parts.

AC power source a device that supplies alternating current.

Figure 9.2 AC motor components.

CREDIT: Normal Life/Shutterstock.

Stators   Coils   Shroud

Shaft

Fan

Frame

Rotor

Bearings

Figure 9.3 A. Example of a fan used inside a motor. **B**. Example of a frame used in a motor. **C**. Example of a rotor and motor.

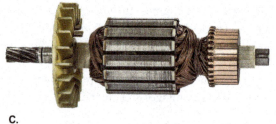

A.   B.   C.

Engines

Camshaft a driven shaft fitted with rotating wheels of irregular shape (cams) that open and close the valves in an engine.

Connecting rod a component that connects a piston to a crankshaft.

Coolant a fluid that circulates around or through an engine to remove the heat of combustion. The fluid can be a liquid (e.g., water or antifreeze) or a gas (e.g., air or freon).

Crankshaft a component that converts the piston's up-and-down or forward-and-backward motion into rotational motion.

Engine block the casing of an engine that houses the pistons and cylinders.

Exhaust port a chamber or cavity in an engine that collects exhaust gases and directs them out of the engine.

Exhaust valve valve in the head at the end of each cylinder. It opens and directs the exhaust from the cylinder to the exhaust port.

Head the component of an engine on the top of the piston cylinders that contains the intake and exhaust valves.

Intake port an air channel that directs the air–fuel mixture to an intake valve.

Intake valve a valve located in the head, at the top of each cylinder, that opens, allowing air–fuel mixture to enter the cylinder.

INTERNAL COMBUSTION ENGINES The major components of a gasoline engine are labeled in Figure 9.4. An internal combustion gasoline engine has many components.

- The **camshaft** is a driven shaft fitted with rotating wheels of irregular shape (cams) that open and close the valves in an engine.

- The **connecting rod** connects the piston to the crankshaft. It swivels at both ends on connecting pins so that its angle can change as the piston moves and the crankshaft rotates.

- **Coolant** is a fluid that circulates around or through an engine to remove the heat of combustion. This fluid can be a liquid, such as water or antifreeze, or a gas, such as air.

- The **crankshaft** converts the piston's up-and-down or forward-and-backward motion into rotational motion.

- The **engine block** is the casing of the engine that houses the pistons and cylinders.

- The **exhaust port** is a cavity or chamber that collects exhaust gases from the exhaust valves and directs them to the exhaust system.

- **Exhaust valves** are valves in the head at the end of each cylinder that open and direct the exhaust from the cylinder to the exhaust port.

- The **head** is the component of an engine on top of the piston cylinders that contains the intake and exhaust valves.

- The **intake port** is an air channel that directs the air–fuel mixture to the intake valves.

- **Intake valves** are located at the top of each cylinder and open to allow the air–fuel mixture to enter the cylinder.

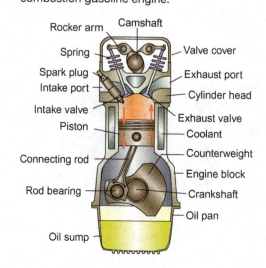

Figure 9.4 Components of an internal combustion gasoline engine.

- The **oil pan/sump** is a component that serves as a reservoir for the oil used to lubricate internal combustion engine parts.
- **Pistons** are components that move up and down or backward and forward inside a cylinder. The pistons are usually equipped with seal and wear rings to isolate the combustion chamber from lubricant oil.
- **Rod bearings** are usually flat, steel rings or sleeves coated with soft metal that are placed between the connecting rod and the crankshaft and help to prevent the rod from breaking under its operating stresses. Pressurized oil prevents metal-to-metal contact between the bearing surface and the crankshaft. Some small engines use splash lubrication instead of pressurized oil because of the light loads they have to handle.
- **Spark plugs** supply the spark that ignites the air–fuel mixture.
- A **valve cover** keeps the valves and camshaft clean and free of dust or debris- and keeps lubricating oil contained.

9.4 Operating Principles of Electric Motors and Engines

Motors

Early pioneers in electricity discovered the principle of electromagnetism. When current is run through a coil of insulated wire that is wrapped around a soft iron bar, a strong magnetic field is created. When the current is removed, the magnetic field collapses. This process is called **electromagnetism** (magnetism produced by an electric current). Electromagnetism plays an important role in electric motors. Whether AC or DC, electric motors operate on the same three electromagnetic principles:

1. Electric current generates a magnetic field.
2. Like magnetic poles repel each other (i.e., positive repels positive and negative repels negative); opposite poles attract each other (i.e., positive attracts negative).
3. The direction of the electrical current determines the magnetic polarity.

A motor consists of two main parts: a stationary magnet, called a stator, and a rotating conductor, called a rotor. The alternating current supplied to the motor flows to the stator, creating a magnetic field using magnets and coiled wire. A typical rotor usually consists of an iron core with copper bars attached to it. These bars conduct electricity easily. The stator creates a rotating magnetic field, which creates a second magnetic field in the rotor. The magnetic fields from the stator and rotor interact, causing the rotor to turn. Figure 9.5 is an AC motor showing the stator, field coils, and rotor.

Electrical meters are used to measure the performance of electrical equipment by showing the amount of electricity drawn (used) by the equipment. Electrical meters can be used to measure the different aspects of the electrical use such as voltage (volts), current (amperes), and power (watts).

Oil pan/sump   a component that serves as a reservoir for the oil used to lubricate internal combustion engine parts.

Piston   a component that moves up and down or backward and forward inside a cylinder.

Rod bearing   a flat steel ring or sleeve coated with soft metal and placed between the connecting rod and the crankshaft.

Spark plug   a component in an internal combustion engine that supplies the spark to ignite the air–fuel mixture.

Valve cover   a cover over the cylinder head that keeps the valves and camshaft clean and free of dust or debris; it also keeps lubricating oil contained.

Electromagnetism   magnetism produced by an electric current.

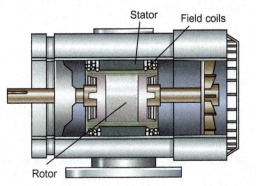

Stator   Field coils

Rotor

Figure 9.5 AC motor stator, field coils, and rotor.

Load the amount of torque necessary for a motor to overcome the resistance of the driven machine.

LOAD Load defines the amount of torque necessary for a motor to overcome the resistance of the machine it drives. The two types of loads are full loads and no loads. A *full load* occurs when the motor is using the full amount of torque for which it is rated. *No load* occurs when the motor is not using torque.

THREE-PHASE VERSUS SINGLE-PHASE MOTORS Most motors in process facilities use three-phase electricity instead of the single-phase electricity found in most homes because three-phase motors are more efficient. Like single-phase motors, three-phase motors can rotate clockwise or counterclockwise, depending on how the wiring is connected to the motor. If the load end is a centrifugal pump, it has a specific flow path for the liquid to enter and exit the pump, and the motor must rotate in a direction that will move the fluid toward the discharge side of the pump. Process technicians should always ensure that the motor rotation is correct before putting the motor in service. This is especially important if any work was performed that involved the motor wiring, or if it is a new installation.

Engines

Internal combustion engines can be classified as two-cycle or four-cycle. Although both engine types contain the same four processes, the way they accomplish them differs.

TWO-CYCLE INTERNAL COMBUSTION ENGINES Two-cycle engines complete the same four processes as four-cycle engines do, but they complete them in two strokes instead of four. Figure 9.6 illustrates the sequence of events in a two-cycle engine.

Because a two-stroke engine fires on every revolution of the crankshaft, it is typically more powerful than a four-stroke engine of equivalent size. In addition, it is lighter and has a simpler construction. For this reason, two-stroke engines are good for portable, lightweight applications (e.g., chainsaws, outboard motors, and some motorcycles), as well as large-scale industrial applications (e.g., locomotives). The down side to two-stroke engines is that they can be less efficient than other types of engines, and they tend to produce a greater volume of emissions because of the unspent fuel that escapes through the exhaust port.

Figure 9.6 Stages of a two-cycle engine.

Stroke 1: Intake and compression   **Stroke 2: Power and exhaust**

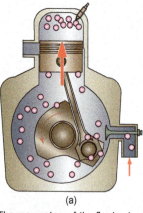 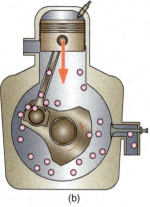 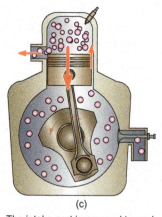

(a)

(b)

(c)

The momentum of the flywheel raises the piston and compresses the fuel mixture in the upper part of the cylinder, while another intake stroke is occurring beneath the piston.

The vacuum created during the upward stroke of the piston draws the fuel/air mixture into the lower part of the cylinder.

A spark plug ignites the fuel mixture when the piston reaches the top of the stroke. This causes the fuel to combust and force the piston downward to complete the cycle.

During the downward stroke, the valve at the bottom of the cylinder is forced closed and the air–fuel mixture is forced into the upper part of the cylinder.

The intake port is exposed toward the end of the power stroke. This allows the compressed fuel/air mixture to escape around the piston and into the upper part of the cylinder. This helps force exhaust gases (along with some of the fresh fuel mixture) out of the exhaust port.

FOUR-CYCLE INTERNAL COMBUSTION ENGINES   Gasoline and diesel engines function using similar four-cycle engines, although there are slight differences and different thermodynamic names. The four strokes in the engine cycle are:

1. Intake

2. Compression

3. Power (ignition)

4. Exhaust

The gasoline engine cycle is called the Otto cycle, and the diesel engine cycle is called the Diesel cycle, each named for its inventor. Each stage of these cycles corresponds with the motion of the engine pistons up and down or backward and forward. The complete cycles have two up and two down (or two backward and two forward) strokes, thus the term *four-cycle engine*. Figure 9.7 illustrates the sequence of events in a four-cycle engine.

Figure 9.7 Stages of a four-cycle engine.

CREDIT: udaix / Shutterstock.

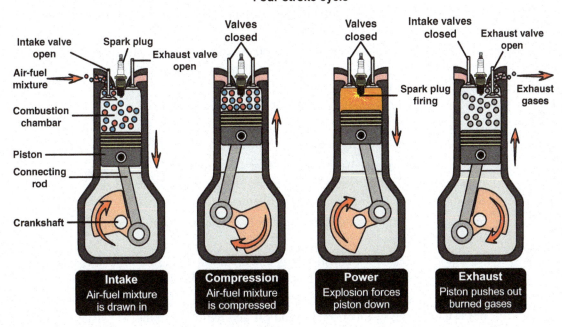

The first stroke is the intake stroke. With the intake valve open, the piston pulls the air–fuel mixture into the cylinder. The second is the compression stroke. During this second stroke, the piston compresses the gas in the cylinder. When a gas is compressed, its temperature increases.

The **compression ratio** is the ratio of the volume of the cylinder at the start of a compression stroke compared to the smaller (compressed) volume of the cylinder at the end of the stroke. Gasoline engines have a compression ratio between 8:1 and 12:1; diesel engines have a compression ratio between 14:1 and 25:1. The higher compression of the diesel engine results in better efficiency. When additional workload is required, an engine can contain multiple cylinders.

In a diesel engine, only air is in the cylinder during the compression stroke. The fuel is injected into the cylinders at the top of the compression stroke and is carefully timed so that it ignites at the right moment. A high-pressure fuel pump is used to directly pump the fuel into the cylinder through an injector that sprays the fuel in as a mist.

In a gasoline engine, ignition timing is controlled by timing the spark. The fuel can be mixed with the air at low pressure when air enters the cylinder, and the fuel and air

Compression ratio   the ratio of the volume of the cylinder at the start of a compression stroke compared to the smaller (compressed) volume of the cylinder at the end of a stroke.

are compressed together during the compression stroke. Older gasoline engines use an external Venturi device (the carburetor) to premix the fuel and air in the proper ratios for optimum combustion. Today, fuel injectors similar to those found in diesel engines meter the gasoline. This results in less air pollution caused by gasoline vapors escaping from the carburetor.

The remaining stages are essentially the same for both types of engine. In the third stroke, the fuel burns and its energy is converted to a much higher pressure and temperature in the cylinder. The expanding combustion gases force the piston back and transfer power to the crankshaft. In the final stage (fourth stroke), the exhaust valves in the head at the top of the cylinder open, and the piston forces the exhaust gas out of the cylinder.

AUXILIARY SYSTEMS OF INTERNAL COMBUSTION ENGINES All engines have a lubrication system for the bearings and moving parts. Because one of the functions of the lubrication system is to remove excess heat, there is a cooling system for the oil. Internal combustion engines also have a cooling system for the engine block. Another component of the internal combustion engine is the fuel and air delivery system (called the injection system).

Combustion Gas Turbines

Some of the major components of a combustion gas turbine are seen in Figure 9.8.

- The *casing* is the housing component of a turbine that contains the high-pressure air and gas and holds the stationary blades.
- The *combustion chamber* is located between the compressor and the turbine. It is where the compressed air is mixed with fuel and the fuel is burned, thereby increasing the temperature and pressure of the combustion gases.
- The *compressor* increases the pressure of the combustion air.
- The *nozzle* directs the flow of fluids to the turbine blades. Nozzles are similar to blades, but they cause the velocity of the steam or gas to increase.
- *Rotating blades* are attached to the shaft of the turbine. The steam or gas causes the turbine to spin by impinging on the rotating blades. The steam or gas slows and is redirected when transferring energy to the rotating blades.
- *Shafts* are rotating metal spindles that hold the rotor and all rotating equipment in place.
- A stage is a set of nozzles or stationary blades, plus a set of rotating blades. Each set is shaped like a disk composed of many narrow fan blades that rotate on the shaft.
- *Stationary blades* are attached to the case and do not rotate, but they change the direction of the flow of the steam or combustion gas and redirect it to the next stage of rotating blades.

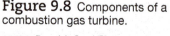

Figure 9.8 Components of a combustion gas turbine.

CREDIT: Fouad A. Saad/Shutterstock.

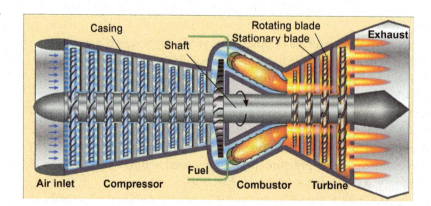

OPERATING PRINCIPLES OF A COMBUSTION GAS TURBINE   During the initial startup of a gas turbine, the compressor is started by an external starting mechanism. When the compressor starts, it delivers pressurized air to the combustion chamber, where fuel is added and the mixture is ignited. This ignition produces expanding combustion gases that drive the turbine. After the turbine is operational, the external starting mechanism can be disconnected from the compressor. This thermodynamic operating principle is called the Brayton cycle.

In the first stage of the Brayton cycle (see Table 8.1), ambient air is drawn into the compressor and its pressure is increased. In the second stage, the hot compressed air enters the combustion chamber, into which fuel is introduced and burned. This further increases the temperature and pressure, and thus the energy, of the combustion gas. In the third stage, the high-pressure, high-temperature gas applies force to the turbine blades, which does work by turning the shaft. In the final stage, the gas exits the turbine at a reduced pressure and temperature through the exhaust, and heat recovery takes place.

A waste heat boiler is an example of a heat recovery system. A waste heat boiler is a heat exchanger that uses turbine exhaust gases to generate steam. This is called a combined cycle and it greatly improves efficiency.

The turbine extracts work from the combustion gas using several sets of rotating and stationary blades and nozzles. Each pair of stationary and rotating elements in a turbine is called a stage. The cone-shaped turbine has blades that increase in size at each stage along the shaft. Gas enters at the small end and exits at a lower pressure at the large end. As nozzles convert some of the pressure energy into velocity energy, the pressure drops and the gas expands.

9.5 Potential Problems of Motors and Engines

Motors

When working with electrical distribution and motors, process technicians should always be aware of potential problems, such as high levels of vibration, high temperatures, and the motor not starting. Table 9.3 lists some potential problems associated with motors.

Table 9.3 Potential Problems and Corresponding Causes Associated with Motors

Condition	Problem	Causes
High vibration	Coupling failures can occur when shafts become misaligned because of high vibration. A coupling failure is often dangerous because of flying shrapnel. Coupling guards must always be in place as a safety measure. Larger motors have high vibration alarms or shutdowns. Excessive vibration can damage machinery.	■ Rotor out of balance (e.g., loss of rotor counterweights) ■ Bearing failure ■ Misalignment
High temperatures	Overheating can cause a number of motor performance problems, such as deterioration of winding insulation. Operation at the wrong current levels, high temperature environments, and loss of lubrication can cause overheating. Larger motors have high temperature alarms or shutdowns.	■ Bearings failure ■ Overload condition; too much load or torque ■ Loss of cooling airflow
Motor not starting	Motor will not start.	■ Interlock protection ■ Frozen bearing because of high temperature ■ Locked rotor ■ Motor overload trip ■ Breaker open or racked out ■ Blown fuse ■ Faulty start/stop relay circuitry ■ No power to the MCC panel

Engines

Problem situations that can occur when operating engines include problems between the driver and the driven equipment, high levels of vibration, high temperatures, turbine trips, and equipment failure. In order to ensure personal safety as well as process safety, process technicians must address these issues as quickly as possible.

DRIVER/DRIVEN EQUIPMENT PROBLEMS   If the driver is operating normally and the driven device is not performing, the shaft coupling must be checked to see if it is connected properly. The clutch should be checked for slippage.

HIGH VIBRATION   Engines can cause high levels of vibration that can damage the machine. Common causes of vibration include an unbalanced or bowed shaft, bearing failures, misalignment, or excessively high speed. Because excessive vibrations are extremely detrimental, large engines have multiple vibration shutdown systems.

MISALIGNMENT   Misalignment is an incorrect alignment between mating components. Proper initial alignment can prevent this problem.

HIGH TEMPERATURES   Typical causes of high temperatures include bearing failures, loss of lubrication, overload from driven machinery, or loss of cooling water or fan airflow. Because high temperatures can damage equipment, many large drivers have high-temperature alarms or shutdowns.

EQUIPMENT FAILURE   Equipment design and strict adherence to operating procedures play a substantial role in preventing equipment failure. In many processes, the operation of the driver represents a significant portion of operating costs, as well as requiring a heightened level of safety considerations. For example, failure of the engine on an emergency firewater pump could lead to disastrous consequences, including spread of fire, explosions, and emissions, possibly extending to other process facility units and the surrounding community. Failure of the engine on a large single-train system could cause a unit shutdown, resulting in substantial production loss, possible flaring, or increased emissions to the environment.

9.6 Safety and Environmental Hazards

Motors

Hazards associated with normal and abnormal motor operations could affect the personal safety of the process technician, process safety, equipment safety, facility operations, and the environment. The most serious hazards associated with normal and abnormal operations include shock or electrocution, injury caused by moving parts, and burns.

The most common personal protective equipment (PPE) used during motor and engine operations includes flame-retardant coveralls (FRCs, shown in Figure 9.9), as well as gloves, safety glasses, and ear plugs. Additional safety precautions that process technicians might be responsible for include lockout/tagout, equipment isolation, and grounding procedures.

Along with electrical hazards, motors also present mechanical hazards to workers. Motors have moving parts that can pull, tear, or rip clothing or skin if they are not properly de-energized before process technicians work on them. Shrouds, coupling guards, and other safety guards should always be in place during normal operation. Figure 9.10 is the equipment hazard symbol that warns of the hazards of moving parts. Motors can also cause a loss of production if they perform inefficiently or if a complete failure occurs.

Motor control centers (MCCs) represent a substantial safety feature for large motors. They are discussed in more detail in Chapter 10. First, they provide protection against short circuits and other electrical faults. Second, they protect the motor against overloading that

Figure 9.9 Using appropriate personal protective equipment (PPE) is an essential part of safe practice when working with motors and engines.

CREDIT: Bonezboyz/Shutterstock.

Figure 9.10 Equipment hazard symbol.

CREDIT: rifkhas/Shutterstock.

could result in overcurrent, which could damage the motor. MCCs can also protect against motor overspeed and can provide interlocks that prevent the motor from running in configurations (e.g., discharge valve closed) that would damage it or other process equipment. And finally, MCCs are used to electrically isolate motors for personnel protection during maintenance.

Studies have shown that the most frequent cause of accidents is human error. Accidents related to electrical energy are no different. If process technicians follow proper safety procedures when working on electrical circuits and electrical devices, the frequency of these accidents can be decreased significantly, keeping people, the equipment, and the environment safe.

Electrical Classification

Because electricity presents a possible ignition source, some areas have been classified by the National Electric Code (NEC) as hazardous locations, according to the potential hazards that may exist. Area classifications are based on classes, divisions, and groups, which, when considered together, are the factors that define an area's hazardous condition.

CLASS   Class designations are grouped by explosive/ignitable materials:

Class I—Locations where flammable gases and vapors may be present in explosive or ignitable mixtures

Class II—Locations where combustible dust may be present in amounts that could produce potentially explosive mixtures

Class III—Locations where fibers that are easily ignitable are present

GROUP   Group designations are used to group the material by relatively similar hazardous characteristics. Groups are assigned letters A through G: A through D are used for Class I, and E through G are used for Class II. Class III has no group designations.

DIVISION   Divisions are used to evaluate the likelihood of the hazardous material being present in ignitable mixtures. Division I means it is normally present, and Division II means it is present during upset conditions. For example, a possible classification for an area with ignitable amounts of acetone present during normal operations would be Class I, Group D, Division I.

The classification of the area dictates the type of motor required for that area. The most protective motor category is explosion-proof, which means that the motor enclosure is capable of withstanding an explosion of a specified vapor or gas, thereby preventing the ignition of additional gas vapors in the surrounding area. Division I areas frequently require the use of explosion-proof motors. Process technicians should know the electrical classifications for their operating areas and should use only the proper electrical equipment in those operating areas.

Engines

Hazards associated with normal and abnormal engine operations can affect personal safety, equipment, and environmental aspects of the process industries. Table 9.4 lists some of the equipment, safety, and environmental concerns associated with engines. Because engines contain rotating parts, process technicians must pay attention to warning signs (see Figure 9.10) and keep loose articles (e.g., clothes, hair, and jewelry) properly secured at all times.

Table 9.4 Equipment, Safety, and Environmental Concerns Associated with Engines

Personal Safety Hazards	Equipment	Environment
■ Burn hazards ■ Fumes from exhaust gases can cause incapacitation if they are not ventilated correctly ■ Noise hazards ■ Danger to hands and arms from rotating equipment ■ Moving parts that can pull, tear, or rip clothing or skin	■ Overspeed, which can cause structure failure and flying debris ■ Extremely high temperatures because of combustion of gases ■ High pressures ■ High vibration levels	■ Exhaust emissions from internal combustion engines and lubrication oil leaks ■ Spills of fuels or lubrication oils and/or process releases from sudden engine failure

9.7 Process Technician's Role in Operation and Maintenance

Motors

Process technicians are a key link in a series of checks and balances that protect people and equipment during both normal operations and shutdown periods. It is the process technician's responsibility to ensure that the equipment is operating properly. Proper training by the employer should cover all of the necessary information. Table 9.5 lists some routine observations that process technicians should make when working around electric motors.

Table 9.5 Process Technician's Role in Motor Operation and Maintenance

Look	Listen	Check
■ Inspect wires to ensure the insulation is not cracked or cut ■ Visually observe for loose covers and shrouds ■ Look for signs of corrosion ■ Monitor lubrication level for bearings ■ Visually observe that the cooling fan is running ■ Monitor lubricant temperature, pressure, and flow	■ Listen for abnormal noises ■ Listen to the bearings; investigate when noisy	■ Check for excessive heat ■ Check for excessive vibration

Scheduled maintenance on electric motors is an essential part of keeping equipment in a condition that minimizes or prevents equipment failure. Process technicians must know how to maintain communications and partnerships with maintenance personnel and coordinate preventive maintenance.

Process technicians have specific roles with regard to the operation and maintenance of motors. This includes the monitoring of lubrication, the condition of seals, and proper housekeeping. Process technicians might also be required to add or replace oil or other lubrication on moving parts and bearings as part of a routine maintenance schedule. Failure to do this can result in equipment damage or failure, or injuries or process incidents.

Housekeeping is another important factor in the maintenance of motors. Pump areas must be cleaned on a regularly scheduled basis and after repair work has been completed.

Instrumentation installed on motors should be checked regularly and recorded on a log sheet (if required by the employer). Recorded data might include vibration monitors, current and power meters, and temperature indications. Abnormal readings should be reported and investigated promptly.

In some facilities, a process technician might use electrical meters to monitor the current that the electric motor is consuming, ensuring that the motor is not operating at more than its full load capacity and that the right size motor is being used for the work being performed.

Many motors cannot withstand frequent stops and starts, which can cause failure of the motor's starter. Because of this, process technicians must adhere to the operating procedures for each motor type.

Engines

Scheduled maintenance activities are an essential part of keeping engines in good condition and preventing their failure. Process technicians must know how the equipment works and how to maintain it in partnership with maintenance personnel.

MONITORING   The process technician is the key person responsible for detecting an abnormal condition in an engine. Table 9.6 describes routine checks that are necessary for the upkeep of engines.

Table 9.6 Process Technician's Role in Engine Operation and Maintenance

Look	Listen	Check
■ Look for loose covers and shrouds ■ Look for leaks ■ Look for excessive vibration	■ Listen for abnormal noise ■ Listen to and check the cooling fan	■ Check for excessive heat ■ Check for excessive vibration

During operation, abnormal sounds are often the first indication of problems. These sounds can be detected if the process technician listens closely to the sounds of the engine each time it is run. Because process technicians spend hours monitoring equipment, abnormal engine sounds are usually easy to detect. Process technicians should also check operating temperatures, pressures, and flows (e.g., coolant, exhaust manifold, oil, and bearings) for abnormalities and check for vibration.

A manual vibration check is often required for small and medium-size internal combustion engines; separate monitoring instrumentation is used for large turbines. The output of an engine-driven device should also be checked (e.g., voltage and current for generators, or pressure and flow for pumps and compressors).

LUBRICATION   Lubrication is critical to the proper operation of all engines. Routine oil sampling, onsite inspection, and offsite analysis are critical parts of a good lubrication program. Onsite checks are used to identify substantial problems, such as water content, the presence of other contaminants, unusual odor or discoloration, foam, and emulsion state. Offsite analysis reports indicate problems such as elemental materials in the oil, and other problems such as bearing wear. Offsite analysis also indicates the condition of the oil and its additives, which should be graphed to ensure the oil is not degrading faster than expected.

If makeup lubrication oil is needed, it is imperative that the proper lubricant is used, as specified in the standard operating procedures or by the manufacturer. Many locations label the required lubricant on the equipment to reduce potential error. Using the wrong lubricant can result in serious and costly equipment damage and, in some cases, catastrophic failure, fire, explosion, and injury to personnel.

OTHER MAINTENANCE ITEMS   In addition to maintaining the equipment components and lubrication, it is also important to keep the fuel system clean so the engine will continue to run smoothly. This is particularly true of diesel engines. Fuel should be sampled and inspected upon delivery, and fuel filters should be regularly inspected, maintained, and changed.

To be able to perform its function, an engine must respond reliably when called upon to do work. Process technicians should also check battery charge and condition or starting air tank pressure for large diesel engines, and conduct regular performance checks for engines that provide emergency backup service.

9.8 Typical Procedures
Motors

Process technicians are responsible for the safe and cost-effective operation of equipment. Typical procedures associated with motors include proper startup, monitoring, shutdown, lockout/tagout, and routine and preventive maintenance.

Before starting any motors or equipment, the process technician must be aware of how this equipment affects other process systems in the unit. After the motor has been started, the process technician should ensure that the other process systems are responding as anticipated.

STARTUP   The typical procedure for starting an electric motor after maintenance work has been completed is as follows:

1. Determine how startup affects the rest of the facility and other facilities that interact with this facility.
2. Confirm with control room personnel that the motor has been returned to operations by the maintenance department if repairs were performed.
3. Notify the control room personnel that the motor is about to be placed on line.
4. Inspect the circuit breaker and other electrical devices in the MCC associated with the motor, and ensure that all are in a proper state of repair.
5. Ensure that the start/stop switch on the motor is in the "off" position.
6. Inspect to make sure that all wiring, insulation, motor connections, and switch covers are secure.
7. Inspect the driven equipment to ensure that the coupling guard is secure; the piping is connected; and all tools, scaffold boards, clothing, gloves, and toolboxes are out of the area.
8. Close all bleeds and vents. Perform the proper valve line-up on the driven equipment per standard operating procedures.
9. Check all lubrication levels, central lubrication systems, oil filters, and coolers to ensure all are ready for use.

10. Turn on the circuit breaker in the MCC.

11. "Bump" (quickly turn on and off) the motor to ensure it is rotating in the proper direction.

12. If the motor is rotating in the proper direction, place the motor in normal operation. If the motor is not rotating in the proper direction, notify appropriate personnel and do not proceed further until proper rotational direction is achieved.

13. Notify the control room that power is being turned on, and place the start/stop switch in the "on" (start) position.

14. Perform a walk-around inspection, looking for leaks, listening for unusual noises, and checking for high temperatures of the equipment.

15. Inform the control room personnel that the motor is online and in normal operation.

MONITORING   Variables that must be monitored to maintain reliable motor operation include motor temperature, motor amps, electrical wiring insulation and shielding, fan operation, bearing lubrication, and vibration. Checking for vibration by touching the electric motor could lead to burns; if done, gloves must be worn. It can also lead to shock, even when wearing leather work gloves. Electrical shock from touching a motor is *always* a reason for shutting down and calling maintenance for repair.

SHUTDOWN   Process technicians should follow the manufacturer, site, and unit-specific standard operating procedures for shutting down the equipment for the process facility.

EMERGENCY   Emergency procedures depend on the nature of the emergency. Process technicians should be aware of all the emergency procedures that are required for their job, and they must be able to follow them in an emergency situation.

LOCKOUT/TAGOUT   When performing maintenance on motors, electrical lockout/tagout procedures must be followed. A generic LOTO procedure can be found in Chapter 6 (Lockout/Tagout), but each facility has specific requirements that must be observed. This lockout/tagout should be performed at both the MCC and the motor. If the driven equipment (e.g., a pump), can also present a hazard by rotating the motor, then it must also be isolated through lockout/tagout.

After the lockout/tagout is complete, before work begins on the equipment, perform the following verification steps (and others as required).

- Perform a walkthrough of the affected work area with all workers.
- Check the de-energization of the electrical circuit by pushing the start button at the start/stop station.
- Check depressurization through bleeder valves.
- Check chained isolation block valves.

Engines

Typical procedures associated with engines include startup, placing in service, normal operation, shutdown, lockout/tagout, and emergency procedures. Operating procedures should be closely followed when starting, operating, and shutting down an engine. It is the process technician's responsibility to speak up if something does not look right.

STARTUP   Startup procedures vary considerably among engine types, but the following steps apply to most engines.

1. When starting an engine, always inspect the lubrication level (with a dipstick in most internal combustion engines or oil bubblers in many gas turbine engines).

2. Ensure that the engine is clear of debris or foreign objects that could get caught in rotating parts.

3. Observe the exhaust for smoke, which can indicate degraded engine performance.

4. Monitor any abnormal phenomena and be prepared to shut down the engine immediately.

Small gasoline and diesel engines have electric starters, and the startup is similar to an automobile engine. Large engines typically use air-driven starters, which use high-pressure air from tanks to provide the motive force to get the engine moving.

Many engines contain interlocks that automatically shut down the engine if the lubrication oil pressure or coolant flow is low or if excessive vibration is detected. The startup process often includes two steps:

1. Get the engine turning by using an electric or air starter.

2. Add gas to start the combustion process.

ENGINE IN SERVICE Procedures for placing an engine in service vary with the engine's function. Process technicians must understand the purpose of the engine and the equipment it drives, as well as how that equipment affects the process system. Before placing the engine and its equipment in service, process technicians should:

1. Know what they expect the response to be for both the engine and the process system.

2. Check the response of the engine, the driven equipment, and the process system after placing the engine in service to verify they are responding as anticipated.

NORMAL OPERATIONS Normal operation requires monitoring key process variables such as oil temperature, oil pressure, engine speed, vibration levels, engine temperature, and other variables. Process technicians must ensure that these variables are controlled as defined in the operating procedures. Testing for vibration should be done with ultrasonic or vibration-sensing equipment.

SHUTDOWN Shutdown of an engine involves two important steps:

1. Cool down the engine by running it under no load or a light load for a brief period (e.g., a few minutes)

2. Monitor, record, and trend the coast down time for some turbines because this can indicate bearing problems. Coast down times can also be monitored for turbochargers of large diesel engines.

LOCKOUT/TAGOUT Lockout/tagout of an engine is similar to other lockout/tagout procedures (see Chapter 6, Lockout/Tagout). During lockout/tagout, process technicians should isolate the starting motive force (e.g., a battery or compressed air), external cooling sources (e.g., circulating water), and the fuel or energy source. It is always necessary to isolate the driven equipment (e.g., a generator, pump, or compressor) before performing maintenance.

EMERGENCY PROCEDURES Emergency procedures are specific to the process facility and, in most instances, include quick isolation of the engine's energy source.

Summary

MOTORS

Motors are important components of industrial manufacturing facilities. Motors are mechanical drivers that convert electrical energy into useful mechanical energy and provide power for a variety of equipment.

Electricity either oscillates (alternating current or AC) or flows in a single direction (direct current or DC). Components of an electric distribution system include three-phase and single-phase systems, transformers, motor control centers (MCCs), circuit breakers, fuses, and switches.

Electric motors can be either alternating current (AC) or direct current (DC). AC motors can be either variable speed or single speed; DC motors are usually variable speed. Single-speed AC motors include synchronous and induction motors. Induction motors, which are the most common design in the process industries, are usually of a squirrel cage design. AC wound rotor designs are variable speed. DC motors are typically variable speed.

The main components of an electric motor include a frame, a shroud, a rotor, a stator, a fan, and bearings. Both AC and DC motors work off the same principles of electromagnetism.

Process technicians should inspect motors and perform scheduled preventive maintenance in order to keep motors in good condition and prevent equipment failure. If the operator is checking for vibration by touching the electric motor, the operator should wear work gloves (to protect against burns) to sense the vibration. Process technicians should also be familiar with proper operating procedures before starting any piece of equipment, and should verify that the actual responses correspond to the anticipated responses. It is crucial to speak up if something does not look right.

ENGINES

Engines are machines that convert energy into useful work. They are used to drive pumps, compressors, electric generators, and other types of equipment. They are also useful in emergencies because they work independently of the facility's electric supply.

Gasoline and diesel engines can provide power in remote and mobile locations. Gas turbines can conserve energy if their waste (exhaust) heat is recovered for use in other processes. Gas turbines are capable of driving large equipment.

Common internal combustion engines are gasoline and diesel, both of which are piston engines. The gasoline engine cycle is called the Otto cycle, and the diesel engine cycle is called the Diesel cycle. Each stage of the cycle corresponds to the motion of the engine's pistons. In a four-cycle engine, a complete cycle consists of two up and two down strokes. Two-cycle engines complete the same four processes as four-cycle engines, but they complete them in two strokes. Internal combustion engines deliver work by converting the linear piston motion into high-speed rotational motion through the use of a crankshaft.

Gas turbines, another type of internal combustion engine, deliver work through high-energy combustion gases that drive turbine blades, causing a shaft to turn at high speeds.

Potential problems related to engines include incorrect fuel or fuel mixtures, lack of lubrication, bearing wear or failure, driver operating but the driven device is not, high levels of vibration, misalignment, high temperatures, overspeed, loss of coolant, or trips.

Equipment malfunction or failure can cause safety and environmental hazards.

Process technicians must be aware of specific procedures associated with engines as they relate to their job requirements. These procedures include startup, shutdown, lockout/tagout, and emergency procedures. When process technicians are involved with the operation and maintenance of equipment and engines located in their operating unit, they must be familiar with the required procedures and know how to access them in a time of emergency. They must understand that speaking up about potential problems is the best way to prevent injury and accidents.

Checking Your Knowledge

1. Define the following terms:
   a. Shroud
   b. Shaft
   c. Bearing
   d. Load
   e. Fan

2. (True or False) AC motors are the most common type of motor used in the process industries because of their simplicity.

3. (True or False) A variable speed motor allows for the rotation speed of the motor to be adjusted.

4. (True or False) Synchronous motors are the most common type of AC motor.

5. A(n) ___ defines the amount of torque necessary for a motor to overcome the resistance of the machine it drives.
   a. watt
   b. amp
   c. volt
   d. load

6. A coupling failure is often dangerous because it may lead to ___.
   a. fire
   b. vibration
   c. a blown fuse
   d. flying shrapnel

7. Which of the following is NOT a potential hazard associated with motors?
   a. Corrosion
   b. Electrocution
   c. Burns
   d. Injury from moving parts

8. Areas where motors are used can be classified as possibly hazardous and include restrictions. Where can a list of restrictions be found?
   a. National Electric Code information
   b. Employee handbook
   c. Motor control center
   d. All of the above

9. What three things must process technicians do as they perform routine tasks on electric motors?
   a. Anticipate, explore, and think
   b. Look, listen, and check
   c. Prevent, understand, and consider
   d. Question, answer, and know

10. In the process industries, engines can also be called _____.

11. What component of internal combustion engines converts the piston's up-and-down or forward-and-backward motion into rotational motion?
    a. Head
    b. Crankshaft
    c. Connecting rod
    d. Engine block

12. List three causes of high temperatures in motors.

13. (True or False) Using the wrong lubricant can result in serious and costly equipment damage and, in some cases, catastrophic failure, fire, explosion, and injury to personnel.

14. List three key process variables that should be monitored during normal operation of engines.

NOTE: Answers to Checking Your Knowledge questions are in the Appendix.

Student Activities

MOTORS

1. Describe the basic principles of electricity, including the difference between AC and DC, and identify which type is most commonly used in the process industry.

2. Write a one-page paper explaining how an electric motor works.

3. Given a picture or a cutaway of an AC motor, identify the following components:

   a. Frame
   b. Shroud
   c. Rotor
   d. Fan
   e. Bearings
   f. Power supply

4. Brainstorm a list of possible problems associated with motors. Be prepared to discuss these problems as a class.

ENGINES

5. In small groups, discuss typical procedures associated with engines. Develop an engine procedure using the following steps as procedure sections. Share as a class when and why these procedures are performed, the possible hazards, and precautions associated with performing them.

   a. Monitoring
   b. Lockout/tagout
   c. Routine and preventive maintenance
   d. Startup/shutdown
   e. Emergency

6. Using a cutaway or drawing of an engine, work with a classmate to identify additional safety precautions that process technicians might be required to take for routine and preventive maintenance.

7. Prepare a two-page report that explains the types of equipment-related operational hazards associated with engines and auxiliary systems during normal and abnormal operations.

Chapter 10
Power Transmission and Lubrication

Objectives

After completing this chapter, you will be able to:

10.1 Describe the elements of electrical power transmission and distribution. (NAPTA Power Transmission and Lubrication 1, 2, 3*) p. 208

10.2 Describe the elements of mechanical power transmission. (NAPTA Power Transmission and Lubrication 1, 2, 3) p. 215

10.3 Explain types of bearings, gears, and seals and their purpose. (NAPTA Power Transmission and Lubrication 4, 5, 6) p. 219

10.4 Describe the principles of lubrication. (NAPTA Power Transmission and Lubrication 1) p. 221

10.5 Identify potential problems associated with mechanical power transmission and lubrication. (NAPTA Power Transmission and Lubrication 10) p. 226

10.6 Describe safety and environmental hazards associated with mechanical power transmission and lubrication. (NAPTA Power Transmission and Lubrication 7) p. 227

10.7 Describe the process technician's role in mechanical power transmission and lubrication procedures. (NAPTA Power Transmission and Lubrication 8, 9) p. 228

*North American Process Technology Alliance (NAPTA) developed curriculum to ensure that Process Technology courses will produce knowledgeable graduates to become entry level employees in process technology. Objectives from that curriculum are named here in abbreviated form. For example "(NAPTA Power Transmission and Lubrication 7)" means that this chapter's objective relates to objective 7 of the NAPTA curriculum about power transmission and lubrication).

Key Terms

Alternating current (AC)—electric current that reverses direction periodically, usually 60 times per second, **p. 212.**

Ammeter—device used to measure the electrical current in a circuit, **p. 209.**

Ampere (amp)—a unit of measure of the electrical current flow in an electrical circuit, comparable to the rate of water flow in a pipe, **p. 209.**

Bearings—mechanical devices used to support shafts and limit radial (up-and-down or side-to-side) and axial (back-and-forth) movement and to hold the rotor in alignment with other parts, **p. 219.**

Belt—a flexible band, placed around two or more pulleys, that transmits rotational energy, **p. 216.**

Chains—mechanical devices used to transfer rotational energy between two or more sprockets, **p. 217.**

Circuit—a system of electrical components that accomplishes a specific purpose, **p. 210.**

Circuit breaker—an electrical component that opens a circuit and stops the flow of electricity when the current reaches unsafe levels, **p. 214.**

Conductor—materials that have electrons that can break free from the flow more easily than the electrons of other materials, **p. 208.**

Couplings—mechanical devices used to connect and transfer rotational energy from the shaft of the driver to the shaft of the driven equipment, **p. 215.**

Direct current (DC)—electrical current that flows in a single direction through a conductor, **p. 212.**

Electricity—the flow of electrons from one point to another along a pathway called a conductor, **p. 208.**

Electrons—negatively charged particles that orbit the nucleus of an atom, **p. 209.**

Fuse—a device used to protect equipment and electrical wiring from overcurrent, **p. 214.**

Gear—a toothed wheel that engages another toothed mechanism in order to change the speed or direction of transmitted motion, **p. 220.**

Gearbox—mechanical device that houses a set of gears, connects the driver to the load, and allows for changes in speed, torque, and direction, **p. 220.**

Generator—a device that converts mechanical energy into electrical energy, **p. 213.**

Lubricant—a substance used to reduce friction between two contact surfaces, **p. 221.**

Motor control center (MCC)—an enclosure that houses the equipment for motor control, including its power source, isolation power switches, lockouts, fuses, overload protection devices, ground-fault protection, and sometimes meters for current (amperes) and voltage, **p. 213.**

Ohm—a measurement of resistance in electrical circuits, **p. 210.**

Rectifier—a device that converts AC voltage to DC voltage, **p. 212.**

Sheave—a V-shaped groove centered on the circumference of a wheel, designed to hold a belt, rope, or cable, **p. 216.**

Sprocket—a toothed wheel used in chain drives, **p. 217.**

Static electricity—electricity that occurs when a number of electrons build up on the surface of a material but have no positive charge nearby to attract them and cause them to flow, **p. 211.**

Switch—an electrical device used to start, stop, or otherwise reconfigure the flow of electricity in a circuit, **p. 215.**

Transformer—an electrical device that takes electricity of one voltage and changes it into another voltage, **p. 213.**

Volt—the unit used to measure the difference in electrical potential between two points in a circuit. One volt is the force that will cause a current of one amp to flow through a resistance of one ohm, **p. 208.**

Voltage—a measurement of the potential energy available to push electrons from one point to another, **p. 208.**

Voltmeter—a device that can be connected to a circuit to determine the amount of voltage present, **p. 209.**

Watt—a unit of measure of electric power; the power consumed by a current of one amp using an electromotive force of one volt, **p. 208.**

10.1 Introduction

In the process industries, operators monitor electrical power transmission, electricity, and mechanical power transmission. Electrical power transmission involves the transfer of electrical power from a source to various pieces of electric dependent equipment in a facility. Mechanical power transmission allows for the to transfer of rotational energy from one point to another in order to power various types of equipment. Without this energy, many processes could not occur.

Lubrication is a key component of mechanical power transmission because it establishes a stable, fluid barrier between equipment components and reduces friction, heat buildup, and equipment wear. The operator monitors and maintains the lubrication systems, which are critical to daily equipment operations.

Electrical Power Transmission and Distribution

Electricity the flow of electrons from one point to another along a pathway called a conductor.

Conductor materials that have electrons that can break free from the flow more easily than the electrons of other materials.

Electricity is the flow of electrons from one point to another along a pathway called a **conductor.** Electricity in a process facility is used to power many pieces of process equipment, including most motors, plant lighting, and instrumentation. Common voltages used in the process industries include low voltage DC (i.e., 12- or 24-volt DC) and AC voltages including 120 volts, 240 volts, and 480 volts. Operators are often responsible for the operation of electric equipment, opening and closing electrical breakers, and managing the source for the process facility electricity.

Electrical transmission in a process facility can involve multiple sources of electrical power. Operators might be responsible for managing the power transmission equipment such as generators, transformers, electrical routing switches, motor control centers (MCCs), breakers, and hand switches. Electrical switches and breakers are *opened* to disconnect power and *closed* to allow electrical flow. Transformers require monitoring to ensure the fluid levels in the transformers are maintained. In the MCCs, electrical power is locked *open* to ensure equipment undergoing maintenance work is de-energized. Many of the controllers in the MCC also have reset buttons, which may be operated if the equipment trips.

To understand electricity, process technicians must be familiar with its principles and know that it is measured in watts, volts, amps, and ohms. To better understand the concept of electric current flow in a circuit, it is often compared to water flowing in a pipe: amps are the measurement of current, which is analogous to the amount of water flowing through the pipe, volts represent the pressure of the water in the pipe (there must be a difference in voltage along the circuit in order for current to flow), and ohms represent anything that could cause resistance to the water flow, like a pinched control valve.

Watt

Watt a unit of measure of electric power; the power consumed by a current of one ampere using an electromotive force of one volt.

A **watt** is the power consumed by a current of one ampere using an electromotive force of one volt. The box on the following page contains the formulas for determining wattage.

Volt

Voltage the potential energy available to push electrons from one point to another.

Volt the unit used to measure the difference in electrical potential between two points in a circuit.

Voltage is the potential energy, or force, available to push electrons from one point to another. The difference in voltage between two points in a circuit is called the potential difference. A **volt** is the unit used to measure this difference in electrical potential. One volt is the force that will cause a current of one ampere to flow through a resistance of one ohm. Process plants use many different levels of electrical voltage. A voltage of 110 is used to supply indoor lighting and varied equipment in offices and labs; 220 and 440 voltage levels supply power to many types of pumps and other associated equipment. Other larger pieces of equipment such as compressor motors might use even higher voltage levels, requiring that the supply come directly from a substation.

Wattage

The formula for determining how many watts a circuit uses, if you know the volts and amperage, is:

$$P = VI$$

P = power in watts
V = potential difference in volts
I = current flow in amperes

For example, a portable electric heater operating at a voltage of 120V and drawing 8.35 amps uses approximately 1,000 watts. (120V × 8.35 amperes = 1002 watts).

The formula for determining how much wattage a circuit uses, if you know the ampere and resistance, is:

$$P = I^2R$$

P = power in watts
I = current flow in amperes
R = resistance to current flow in ohms

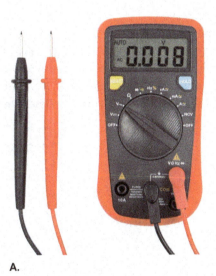

A.   **B.**

Figure 10.1 A. Digital voltmeter/ammeter combination. **B.** Analog faceplate from a voltage meter.

CREDIT: **A.** fotosv/Shutterstock. **B.** Trin Tra/Shutterstock.

Using the water analogy again, imagine electric current, like a river, flowing down a slope (the path of least resistance). The greater the angle of the slope, the faster the water flows. Electric current behaves in a similar way. If the difference between positive and negative charges is low, electrons flow with little force. When the difference is increased, the **electrons** flow with greater force. The force that makes electrons flow is called voltage or electromotive force (EMF), and it is measured in volts (V). Voltage can be measured with a voltmeter (shown in Figure 10.1).

A **voltmeter** is a device that can be connected to a circuit in order to determine the amount of voltage present. The symbol V is used to denote voltage. Sometimes, the letter V is followed by AC or DC (e.g., VAC or VDC) to denote whether the voltage is from alternating current (AC) or direct current (DC).

Electrons negatively charged particles that orbit the nucleus of an atom.

Voltmeter a device that can be connected to a circuit to determine the amount of voltage present.

Amperes

Amperes, or amps, are a unit of measure of the electrical current flow in an electrical circuit, similar to a measurement of gallons of water flow in a pipe. Ampere describes how many electrons are flowing at a given time. The symbol I is used to denote current in equations.

An **ammeter** is a device used to measure the electrical current in a circuit. Ammeters must be connected in series to an electrical circuit in order to display actual amps, since current must go through the ammeter in order to be measured.

Ampere (amp) a unit of measure of the electrical current flow in an electrical circuit, comparable to the rate of water flow in a pipe.

Ammeter device used to measure the electrical current in a circuit.

Figure 10.2 shows an example of an ammeter. Current capacity (in amperes) indicates how much work a circuit can do for a given voltage. In other words, amps register the capacity of a battery or other source of electricity to produce electrons.

Ohm

Ohm a measurement of resistance in electrical circuits.

Ohm is a measure of resistance in electrical circuits. Ohm's law describes how voltage (potential), ohm (resistance), and ampere (current) are related. This law states that the amount of steady current through a conductor is proportional to the voltage across that conductor. This means a conductor with one ohm of resistance (represented by R) has a current of one ampere under the potential of one volt. Simply put, volt equals ampere times ohm (V = IR).

Ohm's Law

The formula for Ohm's law is:

$$V = IR$$

V = voltage
I = current flow in amperes
R = resistance to flow in ohms

When working with Ohm's law, if you know the value of two variables, you can always figure out the third using one of the following calculations:

I = V/R (current = volt ÷ resistance)   **R = V/I** (resistance = volt ÷ current)

Circuits

Circuit a system of electrical components that accomplishes a specific purpose.

A **circuit** is a system of one or more electrical components that accomplishes a specific purpose. Circuits combine conductors with a power supply and usually some kind of electrical component (such as a switch or light) in a continuously conducting path. In a circuit, electrons flow along the path uninterrupted and return to the power supply to complete the circuit. Circuits are connected in either a series or parallel configuration.

In a series circuit, the components are connected in a loop so the electrical current flows in a single path. The same amount of current flows through all components in a series circuit. Any break in the circuit stops the flow of current. In a parallel circuit, components are split into branches so the electrical current follows more than one path. The same amount of voltage flows through all components in a parallel circuit. A break in one branch does not stop current in the other branches. Figure 10.3 shows an example of a series and a parallel circuit.

Figure 10.3 Series and parallel circuit examples.

CREDIT: Designua/Shutterstock.

Series and parallel circuits

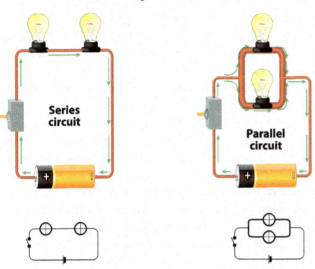

Did You Know?

- Watt is named after James Watt, who invented the steam engine.
- Ohm is named after Georg Ohm, a German mathematician and physicist.
- Ampere is named after André Ampere, the first person to explain the electrodynamic theory.

Grounding

All energized conductors supplying current to equipment are kept insulated from each other, from the ground (earth), and from the equipment user. Many types of equipment have exposed conductive parts, such as metal covers, that are routinely touched during normal operations. If these surfaces become energized, the process technician could complete the circuit and receive a shock. For this reason, non-current-carrying conductive materials should be used to enclose electrical conductors, and the equipment should be properly grounded.

Grounding is the process of connecting an object to the earth using copper wire and a grounding rod to provide a path for the electricity to dissipate harmlessly into the ground. A separate conductor specifically designed for this purpose typically accomplishes grounding. With this conductor in place, if the equipment case becomes energized, a low resistance path for the flow of ground current back to the source is already in place, and it reduces or eliminates the possible **static electricity** and shock hazard to the operator.

Figure 10.4 shows an example of a piece of electrical equipment that is attached to the earth (grounded) through a grounding wire.

Static electricity electricity that occurs when a number of electrons build up on the surface of a material but have no positive charge nearby to attract them and cause them to flow.

Motor ground wires

Figure 10.4 Grounding wires attached to a piece of electrical equipment.

CREDIT: Leonid Eremeychuk/123RF.

Types of Current

Alternating current (AC) electric current that reverses direction periodically, usually 60 times per second.

Alternating current (AC) is electric current that reverses direction periodically, usually 60 times per second. The movement of alternating current is similar to water sloshing backward and forward in a pipe. When a negative charge is at one end of a conductor and a positive charge is at the other end, the electrons move away from the negative charge. But if the charge (polarity) at the end of the conductors is reversed, the electrons switch directions (alternate). In the United States, the AC power supply changes direction 60 times per second. This cycling is called frequency or cycles per second.

Electricity is almost always generated as three-phase electricity because it is more efficient. A circuit consisting of two energized wires, each carrying one phase, and a neutral or ground wire delivers three-phase electricity. The electricity between any one energized wire and the ground, or the electricity between two power wires, is single phase. When any two of the three wires that connect the three phase power are switched, the direction of rotation in rotational equipment such as motors is reversed. It is important to check the rotation for any equipment being returned to service after undergoing maintenance work by conducting a "bump test", where the power switch is quickly "bumped" on and the off again to ensure that the direction of rotation is correct.

Direct current (DC) electrical current that flows in a single direction through a conductor.

Direct current (DC) is electrical current that flows in a single direction through a conductor. Direct current flows like water moving in one direction through a pipe. Batteries are examples of DC electrical power sources.

Did You Know?

Thomas Edison was an inventor who invented the phonograph, among other items, and improved the printing telegraph.

Edison also built the first practical DC (direct current) generator.

CREDIT: Everett Historical/Shutterstock.

DC power is used in limited applications in the process industries. Most frequently, DC power is used where critical equipment must remain functional even during a power outage. Batteries usually provide this type of backup power. Backup batteries do not supply a sufficient amount of power to continue normal unit operation, but will supply enough power to allow personnel to place the facility in a safe condition. DC power also is used to power emergency lighting and for special applications such as high-torque motors.

Commonly, DC motors have large battery banks to provide backup power if the regular power supply fails. These battery banks are kept fully charged by battery chargers that use **rectifiers** (devices that convert AC voltage to DC voltage).

Rectifier a device that converts AC voltage to DC voltage.

AC has an advantage over DC because AC voltages easily can be stepped up (increased) or stepped down (decreased) using transformers. By employing transformers to raise voltage levels, AC systems can distribute electricity economically for hundreds of miles. Table 10.1 summarizes the similarities and differences between AC and DC power.

Table 10.1 Differences and Similarities Between Alternating Current (AC) and Direct Current (DC)

Alternating Current (AC)	Direct Current (DC)
Polarity is switched repeatedly.	Polarity is fixed.
Voltage varies during cycles.	Voltage remains constant.
Voltage can be stepped up or stepped down by a transformer.	Voltage cannot be stepped up or stepped down by a transformer.
More difficult to measure than DC current.	Easier to measure than AC current.
Heating effect is the same as DC current.	Heating effect is the same as AC current.

Electrical Distribution Systems

Electricity is a form of energy that must come from a power source such as batteries or generators. **Generators** are devices that convert mechanical energy into electrical energy. For a power station, the most common way to produce electricity is by burning fuels to power turbines, which rotate magnetic fields inside generators to create electric current. Other methods include hydroelectric, nuclear, thermal, wind, and solar power generation.

Electricity flows in a continuous current from a point of high potential (the power source) to a point of lower potential (e.g., your home or facility) through a conductor such as a wire. High-voltage electricity is transmitted from the power plant to the power grid through a system of wires. During transmission, the high-voltage electricity is routed to a substation that steps down the electricity to a lower, safer voltage. The substation then distributes the electricity through feeder wires to a step-down transformer, lowering the voltage again so residential or commercial customers can use it.

After it is inside the home or facility, the reduced electrical current is used to do work, such as lighting a bulb or operating a motor. To make the system as safe as possible, safety devices such as fuses, protective relays, and ground-fault detectors are used throughout the power transmission process.

When power enters a process facility, it is often at a very high voltage. To make the power usable, the power is routed through a transformer, which adjusts the voltage. From there, the power is routed to a motor control center that contains fuses, circuit breakers, and switches that allow for the activation and control of motors and other electrical equipment.

TRANSFORMERS A **transformer** (Figure 10.5) is an electrical device that takes electricity of one voltage and changes it into another voltage. Transformers that reduce the voltage are referred to as step-down transformers. When voltage is stepped down, there is a corresponding increase in current. The power (the product of voltage and current) remains constant. Transformers that raise voltage are called step-up transformers.

MOTOR CONTROL CENTERS (MCCs) **Motor control centers (MCCs)** are enclosures that house the equipment for motor control, including isolation power switches, circuit breakers, motor starters, lockouts, fuses, overload protection devices, ground-fault protection, and sometimes meters for current (amperes) and voltage. Most industrial motors (except small motors) receive their power from a central MCC. MCCs are sometimes called electrical substations or *electrical rooms*. Many MCC units can be contained in one electrical room or substation (Figure 10.6).

MCCs are maintained at cool temperatures and are usually isolated from the rest of the facility so they do not serve as an ignition source in the event of a spill or release of flammable material.

While a typical motor has a local start/stop switch located on or near the equipment in the field, a low-voltage control circuit powers that switch, which in turn operates a control relay in the MCC. The control relay is what actually provides current to the equipment. Its breaker in the MCC is used to isolate and lockout the electricity source when maintenance is performed on that equipment.

Generator a device that converts mechanical energy into electrical energy.

Transformer an electrical device that takes electricity of one voltage and changes it into another voltage.

Figure 10.5 Electrical transformer.

CREDIT: only kim/Shutterstock.

Motor control center (MCC) an enclosure that houses the equipment for motor control, including its power source, isolation power switches, lockouts, fuses, overload protection devices, ground-fault protection, and sometimes meters for current (amperes) and voltage.

Figure 10.6 Motor control centers (MCC). **A.** An MCC unit. **B.** An MCC, also called an electrical substation or electric room.

CREDIT: A. Seksun Yodmauk/Shutterstock. B. Lemau Studio/Shutterstock.

A.

B.

All motors should have controllers that protect them from instantaneous overload (e.g., a short circuit), as well as thermal overload from working beyond design limits. Motors also should have a way to isolate the circuitry during maintenance work. Typically, the motor's overload, starting and stopping devices, and disconnecting device are built together in one motor controller. Safety features of MCCs are addressed later in this chapter.

FUSES   A **fuse** is used to protect equipment and electrical wiring from overcurrent. Fuses are designed to open or "blow" to stop current flow if the current becomes too great. After the fuse is blown, it must be replaced in order to re-energize the circuit. All fuses are rated for a specific current rating above which they will blow. Fuses are also rated for the maximum voltage of the circuit in which they can be used and are designated as either fast acting or slow acting. (Note: Slow-acting fuses must be used with motors, because there is a large current in-rush when a motor starts.)

Fuses are usually limited to DC and AC single-phase applications. In three-phase applications, or where ground-fault protection is needed, a circuit breaker is typically used instead of a fuse.

Fuse   a device used to protect equipment and electrical wiring from overcurrent.

CIRCUIT BREAKERS   A **circuit breaker** (shown in Figure 10.7) is an electrical component that opens a circuit and stops the flow of electricity when the current reaches unsafe levels. Circuit breakers act as a reusable form of protection from overcurrent. When overcurrent occurs, the circuit breaker trips and interrupts the flow through the circuit. The circuit breaker then must be reset to re-energize the circuit.

Circuit breaker   an electrical component that opens a circuit and stops the flow of electricity when the current reaches unsafe levels.

Figure 10.7 Large circuit breaker used in process industries.

CREDIT: Vachagan Malkhasyan/Shutterstock.

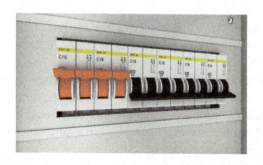

Many electrical components have a circuit breaker reset button that is used to reset the device for operation. An MCC might have one or more circuit breakers to protect each motor from current overload. These breakers have different response speeds and sensitivities.

SWITCHES   A **switch** is an electrical device used to start, stop, or otherwise reconfigure the flow of electricity in a circuit. Light switches are the most common household application of switches.

In large motors, switches send a control signal to an MCC. Large electrical switches used for isolation during system repairs and maintenance are called disconnects. Disconnects cannot be operated under load (while the equipment is in operation), so the power must first be interrupted using the start/stop switch and circuit breakers.

Switch   an electrical device used to start, stop, or otherwise reconfigure the flow of electricity in a circuit.

10.2 Mechanical Power Transmission

Mechanical power transmission transfers rotational energy from a driver to driven equipment with a minimal loss of energy to friction. Lubrication systems provide a fluid film between two surfaces that move relative to each other. Lubrication also is used to remove heat produced by friction in rotating equipment. To understand the operating principles of mechanical transmission, a process technician must understand the components of power transmission, which include couplings, gearboxes, speed reducers and increasers, belts, chains, magnetic and hydraulic drives, bearings, and gears.

Couplings

Couplings are mechanical devices used to connect and transfer rotational energy from the shaft of the driver to the shaft of the driven equipment. Couplings have several functions. They connect two rotating shafts so the rotating movement from one shaft can be transferred to another shaft. Couplings also align the rotating shafts that are coupled together. Figure 10.8 shows an example of a shaft coupling.

Couplings   mechanical devices used to connect and transfer rotational energy from the shaft of the driver to the shaft of the driven equipment.

Figure 10.8 Example of a coupling.

CREDIT: Pnor Tkk/Shutterstock.

While equipment can operate with a very slight degree of misalignment, the better the alignment between the rotating shafts, the longer the coupling will operate without wearing down.

The process industries use several types of couplings, including flexible, rigid, and fluid. Flexible couplings are used most frequently in general mechanical power transmission service because they allow some misalignment without affecting the transmission of power. Flexible couplings also provide some isolation of vibration between drivers and the driven equipment. The rotating shafts can expand and contract because of temperature changes, so the coupling is designed to accommodate this movement while transmitting the rotational energy.

Rigid couplings are used in applications where only small amounts of misalignment can be tolerated without damage to equipment or decreased performance.

Fluid couplings are used in applications where variable speed operations occur. Fluid couplings cushion the shock from equipment overloads, machinery jams, reversing operations, or sudden speed changes by increasing slip.

Belts

Belt a flexible band, placed around two or more pulleys or sheaves, that transmits rotational energy.

Sheave a V-shaped groove centered on the circumference of a wheel designed to hold a belt, a rope, or a cable.

Belts are flexible bands, placed around two or more pulleys or sheaves, that transmit rotational energy. V-belt drives are the most commonly used in an industrial setting. V-belts, which are usually designed to move in one direction, use traction to transfer motion between two **sheaves** (V-shaped grooves centered on the circumference of a wheel designed to hold a belt, a rope, or a cable). V-belt sheaves (shown in Figure 10.9) exhibit very low wear.

Figure 10.9 Examples of V-belt sheaves. **A.** Web type. **B.** Solid type.

CREDIT: Viktor Chursin/Shutterstock.

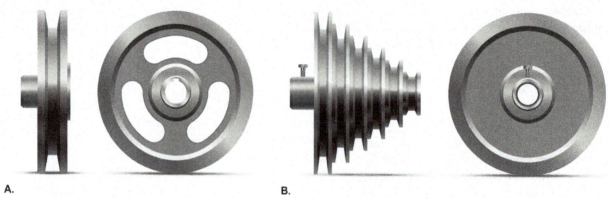

A.

B.

The three categories of belts—standard, light-duty, and high-capacity—are determined by the size, shape, and width of the belt. Standard belts are the most commonly used in industrial settings. Light-duty belts are used with smaller drive pulleys. High-capacity belts are used when the equipment has higher horsepower or loading conditions.

It is important to create and maintain proper tension when installing or working with belts because improper tension can decrease the life of the belt and prevent the system from operating properly. If a belt is too loose, it might slip or vibrate up and down. If the belt is too tight, it might prevent proper rotation of the sheaves or damage sheave bearings.

As a general rule, when adjusting the tension on a belt, there should be $1/4$ inch of slack per foot of distance between pulleys. Figure 10.10 illustrates the proper amount of tension on a belt.

Figure 10.10 Correct tension on a belt.

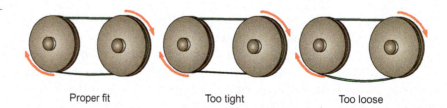

Proper fit   Too tight   Too loose

Did You Know?

The gear ratio is the relationship between the number of teeth on two gears, the driven gear and the driving gear.

For example, if the larger, driven gear has 26 teeth and the smaller, driving gear has 13 teeth, the gear ratio is 2:1. This means that for every two rotations of the driving gear, the driven gear makes one rotation.

CREDIT: Aerodim/Shutterstock.

Chains

A **chain** is a mechanical device used to transfer rotational energy between two or more sprockets. Chains perform the same function of transferring rotational energy as belts and gears, but they do not use friction to transfer motion. Instead, the links of the chain interlock with the teeth on a **sprocket** (a toothed wheel used in chain drives).

Figure 10.11 shows an example of a roller chain (one of the most common types of chain) engaged with sprocket teeth.

Chains mechanical devices used to transfer rotational energy between two or more sprockets.

Sprocket a toothed wheel used in chain drives.

Figure 10.11 Roller chain and sprocket.

CREDIT: cherezoff/Shutterstock.

The load capacity of a chain drive can be increased with multiple-strand chains. Each link in a chain drive transmits load in tension to and from sprocket teeth. Because of this, sprockets (see Figure 10.12) should be replaced whenever the teeth are worn.

Figure 10.12 Example of a typical sprocket.

CREDIT: Sharomka/Shutterstock.

Chain drives have many advantages over belt drives. For example, because a chain drive requires only a few sprocket teeth for effective engagement, it allows higher reduction ratios than are usually permitted with belts. Chains also have some disadvantages, however. These advantages and disadvantages are listed in Table 10.2.

Table 10.2 Advantages and Disadvantages of Chains over Belts

Advantages	Disadvantages
■ No slippage between chain and sprocket teeth. ■ Minimal stretch, allowing chains to carry heavy loads. ■ Long operating life expectancy because flexure (bending) and friction contact occur between hardened bearing surfaces separated by lubrication (oil film). ■ Capable of operating in extreme environments, especially if high-alloy metals and other special materials are used (e.g., areas with extreme temperature and/or moisture levels, or areas that are extremely oily, dusty, dirty, or corrosive). ■ Long shelf life because metal chains ordinarily do not deteriorate with age and are less affected by sun, moisture, and temperature. ■ Certain types can be replaced without disturbing other components mounted on the same shafts as sprockets.	■ Noise level is usually higher than with belts or gears (although silent chain drives are relatively quiet). ■ Chains can elongate because of wearing of link and sprocket teeth contact surfaces. ■ Usually limited to applications with lower speeds than belts or gears.

Magnetic and Hydraulic Drives

A magnetic drive (also referred to as a "mag drive") is a general term for a coupling used to transmit power through magnetic forces between the driver and the driven element (Figure 10.13). In this type of driver, the driver and driven sides might have no physical contact. Instead, they use a standard electric motor to drive a set of permanent magnets (which are mounted on a carrier or drive assembly) to drive an inner rotor connected to a pump's impeller. Because the two devices are connected magnetically, a pump seal is eliminated. Thus, these types of pumps are often referred to as sealless pumps.

Hydraulic drives transfer energy through closed circuit movements of a pressurized liquid in a conduit. In this type of drive, the pump section supplies pressurized liquid and

Figure 10.13 Magnetically driven pump.

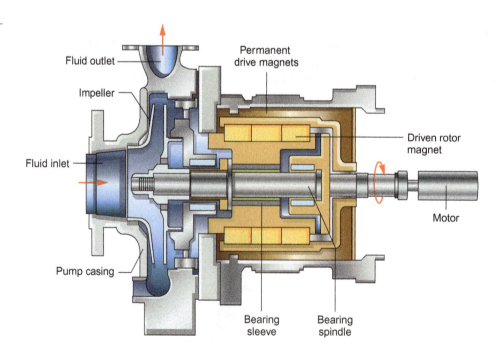

the motor section converts the liquid flow into shaft power. The automatic transmission system in an automobile is an example of a hydraulic drive.

10.3 Bearings, Gears, and Seals

Bearings (see Figure 10.14) are mechanical devices used to support shafts, limit radial (up-and-down or side-to-side) and axial (back-and-forth) movement, and hold the rotor in alignment with other parts. Lubricated bearings also help reduce friction and allow smooth movement of rotary equipment.

Thrust bearings, such as those used in turbines, centrifugal compressors, and large centrifugal pumps, prevent excessive axial (back-and-forth or end-to-end) movement parallel to the shaft and hold the rotor in correct alignment with nonmoving parts. Radial bearings support and prevent a rotor from moving from side to side or up and down.

Excessive vibration could be an indication that bearings are beginning to fail. To help monitor vibration, sensors can be used to measure the frequency and amplitude of the vibration. The results of these measurements can be displayed on a local control panel or in a control room, with limits set for alarms at high frequency.

Bearings require lubrication. Even with proper lubrication, bearings can wear out. Checking the temperature of a bearing is one of the tasks an operator may have to perform while monitoring equipment. High temperatures may indicate lack of lubrication or bearing wear.

Bearings   mechanical devices used to support shafts and limit radial (up-and-down or side-to-side) and axial (back-and-forth) movement and to hold the rotor in alignment with other parts.

Figure 10.14 Bearings help support shafts and reduce friction.

CREDIT: Kalabi Yau/Shutterstock.

There are many different types of bearings. The most common, however, are roller, ball, sleeve, and thrust bearings. Rolling element bearings include ball, cylindrical, tapered, and barrel-shaped rolling elements.

Roller bearings (see Figure 10.15A) are used in moderate-pressure applications and at low speeds (less than 1,000 rpm). The principal parts of roller bearings include inner surface (races) and outer rings, rolling elements, and the cage that spaces the rolling elements.

Ball bearings (see Figure 10.15B) are used in smaller machinery at low speeds. The principal components of ball bearings are the same as roller bearings. The difference is the shape of the rolling element (round versus cylindrical).

Tapered roller bearings (see Figure 10.15C) are used to prevent axial and radial movement of the shaft. Sleeve (fluid film or journal) bearings (see Figure 10.15D) are used for large equipment and high speeds. This type of bearing usually has a softer lining plate or pad that aligns with a polished inner shaft surface. The plate or pad can be split for ease of removal. Thrust bearings (see Figure 10.15E) are used to prevent axial movement of the shaft.

Figure 10.15 A. Roller bearings. **B.** Ball bearings. **C.** Tapered roller bearings. **D.** Sleeve (fluid film) bearing. **E.** Fluid film thrust bearing.

CREDIT: **A.** loraks/Shutterstock. **B.** Billion Photos/Shutterstock. **C.** pelfophoto/Shutterstock **E.** Vasyl S/Shutterstock.

A.

B.

C.

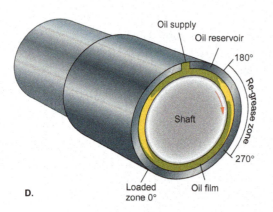
D.

E.

Gears

Gear a toothed wheel that engages another toothed mechanism in order to change the speed or direction of transmitted motion.

A **gear** is a toothed wheel that engages another toothed mechanism in order to change the speed or direction of transmitted motion. Figure 10.16 shows some of the more common types of gears.

Spur gears (see Figure 10.16A), which are the most commonly used gears in the process industries, have teeth that are cut parallel to the shaft. Spur gears are used primarily in low-power applications.

Helical gears (see Figure 10.16B) are similar to spur gears, but their teeth are cut on an angle. Because of their design, helical gears can handle more force than simple spur gears. Therefore, they are used in applications that require greater power (torque) and equipment size. Double helical gears can be beneficial because of their design. Each row of teeth cancels the thrust of the other, lessening the thrust load. This can eliminate the need for a thrust bearing.

Worm gears (see Figure 10.16C) are gears with slanted teeth, placed at a 90-degree angle, which are designed to mesh with another gear (referred to as a worm). Worm gears are used to obtain large speed reductions between nonintersecting shafts, and they are generally nonreversible (i.e., they can turn in only one direction).

A bevel gear (see Figure 10.16D) connects two intersecting shafts in any given speed ratio. Bevel gears are connected at a 90-degree angle and transmit power smoothly through a low power range.

Gearbox a mechanical device that houses a set of gears, connects the driver to the load, and allows for changes in speed, torque, and direction.

Gears also can be found in a gearbox, where they are used to increase or reduce the speed of driven equipment. A **gearbox** (see Figure 10.17) houses a set of gears, connects the driver to the load, and allows for changes in speed, torque, and direction. Gearboxes use a

Figure 10.16 A. Spur gears. **B.** Helical gear. **C.** Worm gear. **D.** Bevel gear.

CREDIT: A. Milos Luzanin/Shutterstock. **B.** Aerodim/Shutterstock. **C.** maxuser/Shutterstock. **D.** Bill Haag/Shutterstock.

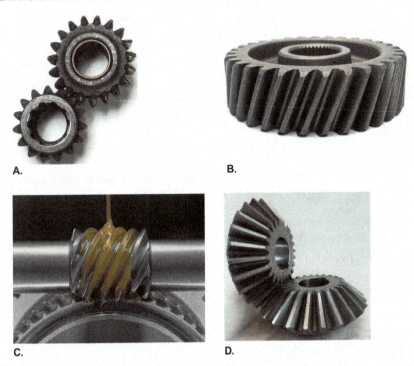

Figure 10.17 Gearbox.

CREDIT: asharkyu/Shutterstock.

series of large and small gears to adjust speed. For example, if the driver speed output is too high, a gearbox might be used to decrease the speed.

10.4 Principles of Lubrication

Lubrication is the application of a lubricant between moving surfaces in order to reduce friction and minimize heat generation. **Lubricants** can come in many different thicknesses and forms. For example, lubricants can be thin oils with low viscosity (resistance to flow) or thick, high viscosity greases. Figure 10.18 shows examples of various types of lubricants.

Lubricant   a substance used to reduce friction between two contact surfaces.

Figure 10.18 Examples of the various types of lubricants.

CREDIT: **A.** wk1003mike/Shutterstock. **B.** studiovin/Shutterstock. **C.** Evgenii Kurbanov/Shutterstock.

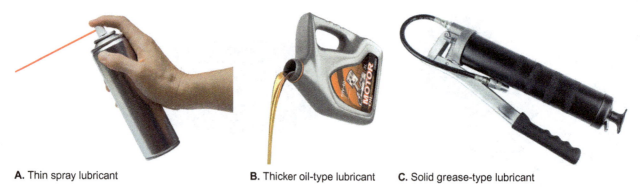

A. Thin spray lubricant   **B.** Thicker oil-type lubricant   **C.** Solid grease-type lubricant

Lubricants consist of a base (e.g., a natural or synthetic hydrocarbon) and appropriate additives. The most important additives are corrosion inhibitors and antioxidants. These additives provide protection to prolong equipment life. Other additives might include anti-foaming agents, emulsifiers, antiwear components, viscosity range extending additives, and extreme pressure chemical components.

Did You Know?

The Society of Automotive Engineers (SAE) has an established numerical system for rating motor oils according to their viscosity.

Some of the numbers contain the letter W, which stands for "winter." With 10W30 motor oil, your engine is protected with 10-weight oil in colder temperatures and 30-weight oil in warmer temperatures.

CREDIT: Ksander/Shutterstock.

Appropriately designed packing and seals hold lubricants in place and keep foreign materials out. Seals are used between sections in motors and turbine rotors. Figures 10.19 and 10.20 show examples of stuffing boxes, packing, and seals.

Figure 10.19 Stuffing box assembly.

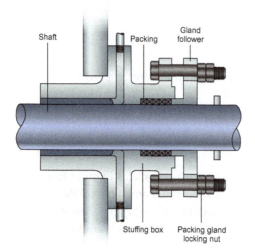

Lip seal

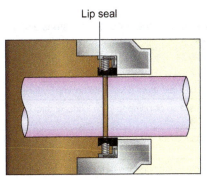

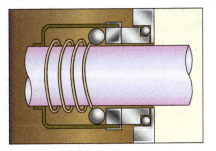

Lip seal **Mechanical seal**

Figure 10.20 Examples of packing and seals.

With the lubricant contained by packing and seals, most devices use capillary action to draw the lubricant film into the narrow spaces between components. Lubricants reduce friction and serve as heat transfer agents (i.e., they circulate through the bearing and back to a reservoir with a cooling system in order to remove heat). In general, high-speed applications use the less-viscous oils. Slower devices, such as vehicle wheels with ball or roller bearings, use more viscous oils or greases. Frictional resistance of greases is higher than that of oils, but grease is not lost from the lubricated area as readily because of its viscosity.

Examples of industrial equipment that uses lubricants include bearings, engines, chains, and gears. The lubricators are classified into several types: automatic, controlled feed, air-line, and electrically operated. Lubricating equipment regulates the quantity of the lubricant that is applied to the lubricated machine. Some other types of lubrication equipment for liquid lubricants include wick-feed oilers, centralized oil systems, and splash-oiling devices. Table 10.3 lists and describes different types of lubricating devices.

Did You Know?

Capillary action is what allows paper towels to wick up liquid.

Capillary action also allows lube oil to be drawn into the narrow spaces between two machine components.

Many lubricating devices use a centralized lubricant delivery system (see Figure 10.21) to introduce lubricant into specific areas of a piece of equipment. A centralized oil system is made up of a reservoir, pumps, heat exchangers for heating or cooling, lubricant filters, valves, and instruments.

Bottle oilers (see Figure 10.22) consist of a see-through container mounted next to an oil reservoir on a pump or other equipment. In this type of system, the oil is gravity-fed when air enters the container due to a decrease in the reservoir level. This provides a constant and consistent level of lubrication for equipment.

Force feed lubricators can be automatic or mechanical systems. They use a pump to force oil into machine components.

Table 10.3 Types of Lubricating Devices Used in the Process Industries and Their Descriptions

Types	Description
Centralized systems	Includes a central oil conditioning and pumping system.
Drip feed	Delivers lubricant through a drip that flows down onto bearing surfaces.
Force feed	Found in larger machinery; uses an oil pump to force lubricant through ducts drilled into the rotating members and casing.
Grease fittings	Receptacles that contain a ball check valve and receive grease from a portable dispensing gun.
Oil mist	Uses an oil distribution system to transport the oil through a mist generator to the lubricated parts.
Recycling system	Returns the oil to a reservoir where filtration and cooling occurs before redistribution back through the system.
Ring oilers	Metal rings that fit around and rotate off a shaft. Oilers have a larger diameter than the shaft, and the lower end dips into an oil pool. Oil is carried on the ring up to the shaft. As it contacts the shaft, it distributes the oil.
Splash feed	Uses a moving member, such as a crankshaft, to distribute the liquid by splashing droplets out of a pool and into the general bearing areas.
Wick feed	Similar to a drip feed oiler; supplies lubricant to the bearing through the capillary action of fibrous material.
Gravity feed	Commonly called a bottle oiler, oil resides in a bowl-type reservoir connected to the bearing housing. Oil flows by gravity to the casing to lubricate bearings.

Figure 10.21 Centralized lubrication system.

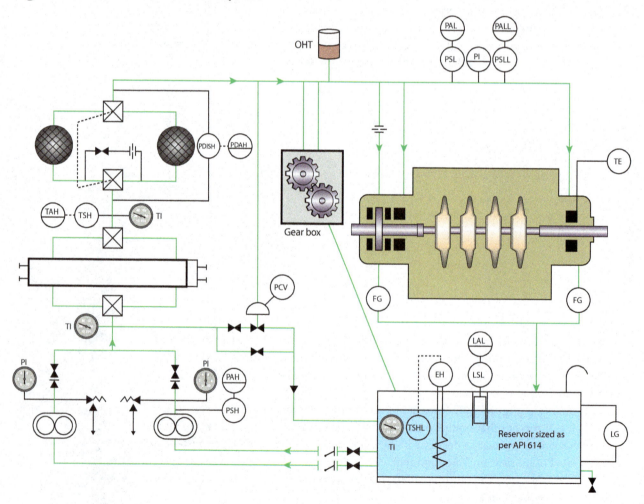

Figure 10.22 Bottle oiler.

CREDIT: Oil and Gas Photographer/ Shutterstock.

Grease fittings are receptacles that contain a ball check valve and receive grease from a portable dispensing gun. Figure 10.23 shows an example of a grease dispenser attached to a grease fitting.

Oil mist lubrication (see Figure 10.24) uses an oil distribution system to transport the oil through a mist generator and then through piping to the lubricated parts. A two-cycle motor engine is an example of a device that uses oil mist lubrication. In a two-cycle engine, oil is mixed with the fuel; as the fuel is vaporized, the oil is carried into the engine.

Recycling lubrication systems return the oil to a reservoir, where filtration and cooling occur before the lubricant is redistributed back through the system.

Figure 10.23 Grease dispenser (grease gun) attached to grease fitting.

CREDIT: Patnaree Asavacharanitich/ Shutterstock.

Figure 10.24 Oil mist lubrication system.

CREDIT: Dale Stagg/Shutterstock.

Ring oilers (see Figure 10.25) are metal rings that fit and rotate around a shaft. These types of oilers have a larger diameter than the shaft so the bottom of the ring dips into an oil pool. Oil is carried on the ring up to the shaft and the bearings.

Figure 10.25 Ring oiler.

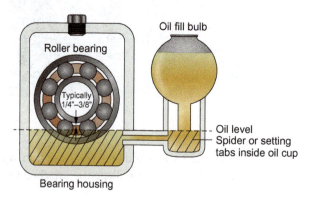

Splash feed uses a moving member, such as a crankshaft (see Figure 10.26), to distribute the lubricant by splashing droplets of the oil out of a pool and into the bearing areas.

A wick feed oiler (Figure 10.27) is similar to a drip feed oiler. Instead of dripping the oil, however, wick feed oilers supply lubricant to bearings through the capillary action of a fibrous material.

Figure 10.26 Crankshaft in a crankcase.

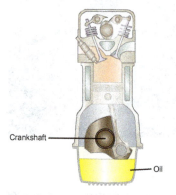

Figure 10.27 Wick feed oiler.

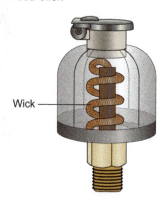

Proper equipment lubrication must be maintained at all times. The process technician's responsibilities regarding equipment lubrication can include observation of the equipment, simple maintenance tasks of adding oil to the equipment, and working with the maintenance department on dedicated procedures specific to the equipment. Specific procedures for lubrication maintenance, from filter changes to oil changes, can be scheduled as weekly or monthly tasks, depending on the size, type, and criticality of the equipment. The process technician will be trained accordingly on procedures specific to the operating equipment in their process unit or area.

10.5 Potential Problems

It is important to maintain proper equipment lubrication because failure to do so can lead to premature wear, heat buildup, and equipment failure that could affect other parts of the process. The entire process facility could be shut down if a gearbox located on a critical piece of equipment fails because of improper lubrication. Because of this, process technicians should always be alert and able to detect unusual noises, excessive vibrations, changes in lubricant color and consistency, and unusual smells that might indicate that the lubricant temperature is getting too high. Process technicians should also be aware of associated protection devices such as interlock systems and vibration monitors, as well as startup and shutdown procedures. If something unusual occurs, it is important to speak up about it to ensure personal safety and safety of other workers and the equipment.

Interlock Protection

Some interlock systems are designed to keep large, single-train, rotating equipment from sustaining serious damage when upset conditions or component failures occur. Process technicians must be aware of how interlock systems work so that they can help prevent emergency shutdowns and promptly recover and restart the process should a shutdown occur. Process technicians must also know the various conditions that can cause equipment shut down, such as the trip points, and how to determine which condition initiated the trip event.

Vibration Monitoring

Excessive vibration, particularly on large rotating equipment, can indicate a potentially serious problem. Portable and/or permanently mounted vibration sensors are often used to detect vibrations that are outside the normal range. Process technicians are responsible for monitoring these vibration sensors and recording data trends so potential problems can be identified and resolved.

Startup and Shutdown Procedures

Improper startup can cause sudden surges that can shorten the life of the rotating equipment and cause excessive stress on bearings, gears, seals, and other components. Because of this, specific operating procedures must be closely followed when rotating equipment is started up or shut down.

Improper Seal Procedures

In many seal installations, clean liquid purges lubricate and cool rotating components. Process technicians operate the pumps (if in place) that supply the seal liquids. They might encounter improper levels of lubrication oil in a gearbox, the use of the wrong type of oil or grease, or water in the lubricant. Each of these can cause mechanical problems in the pump, and all three can cause worn teeth on a gearwheel.

10.6 Safety and Environmental Hazards

Hazards associated with normal and abnormal transmission and lubrication operations can negatively affect the personal safety of the process technician, equipment reliability, and the environment. For example, seal failure can allow process fluid to escape into the atmosphere and negatively affect personnel and the environment.

Personal Safety

Process technicians should always be aware of the potential hazards and the necessary safety precautions associated with transmission and lubrication operations. Couplings pose a risk because of the possibility of failure (disintegration and flying debris). All couplings must be covered with a coupling guard to keep people from accidentally coming into contact with rotating equipment and to protect workers in case of flying debris.

Although the lubricating oils are below their flashpoint, fire danger still exists because of potential exposure to hot surfaces (e.g., steam lines).

Process technicians might be required to make routine and preventive maintenance checks. Personal protective equipment, such as hard hats, fire-retardant clothing, gloves, hearing protection, and safety glasses must be worn when working in process areas. Process technicians should also be aware of the danger of loose garments, long hair, gloves, and jewelry getting caught in rotating equipment and avoid these hazards when working around rotating equipment. Many process facilities have personal protective equipment dress codes to reduce this hazard.

Equipment Operations

Equipment operation hazards are associated with power transmission during normal and abnormal operations. The possibility of failed transmission, gears, and lubrication systems exists at all times. Any of these failures directly affect the operation of plant equipment.

The coupling is designed to be the "weak link" between the driver and driven equipment, to prevent serious damage in the event of a malfunction. If the coupling does fail, there can be a danger of flying debris, depending on the type of coupling used. Noisy gearboxes might also be an indication of potential problems. Process technicians must be alert to changes in noise levels and also check the temperatures, smells, and color changes, as appropriate, to prevent personnel injury.

As with any high-pressure fluid, hydraulic fluid presents a potential safety hazard. Bearing failures in centrifugal pumps or other rotating equipment pose a hazard because it is not uncommon under abnormal conditions for heat buildup to create an ignition source and result in a fire.

Environmental Impact

Process technicians are responsible for maintaining equipment operation to prevent leaks and emissions. Releases to the environment include liquid spills to the ground as well as vapor releases to the air. Leaks, unusual noises, and vibrations should all be investigated promptly. Appropriate maintenance or technical support personnel might need to be involved to correct the problem.

10.7 Process Technician's Role in Operation and Maintenance

Process technicians might be required to complete a number of tasks associated with the maintenance of power transmission and lubricating equipment. These tasks can include:

- Swinging or replacing oil filters
- Swinging oil coolers or backflushing the waterside of oil coolers
- Collecting and recording vibration levels and readings
- Checking oil levels and adding oil if necessary
- Checking the grease supply and adding grease if necessary
- Changing oil
- Sampling oil for water and other contaminants
- Checking bearing surface temperatures with a skin or surface pyrometer
- Monitoring and ensuring proper operation of oil centrifuges

Lubrication Maintenance

Process technicians are responsible for maintaining oil levels, changing system filters, and/or adding grease to bearings.

Oil levels in lube oil reservoirs must be checked frequently and maintained at designed operating levels. Vents in the lube oil reservoir should be monitored to ensure they are not venting process materials or vapors. In some systems, water contamination of the oil can occur in several ways. Larger rotating equipment has sight flow windows through which the color of the lube oil can be checked. A milky or frothy appearance should be immediately investigated because it indicates contamination of the oil. Water contamination can be removed using a portable centrifuge with a slipstream of oil from the pump.

Filter changes are crucial because a clean lube oil supply ensures that the lubrication system will operate properly. To keep it clean, the lube oil usually flows through filters before going into the machinery. Filter pressure drop is a variable that should be closely monitored to help avoid problems. An excessive differential pressure across the filter (from inlet to outlet) is an indication of a dirty or plugged filter. Filter elements are replaced before the pressure drop reaches a point that would prevent oil from circulating through it. Extreme care must be taken when swinging filter elements to avoid shutting off the oil flow. Process technicians must also ensure that the flow does not bypass the filter elements. Unfiltered oil might contain contaminants that could damage the lubricated equipment.

In many process facilities, rotating equipment components have grease fittings. It may be the responsibility of the process technician to add grease to these fittings at intervals specified by equipment procedures. Selecting the proper lubricant, adding the correct amount, and recognizing contamination are important responsibilities of a process technician. In a large process facility or refinery, many different kinds of greases and lubricating oils are used. The process technician is responsible for using the proper lubricant to service each piece of equipment.

Summary

Electricity either oscillates (alternating current or AC) or flows in a single direction (direct current or DC). Components of an electric distribution system include three-phase and single-phase systems, transformers, motor control centers (MCCs), circuit breakers, fuses, and switches. For safety reasons, when operating large electrical breakers, personnel should not stand directly in front of the breaker.

Mechanical power transmission and lubrication systems are critical to the daily operation of process equipment. The purpose of mechanical power transmission is to transfer rotational energy from one point to another with minimum loss of energy to friction. Rotating equipment coupling guards must always be secured in place. Loose-fitting clothes, hair, and jewelry must always be kept away from rotating equipment.

Lubrication is the process of establishing a stable fluid film between two hard metal surfaces moving relative to each other. The objective of lubrication systems is to transport the lubricating fluid into the area between the metal surfaces of the equipment in order to minimize contact friction. Lubrication prevents metal-to-metal moving contact and removes heat caused by friction, both of which are generally destructive to equipment. Lubrication of valves makes them easier to open and close. Process technicians must wear the proper PPE when handling lubricants.

Mechanical power transmission consists of various components that connect drivers to driven equipment in order to transfer power. Couplings are power transfer devices that connect the shafts of drivers to driven equipment. Types of couplings are flexible, rigid, and fluid couplings.

Gearboxes contain gears used to increase or decrease the speed of driven equipment relative to the speed of a driver. Belts use traction to transfer rotational energy to pulleys. Chains are components that perform the same function but transfer rotational energy through sprockets.

Magnetic drive pumps transmit power through magnetic forces between the driver and the driven elements. Hydraulic drives transfer energy through closed-circuit movements of a pressurized liquid in conduits.

Bearings are used to support shafts and reduce friction of radial (up-and-down or side-to-side) and axial (back-and-forth) movement. They allow smooth movement between two surfaces on rotary or linear moving equipment. Roller, ball, sleeve (fluid film), and thrust are the four main types of bearings. When inspecting bearings for vibration or wear, it is common to feel the bearing housing. The operator should wear gloves to safely feel a bearing housing.

Gears are used in most mechanical devices to transmit motion and control the speed of equipment. Gears come in many sizes and shapes. The most common types of gears are spur, helical, worm, and bevel.

Problems that can occur as a result of improper mechanical energy transmission or lubrication can affect equipment life, process facility operations, interlock protection, vibration monitoring, startup and shutdown procedures, packing and seals, and improper seal procedures. Ultimately, problems can result in the necessity of replacement costs much sooner than anticipated and might affect personal or process safety.

Appropriately designed seals hold lubricants in place and keep foreign materials out. These seals are placed at shaft exit points just beyond the bearings. In many seal installations, clean fluid purges lubricate and cool rotating components.

Interlock systems are designed to keep large (and often single-train) rotating equipment from sustaining serious damage when upset conditions or component failures occur. Leaks, noise, unusual smells, color changes, and unusual vibrations should all be investigated promptly by appropriate support personnel.

Hazards associated with normal and abnormal transmission/lubrication operations can affect the personal safety of process technicians, equipment reliability, and the environment.

Process technicians must perform various types of maintenance, including transmission system maintenance, lubrication maintenance (including maintaining oil levels), changing filters, and adding grease to bearings.

Checking Your Knowledge

1. Define the following terms:
   a. Alternating current (AC)
   b. Direct current (DC)
   c. Ammeter
   d. Circuit breaker
   e. Fuse
   f. Motor control center (MCC)
   g. Switch
   h. Ampere
   i. Ohm
   j. Volt
   k. Watt

2. A rigid coupling is used:
   a. to allow a motor to accelerate to full speed before the load is increased.
   b. in applications where small amounts of misalignment can be tolerated.
   c. to cushion the shock from equipment overloads.
   d. more frequently in general power transmission service and permits some misalignment.

3. (True or False) One advantage of a chain over a belt driver is that there is no slippage between chain and sprocket teeth.

4. Which of the following definitions best describes sleeve (fluid film) bearings?
   a. They are used in moderate-pressure applications in small to midsize operations (less than 1,000 rpm).
   b. They are used in smaller machinery at low speeds.
   c. They are used in larger machinery at low speeds.
   d. They are used in larger machinery at high speeds.

5. _____ gears are used to obtain large speed reductions between nonintersecting shafts placed at 90-degree angles to one another.
   a. Spur
   b. Worm
   c. Helical
   d. Bevel

6. _____ facilitates the spreading of lubricant film into narrow spaces between equipment components.
   a. Lubrication
   b. Capillary action
   c. Corrosion inhibitors
   d. Antifoams

7. Which of the following are the most important additives used in lubricants? (Select all that apply.)

   a. Corrosion inhibitors

   b. Antifoams

   c. Emulsifiers

   d. Antioxidants

8. Which type of lubricator is used in larger machinery and uses an oil pump to inject lubricant through ducts drilled into rotating members and casings?

   a. Ring oilers

   b. Wick feed

   c. Force feed

   d. Drip feed

9. (True or False) Interlock systems are designed to prevent rotating equipment from sustaining serious damage when upset conditions or component failures occur.

10. (True or False) Equipment vibration on large rotating equipment can be an indication of a potentially serious problem.

11. _____ is the process of connecting an object to the earth using copper wire and a rod to provide a path for electricity to dissipate harmlessly into the ground.

12. (True or False) For motors, fast-acting fuses must be used because there is a large current in-rush when a motor starts.

13. What type of bearings are used to prevent axial movement of the shaft?

   a. Roller

   b. Ball

   c. Sleeve

   d. Thrust

14. The _____ is designed to be the "weak link" between the driver, gearbox, or machinery train to prevent serious damage in the event of a malfunction.

15. List four tasks that process technicians might be required to complete to maintain power transmission and lubricating equipment.

16. (True or False) Process technicians perform tasks such as maintaining oil levels, changing system filters, and adding grease to bearings.

NOTE: Answers to Checking Your Knowledge questions are in the Appendix.

Student Activities

1. Describe the basic principles of electricity, including the difference between AC and DC, and identify which type is most commonly used in the process industry.

2. Write a one-page paper about the purpose and principle of mechanical power transmission.

3. With a classmate, research couplings and explain the three main types of couplings.

4. With a pulley system in a lab setting, show how to properly adjust belt tension.

5. Given a drawing, identify the location of the bearings on a piece of equipment and identify the type of each bearing.

6. Given drawings or actual gears, identify and describe spur, helical, worm, and bevel gears.

7. Explain the process technician's role in the following tasks:

   • Power transmission system maintenance

   • Maintaining oil levels

   • Changing system filters

   • Adding grease to bearings

8. With a classmate, prepare to exhibit or demonstrate different types and grades of lubricants, lubricating devices, and/or typical lubricating procedures or requirements.

Chapter 11

Heat Exchangers

Objectives

After completing this chapter, you will be able to:

11.1 Identify the purpose, common types, and applications of heat exchangers. (NAPTA Heat Exchangers 1, 2*) p. 234

11.2 Identify the components of heat exchangers and the purpose of each. (NAPTA Heat Exchangers 3, 4) p. 242

11.3 Describe the operating principles of heat exchangers. (NAPTA Heat Exchangers 5, 10) p. 243

11.4 Identify potential problems associated with heat exchangers. (NAPTA Heat Exchangers 9) p. 247

11.5 Describe safety and environmental hazards associated with heat exchangers. (NAPTA Heat Exchangers 6) p. 249

11.6 Describe the process technician's role in heat exchanger operation and maintenance. (NAPTA Heat Exchangers 8) p. 249

11.7 Identify typical procedures associated with heat exchangers. (NAPTA Heat Exchangers 7) p. 250

* North American Process Technology Alliance (NAPTA) developed curriculum to ensure that Process Technology courses will produce knowledgeable graduates to become entry-level employees in process technology. Objectives from that curriculum are named here in abbreviated form. For example, "(NAPTA Heat Exchangers 1, 2)" means that this chapter's objective relates to objectives 1 and 2 of the NAPTA curriculum about heat exchangers.

Key Terms

Aftercooler—a heat exchanger, located after the final discharge stage of a compressor, **p. 240.**

Backwashing (backflushing)—a procedure in which the direction of flow through the exchanger is reversed to remove solids, **p. 251.**

Baffles—metal plates that are placed inside a vessel or tank and are used to alter the flow, facilitate mixing, or cause turbulent flow, **p. 247.**

British thermal unit (BTU)—a measure of energy in the English (Imperial standard) system; the heat required to raise the temperature of 1 pound of water 1 degree Fahrenheit at sea level, **p. 234.**

Calorie (cal)—gram calorie; the amount of heat energy required to raise the temperature of 1 gram of water by 1 degree Celsius at sea level, **p. 234.**

Chiller—devices used to cool a fluid to a temperature below ambient temperature. Chillers generally use a refrigerant as a coolant, **p. 240.**

Condenser—an exchanger used to convert a substance from a vapor to a liquid, **p. 240.**

Conduction—the transfer of heat from one substance to another by direct contact, **p. 234.**

Convection—the transfer of heat as a result of fluid movement, **p. 234.**

Countercurrent flow—flow movement that occurs when two fluids are moving in opposite directions, **p. 246.**

Crossflow—movement that occurs when two streams of fluid flow perpendicular (at a 90-degree angle) to each other, **p. 246.**

Economizer—a type of heat exchanger that preheats a fluid with waste heat, such as boiler feedwater being preheated with flue gases exiting a boiler, **p. 244.**

Exchanger head—device located on the end of a heat exchanger that directs the flow of fluids into and out of the tubes; also called a channel head, **p. 242.**

Finfan coolers—air coolers consisting of banks of finned tubes, connected by an external tubesheet, that transfer heat from process streams flowing through the tube banks to the atmosphere, **p. 240.**

Fouling—accumulation of deposits that build up on the surfaces of processing equipment, **p. 247.**

Heat—a measure of the transmission of energy from one object to another as a result of the temperature difference between them, **p. 234.**

Heat exchangers—devices that are used to transfer (exchange) heat from one substance to another, usually without the two substances physically contacting each other, **p. 234.**

Heat transfer—the transmission of energy from one object to another as a result of a temperature difference between the two objects, **p. 234.**

Hybrid flow—movement of fluid in a heat exchanger that is any combination of counterflow, parallel flow, and crossflow, **p. 246.**

Interchanger—a process-to-process heat exchanger that uses hot process fluids on the tube side and cooler process fluids on the shell side. Also known as a cross exchanger, **p. 240.**

Intercooler—a heat exchanger used between the stages of a compressor to cool the gas and remove any liquid before it reaches the next stage, **p. 240.**

Laminar flow—(also called *streamline flow*) movement of fluid that occurs when the Reynolds number is low (at very low fluid velocities), **p. 246.**

Parallel flow—movement of fluid that occurs in a heat exchanger when the shell flow and the tube flow are parallel to one another, **p. 246.**

Preheater—a heat exchanger that adds heat to a substance prior to a process operation, **p. 240.**

Radiation—the transfer of heat through electromagnetic waves (e.g., warmth emitted from the sun), **p. 234.**

Reboiler—a tubular heat exchanger, usually located at the bottom of a distillation column or stripper, that is used to supply the necessary column heat, **p. 240.**

Reynolds number—a numeric designation used in fluid mechanics to indicate whether a fluid flow in a particular situation will be laminar (smooth) or turbulent, **p. 246.**

Shell—the outer casing, or external covering, of a heat exchanger, **p. 242.**

Temperature—a measure of the thermal energy of a substance (e.g., the "hotness" or "coldness") that can be determined using a thermometer, **p. 234.**

Tube bundle—a group of fixed, parallel tubes though which process fluids are circulated, **p. 242.**

Tube sheet—a formed metal plate with drilled holes that allow process fluids to enter the tube bundle, **p. 242.**

Turbulent flow—movement of fluids that occurs when the Reynolds number is high, characterized by irregular flow patterns and high velocities, **p. 246.**
Vaporizer—a device that converts a liquid into a vapor, **p. 241.**

11.1 Introduction

Heat exchangers are devices that are used to transfer (exchange) heat from one substance to another, usually without the two substances physically contacting each other. Without heat exchangers, many processes could not occur properly. The primary modes of heat transfer used in heat exchangers are conduction and convection. (Heat also can be transferred by radiation, but this is not usually found in heat exchanger systems.)

When there is a temperature difference between two fluids, heat is transferred from the fluid with the higher temperature to the fluid with the lower temperature. In other words, the heat travels from hot to cold.

Heat exchangers come in a variety of types, including double-pipe, spiral, plate and frame, and shell and tube. The most common type, however, is the shell and tube exchanger. Although the designs of different heat exchangers vary, all have similar components and use the same principles of heat transfer. When working with heat exchangers, process technicians should be aware of the operational aspects of the exchanger and the factors that could affect heat exchange.

HEAT TRANSFER OVERVIEW When working with **heat transfer,** it is important to know that temperature and heat are not the same thing. **Temperature** is a measure of the average thermal energy of a substance (e.g., the "hotness" or "coldness"), which can be determined using a thermometer. **Heat** is a measure of the transmission of energy from one object to another as a result of the temperature difference between them.

According to the principles of physics, heat energy always moves from hot to cold. Heat, a form of energy, cannot be created or destroyed; it can only be transferred from one object to another.

Many different units are used to measure energy, and more specifically heat energy, depending on the measurement system in use. Heat can be measured in British thermal units or joules; it may also be measured in calories. A **British thermal unit (BTU)** is the amount of heat energy required to raise the temperature of 1 pound of water by 1 degree Fahrenheit. A **calorie (cal)**, or gram calorie, is the amount of heat energy required to raise the temperature of 1 gram of water by 1 degree Celsius. One BTU is equivalent to 252 calories. The term *Calorie* (Cal), or kilogram calorie with a capital "C" can be found on food labels. The term Calorie is really 1,000 times larger than the calories referred to in this chapter. That means one BTU is equivalent to 0.252 of the Calories we eat. Table 11.1 provides some equivalencies.

Heat exchangers devices that are used to transfer (exchange) heat from one substance to another, usually without the two substances physically contacting each other.

Heat transfer the transmission of energy from one object to another as a result of a temperature difference between the two objects.

Temperature a measure of the thermal energy of a substance (e.g., the "hotness" or "coldness") that can be determined using a thermometer.

Heat a measure of the transmission of energy from one object to another as a result of the temperature difference between them.

British thermal unit (BTU) a measure of energy in the English (Imperial standard) system; the heat required to raise the temperature of 1 pound of water 1 degree Fahrenheit at sea level.

Calorie (cal) gram calorie; the amount of heat energy required to raise the temperature of 1 gram of water by 1 degree Celsius at sea level.

Table 11.1 Comparison of Heat Units

	BTUs	calories (gram)	joules	Calories (food)
1 BTU =	----	252	1055	0.252
1 calorie (gram) =	0.00397	----	4.186	0.001
1 joule =	0.000948	0.239	----	0.000239
1 Calorie (food) =	3.97	1000	4186	----

Since the adoption of the SI system, the unit of energy in the International System of Units is the joule. One small calorie is about 4.2 joules, so one Calorie is about 4.2 kilojoules.

As mentioned, heat can be transferred by conduction, convection, and sometimes radiation. **Conduction** is the transfer of heat from one substance to another by direct contact (e.g., transferring heat energy from a frying pan to an egg or from heat tracing attached to a pipe, through the pipe, to a process fluid). **Convection** is the transfer of heat as a result of fluid movement (e.g., warm air circulated by a hair dryer or heat transferred from hot water to cold water in a pot). **Radiation** is the transfer of heat by electromagnetic waves (e.g., warmth emitted from the sun). Figure 11.1 shows examples of different types of heat transfer.

Conduction the transfer of heat from one substance to another by direct contact.

Convection the transfer of heat as a result of fluid movement.

Radiation the transfer of heat through electromagnetic waves (e.g., warmth emitted from the sun).

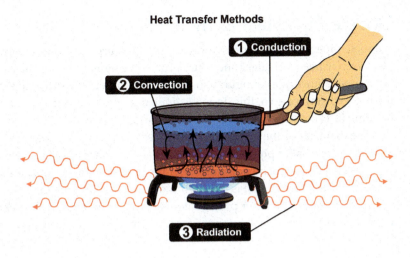

Heat Transfer Methods

❶ Conduction

❷ Convection

❸ Radiation

Figure 11.1 Heat may be transferred by conduction, convection, or radiation.

CREDIT: udaix/Shutterstock.

Types of Heat Exchangers

The basic purpose of all heat exchangers is to heat or cool fluids, but the design of the exchangers can differ. Types of heat exchangers include double-pipe, spiral, cold box, air cooled, plate and frame, and shell and tube. Exchangers also can be classified by their function (e.g., condenser, reboiler, preheater, vaporizer, chiller, intercooler, and after-cooler). Various symbols representing different types of heat exchangers are shown in Figure 11.2.

Figure 11.2 Some common heat exchanger symbols.

Heat exchanger symbols				
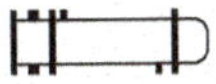 Floating head heat exchanger 1	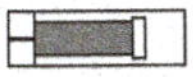 Floating head heat exchanger 2	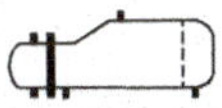 Kettle reboiler	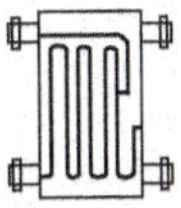 Plate exchanger	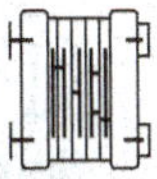 Plate and frame heat exchanger
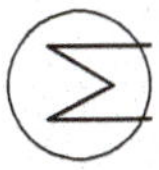 Heat exchanger 1	 Heat exchanger 2	 Condenser	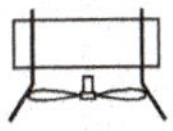 Forced draft air cooler	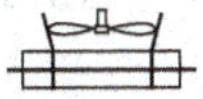 Induced draft air cooler
 Finfan cooler	 Finned tubes heat exchanger	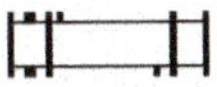 Shell & tube multi-pass heat exchanger	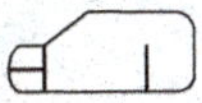 Reboiler heat exchanger	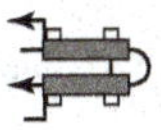 Hairpin exchanger
 Spiral heat exchanger 1	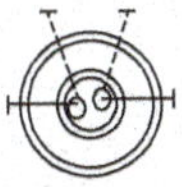 Spiral heat exchanger 2	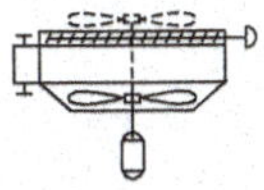 Air cooled exchanger	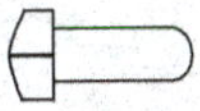 U-tube heat exchanger 1	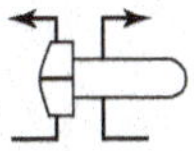 U-tube heat exchanger 2

Double-Pipe Heat Exchangers

A double-pipe heat exchanger, sometimes called a hairpin, (shown in Figure 11.3) is the simplest form of heat exchanger and consists of a pipe within a pipe. The strength of inner and outer pipes allows this configuration to be suited for high-pressure applications. Because of the reduced surface area, this design is limited to low heat duty, where the change in temperature (ΔT) between the two fluids is not large.

The diameter of the larger pipe is reduced by a flange or other transition piece and is welded to the smaller pipe to create an annular (ring-shaped) space between the two pipes. Nozzles attached to each end of the outer pipe allow a stream to pass through the annular space, while another stream is passed through the smaller pipe for heat exchange.

In some cases, the inner pipe has louvers or fins to increase the heat transfer capacity. Finned tubes improve heat exchange for low-density fluids, such as gases.

Figure 11.3 Double-pipe heat exchanger.

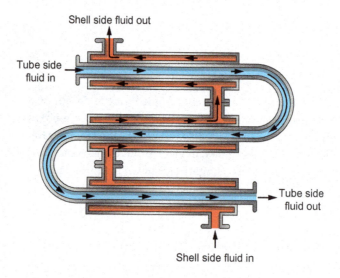

Spiral Heat Exchangers

Spiral heat exchangers (shown in Figure 11.4) consist of spiral plates or helical tube cores fixed in the shell. Spiral heat exchangers offer high reliability and are good for operations with high fouling potential, such as slurries. Because there are no *dead spaces* (areas without movement), the helical flow pattern creates a high-turbulence area that is self-cleaning.

Figure 11.4 Spiral heat exchanger.

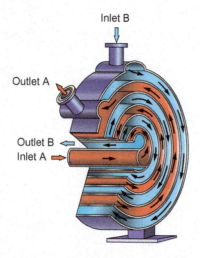

Cold Box Heat Exchangers

A cold box heat exchanger is a piece of equipment used in low-temperature and cryogenic processes such as the purification and liquefaction of natural gas, the production of ethylene, hydrogen recovery, and the separation of air components. The cold box is actually the enclosure, usually an insulated rectangular housing, for one or more heat exchange sections for multiple streams, in addition to other equipment related to the process such as knockout drums, piping, and instrumentation. The exchanger is a plate and fin design, most often made of brazed aluminum, consisting of layers of corrugated fins separated by aluminum plates (Figure 11.5). Aluminum is used because of its ability to withstand the low temperature requirements and its excellent heat transfer capabilities. Heat is transferred from a process stream, through the fins and separator plates, into the next set of fins and then to the adjoining process stream. A nitrogen purge is used to maintain an inert environment in the cold box and to protect it from developing a potentially explosive condition if a leak develops.

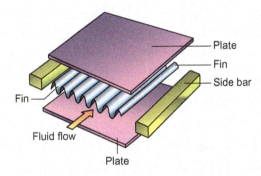

Figure 11.5 Plate-fin heat exchanger.

Air Cooled Heat Exchangers

Similar to a car radiator, air cooled heat exchangers, or finfans, use ambient air to cool a process stream. They are used in areas where water is not available as a cooling medium or when ambient air temperature can provide sufficient cooling to achieve the required outlet temperature. A header box distributes the process stream through multiple rows of finned tubes, over which air flows. The tube fins greatly increase the surface area over which heat transfer occurs, conducting heat from the tubes and transferring it to the air. Air cooled heat exchangers can be of a forced draft design (with the fan located beneath the tubes as shown in Figure 11.6) or an induced draft design (with the fan located above the tubes).

Figure 11.6 Forced draft air cooled heat exchanger.

CREDIT: Oil and Gas Photographer/ Shutterstock.

Shell and Tube Heat Exchangers

Shell and tube heat exchangers, the most common type of exchangers in the process industries, consist of a tube bundle inside a shell. They are used for heat removal in chillers, condensers, coolers, and in evaporation and refrigeration systems. They can also add heat to process fluid in preheaters, waste heat boilers, thermosiphon and kettle reboilers, and process heaters.

The main methods of heat transfer in a shell and tube heat exchanger are conduction (through the tube wall and tube surface to the shell fluid) and convection (fluid movement within the shell and the tubes). For example, hot process fluids can be passed through the tubes while cool water is pumped into the shell. As the cool water circulates around the tubes, heat is transferred from the fluid in the tubes to the water in the shell.

Drawings of shell and tube heat exchangers are shown in Figures 11.7, 11.8, and 11.9. Each of these shell and tube heat exchangers consists of a cylinder with flanged, removable head(s) attached to end(s) that direct flow through the exchanger tubes and allow easy access for cleaning the tubes.

Figure 11.7 shows a fixed tubesheet type of shell and tube heat exchanger. The advantage of this design is its low cost. The disadvantage is that, because of stress created by thermal expansion on the fixed tubesheets, the temperature differential (ΔT, the difference between the coldest possible temperature of the cold fluid and the hottest possible temperature of the hot fluid) cannot be too great. Fixed tubesheet heat exchangers can be designed for single-pass or multipass arrangements based on the design of the channel heads.

Figure 11.7 Shell and tube heat exchangers (fixed tubesheet type). **A.** Single pass. **B.** Multipass.

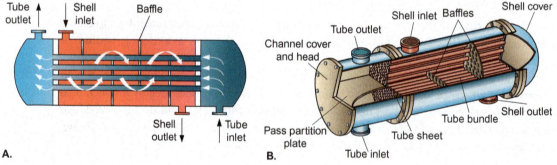

Figure 11.8 Shell and tube heat exchanger (floating head type).

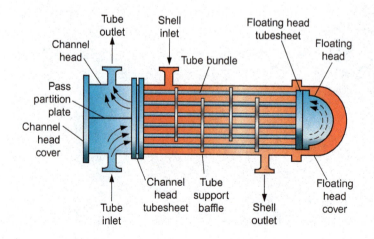

Figure 11.8 shows a floating head type of shell and tube heat exchanger. A *floating head* is an arrangement in which the process fluid coming out of the tubes at one end is captured and directed by a head that is not fixed to the shell of the heat exchanger. The floating head is used to address thermal stress caused by thermal expansion. Advantages of this design are midrange cost and the ability to handle a larger ΔT than a fixed tubesheet design. The limiting factor for ΔT in this type of exchanger is created by the thermal expansion of the studs securing the floating head bonnet.

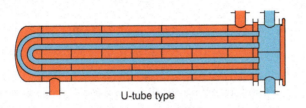

U-tube type

Figure 11.9 Shell and tube heat exchanger (U-tube type).

Figure 11.9 shows the U-tube type of shell and tube heat exchanger. A U-tube heat exchanger is a type of shell and tube heat exchanger that uses U-shaped tubes. In this type of heat exchanger, baffles direct the shell side flow back and forth across the tubes. The benefit of this design is that the tubes are better able to deal with thermal expansion. Because they start and end on the same side of the exchanger, they are able to handle a very large ΔT. All the tubes are allowed to expand without putting stress on the tubesheet. The disadvantage of this design is that each tube must be bent differently, resulting in high construction costs.

The shell side of any type of shell and tube exchanger can be designed with baffles (shaped metal dividers). These baffles help exchangers increase fluid turbulence, thereby increasing the overall amount of heat transferred. This results in an exchanger that requires less surface area.

A channel head with partition plates (pass dividers) changes a shell and tube heat exchanger from a single to a multipass heat exchanger (see Figures 11.7B and 11.8). This makes more efficient use of the cooling medium and reduces fouling rates.

Kettle Reboilers

Kettle reboilers are a special type of heat exchanger (shown in Figure 11.10) that have a vapor-disengaging space built into the shell. Reboilers are commonly used with distillation towers to generate the vapor which drives the fractionation.

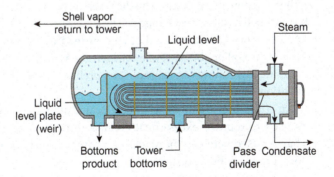

Figure 11.10 Kettle type reboiler heat exchanger.

Plate and Frame Heat Exchangers

Plate and frame heat exchangers (Figure 11.11) consist of a series of thin plates that are mounted on a frame. The plates are sealed with and separated by gaskets, which prevent leakage between them. Two fluids (hot and cold) are each directed through alternating plates. The high fluid velocities and the thin wall of the plates result in a very high overall heat transfer rate (two to three times that of a shell and tube exchanger).

Figure 11.11 A. Diagram of plate and frame heat exchanger. **B.** Photo of plate and frame heat exchanger.

CREDIT: B. NavinTar/Shutterstock.

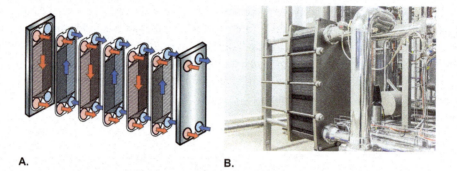

A.   B.

The high transfer rate allows a typical plate and frame exchanger to be relatively small, saving considerable space in a process plant.

Because of their design, plate and frame exchangers usually are used with clean (solids-free), nontoxic materials, and they operate at fairly low pressures and temperatures. Their efficiency makes plate and frame exchangers useful where very high overall heat transfer rates are required.

Exchanger Applications and Services

Heat exchangers are used to heat or cool process fluids. They can be used alone or coupled with other heat-exchanging devices such as cooling towers. The name used for a heat exchanger often is associated with its application, such as preheater, reboiler, aftercooler, condenser, chiller, intercooler, interchanger, fin fan cooler, heater, and vaporizer.

A **preheater** is a heat exchanger that adds heat to fluids prior to a process operation. A preheater can be used to heat a liquid to near its boiling point before it enters a distillation tower or to heat the feedstock before it enters a reactor.

A **reboiler** is a tubular heat exchanger, usually placed near the base of a distillation column or stripper, which is used to supply necessary column heat. The purpose of a reboiler is to produce vapor from the liquid that enters it. A reboiler provides the distillation column with the vapor necessary for separation inside the column. The control temperature is based on the boiling point of the fluid being converted to vapor. The heating medium for a reboiler can be steam or hot fluids from other parts of the facility.

A **condenser** is a heat exchanger used to convert a vapor to a liquid, often found in the overhead system of a column. The design of a condenser can be the same as a preheater (e.g., a typical shell and tube exchanger). The difference, however, is the temperature of the fluid being used for heat exchange (i.e., cool instead of warm). An everyday example of a condenser is the coil on the back of a refrigerator, which condenses the refrigerant.

Figure 11.12 shows an example of a preheater, a reboiler, and a condenser attached to a distillation column.

An **aftercooler** is a heat exchanger, located after the final stage discharge of a compressor. It is used to remove excess heat created during compression. An **intercooler** is a heat exchanger used between the stages of a compressor to cool the gas and remove any liquid before it reaches the next stage.

An **interchanger** (cross exchanger) is a process-to-process heat exchanger. Interchangers use hot process fluids on the tube side and cooler process fluids on the shell side. An example of interchangers used on columns or towers is where the hot bottom fluids are pumped through the tube side, and the feed to the column is put through the shell side. This action preheats the feed to the column and cools the bottom material.

Chillers (coolers) are devices used to cool a fluid to a temperature below ambient temperature. Chillers generally use a refrigerant as a coolant.

Finfan coolers are air coolers consisting of banks of finned tubes connected by an external tubesheet. The finned tubes transfer heat from process streams flowing through the tube

Preheater a heat exchanger that adds heat to a substance prior to a process operation.

Reboiler a tubular heat exchanger, usually located at the bottom of a distillation column or stripper, that is used to supply the necessary column heat.

Condenser an exchanger used to convert a substance from a vapor to a liquid.

Aftercooler a heat exchanger, located after the final discharge stage of a compressor.

Intercooler a heat exchanger used between the stages of a compressor to cool the gas and remove any liquid before it reaches the next stage.

Interchanger a process-to-process heat exchanger that uses hot process fluids on the tube side and cooler process fluids on the shell side. Also known as a cross exchanger.

Chiller devices used to cool a fluid to a temperature below ambient temperature. Chillers generally use a refrigerant as a coolant.

Finfan coolers air coolers consisting of banks of finned tubes, connected by an external tubesheet, that transfer heat from process streams flowing through the tube banks to the atmosphere.

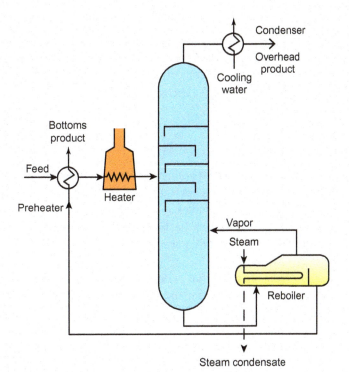

banks to the atmosphere. Heat transfer is assisted by fans that push or pull a high volume of air across the bank of tubes. Finned tubes provide additional heat transfer capacity by increasing the heat transfer surface.

A *heater* is a heat exchanger used to heat process materials. A *cooler* is a heat exchanger used to cool process materials.

A **vaporizer** (a device that converts a liquid into a vapor) is constructed by combining a heat exchanger with a separate vapor/liquid separator. Figure 11.13 shows a vaporizer that would be used to heat liquid nitrogen to make nitrogen gas.

Vaporizer   a device that converts a liquid into a vapor.

Figure 11.13 Vaporizer used to convert liquid nitrogen to nitrogen gas.

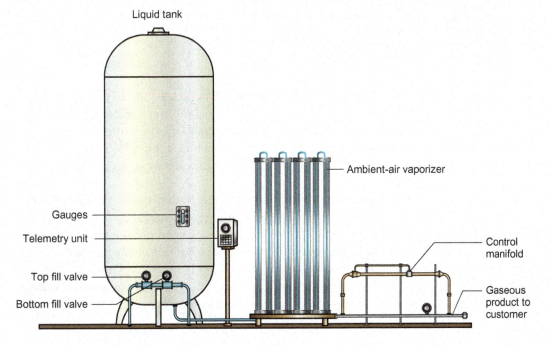

A *thermosiphon* heat exchanger is a form of heat exchanger that creates fluid circulation that is not mechanical but is based on convection. This circulation is caused by the difference in densities of the fluids. Circulation depends on the difference in density between the very hot material exiting the top of the thermosiphon heat exchanger and the cooler material entering the bottom of the exchanger.

11.2 Heat Exchanger Components

The components of a heat exchanger vary based on the design and purpose of the exchanger, but some commonalities are found among them. For example, the components of a typical shell and tube heat exchanger include the tubes, tube sheets, segmental baffles, and tie rods that make up the tube bundle. Figure 11.14 shows a photo of tubes, a tube bundle, and tube bundle puller (a piece of equipment used to remove a tube bundle from a shell for maintenance or cleaning).

The **tube sheet** is a flat plate to which the tubes in a heat exchanger are fixed. The tube sheet consists of a formed metal plate with holes drilled through it. Tubes then are fitted into the holes and attached to the tube sheet by welding or flaring. This group of tubes is called a tube bundle.

A **tube bundle** is a group of fixed or parallel tubes through which process fluids are circulated. Each tube increases the surface area available for heat transfer. The **shell** of a heat exchanger is the outer covering that retains the shell-side fluid. The tubes are housed inside the shell and the tube bundle is sealed to prevent leakage between the tube side and the shell side fluids. Fluid moves into and out of the shell and tubes through inlet and outlet nozzles. The **exchanger head** (also called the *channel head*) is located on the end of a heat exchanger and is where the flow of fluids is directed into and out of the tubes.

Tube sheet a formed metal plate with drilled holes that allow process fluids to enter the tube bundle.

Tube bundle a group of fixed, parallel tubes though which process fluids are circulated.

Shell the outer casing, or external covering, of a heat exchanger.

Exchanger head device located on the end of a heat exchanger that directs the flow of fluids into and out of the tubes; also called a channel head.

Figure 11.14 Tube bundle, tubes, tube bundle puller.

CREDIT: Courtesy of Brian Ellingson.

Finned tubing (shown in Figure 11.15) consists of thin plates of metal that are attached to the outside of a tube to provide additional surface area for heat transfer. Because of the increased surface area, a finned tube allows more heat transfer to occur than an equivalent length of plain tubing. A car radiator is a typical finned tube application in everyday life. In the process industries, finned tubes often are used in air cooled exchangers (fin fans), air preheaters, and tank heaters.

The *channel* is the flanged section of the heat exchanger head that contains tube side flow nozzles and any baffles used to create a tube side multipass exchanger.

The *channel cover* is a plate that is bolted to the outside end of the channel. It retains exchanger fluid and provides access for cleaning. (Note: Some exchangers are built with shell bonnets instead of channel covers. Bonnets are built in one piece, with a flanged end bolted onto the shell so that they can be removed easily.) A gasket is compressed between the flanged metal pieces of the channel cover and the exchanger to prevent leakage to the atmosphere.

Figure 11.15 Example of finned tubing.

CREDIT: Sutham Booranaphol/Shutterstock.

Baffles, or dividers, are metal plates spaced along the tube bundle that hold and support the tubes in place and direct the shell side fluid flow across the outside of the tubes. Baffles contain the same tube-hole pattern as the tube sheet, with a section cut off to form an opening for fluid flow. The baffles cause fluid to flow back and forth across the tube bundle to improve heat transfer.

Baffles are fastened together by *tie rods*, which are small-diameter metal rods installed along the tubes. By spacing baffles properly and using tie rods to hold them in position, tube vibration that results from fluid flow, condensing vapors, and piping pulsations can be minimized.

Some tube bundles must be allowed to expand when they are heated. These bundles either have expansion joints in the shell, with fixed tube sheets welded to the shell on both ends, or they have one fixed tube sheet and a floating head on the other end. Expansion joints allow the tube bundle to expand without touching the shell.

A *floating head* is a component which allows the process fluid coming out of the tubes at one end to be captured and directed by a head that is not fixed to the shell of the heat exchanger. The floating head is used to address thermal stress due to thermal expansion. A U-tube heat exchanger also solves the problem of thermal stress.

A *weir* is a flat or notched dam that functions as a barrier to flow. In a kettle reboiler, the weir is located at the end of the tube bundle and ensures that all of the tubes remain covered by the shell side fluid. In this arrangement, the weir acts as a dam to help contain the liquid at a constant level within the exchanger. This prevents overheating of the tubes and promotes better heat transfer. When excess flow enters the exchanger, it flows over the weir and can be circulated back into the system or removed as bottoms product.

11.3 Principles of Operation of Heat Exchangers

In the process industries, many chemical reactions generate heat or require the addition of heat. Heat also can be produced by friction in rotating equipment (e.g., pumps and compressors). Some of this heat is transmitted to process fluids and must be removed. The only way to remove heat from a process stream is to transfer it to another process stream or to some other heat-absorbing material.

A practical example of a heat exchanger is a car radiator. In this type of exchanger, hot liquid flows through the radiator tubes. As the car moves forward, cool air is drawn into the radiator grill and over the radiator tubes. As the air passes over the tubes, the heat from the liquid is transferred to the air via conduction and convection.

To improve efficiency, reduce the waste of heat energy, and reduce energy costs, heat in the process industries is often recycled. For example, the material leaving a distillation column (tower) can be used to preheat the feedstock coming into the tower by means of a heat exchanger. In other process operations, heat leaving a process heater (furnace) as flue gas can be used to preheat air entering the furnace in an air preheater, rather than allowing it to

Did You Know?

It takes 970 BTUs of latent heat (the heat required to induce a phase change) to change 1 pound of 212-degree Fahrenheit water into steam.

It takes only 144 BTUs to change ice at 32 degrees into water.

212 °F
Steam

970 BTUs
Latent Heat

1 pound
of water

Economizer a type of heat exchanger that preheats a fluid with waste heat, such as boiler feedwater being preheated with flue gases exiting a boiler.

escape out the furnace stack to the atmosphere. The heat exiting a boiler is used in the same way, as it is used to preheat water entering the boiler in a piece of equipment called an **economizer.**

Another way to reduce energy costs is to use the most efficient heat exchanger possible. The more efficient the exchanger, the less energy required for heat transfer.

Heat exchangers can be used for a variety of applications. The amount of heat transfer that occurs in a heat exchanger is a function of several factors, including the temperature difference between the streams, the surface area available for heat transfer, and the overall heat transfer coefficient.

Heat Transfer Coefficient

The overall heat transfer coefficient is a calculated variable that indicates the rate at which heat can be transferred. As mentioned, heat transfer rate is determined by three variables: the temperature difference between the two streams, the surface area of the exchanger, and the overall heat transfer coefficient. The heat transfer rate increases when:

- The temperature difference between the fluids increases
- The surface area increases
- The heat transfer coefficient increases.

This chart shows some common fluids and their specific heat capacities (the amount of heat required to change their temperature by one degree).

Product	Specific Heat Capacity - C_p *(BTU/lb.°F)*
Air	0.024
Ammonia (liquid)	1.12
Ammonia (vapor)	0.497
Gasoline	0.53
Heptane	0.535
Hexane	0.54
Kerosene	0.48
Octane	0.51
Oil, mineral	0.4
Olive oil	0.47
Petroleum	0.51
Propane, vapor	0.576
Propylene glycol	0.6
Water, fresh	1
Water, sea 36°F	0.938

The cold fluid in a heat exchanger gains the BTUs removed from the warmer fluid. This exchange is explained in the formula:

$$M_1 \times \Delta T_1 \times C_{p1} = M_2 \times \Delta T_2 \times C_{p2}$$

Where M = Mass, ΔT = the change in a fluid's temperature (exchanger inlet temperature vs. outlet temperature), and C_p is the fluid's specific heat capacity.

Example: A countercurrent shell and tube heat exchanger has gasoline on the shell side and propane vapor on the tube side. The tube side inlet propane vapor flow is 2,500 lb/hour at a temperature of 320°F (160°C). The outlet propane temperature is 110°F (43°C). On the shell side, the inlet flow of gasoline is 5,300 lb/hour at a temperature of 40°F (4.4°C). What is the temperature of the gasoline exiting the shell side? The units are calculated in degrees Fahrenheit.

Propane is fluid 1. Gasoline is fluid 2.

$M_1 \times \Delta T_1 \times C_{p1} = M_2 \times \Delta T_2 \times C_{p2}$
$2500 \times (320° - 110°) \times 0.576 = 5300 \times \Delta T \times 0.53$
$2500 \times 210° \times 0.576 = 5300 \times \Delta T \times 0.53$
$302,400 = 5300 \times \Delta T \times 0.53$
$302,400 = 2809 \times \Delta T$
$302,400/2809 = \Delta T$
$107.654 = \Delta T$
$40° + 107.654 = 147.654°$
Inlet temperature + ΔT = Outlet temperature

There are also factors that decrease the heat transfer rate, including heat exchanger fouling (discussed in the Potential Problems section).

Did You Know?

Heat exchange systems also occur in nature. Penguins and wading birds use a heat exchange system to reduce the amount of body heat lost through their legs as they stand on the ice or move through cold water. The heat exchangers in the body are composed of blood vessels that run parallel to each other, with blood flowing in opposite directions.

Warm blood flows from the body into the legs. Cooled blood runs from the legs up into the body alongside the warm blood. As the two flows move parallel to one another, the cooler blood is warmed, thereby conserving body heat. In other words, the blood vessels act like a countercurrent heat exchanger. Such an arrangement can prevent hypothermia (excessive loss of body heat to the environment).

CREDIT: vladsilver/Shutterstock.

Types of Flow

In a shell and tube exchanger, liquids or gases of varying temperatures are introduced into the shell and tubes to facilitate heat exchange. For example, hot process fluids can be pumped into the tubes while cool water is pumped into the shell. As the cool water circulates around the outside of the tubes, heat is transferred through the tube walls to the water in the shell.

The flow pattern of fluids through a heat exchanger can be parallel, countercurrent, crossflow, or a hybrid flow, exhibiting characteristics of two flow patterns. Flow also is characterized as *turbulent* (disrupted or unorganized) or *laminar* (smooth and streamlined). The different flow types are important because they affect heat exchanger operations. Figure 11.16 shows types of flow.

Parallel flow (shown in Figure 11.16A) occurs when the streams on both the shell and tube sides move in the same direction. Parallel flow is most often used with heat-sensitive fluids that must be kept below a certain temperature.

Countercurrent flow occurs when the tube side stream and the shell side stream flow in opposite directions. Because it is the most efficient flow type, most shell and tube exchangers have countercurrent flow design. Figure 11.16B shows an example of a countercurrent flow.

Crossflow (shown in Figure 11.16C) occurs when two streams move perpendicular to each other. This is commonly achieved by installing baffles, or dividers, in the shell side of the shell and tube heat exchanger, which direct the flow. In an exchanger where steam is condensed, the tubes with cooling water run parallel to the ground. Steam enters from the top, flows perpendicularly across the outside of the tubes, is condensed into its liquid state, and then exits at the bottom.

Parallel flow movement of fluid that occurs in a heat exchanger when the shell flow and the tube flow are parallel to one another.

Countercurrent flow flow movement that occurs when two fluids are moving in opposite directions.

Crossflow movement that occurs when two streams of fluid flow perpendicular (at a 90-degree angle) to each other.

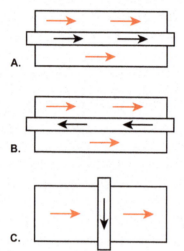

Figure 11.16 A. Parallel flow. **B.** Countercurrent flow. **C.** Crossflow.

Hybrid flow movement of fluid in a heat exchanger that is any combination of counterflow, parallel flow, and crossflow.

Laminar flow (also called *streamline flow*) movement of fluid that occurs when the Reynolds number is low (at very low fluid velocities).

Reynolds number a numeric designation used in fluid mechanics to indicate whether a fluid flow in a particular situation will be laminar (smooth) or turbulent.

Turbulent flow movement of fluids that occurs when the Reynolds number is high, characterized by irregular flow patterns and high velocities.

Hybrid flow occurs when the heat exchanger has multiple passes and baffles. This causes portions of the flow path to be parallel flow, crossflow, and/or countercurrent flow patterns.

As fluids move through a pipe or heat exchanger, the flow can be either laminar or turbulent. **Laminar flow,** or streamline flow (shown in Figure 11.17A), occurs when the fluid flow velocities are low, resulting in regular flow patterns. In other words, the fluid flow is smooth and unbroken, like a series of laminations or thin cylinders of fluid slipping past one another inside a tube. Laminar flows occur when the **Reynolds number** (a number used in fluid mechanics to indicate whether a fluid flow in a particular situation will be smooth or turbulent) is low.

Turbulent flow (shown in Figure 11.17B) is a condition in which fluid flow patterns are disturbed so there is considerable mixing. Turbulent flow occurs when the fluid flow velocities and the Reynolds number are high.

Figure 11.17 A. Laminar flow. **B.** Turbulent flow.

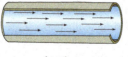

Laminar
flow

Turbulent
flow

The ideal flow type in a heat exchanger is turbulent flow because it provides more mixing and better heat transfer for heating and cooling. **Baffles** are metal plates used to alter the flow of materials or facilitate mixing in an exchanger. Baffles support the tubes, increase turbulent flow, increase heat transfer rates, and reduce hot spots.

Baffles metal plates that are placed inside a vessel or tank and are used to alter the flow, facilitate mixing, or cause turbulent flow.

Did You Know?

Turbulent flow (shown in Figure 11.17B), the most desired flow type in most heat exchangers, can be achieved only when the flow rate is high enough. Too little flow for a pipe size results in a laminar flow.

What Goes Where? Rules of Thumb for Heat Exchangers

Which fluid is put in the tube side of a shell and tube heat exchanger? Although there are many considerations, these three are the most common.

1. Fluids that have a greater potential for fouling. Cleaning the inside of the tubes in a heat exchanger is usually easier than cleaning the shell side. Cooling water is the most common example of a fluid that has a high probability of fouling; therefore, it is usually in the tubes. After the heads of the exchanger are removed, the tubes can be cleaned with high-pressure water from a hydroblaster.

2. Higher pressure material. It is more expensive to make a shell thick enough to hold high pressure material. Compare the MAWP (maximum allowable working pressure) for each fluid and if the ΔP is greater than ~200 PSI, the higher pressure fluid is usually in the tubes. If neither of the fluids is more likely to foul and require frequent cleaning, the pressure rule applies.

3. Higher temperature material. If neither fluid is more prone to fouling and the pressures of the two fluids are relatively close, the higher temperature material will usually be on the tube side.

11.4 Potential Problems

Heat exchangers are designed to perform a specific function. Deviations from normal operating conditions can cause process upsets. For example, when a distillation column reboiler provides insufficient heat, it often results in improper separation of the materials in the column and thus poor product quality and higher production cost.

Consequences of Deviation from Normal Operation

A certain temperature differential must exist between the two fluids in a heat exchanger in order for sufficient heat transfer to take place. Another problem can be an excessive pressure drop, which can cause loss of flow, reduce the amount of heat transfer, and increase the likelihood of fouling.

Fouling is the accumulation of deposits (such as sand, silt, scale, sludge, fungi, and algae) that build up on the internal surfaces of processing equipment. Fouling can be caused by a variety of factors, including mineral deposits or other impurities (e.g., corrosion products) in the fluid. These contaminants or impurities form a thin film layer of scale on the walls or tubes of the exchanger. The scale acts like an insulating barrier, increasing the resistance to heat transfer and reducing the effectiveness of the exchanger. It also can reduce

Fouling accumulation of deposits that build up on the internal surfaces of processing equipment.

the interior diameter of the tubes, causing a higher than normal drop in pressure across the exchanger.

A symptom of fouling is an undesirable change in temperature (or delta T) and change in pressure (or delta P). When the buildup (fouling) reaches a certain point, the exchanger must be taken out of service and have its internal surfaces cleaned.

Process technicians should be familiar with the troubleshooting steps (causes and solutions) for the following situations: fouling, excessive steam shell side pressure (*chest pressure*), leaks, system upsets, corrosion, erosion, reduced fluid velocity, backwashing, and vapor-locked conditions. Table 11.2 lists potential problems and their solutions when troubleshooting heat exchangers.

Table 11.2 Potential Problems and Their Solutions When Troubleshooting Heat Exchangers

Potential Problem	Cause/Solution	Description of Cause/Solution
Fouling	Cause	Plugging of the tube side caused by a buildup of contaminants at the tube sheet inlet and inside the tubes can lead to deposition of solids such as tarlike materials on heat transfer surfaces. Scaling and carbon buildup (coking) are also common forms of fouling.
	Solution	In the case of contaminant buildup such as scaling and sludge, backflushing, hydroblasting (high-pressure water spray), or chemical cleaning is usually required to clean tube deposits. A fouled tube bundle can be cleaned by a machine with a rotating rod-out device. If shell side fouling is caused, chemicals can be used to clean the bundle.
Excessive steam chest (shell side) pressure	Cause	In heat exchangers that use steam as the heating medium, an increase in chest pressure (increased shell side pressure) indicates reduced heat transfer. The high pressure typically results because of a decrease in the rate of steam condensing on the shell side. Common causes of excessive chest pressure are: ■ Low liquid levels ■ Malfunctioning steam traps ■ Tube fouling ■ Noncondensable compounds in the steam ■ Excessive steam temperature ■ Closed valve on the steam outlet line
	Solution	Monitor the pressure for any loss of heat transfer and watch for tube fouling. Monitor liquid level, maintain proper steam temperatures, and check for proper steam trap operation.
Leaks	Cause	Leaks around flange gaskets and valve packing can result from worn gaskets, corrosion, loose bolts, or normally occurring wear. Leaks across tube sheets or through ruptured tubes are caused by erosion, corrosion, or substandard construction. These structural abnormalities result in the contamination of the heating/cooling medium or the product, depending on relative pressure levels in the tubes/shell of the exchanger. Leaks also can be caused by thermal shock, high pressure differential, and/or incorrect design.
	Solution	Frequent inspection of the packing material around flanges and valve gaskets for corrosion. Replace any materials that appear to be damaged from erosion, corrosion, or substandard construction. Inspect for and tighten any loose bolts. Repair tube leaks by plugging leaking tubes or rolling tubes if the leak is at the tube sheet.
Corrosion, erosion	Cause	Corrosion is the removal of metal as a result of a chemical reaction within a process fluid or other material. Erosion is the removal of metal by impingement caused by an abrasive and/or high-velocity fluid stream.
	Solution	Tube corrosion should not occur if the proper material of construction is employed in the heat exchanger tube bundle. Erosion can be prevented if suspended solids are removed from the fluid stream or the velocity of the stream can be lowered without significantly affecting the heat transfer.
Reduced fluid velocity	Cause	Reduced fluid velocity can occur as a result of fouling (scaling), restrictions (e.g., partially opened valves), or improper pump operations (internal wear on a pump that reduces the performance, plugging of pump screens, or a low-suction head on a pump).
	Solution	To increase reduced fluid velocity caused by fouling, restrictions, or improper pump operations, process technicians should clean screens on the pump, switch and repair failing pumps, and adjust levels to provide proper suction head.
Vapor-locked conditions	Cause	Vapor-locked conditions can occur when vapors become trapped in spaces where heat transfer normally occurs. For example, a thermosiphon reboiler can become vapor-locked when the liquid level is above the return nozzle of the reboiler.
	Solution	Necessary adjustments to steam, process flows, and levels are needed to correct the vapor-lock condition.

11.5 Safety and Environmental Hazards

Hazards associated with normal and abnormal heat exchanger operation can affect personnel safety and the environment. For example, opening and closing valves too quickly can be hazardous when high-pressure, high-temperature, and/or hazardous materials are involved. Also, draining and cleaning an exchanger for maintenance exposes the process technician to the potential for burns and contact with chemicals. During process upsets and emergencies, additional hazards are present, such as chemical contact with the skin and eyes. Burns are possible when working around steam piping. Thermal shock and overheating of fouled tubes can result in ruptured tubes that can cause injury to personnel and damage to equipment.

Personal protective equipment (PPE) should be worn during normal and abnormal heat exchanger operations. In addition to basic PPE (gloves, safety shoes and glasses, hardhat, hearing protection, and flame-resistant clothing), additional PPE that can be required includes:

- Goggles, when sampling around exchangers
- Breathing protection (supplied fresh air) if toxic gases could be present
- Face shields when using chemicals, steam, or high-pressure water.

Many chemicals are toxic and harmful to the environment if leaked to the atmosphere. Because of this, the government regulates the amounts that can be released. Exceeding the regulated amounts requires reporting, and fines are possible if reporting and regulations are not followed. Process fluids also can leak into other systems such as cooling tower water causing corrosion, foaming, and release to the atmosphere.

Many hazards are associated with improper heat exchanger operation. Table 11.3 lists some of these hazards and their effects.

Table 11.3 Hazards Associated with Improper Heat Exchanger Operation

Improper Operation	Possible Effects			
	Individual	Equipment	Production	Environment
Attempting to place the exchanger in service with the bleed valves open	Exposure to chemicals or steam	Adverse effects on adjacent equipment	Loss of production	Spill to the environment
Applying heat to the exchanger without following the proper warmup procedure	Exposure to chemicals or steam if the tubes rupture	Tube rupture because of thermal shock	Downtime because of repair	Spill to the environment if the tubes rupture
Improper alignment of the inlet and outlet valves		Overheating, tube fouling	Downtime because of repair, product contamination	
Operating a heat exchanger with ruptured tubes	Exposure to chemicals or steam	Possible explosive condition resulting in severe damage	Product ruined by contamination; downtime because of repair	Potential environmental release
Opening a cold fluid to a hot heat exchanger	Exposure to chemicals or steam	Overpressurizing the heat exchanger because of rapid expansion of the liquid; potential for thermal shock	Heat exchanger damage resulting in downtime because of repair	Potential environmental release

11.6 Process Technician's Role in Operation and Maintenance

The role of the process technician is to ensure the proper operation of heat exchangers, including routine and preventive maintenance. Process technicians should always monitor heat exchangers for abnormal noises, leaks, excessive heat, pressure drops, temperature drops, or other abnormal conditions in order to ensure personal and process safety. They should conduct preventive maintenance as needed. Table 11.4 lists some of the items a process technician should monitor when working with heat exchangers.

Table 11.4 Process Technician's Role in Operation and Maintenance

Look	Listen	Check
■ Inspect equipment for external leaks. ■ Check for internal tube leaks by collecting and analyzing samples. ■ Look for abnormal pressure changes (could indicate tube plugging). ■ Check temperature gauges for high temperature readings. ■ Inspect insulation.	■ Listen for abnormal noises (e.g., rattling or whistling). High-pressure leaks often are invisible but emit a high-pitched sound and can be very dangerous.	■ Check for excessive vibration (vibration can loosen the tubes in the tube sheet, or wear through tube walls). ■ Check inlet and outlet lines for unusually high or unusually low temperatures.

11.7 Typical Procedures

Process technicians should be aware of typical heat exchanger procedures, which include monitoring, inspecting for leaks, lockout/tagout, routine and preventive maintenance, sampling, startup, shutdown, and emergency procedures. Failure to perform proper maintenance and monitoring could affect the process and result in equipment damage or injury.

Startup

The following steps represent a general startup procedure for a shell and tube heat exchanger that has been taken out of service for maintenance activities. In this example, the procedure requires that the low-temperature side of the heat exchanger be filled first. This controls the heat-up of the exchanger components and reduces the amount of thermal shock to which the exchanger is subjected. It is important to note that facility personnel must always follow specific operating procedures developed for their facility.

1. Inspect the heat exchanger to ensure that it is ready for startup.
2. Verify that all the drains and bleeders are shut and plugged.
3. Establish the cold stream flow to the heat exchanger.
4. Open the cold side vent valve slightly to allow any trapped air to escape; slowly crack open the cold side inlet valve.
5. Shut off the vent valve when the cold side is completely filled; open the cold side inlet valve fully.
6. Slowly open the cold side outlet valve.

After filling the low temperature side, the hot side of the exchanger must be lined up.

1. Open the hot side vent valve.
2. Slowly crack open the hot side outlet valve to allow the hot side to back fill with process fluid and to remove any trapped air.
3. After the hot side is filled, close the vent valve and slowly open the outlet valve fully.
4. Open the inlet valve on the hot side slowly until it is fully open.
5. As the exchanger comes up to normal operating conditions, monitor and check for leaks or abnormal conditions.

Shutdown

The following steps represent a general procedure for shutting down a shell and tube heat exchanger. Individual facilities have specific procedures that might vary from this procedure. In this example, the side with the hot stream is shut down first.

1. Shut off the hot side inlet valve.
2. After the hot side has cooled, shut off the hot side outlet valve and open the hot side vent.
3. Shut off the cold side inlet valve.
4. Shut off the cold side outlet valve and open the cold side vent.
5. Allow the exchanger to cool.
6. Follow procedures to open drain valves and empty exchanger.

General Maintenance Tasks

General maintenance tasks performed on heat exchangers may include backwashing, water blasting, sand blasting, and acidizing.

- **Backwashing (backflushing)** is a procedure in which the direction of flow through the exchanger is reversed to remove solids that have accumulated in the heat exchanger. Heat exchangers can be equipped with special valves so back flushing can be performed. While backwashing can be performed by process technicians, water blasting, sand blasting, and acidizing require specialized training to perform.

- Water blasting is a surface cleaning technique used on heat exchangers that uses high-pressure jets of water to remove scale or other deposits from the inside of exchanger tubes.

- Sand blasting is a surface cleaning technique involving entraining sand into a stream of high-pressure air to remove scale or other deposits from accessible surfaces.

- Acidizing is a surface cleaning technique that uses acid to remove scale or other deposits from accessible surfaces.

Backwashing (backflushing) a procedure in which the direction of flow through the exchanger is reversed to remove solids.

Summary

Heat exchangers are devices used to transfer heat from one fluid to another, usually without the two fluids physically contacting each other. Heat exchangers usually accomplish this heat transfer through conduction and convection.

There are many types of heat exchangers, including double-pipe, spiral, shell and tube, U-tube, and plate and frame. The most common type of heat exchanger is the shell and tube exchanger.

Double-pipe heat exchangers consist of one pipe within a larger diameter pipe. The double-pipe provides extra strength in the pipe walls, reducing the likelihood of rupture.

Spiral heat exchangers consist of spiral plates, or helical tube cores fixed in the shell.

In a shell and tube exchanger, liquids or gases of varying temperatures are directed into the shell and tubes. As the fluids move through the shell, baffles cause turbulence (mixing) in the fluid. Turbulence is desired because it facilitates heat transfer and creates more surface area for heating and cooling.

U-tube exchangers are shell and tube heat exchangers that use U-shaped tubes. The tubes start and end on the same side of the exchanger, which enables them to handle thermal expansion better.

The basic components of a heat exchanger vary based on design and purpose, but some commonalities exist. The basic components of a shell and tube heat exchanger include the tube bundle, tube sheets, baffles, tube inlet and outlet, shell, shell inlet and outlet, exchanger cover, and pass partition.

Heat exchangers, including reboilers, preheaters, aftercoolers, condensers, chillers, or interchangers, can be used in a variety of applications. An exchanger service is the actual functional application of the exchanger such as cooler, heater, condenser, vaporizer, thermosiphon, reboiler, or air cooler.

Flow through an exchanger is characterized by the pattern the flow takes, which can be countercurrent, parallel, cross, or hybrid. Flow also can be characterized as turbulent (agitated) or laminar (smooth).

Improperly operating a heat exchanger can lead to safety and process issues, including possible exposure to chemicals or hot fluids, or equipment damage because of thermal shock.

Process technicians should always monitor heat exchangers for abnormal noises, leaks, excessive heat, pressure drops, temperature drops, or other abnormal conditions. They should conduct preventive maintenance as needed.

Checking Your Knowledge

1. Define the following terms:

  a. Backwashing

  b. Baffle

  c. BTU

  d. Fouling

  e. Counterflow

  f. Crossflow

  g. Laminar flow

  h. Parallel flow

  i. Turbulent flow

  j. Convection

  k. Conduction

  l. Radiation

2. Heat exchangers are used to transfer heat between:

  a. separation equipment

  b. hot and cold streams

  c. chemical reactions

  d. temperature and pressure

3. (True or False) One BTU is equivalent to 252 calories.

4. Which of the following terms describes when the tube side stream and the shell side stream flow in opposite directions?

  a. Crossflow

  b. Parallel flow

  c. Turbulent flow

  d. Countercurrent flow

5. Select the components that are found in a heat exchanger. (Select all that apply.)

  a. Channel cover

  b. Baffles/divider

  c. Expansion joint

  d. Tube sheet

6. A _____ is a part of a heat exchanger that is bolted to the outside end of the channel to retain exchanger fluid and provide access for cleaning.

  a. tube sheet

  b. channel cover

  c. tie rod

  d. baffle

7. Which of the following pieces of equipment is used to convert a vapor to a liquid?

  a. Preheater

  b. Reboiler

  c. Chiller

  d. Condenser

8. Which of the following is NOT a common potential problem in heat exchangers?

  a. Chemical balance

  b. Excessive chest pressure

  c. Fouling, erosion, and corrosion

  d. Vapor-locked conditions

9. Match the items on the left with the correct description on the right.

I. Thermosiphon	**a.** Produces vapor from the liquid that enters it and provides the distillation column with the vapor necessary for separation inside the column.
II. Cooler	**b.** Uses cooling to convert a substance from a vapor to a liquid
III. Reboiler	**c.** Is used to cool hot liquids
IV. Condenser	**d.** Provides circulation based on density differences

10. When should breathing protection be used with heat exchangers?

    a. During normal operations

    b. During abnormal situations

    c. When toxic gases might be present

    d. At all times

11. (True or False) Thermal shock and overheating of fouled tubes can result in ruptured tubes.

12. List three things a process technician should look for during the operation and maintenance of heat exchangers.

13. _____ is a surface cleaning technique to remove scale or other deposits from accessible surfaces.

    a. Acidizing

    b. Air jetting

    c. Backwashing

    d. High-point bleeding

14. (True or False) Backwashing requires specialized training to perform.

NOTE: Answers to Checking Your Knowledge questions are in the Appendix.

Student Activities

1. Using a diagram or a cutaway of a multipass heat exchanger, identify the various components and explain the purpose of each.

2. If a working heat exchanger unit is available, vary the flow rates and temperatures of both shell and tube side fluids and record the changes in pressure, flow, and temperature of the tube side inlet and outlet and the shell side inlet and outlet.

3. With a group, determine how the heat transfer coefficient is calculated and make a short presentation to the class.

4. List the three variables that determine the heat transfer rate.

5. Describe the different applications of a typical shell and tube exchanger. Be sure to include the following:

    a. Reboiler

    b. Thermosiphon

    c. Preheater

    d. Aftercooler

    e. Condenser

Chapter 12
Cooling Towers

Objectives

After completing this chapter, you will be able to:

12.1 Identify the purpose, applications, and common types of cooling towers. (NAPTA Cooling Towers 1, 2*) p. 256.

12.2 Identify the components of cooling towers and explain their purpose. (NAPTA Cooling Towers 3, 4, 5) p. 259.

12.3 Describe the operating principles of cooling towers. (NAPTA Cooling Towers 6) p. 261.

12.4 Identify potential problems associated with cooling towers. (NAPTA Cooling Towers 11) p. 263.

12.5 Describe safety and environmental hazards associated with cooling towers. (NAPTA Cooling Towers 7) p. 266.

12.6 Describe the process technician's role in typical procedures associated with cooling towers. (NAPTA Cooling Towers 8, 9, 10) p. 268.

* North American Process Technology Alliance (NAPTA) developed curriculum to ensure that Process Technology courses will produce knowledgeable graduates to become entry level employees in process technology. Objectives from that curriculum are named here in abbreviated form. For example, "(NAPTA Cooling Towers 1, 2)" means that this chapter's first objective relates to objectives 1 and 2 of the NAPTA curriculum about cooling towers).

Key Terms

Ambient air—any part of atmospheric air that is breathable, **p. 264.**

Approach—a term that describes how closely a cooling tower can cool the water to the wet bulb temperature of the air, **p. 263.**

Basin—a reservoir at the bottom of the cooling tower where cooled water is collected so it can be recycled through the process, **p. 260.**

Biocides—chemical agents that are capable of controlling undesirable living microorganisms (such as in a cooling tower), **p. 268.**

Blowdown—the process of removing small amounts of water from the cooling tower to reduce the concentration of impurities, **p. 269.**

Circulation rate—the rate at which cooling water flows through the tower and through the cooling water system, **p. 265.**

Closed circuit cooling towers—cooling towers that use a closed coil over which water is sprayed and a fan provides air flow, and in which there is no direct contact between the water being cooled and the water and air used to cool it, **p. 256.**

Coolers—heat exchangers that use cooling tower water to lower the temperature of process materials, **p. 259.**

Cooling range—the difference in temperature between the hot water entering the tower and the cooler water exiting the tower, **p. 265.**

Cooling towers—structures designed to lower the temperature of water using latent heat of evaporation, **p. 256.**

Counterflow—the condition created when air and water flow in opposite directions, **p. 258.**

Crossflow—flow of water downward with air forced horizontally across the water's flow path, **p. 258.**

Cycles of concentration—the maximum multiplier for the miscellaneous substances in the circulating cooling water as compared to the amount of those substances in the makeup water, **p. 262.**

Dew point—the temperature at which air is completely saturated with water vapor (100 percent relative humidity), **p. 263.**

Drift—carrying of water with the air stream out of the cooling tower; also known as *windage* (a force created on an object by friction when there is relative movement between air and the object), **p. 260.**

Drift eliminators—devices that prevent water from being blown out of the cooling tower; used to minimize water loss, **p. 260.**

Dry bulb temperature—the actual temperature of the air that can be measured with a thermometer, **p. 263.**

Erosion—the degradation of tower components (fan blades, wooden portions, and cooling water piping) by mechanical wear or abrasion by the flow of fluids (often containing solids). Erosion can lead to a loss of structural integrity and process fluid contamination, **p. 265.**

Evaporation—the process in which a liquid is changed into a vapor through the latent heat of evaporation, **p. 261.**

Fill—material inside the cooling tower, usually made of plastic, that breaks water into smaller droplets and increases the surface area for increased air-to-water contact, **p. 260.**

Foaming—the formation of a froth caused by mixing of water contaminants with air; impairs water circulation and can cause pumps to lose suction or cavitate, **p. 266.**

Forced draft cooling towers—cooling towers that contain fans or blowers at the bottom or on the side of the tower to force air through the equipment, **p. 257.**

Heat duty—the amount of heat energy a cooling tower is capable of removing, usually expressed in MBTU/hour, **p. 265.**

Humidity—moisture content in ambient air measured as a percentage of saturation, **p. 263.**

Induced draft cooling towers—cooling towers in which air is pulled through the tower internals by a fan located at the top of the tower, **p. 257.**

Louvers—movable, slanted slats that are used to adjust the flow of air, **p. 260.**

Makeup water—the amount of water that must be added to compensate for the water leaving the cooling water system through evaporation, drift, and blowdown, **p. 260.**

Natural draft cooling towers—cooling towers in which air movement is caused by wind, temperature difference, or other nonmechanical means, **p. 257.**

Psychrometer—an instrument for measuring humidity, consisting of a wet bulb thermometer and a dry bulb thermometer; the difference in the two thermometer readings is used to determine atmospheric humidity, **p. 264.**

Relative humidity—a measure of the amount of water in the air compared with the maximum amount of water the air can hold at that temperature, **p. 263.**

Tube leaks—leaks in a tube that can result in process chemical entering the circulating water, possibly resulting in a fire, explosion, or environmental or toxic hazard, **p. 266.**

Water distribution header—a pipe that provides water to a distributor box located at the top of the cooling tower so the water can be distributed evenly onto the fill, **p. 260.**

Wet bulb temperature—the lowest temperature to which air can be cooled through the evaporation of water, **p. 263.**

12.1 Introduction

Cooling towers are heat transfer devices used to cool water. Until the latter part of the 20th century, many facilities employed once-through cooling water, and the use of cooling towers was the exception, rather than the rule. With the emphasis on minimal environmental impact and water conservation, many facilities now include cooling towers in their plant design, despite the large capital investment and high operating costs they involve.

In cooling towers, water usually flows over internal components made of plastic or wood. These internal components are designed to break the water into tiny droplets, thereby increasing surface area and promoting maximum water-to-air contact. Many different types of cooling towers are in use today, but all have similar components and use the same principles of heat transfer.

The main purpose of cooling towers is to remove heat from process cooling water so the water can be recirculated through the process. Cooling towers reduce the temperature of water through convection and evaporation. Cooling tower water can be supplied from municipal water systems or from other bodies of water, such as rivers or cooling ponds.

Cooling towers can be categorized as either open circuit (wet) design or closed circuit (dry) design.

Types of Cooling Towers

Cooling towers structures designed to lower the temperature of water using latent heat of evaporation.

Cooling towers are structures designed to lower the temperature of a circulating cooling water stream. They accomplish this through the thermodynamic principle of evaporation and sensible heat loss. Evaporation is the process by which a liquid turns into a vapor without its temperature reaching its boiling point. When water changes from a liquid into a vapor, it absorbs a substantial amount of heat from the surrounding water. This process is called latent heat of vaporization. Cooling towers (Figure 12.1) come in many shapes, sizes, and air flow classifications (e.g., single-cell, and multicell).

Closed circuit cooling towers cooling towers that use a closed coil over which water is sprayed and a fan provides air flow, and in which there is no direct contact between the water being cooled and the water and air used to cool it.

Closed circuit cooling towers are used in refrigeration and other HVAC systems. In **closed circuit cooling towers**, the working fluid passes through a heat exchange coil and water is sprayed over it. A fan draws air through the water spray, removing heat from the coil by the process of convection. There is no direct contact between the water being cooled and the spray water.

Figure 12.1 Open and closed cooling towers. **A.** Natural draft open cooling tower. **B.** Induced draft open cooling tower. **C.** Forced draft open cooling tower. **D.** Closed circuit cooling tower.

CREDIT: **A.** Martin Lisner/Shutterstock. **B.** Chatchawal Kittirojana/Shutterstock. **C.** Chris Rout /Alamy Stock Photo.

A. Natural draft.

B. Induced draft.

C. Forced draft.

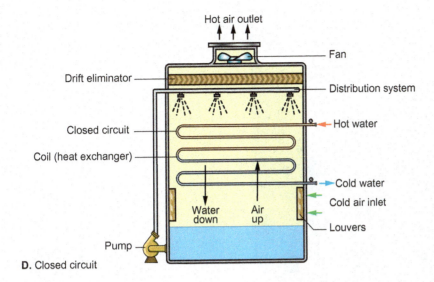

D. Closed circuit

This chapter focuses on *open* cooling towers, which are the type most commonly found in industry for process cooling purposes.

Open cooling towers are classified by their air flow method as well as by the relationship between their air and water flow patterns. Air flow classifications are natural draft, induced draft, and forced draft. Air-to-water flow patterns can be either crossflow or counterflow.

Air Flow Classifications

Natural draft cooling towers, which are most frequently seen in nuclear and coal-fired power plants, use temperature differences inside and outside the stack to facilitate air movement. In a hyperbolic, natural draft cooling tower, there is no fan. Instead, the hyperbolic design creates a chimney effect that induces air flow through natural air currents (draft). The structures of natural draft cooling towers are very tall (typically 300 feet, or almost 100 meters).

Instead of forced movement to facilitate draft, air flow occurs naturally because of density differences between the warm, moist air inside the tower and the cooler, drier air outside the tower. These cooling towers remove massive amounts of heat. Depending on the atmospheric conditions, a cloud of water vapor may be seen exiting the stack. On hot, dry days this cloud is less visible because the water evaporates into the warm, dry air more quickly than it does into cold or humid air. Figure 12.2 shows examples of a natural draft cooling tower.

Induced draft cooling towers (Figure 12.3) have fans at the top of the tower that pull air through the tower. A shroud surrounds the fan, and air drawn through the cooling tower is exhausted through this shroud. In this type of cooling tower, the fan at the top induces (pulls) hot moist air out of the tower and into the atmosphere. The result is that cool air enters at a low velocity and warm air exits at a high velocity. This is usually the most efficient arrangement for cooling towers. Induced draft towers can consist of several connected cells, with separate fans and water inlet lines.

Forced draft cooling towers (Figure 12.4) have fans or blowers at the bottom side of the tower that force air through the equipment. The pressure produced by the fan helps move the air through the tower and out the top. This results in cool air entering at a high velocity and warm air exiting at a low velocity. One of the disadvantages of the forced draft design is that the low exiting velocity makes the tower more likely to experience recirculation - where the warm, humid discharged air flows back into the air intake and is recirculated. Forced draft towers are usually built in units or cells. Several cells can be connected together to create the appearance of a single unit, but each cell has its own fans and water feed lines.

Single-cell and multicell towers are both used in the industry. In a multicell cooling tower, individual cells can be taken out of service or isolated for maintenance and temperature control purposes. Multicell towers are used most often when the heat removal demand varies.

Natural draft cooling towers cooling towers in which air movement is caused by wind, temperature difference, or other nonmechanical means.

Induced draft cooling towers cooling towers in which air is pulled through the tower internals by a fan located at the top of the tower.

Forced draft cooling towers cooling towers that contain fans or blowers at the bottom or on the side of the tower to force air through the equipment.

Figure 12.2 A. Natural draft cooling tower. **B.** Diagram of a natural draft cooling tower.

CREDIT: A. Bildagentur Zoonar GmbH/Shutterstock.

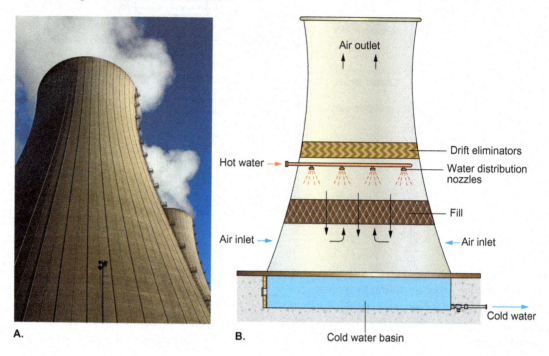

A.

B.

Figure 12.3 Diagram of an induced draft tower showing the flow inside the equipment.

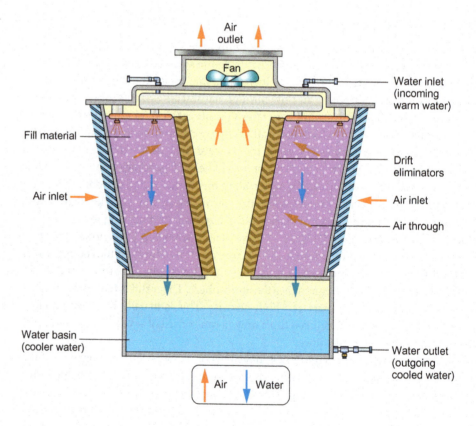

Crossflow   flow of water downward with air forced horizontally across the water's flow path.

Counterflow   the condition created when air and water flow in opposite directions.

Air to Water Flow Patterns

Crossflow and counterflow are two types of air flow patterns used in cooling towers. In a **crossflow** cooling tower (Figure 12.5A), water flows downward and air is forced horizontally across the water's flow path. In a **counterflow** cooling tower (see Figure 12.5B), air and water flow in opposite directions. The water flows downward while the air is forced upward by induced, forced, or natural draft.

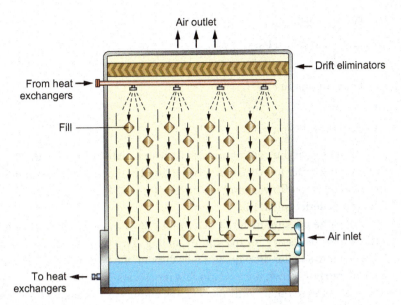

Figure 12.4 Diagram of a forced draft cooling tower.

Figure 12.5 A. Crossflow cooling tower. **B.** Counterflow cooling tower.

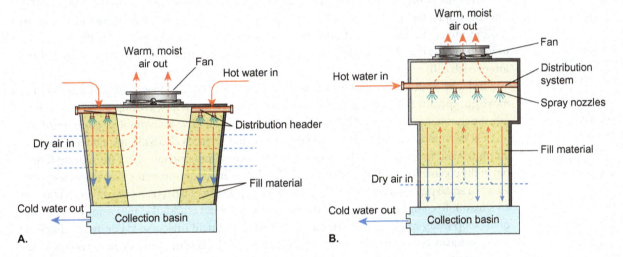

Applications of Cooling Towers

Cooling tower water can be used in a variety of heat exchanger applications, including process condensers and coolers. Cooling water also can be used to heat and vaporize cryogenic (very low temperature) liquids such as anhydrous ammonia or liquid propane.

Condensers are heat exchangers that cool hot vapors and convert them into a liquid. As the vapors flow through the heat exchanger, the cool water is heated and the hot vapors are cooled. As the vapors cool, they condense and are pumped out for storage or are returned to the process. The cooling water (which is now hotter) is returned to the cooling tower to be cooled.

Coolers are heat exchangers that use cooling tower water to lower the temperature of process materials. In lubricating systems (found in rotating equipment), coolers are used to cool lubrication oils.

Coolers   heat exchangers that use cooling tower water to lower the temperature of process materials.

12.2 Components of Cooling Towers

Although cooling tower designs vary, most have similar components (Figure 12.6), many of which are made using plastic or wood. These components include:

- Fans and motors
- Fan shrouds or chimney

- Water distribution system
- Fill (slats, splash bars)
- Louvers
- Drift eliminators
- Basin
- Pumps and sumps.

Water distribution header a pipe that provides water to a distributor box located at the top of the cooling tower so the water can be distributed evenly onto the fill.

Fill material inside the cooling tower, usually made of plastic, that breaks water into smaller droplets and increases the surface area for increased air-to-water contact.

Louvers movable, slanted slats that are used to adjust the flow of air.

Drift eliminators devices that prevent water from being blown out of the cooling tower; used to minimize water loss.

Drift carrying of water with the air stream out of the cooling tower; also known as *windage* (a force created on an object by friction when there is relative movement between air and the object).

Makeup water the amount of water that must be added to compensate for the water leaving the cooling water system through evaporation, drift, and blowdown.

Basin a reservoir at the bottom of the cooling tower where cooled water is collected so it can be recycled through the process.

At the top of the cooling tower is a **water distribution header** that provides returned (hot) cooling water to a distributor box located at the top of forced and induced draft cooling towers. The water flows out of the distributor box through a series of nozzles and begins to fall onto material called fill.

Fill is the packing material in a cooling tower which distributes the water over a large surface area, necessary for maximum cooling. Two types of fill are splash fill, which breaks the water into smaller droplets as it falls through the cooling tower, and film fill, which allows the water to flow down the tower in thin film layers. The fill materials direct the flow of water and maximize its contact with air along the falling path. Breaking the water into smaller droplets helps increase the surface area of the water. This, coupled with increased air contact along its falling path, facilitates the process of evaporation and increases the amount of cooling.

Louvers, which are movable, slanted slats, are installed on the outer sides of the cooling tower to adjust the quantity and direction of the air flow into the tower. In an induced draft cooling tower, air flow is facilitated by a large motor-driven fan located at the top of the tower. This fan usually is equipped with speed-reducing gear boxes that are located at the top of each cell.

Induced draft towers and forced draft cooling towers utilize mechanical means (fans) to move air upward through the tower. In a hyperbolic, natural draft cooling tower, there is no fan; the chimney effect created by the stack itself induces air flow via natural air currents (draft).

On the interior surfaces of the cooling tower (the surfaces closest to the center where the air movement from the fan is the greatest), **drift eliminators** prevent water from being blown out of the cooling tower. The main purpose of a drift eliminator is to minimize water loss. Some water (a small percentage), however, is always lost because of **drift** (carrying of water with the air stream). To compensate for this water loss, **makeup water** is added to the basin. The addition of makeup water might require increased chemical treatment (discussed later in this chapter).

The **basin** is a reservoir (sump) at the bottom of the cooling tower where cooled water is collected so that it can be recycled. The basin is also where chemicals are injected into the cooling water to reduce bacterial growth and adjust water chemistry.

Water inside the basin is pumped out of the cooling tower by water circulation pumps and returns to the heat exchangers and other process components. After the water has passed through the exchangers and heat transfer has occurred, the heated water is returned to the top of the cooling tower to again be cooled and recycled.

Figure 12.6 Induced draft, crossflow cooling tower components.

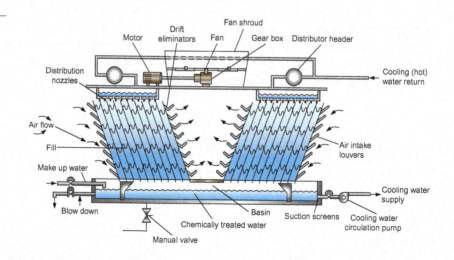

12.3 Principles of Operation

Cooling towers work primarily based on the process of **evaporation**, a process in which a liquid is changed into a vapor. When water changes from a liquid into a vapor, it absorbs a considerable amount of heat from the surrounding water. This evaporative process is what cools the water.

The process of evaporation is much more efficient than convection at removing heat, although both play a role in the process. About two-thirds of the heat is transferred by evaporation, but convection (the transfer of heat by circulation or movement of liquid or gas) is also important. The circulation of air and water makes the heat transfer process possible and works extremely well.

Water is an excellent cooling medium because it is 50 to 100 times more efficient at conducting heat than air alone. Water has the ability to accept more heat than virtually any other available cooling fluid. Lower humidity allows evaporation (latent heat of vaporization) to make up 80 percent to 90 percent of the cooling, while sensible heat loss accounts for the remaining 10 percent to 20 percent of the cooling.

Evaporation   the process in which a liquid is changed into a vapor through the latent heat of evaporation.

Overview of the Cooling Process

The cooling process in a cooling tower has eight steps.

1. Cool water is pumped from the cooling tower basin to the process heat exchangers.
2. In the heat exchangers, heat is transferred to the circulating water.
3. The heated water is returned to the cooling tower distribution header for cooling.
4. The distribution header fills the distribution boxes (or trough), and the hot water flows through spray nozzles downward onto the fill.
5. The fill provides air-to-water contact by breaking the water into smaller droplets or thin film layers, increasing the surface area of the water and promoting heat exchange.
6. As the water falls through the tower, it is exposed to air that removes heat through latent heat of vaporization and convection.
7. As the cool water reaches the bottom of the tower, it is collected in the basin.
8. The water in the basin is then pumped back to the process heat exchangers, and steps 2 through 8 are repeated.

Calculating Heat Removal in BTUs

Open cooling towers move a tremendous amount of water. It is common to move thousands of gallons (each about 3.8 cubic meters) or more per minute. The temperature of the water leaving the cooling tower should be constant throughout the year. A cooling water system usually is designed so the returning hot water is about 15 to 30 degrees Fahrenheit (-9.4 to -1.1 degrees C) warmer than the water exiting the tower. Cooling towers can remove a large number of BTUs from a process.

Example: Calculate the heat load (BTUs removed) in a cooling tower that circulates 40,000 GPM with cooling water exiting the tower at 85 degrees F (29.4 degrees C) and the warm cooling water returning to the tower at 115 degrees F (46.1 degrees C).

One BTU changes the temperature of 1 pound of water 1 degree F, so 30 BTUs must be removed from every pound of water to effect a 30-degree F change.

The 40,000 gallons of water must first be converted to pounds. One gallon of water is equal to ~8.33 pounds, so 40,000 GPM $\times$ 8.33 = 333,200 pounts per minute (ppm).

The formula for calculating heat duty is:

$$\text{BTU/min} = \text{Flow rate (in pounds per minute)}$$
$$\times \text{ Temperature difference} \times \text{Heat transfer coefficient}$$

The temperature difference between water exiting and returning to the tower is 30 degrees F (85 degrees F to 115 degrees F).

We know from Chapter 11, Heat Transfer Coefficients, that the heat transfer coefficient of water is 1.

So:

333,200 (pounds of water per min) $\times$ 30 (temperature difference) $\times$ 1 (heat transfer coefficient of water) = 9,996,000 BTU/min, or 9,996 MBTUs/min (999.6 MMBTUs). This is the approximate amount of heat removed from 40,000 GPM of water to lower the temperature 30 degrees F.

Another formula used to approximate the amount of heat rejected by a cooling tower is:

$$\text{BTU/hr} = \text{GPM} \times 500 \times \Delta T$$

where 500 is a constant representing the weight of water in pounds per gallon (~8.33) $\times$ the heat transfer coefficient of water (1) $\times$ a conversion factor (60) for changing minutes to hours.

Cycles of Concentration

The makeup water to a cooling tower often comes from lakes, rivers, or wells. Small amounts of dissolved solids (minerals, salts, metals) in the makeup water usually will not have a significant effect on the cooling tower's operation, but in large enough concentrations they can cause scaling, fouling, and corrosion in process equipment. As water in a cooling tower evaporates, these impurities are left behind, and their concentration in the circulating water increases. Makeup water is analyzed to determine the amount of dissolved solids the water contains. The circulating water or blowdown is analyzed for the same impurities, and the results are compared. The **cycles of concentration** in a cooling tower reflect the comparison of these two numbers. If the amount of impurities in the circulating water is three times the amount in the makeup water, then the cycles of concentration are 3. Generally, the cycles of concentration are maintained in the range of 4–8, although there are instances where cooling towers are operated at much higher numbers. Operators control this concentration by adjusting the blowdown. Raising the cycles of concentration will result in a reduction in the amount of blowdown and makeup water, leading to cost savings, but can result in higher chemical treatment costs.

Cycles of concentration   the maximum multiplier for the miscellaneous substances in the circulating cooling water as compared to the amount of those substances in the makeup water.

Calculating Quantity of Makeup Water

Cooling towers are always losing water due to evaporation, drift, and blowdown. Fresh makeup water needs to equal the water lost to maintain the correct level in the basin.

$$\text{Makeup} = \text{Evaporated loss} + \text{Drift loss} + \text{Blowdown}$$

Example: A cooling tower circulates 40,000 GPM water. The hot water entering the top of the tower is 115 degrees F (46.1 degrees C) and is cooled to 85 degrees F (29.4 degrees C) by the time it enters the cooling tower basin. Assume drift loss is 0.35% of circulation rate, evaporative losses are 1% for each 10 degree F (5.5 degrees C) water temperature decrease, and blowdown rate is 0.25% of circulation rate. What is the required makeup in GPM?

$$\text{Makeup water} = [40{,}000\,(0.01 \times 3) + 40{,}000\,(0.0035) + 40{,}000\,(0.0025)] = 1440 \text{ GPM}$$

Evaporative loss Drift loss Blowdown loss

12.4 Potential Problems Affecting Cooling Tower Operation

Many factors affect cooling tower operation, including relative humidity, outside air temperature, wind velocity, and tower design. Other factors include heat exchanger tube leak or inadequate circulation because of pump malfunction. Water temperature also can be a factor if the cooling tower is located in an area with extremely low temperatures. For this reason, cooling towers in cold areas usually are equipped with temperature control devices to protect from possible freezing.

Humidity

Humidity is the moisture content in the air. Local humidity levels must be considered when designing a cooling tower because evaporation is much more difficult in areas with high levels of humidity. For example, a cooling tower in the hot, dry conditions of Arizona could be significantly smaller than one in the hot, humid climate of south Texas and still provide the same amount of cooling. The reason evaporation is more difficult in humid areas is that the ambient air is already partially saturated with moisture, so less water can be absorbed.

Relative humidity is the amount of water in a given amount of air at a given temperature, compared to the maximum amount of water that the air can hold at that same temperature. Relative humidity and the temperature of the air determine the amount of heat that can be removed.

If the humidity is high, the air can absorb only a small amount of water vapor, so the evaporation process will be slow. If the humidity is low, however, the air can hold a much larger amount of water vapor, so the evaporation rate will be much quicker.

Dew point is the temperature at which air is completely saturated with moisture. As the temperature in the cooling tower approaches the dew point, the evaporation rate declines and reduces the cooling rate. When the dew point is reached, the amount of evaporation drops to zero.

The **approach** is a measure of the temperature difference between the air inlet temperature (wet bulb) and the water outlet temperature and refers to the cooling tower's ability to evaporate and cool water. For example, a 5-degree approach means that, if everything is working as designed, the cooling tower should cool the water temperature to within 5 degrees of the wet bulb temperature of the air. The approach range depends on the design of the tower, the relative flow rates of the air and the water, and the contact time.

Temperature

Air temperature affects cooling towers by increasing or decreasing the rate of evaporation. If the temperature is cold, the molecules of air move much more slowly, so the rate of evaporation is decreased. If it is hot, the molecules of air move much more rapidly, so the process of evaporation occurs more quickly.

Air temperature can be evaluated through wet or dry bulb measurements. **Dry bulb temperature** is the actual temperature of the air that can be measured with a thermometer and does not take humidity into consideration. **Wet bulb temperature** is the lowest theoretical temperature that air can be cooled to through the evaporation of water. If the humidity in the air is less than 100 percent (e.g., the air is not completely saturated with water), the process of evaporation will take place as the water comes into contact with the air.

Humidity moisture content in ambient air measured as a percentage of saturation.

Relative humidity a measure of the amount of water in the air compared with the maximum amount of water the air can hold at that temperature.

Dew point the temperature at which air is completely saturated with water vapor (100 percent relative humidity).

Approach a term that describes how closely a cooling tower can cool the water to the wet bulb temperature of the air.

Dry bulb temperature the actual temperature of the air that can be measured with a thermometer.

Wet bulb temperature the lowest temperature to which air can be cooled through the evaporation of water.

Psychrometer an instrument for measuring humidity, consisting of a wet bulb thermometer and a dry bulb thermometer; the difference in the two thermometer readings is used to determine atmospheric humidity.

Wet and dry bulb temperatures, measured with a **psychrometer** (Figure 12.7A), are used to evaluate the effectiveness of a cooling tower at current atmospheric conditions. Because the evaporation of water from the surface of the thermometer has a cooling effect, the temperature indicated by a wet bulb thermometer is less than the temperature indicated by a dry bulb thermometer. At 100% humidity, however, both wet bulb and dry bulb temperatures are equal. As mentioned, the rate of evaporation from the wet bulb thermometer depends on humidity because the rate of evaporation is slower when the air is already full of water vapor. The difference in the temperatures indicated by the two thermometers (wet and dry bulb) gives a measure of atmospheric humidity. A psychrometric chart (see Figure 12.7B) also can be used to find the wet bulb temperature.

Figure 12.7 A. Psychrometer. **B.** Psychrometric chart. To read it, start at the dry bulb temperature on the X-axis. Go up to the humidity line (curved lines) and then follow the diagonal line up to the left to locate the wet bulb temperature.

CREDIT: **A.** Fouad A. Saad/Shutterstock. **B.** Slave SPB/Shutterstock.

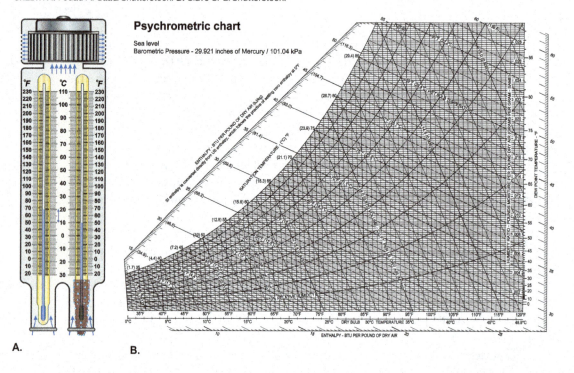

A.

B.

Ambient air any part of atmospheric air that is breathable.

High cooling water temperature occurs as a result of improper system balance or ambient air conditions. (**Ambient air** is any part of atmospheric air that is breathable.) Worn-out fan blades, nonworking fans, improper water/air distribution, foaming, fouling, scaling, and corrosion can all lead to improper heat balance that results in inadequate cooling in the cooling tower and reduces the heat transfer efficiency.

Low cooling water temperature can be the result of improper system balance or ambient air temperature. For example, cold weather can make it difficult to control the temperature of the tower. Sometimes controlling the number of fans or banks of fans in operation helps alleviate this problem. In geographic areas where cold weather is prevalent, cooling towers might be equipped with bypasses that divert return water to the basin in order to control temperature. Also during cold weather, fans might be shut down and the towers closed to prevent ice accumulation. Factors affecting cooling tower performance include:

- Humidity and ambient temperature
- Cooling tower design
- Wind speed and direction (in natural draft cooling towers)
- Contamination of cooling water.

Table 12.1 lists how these variables affect cooling tower performance.

As evaporation occurs, the air absorbs water. Energy is required to change this water from a liquid to a vapor. This energy comes in the form of latent heat transfer. Air continues to absorb water until there is no more water for evaporation or the air becomes saturated (i.e., cannot hold any more water).

Table 12.1 Variables That Affect Cooling Tower Performance

Variable	Effect of Variable on Cooling Tower Performance
Temperature	High ambient temperature = high cooling tower loads
Tower design	High air flow = high cooling rate
Humidity	High humidity = high water vapor in the air = low cooling rate
Wind velocity and direction	High wind velocity = high air flow = high cooling rate Wind direction opposite of cooling tower orientation = decreased evaporation (Note: Induced or forced draft towers have greater air flow than similar-size natural draft towers and do not depend on wind currents).
Water contamination	High contamination level (e.g., algae buildup) = low cooling rate (e.g., fouling)

Wind Velocity

Wind velocity is the speed of the wind outside the tower. Wind velocity affects cooling towers by increasing or decreasing the rate of evaporation.

Tower Design

Cooling tower design can affect heat transfer in a variety of ways. For example, the design of the fill can increase or decrease heat transfer based on the amount of water-to-air contact. The amount and type of draft also can increase or decrease the amount of cooling based on the amount of water-to air contact.

The **cooling range** is the difference in temperature between the warm entering water and the cool exiting water. **Circulation rate** is the rate at which water flows through the tower. Circulation rate can affect heat transfer (increased circulation equals increased cooling).

Heat duty is the amount of heat energy a cooling tower is capable of removing, usually expressed in MBTU/hour.

Cooling range the difference in temperature between the hot water entering the tower and the cooler water exiting the tower.

Circulation rate the rate at which cooling water flows through the tower and through the cooling water system.

Heat duty the amount of heat energy a cooling tower is capable of removing, usually expressed in MBTU/hour.

Corrosion, Erosion, Fouling, and Scale

CORROSION Corrosion is the deterioration of a metal by a chemical reaction (e.g., iron rusting). Over time, corrosion can cause a phenomenon known as *stress corrosion cracking* in some stainless steel materials. In cooling tower systems, this can result in decreased heat transfer and loss of structural integrity. It may also allow process fluids in heat exchangers to leak into the cooling water if the corrosion is severe enough.

EROSION **Erosion** is the mechanical degradation or wearing away of tower components (e.g., fan blades, wooden elements, and cooling water piping). Erosion also can lead to a loss of structural integrity and process fluid contamination.

FOULING Fouling is the accumulation of deposits (such as sand, silt, scale, sludge, fungi, and algae) on the surfaces of processing equipment. Fouling of exchangers can result from contamination of the cooling water at the tower. Fouling of cooling tower components such as the packing/fill, tower basin, or drift eliminators can result from improper chemical addition and control. As fouling increases, heat exchanger tubes become blocked and water flow rate

Did You Know?

Cooling towers that are common in nuclear plants employ natural draft instead of fans to facilitate air movement.

CREDIT: Meryll/Shutterstock.

Erosion the degradation of tower components (fan blades, wooden portions, and cooling water piping) by mechanical wear or abrasion by the flow of fluids (often containing solids). Erosion can lead to a loss of structural integrity and process fluid contamination.

decreases. As the flow rate decreases, heat transfer in the cooling tower between the water and air flow decreases. Fouling of pumps, which is commonly caused by plugged suction screens, can lead to low pump discharge pressure and can prevent water from circulating properly through the cooling water system.

SCALE   Scale occurs as a result of dissolved solids depositing on the inside surfaces of hot equipment. As scale is deposited, an insulating layer is formed. As with fouling, the presence of scale can prevent water from flowing properly. It reduces the efficiency of the equipment and reduces heat transfer capability.

Foaming   the formation of a froth caused by mixing of water contaminants with air; impairs water circulation and can cause pumps to lose suction or cavitate.

FOAMING   Foaming is a continuous formation of bubbles on the surface of cooling tower water. Foaming generally is caused when the alkalinity of the tower water is high. It is visible at the top of the tower cells where the distribution takes place and in the bottom basin. The resulting frothy mixture reduces the effectiveness of the cooling tower heat transfer process. It also can cause poor equipment performance, and it can overflow tanks or sumps. Foam from the distribution header at the top of the tower can fall onto personnel, vehicles, or other equipment in the area.

Foaming can result from excessive agitation, improper levels, air leaks, cavitation, or contamination. The potential for foaming is increased when impurities are present. Foaming can be caused by:

- Chemical overfeed
- Overcycling
- Excessive suspended solids
- Process contamination
- Protein byproducts from microbiological growth
- Surfactants
- High alkalinity.

Foaming can impair water circulation and cause pumps to lose suction, can ruin paint on vehicles and other equipment, and can be an inhalation source of *Legionella* bacteria. Foaming can be controlled or eliminated with antifoam agents.

12.5 Safety and Environmental Hazards

Many hazards are associated with normal and abnormal cooling tower operations. These hazards can affect equipment, unit operations, the environment, and the personal safety of process technicians. Table 12.2 lists some of these hazards and their possible effects.

In addition to the hazards caused by improper operation, hazards also are associated with normal operations. For example, slip hazards exist around towers because of water leakage and moisture collection caused by drift. Typical personal protective and safety equipment used during normal and abnormal cooling tower operations includes goggles, rubber gloves, rain suits, and respirators. In instances where problems are found or suspected, full-face respirators with organic cartridges are required. Full-face respirators with organic cartridges protect workers in areas where there is exposure to cooling water chemicals and contaminants.

All precautions and safeguards must be taken to ensure that cooling towers are in proper working condition to prevent potential injury to individuals or environmental hazards.

Heat Exchanger Tube Leaks

Tube leaks   leaks in a tube that can result in process chemicals entering the circulating water, possibly resulting in a fire, explosion, or environmental or toxic hazard.

Heat exchanger **tube leaks** can result in process chemicals entering the circulating water, possibly resulting in a fire, explosion, or an environmental or toxic hazard. In some cases, air-monitoring instruments are added to detect toxic or flammable chemicals entering the cooling tower through leaks in heat exchanger tubing.

Table 12.2 Hazards Associated with Improper Cooling Tower Operation

Improper Operation	Possible Effects			
	Individual	Equipment	Production	Environment
Failure to treat cooling water chemically	Exposure to harmful microorganisms (e.g., *Legionella* bacteria, which causes Legionnaires disease)	Algae growth or sludge buildup can foul process equipment such as exchangers	Reduced production Off-spec products	Extremely high contaminant levels in the blowdown
Improper cooling tower lineup	Eye and skin irritation if exposure occurs	Pump damage because of blocked discharge or suction	Off-spec products	Spills to the environment of effluent water intended for water treatment facilities
Circulating cooling water through a heat exchanger with ruptured tubes	Exposure to process chemicals	Tower contamination; depletion of treatment chemicals; possible fire or explosive conditions; possible damage to tower components	Product ruined by contamination	Hazardous chemical spill
Exposing skin to or ingesting chemically treated cooling water	Eye and skin irritation and/or severe illness			
Improper chemical addition to cooling tower	Eye and skin irritation and/or severe illness	Corrosion; erosion	Loss of tower performance	Hazardous chemical spill

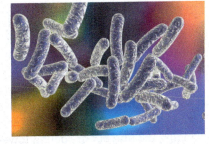
Fires

Many older cooling towers are built of wood and can act like kindling (thin, dry wood used to start a fire) in certain situations. For example, if a tube leak develops when a heat exchanger is being used to cool a flammable substance, the cooling tower can be saturated with flammable gas in a very short time. Static electricity and lightning are among possible causes of ignition. Fans that provide an extremely large amount of air flow can turn the tower into a huge fire hazard. Cooling towers generally are equipped with sprinkler and trip systems that shut down fans in the event of a fire.

A lengthy shutdown of an operating unit can be particularly dangerous. During this downtime, a wooden structure can dry out completely and become susceptible to ignition sources. To prevent fire, the tower must be kept wet at all times. This is achieved with a permanent fire protection sprinkler system or with a simple wetdown system consisting of a small pump, piping, and nozzles that pump water from the cold water basin to the top areas of the tower.

12.6 Process Technician's Role in Operation and Maintenance

A process technician working in an area with a cooling tower must understand the operating procedures associated with safe operation of the entire unit, including the cooling tower's operating area. In the event of an emergency or shutdown of the unit, the process technician must be able to troubleshoot the problem and communicate with the control room about the activity in the field.

Typical procedures associated with operating cooling towers safely include monitoring, adjusting to weather conditions, temperature control, blowdown, adding makeup water, testing, adding chemicals, and maintenance.

Component and Concentration Monitoring

Cooling tower components must be monitored during normal operation. This monitoring includes checking for overloading of electric motors, cycles of concentration, vibration, water basin level, and proper operation of the liquid level (sump) and water addition system.

When water passes through a cooling tower, some water is evaporated and minerals are left behind in the recirculating water. As evaporation continues, the concentration of solids in the water becomes much greater than in the original makeup water. This can eventually lead to saturated conditions. If the cooling tower contains an excessive amount of solids, then the process technician must perform blowdown operations to reduce the concentration.

Temperature Monitoring

Cooling towers are usually designed with some excess capacity because additional heat loads are often added to the process unit, requiring additional cooling tower capacity. Conditions can affect water temperature, which in turn affects the capacity for heat removal.

The capacity of a tower during hot summer weather is less than during the cold days of winter. Any decline in cooling tower capacity limits the manufacturing capacity of the facility. Because of this, cooling water temperature is monitored carefully. It is controlled by changing the number of cooling tower cells and fans operating, by changing the number of pumps running, and by adjusting valves to ensure that water is circulating effectively.

Chemical Treatment

Chemical treatment is important in cooling towers because they are open systems. An untreated cooling tower system is susceptible to algae growth, sludge, scale buildup, and fouling of exchangers and cooling equipment.

To determine the proper chemical treatment, the circulating system water chemistry must be monitored on a regular basis for parameters such as conductivity, pH, total dissolved solids, and the presence of process chemicals. The most common chemical additives are biocides, corrosion inhibitors, dispersants, and pH control chemicals.

Biocides control biological growth (e.g., bacteria and algae) in the tower and in circulating water. Biocides often are added at night when evaporation is lowest, allowing the biocides the to have maximum effect.

Chemicals are added to cooling tower water to prevent corrosion in piping and heat exchange equipment. These corrosion inhibitors also reduce the corrosive effect of the cooling water, both in the cooling tower and in the facility equipment where cooling water is used.

Dispersants are used to clean the cooling tower and other equipment by keeping foreign material distributed in the water until filters can remove it.

A cooling tower requires pH adjustment for the health of the tower and equipment. These pH controls assist in corrosion control and suppress organic growth (e.g., algae).

Biocides chemical agents that are capable of controlling undesirable living microorganisms (such as in a cooling tower).

As water is lost to evaporation, the accumulation of dissolved solids changes the pH and the chemical concentrations in the water. Additives and/or blowdown are required to counteract these effects.

Blowdown

Blowdown is the process in which a certain amount of water is discharged and removed from the system to maintain the allowable cycles of concentration. The allowable cycles of concentration are based on the miscellaneous elements in the source water. Increasing the blowdown lowers the concentration of these elements. Evaporative loss, drift loss, and blowdown are replaced with fresh makeup water. This is usually an automatic process, but process technicians should still monitor these levels to make sure the system is working properly.

Blowdown the process of removing small amounts of water from the cooling tower to reduce the concentration of impurities.

Maintenance

Some work on cooling towers requires that lockout/tagout procedures be implemented. When that occurs, unit technicians must be able to isolate the unit completely from all energy sources to ensure the safety of themselves and their coworkers.

Because of the fire danger presented by wooden cooling towers, hot work around cooling towers should be approved by supervision, limited, and closely monitored. In addition to the fire hazard of the wooden construction material, the natural oils in cedar planks can ignite quickly and the design of the fill adds to the likelihood that flames will spread. The fire hazard increases exponentially if tower components are allowed to dry out during outages.

When working with cooling towers, process technicians should be aware of problems that can occur and be able to perform preventive maintenance. Typical scheduled maintenance activities for process technicians include:

- Sampling for chemical control (daily)
- Performing blowdown for solids control and removal (as required)
- Pulling and cleaning pump suction screens and washing the stairway and deck (as required)
- Disinfecting the cooling tower (as required).

Table 12.3 lists some additional monitoring and maintenance tasks that process technicians must perform when working with cooling towers.

Like other elements in a process system, cooling towers require routine preventive maintenance. Mechanical and chemical cleaning is required per manufacturer and unit specifications. The cooling tower must be disinfected periodically. Biocide treatments must be maintained

Table 12.3 Process Technician's Role in Operation and Maintenance

Look	Listen	Check
■ Observe for leaks. ■ Check basin water levels to make sure they are adequate. ■ Monitor chemical balance (pH and conductivity). ■ Check filter screens for plugging. ■ Check temperature differentials. ■ Look for broken fill materials (to fix at next turnaround). ■ Look for ice buildup in cold climates. (Note: The weight of ice buildup can collapse the internal components of a cooling tower.) ■ Check for proper water distribution on top of the tower. ■ Determine cycles of concentration.	■ Listen for abnormal noises (e.g., grinding sounds associated with pump cavitations, or high-pitched sounds associated with improperly lubricated fan bearings).	■ Use the correct device to check for excessive heat in fan and pump motors. ■ Check for excessive vibration in fans and pumps.

on a regular basis to ensure that the system is within its acceptable chemical range. Full-face respirators, and possibly protective suits, might be required for these tasks.

Process technicians should also routinely sample cooling water to ensure that the pH is within an acceptable range. If the readings are not in alignment with specs, then the process technician needs to make changes to the system as required.

Process technicians should monitor the temperature of the cooling water supply and run only the number of fans required to produce the proper cooling water supply temperature. Technicians also should take additional safety precautions for routine and preventive maintenance, such as the use of full-face respirators when adding treatment chemicals to the cooling tower water. Most highly toxic chemicals have been replaced with more environmentally friendly ones.

Failure to perform proper maintenance and monitoring can affect the process and result in personnel injury or equipment damage. Equipment operational hazards are also associated with cooling towers during normal and abnormal operation. For example, cooling towers use large, motor or turbine-driven pumps. At times, emergency cooling water is needed during power outages. In these cases, circulating pumps with other drivers, engines, or turbines are used, so hazards associated with rotating equipment are also present.

Summary

Cooling towers are structures designed to lower the temperature of a circulating cooling water stream. They accomplish this through the thermodynamic principle of evaporation and sensible heat loss. When water changes from a liquid into a vapor, it absorbs a substantial amount of heat from the surrounding water. This process is called latent heat of vaporization.

The main purpose of cooling towers is to remove heat from process cooling water so the water can be recycled and recirculated through the process. Cooling tower water can be used in a variety of applications, including process condensers and equipment coolers.

Cooling towers may be open (wet) or closed (dry) in design. In closed circuit cooling towers, the working fluid passes through a heat exchange coil and water is sprayed over it. A fan draws air through the water spray, pulling heat away from the coil by the process of convection. There is no direct contact between the water being cooled and the sprayed water. Most cooling towers used in process industries are of the open cooling design, which require a considerable amount of attention for a variety of reasons, including the potential for contact with hot water and exposure to toxic or corrosive chemicals.

Heat exchangers remove heat from process fluids. Cooling towers remove heat from heat exchanger cooling water. Maintaining the heating and cooling relationship between exchangers and towers is a continuous process. Proper operation of a cooling tower involves maintaining a constant cool water temperature throughout the year. On cold days, fans are slowed or turned off to maintain the constant temperature.

Cooling towers are classified as natural draft, forced draft, or induced draft. They also come in many shapes and sizes and can be single-cell or multicell. The two types of flow patterns in cooling towers are crossflow and counterflow. In a crossflow tower, air flows generally perpendicular (at a 90-degree angle) to the direction of the water flowing downward. In counterflow towers, water and air generally flow on the same plane (e.g., vertically) but in opposite directions.

Cooling towers include a water distribution header and water boxes that distribute the water while fill redirects the flow. A basin stores water, and makeup water replaces water lost to evaporation, blowdown, and drift. Drift eliminators minimize water loss. Suction screens filter out debris. Fans with a shroud force, or induce, air flow in induced or forced draft towers.

Many factors affect cooling tower performance. These factors include temperature, humidity, wind velocity, water contamination, and tower design.

Potential problems associated with cooling towers include heat exchanger tube leaks, other process equipment leaks, and inadequate circulation because of pump malfunction. A heat exchanger tube leak can affect cooling tower operation by allowing process chemicals to enter the circulating water system and to be exposed to the atmosphere in the cooling tower.

Safety and environmental hazards associated with cooling towers include normal and abnormal cooling tower operations and operation of the circulating water system. Hazards include slip hazards, biological hazards, chemical exposure, and fire or explosions. Appropriate PPE (including full-face respirators) is used to protect employees working in areas where they might be exposed to cooling water chemicals and contaminants.

When monitoring and maintaining cooling towers, process technicians must remember to check pumps for excessive vibrations, noise, or heat. They also must ensure that water levels are adequate, there are no leaks, filter screens are not plugged, and proper heat exchange is occurring. In cold climates, they should check for excessive ice buildup inside the tower because ice can damage the internal structures of the tower. They also must monitor and adjust water chemistry.

Checking Your Knowledge

1. Define the following terms:

   a. Dew point

   b. Basin

   c. Biocides

   d. Blowdown

   e. Cooling towers

   f. Drift eliminators

   g. Evaporation

   h. Fill

   i. Forced draft cooling towers

   j. Induced draft cooling towers

   k. Natural draft cooling towers

   l. Relative humidity

   m. Wet bulb temperature

   n. Dry bulb temperature

   o. Circulation rate

   p. Louvers

   q. Counterflow

2. (True or False) Cooling towers are designed to raise the temperature of a water stream.

3. (True or False) Natural draft towers have fans or blowers at the bottom of the tower that force air through the equipment.

4. _____ are movable plates that adjust the quantity and direction of the air flow into the tower.

   a. Spray nozzles

   b. Fans

   c. Louvers

   d. Basins

5. Drift is water carried over with _____ from the tower.

   a. air

   b. heat transfer

   c. entrainment

   d. evaporation

6. The _____ and temperature of the air determine the amount of heat removed in a standard amount of air.

   a. steam

   b. heat transfer

   c. entrainment

   d. relative humidity

7. The lowest temperature to which air can be cooled through the evaporation of water is _____.

   a. humidity

   b. dry bulb temperature

   c. wet bulb temperature

   d. ambient temperature

8. Algae can accumulate and other plant and animal life can appear if cooling tower _____ is not controlled.

   a. thermodynamics

   b. water chemistry

   c. temperature

   d. flow

9. (True or False) The process of evaporation is less efficient than convection at removing heat.

10. (True or False) The temperature of the water leaving the cooling tower should be constant year-round.

11. Which of the following hazards can cause a hazardous chemical spill? (Select all that apply.)

    a. Improper cooling tower lineup

    b. Failure to treat cooling water chemically

    c. Circulating cooling water through a heat exchanger with ruptured tubes

    d. Improper chemical addition to a cooling tower

12. (True or False) Legionnaires disease bacteria (LDB), which can exist in cooling towers, causes a form of pneumonia.

NOTE: Answers to Checking Your Knowledge questions are in the Appendix.

Student Activities

1. Given a model or diagram of a cooling tower, identify the following components and explain the purpose of each:

   a. Water distribution header and water boxes

   b. Fill

   c. Basin

   d. Makeup water

   e. Suction screens

   f. Louvers

   g. Drift eliminators

   h. Fan

2. Prepare a report about the three main types of cooling towers used in industry. In the report, describe how water is cooled in the tower. Include the flow description and a drawing with the components labeled.

3. Work with a team member to prepare a presentation about the cycles of concentration, the importance of chemical treatment, water chemistry (include types of common chemical additives), and blowdown. Include in your presentation the importance of each, how they interact, and how they can improve productivity within a process facility.

4. Research and explain why nuclear power facilities use hyperbolic cooling towers.

Chapter 13
Furnaces

Objectives

After completing this chapter, you will be able to:

13.1 Identify the purpose, common types, and applications of furnaces. (NAPTA Furnaces 1, 2*) p. 275

13.2 Identify the components of furnaces and the purpose of each. (NAPTA Furnaces 3, 4, 5) p. 277

13.3 Explain the operating principles of furnaces. (NAPTA Furnaces 6, 10) p. 280

13.4 Identify potential problems associated with furnaces. (NAPTA Furnaces 11) p. 282

13.5 Describe safety and environmental hazards associated with furnaces (including stack and emissions controls). (NAPTA Furnaces 8) p. 284

13.6 Describe the process technician's role in furnace operation and maintenance. (NAPTA Furnaces 9) p. 286

13.7 Identify typical procedures associated with furnaces. (NAPTA Furnaces 7) p. 287

* North American Process Technology Alliance (NAPTA) developed curriculum to ensure that Process Technology courses will produce knowledgeable graduates to become entry level employees in process technology. Objectives from that curriculum are named here in abbreviated form. For example, "(NAPTA Furnaces 1, 2)" means that this chapter's objective relates to objectives 1 and 2 of the NAPTA curriculum on furnaces).

Key Terms

Air register—also called the burner damper, a device located on a burner that is used to adjust the primary and secondary air flow to the burner; it provides the main source of air entry to the furnace, **p. 278.**

Assisted draft furnace—a system that uses electric motor, steam turbine-driven rotary fans, or blowers to push air into the furnace (*forced draft*) or to draw flue gas from the furnace to the stack (*induced draft*), **p. 276.**

Balanced draft furnace—a furnace that uses both forced and induced draft fans to facilitate air flow; fans induce flow out of the firebox through the stack (*induced draft*) and provide positive pressure to the burners (*forced draft*), **p. 277.**

Box furnace—a square, box-shaped furnace designed to heat process fluids or to generate steam, **p. 275.**

Burners—devices with open flames used to provide heat to the firebox; number of burners needed depends on the furnace's size and capacity, **p. 278.**

Cabin furnace—a cabin-shaped furnace designed to heat process fluids or to generate steam, **p. 275.**

Coke—carbon deposits inside process tubes (*coking*), which can result in poor heat transfer to process fluid and in overheating and failure of the tube metal, **p. 282.**

Convection section—the upper and cooler section of a furnace where heat is transferred and convection tubes absorb heat, primarily through the method of convection; the part of the furnace where process feed and/or steam enters, **p. 277.**

Convection tubes—furnace tubes, located above the shock bank tubes, that receive heat through convection, **p. 279.**

Damper—a device that regulates the flow of air into a furnace and flue gases leaving a furnace, **p. 279.**

Firebox—the combustion area and hottest section of the furnace where burners are located and radiant heat transfer occurs, **p. 278.**

Forced draft furnace—a furnace that uses fans or blowers to force the air required for combustion into the burner's air registers, **p. 277.**

Fuel lines—supply lines that provide the fuel required to operate the burners, **p. 277.**

Furnace—a piece of equipment in which heat is released by burning fuel and is transferred directly or indirectly to a fluid mass for the purpose of increasing the temperature of fluids flowing through tubes to effect a physical or chemical change; also referred to as a *process heater*, **p. 275.**

Heater—a term used to describe a fired furnace, **p. 275.**

Induced draft furnace—a furnace that uses fans, located between the convection section and the stack, to pull air up through the furnace and induce air flow by creating a lower pressure in the firebox, **p. 277.**

Natural draft furnace—a furnace that has no fans or mechanical means of producing draft; instead, the heat in the furnace causes natural draft, **p. 276.**

Nitrogen oxides (NOx)—undesirable air pollution produced from reactions of nitrogen and oxygen. The primary NO_x species in process heaters are nitric oxide (NO) and nitrogen dioxide (NO_2), **p. 280.**

Pilot—an initiating device used to ignite the burner fuel, **p. 278.**

Radiant section—the lower portion of a furnace where heat transfer is primarily through radiation, **p. 277.**

Radiant tubes—tubes located in the firebox that receive heat primarily through radiant heat transfer, **p. 278.**

Radiation—the transfer of heat by electromagnetic waves, **p. 277.**

Refractory lining—a form of insulation used inside high temperature boilers, incinerators, heaters, reactors, and furnaces; a common type is called fire brick, **p. 278.**

Shock bank (shock tubes)—also called the *shield section*, a row of tubes located directly above the firebox (radiant section) in a furnace that receives both radiant and convective heat; protects the convection section from exposure to the radiant heat of the firebox, **p. 279.**

Stack—a cylindrical outlet, located at the top of a furnace, through which flue (combustion) gas is removed from the furnace, **p. 279.**

Stack damper—a device that regulates the flow of flue gases leaving a furnace, **p. 279.**

Vertical cylindrical furnace—a furnace design in which the radiant section tubes are laid out in a vertical configuration along the walls of the cylinder; it has a smaller footprint than a cabin furnace and is used in operations where less heat capacity is required, **p. 275.**

13.1 Introduction

A **furnace**, also referred to as a process **heater**, is a piece of equipment in which heat is released by burning fuel and is transferred directly or indirectly to a fluid mass for the purpose of increasing the temperature of fluids flowing through tubes to effect a physical or chemical change. In some furnaces, convection section tubes are also used to preheat boiler feedwater. Furnaces are an important part of many processes, both because they provide the heat necessary to facilitate chemical reactions and physical changes and because they can be used to incinerate unwanted waste streams.

Several types of industrial furnaces exist, all with similar principles of operation. They differ, however, in shape, heat source, function, process cycle, draft type, mode of heat application, and the atmosphere inside the furnace.

Furnace a piece of equipment in which heat is released by burning fuel and is transferred directly or indirectly to a fluid mass for the purpose of increasing the temperature of fluids flowing through tubes to effect a physical or chemical change; also referred to as a *process heater*.

Heater a term used to describe a fired furnace.

Did You Know?

Burning municipal waste in an incinerator produces heat that can be used to make steam. This steam can be used to power turbines that drive generators that produce electricity.

Metals that might be present during the incineration process also can be reclaimed and sent for recycling. The reuse of both the materials and the heat energy is compliant with environmental protection regulations.

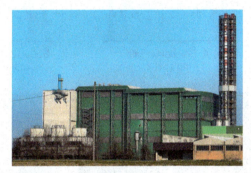

CREDIT: CervelliInFuga/Shutterstock.

Furnace Types and Applications

Furnaces can be used in a wide variety of applications. In mining and metallurgical operations, furnaces can be used to extract metal from ore, facilitate the casting and shaping of metal, and improve the properties of metal through heat treatment and hardening. In glass and plastics manufacturing, furnaces can be used to melt glass or plastic so it can be shaped into containers, formed into light bulbs, extruded into fibers, pressed into sheets, or molded into other shapes. In the waste treatment industry, furnaces can be used to *incinerate* (burn at a high temperature and convert to ash) medical and municipal solid wastes so they take up less space and are less hazardous. In other industrial applications, such as oil and gas refining or chemical processing, furnaces are used to heat fluids in order to change the physical properties (e.g., lower the viscosity or convert from a liquid to a gas), and facilitate chemical reactions. Furnaces addressed in this chapter are used in the oil and gas refining process.

Common Furnace Designs

Furnace designs vary with regard to function, heating duty, type of fuel, and method of introducing combustion air. The most common types of industrial furnaces are box furnaces, vertical (cylindrical) furnaces, and cabin furnaces. Figure 13.1 shows examples of each of these types of furnaces with some of the different coil arrangements that are used.

Box furnaces (see Figure 13.1A) can be used to heat process fluids or generate steam. They also can be used for heat treatment applications such as tempering, hardening, or firing. **Vertical cylindrical furnaces** (see Figure 13.1B), which are also referred to as *vertical cylindrical heaters* or VCs, are similar to box furnaces but are cylindrical (tube like) in shape and stand upright (vertical), so they require less space. **Cabin furnaces** (see Figure 13.1C) are used in high-temperature processes. Cabin furnaces derive their name from their shape, which resembles a log cabin with a chimney.

Box furnace a square, box-shaped furnace designed to heat process fluids or to generate steam.

Vertical cylindrical furnace a furnace design in which the radiant section tubes are laid out in a vertical configuration along the walls of the cylinder; it has a smaller footprint than a cabin furnace and is used in operations where less heat capacity is required.

Cabin furnace a cabin-shaped furnace designed to heat process fluids or to generate steam.

Figure 13.1 A. Box furnace.
B. Vertical (cylindrical) furnace.
C. Cabin furnace.

CREDIT: *(Bottom row, from left)* eakka-luktemwanich/Shutterstock. tamapapat/Shutterstock. meepoohfoto/Shutterstock.

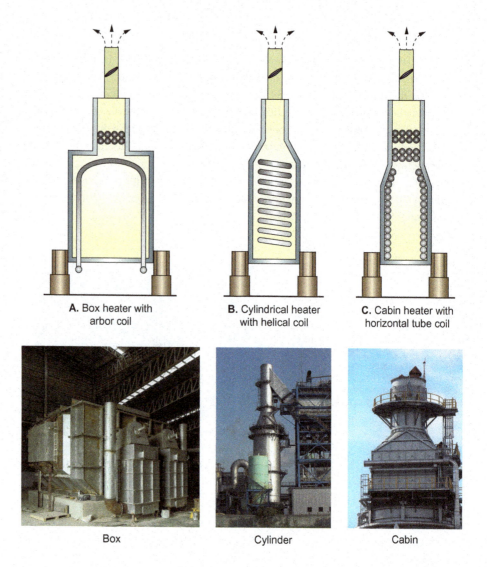

A. Box heater with arbor coil

B. Cylindrical heater with helical coil

C. Cabin heater with horizontal tube coil

Box   Cylinder   Cabin

Furnace Draft Types

Like cooling towers, air flow (draft) inside furnaces can be provided by natural air currents (natural draft) or mechanical air currents (assisted draft).

Natural draft furnaces (Figure 13.2) have no fans or mechanical means of producing draft. Because they lack mechanical air movers, the stacks on natural draft furnaces must be taller than on other furnaces in order to achieve adequate draft.

Because hot furnace gases are not as dense as the ambient air, they rise, creating a differential pressure between the top and bottom of the furnace. This differential pressure (referred to as the *chimney effect*, *furnace draft*, or *thermal head*) creates a slight vacuum that pulls atmospheric air into the furnace and causes draft. As hot flue gas rises through the stack, a negative pressure is created inside the firebox. This negative pressure pulls air into the burner air registers at the bottom of the furnace.

Assisted draft furnaces use rotary blowers, driven by electric motors or steam turbines. The three main types of air flow in assisted draft furnaces are induced draft, forced draft, and balanced draft. Induced draft draws flue gas from the furnace to the stack. Forced draft pushes air into the furnace. Balanced draft has fans to both force and induce air movement. All three of these draft types are shown in Figure 13.3.

The fan speed of assisted draft furnaces can be either fixed or variable, and it can be controlled manually or automatically from a control panel. While assisted draft systems require more energy to operate than natural draft systems, because of the energy required

Natural draft furnace a furnace that has no fans or mechanical means of producing draft; instead, the heat in the furnace causes natural draft.

Assisted draft furnace a system that uses electric motor, steam turbine-driven rotary fans, or blowers to push air into the furnace (*forced draft*) or to draw flue gas from the furnace to the stack (*induced draft*).

Figure 13.2
Natural draft furnace
(cabin-type).

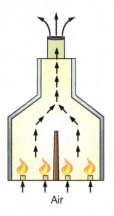

Figure 13.3 **A.** Induced draft cabin furnace. **B.** Forced draft cabin furnace. **C.** Balanced draft cabin furnace.

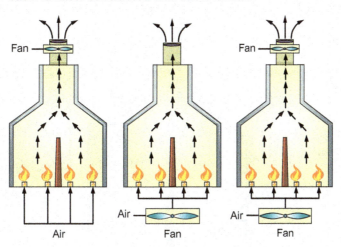

to operate the fans, they are actually more efficient because they provide more control and require smaller burners for the same firing capacity.

An **induced draft furnace** (shown in Figure 13.3A) uses a fan located at the top of the furnace to draw flue gas from the furnace body into the stack. As the gas moves into the stack, the pressure in the firebox is decreased, thereby inducing draft and drawing ambient air into the furnace.

A **forced draft furnace** (shown in Figure 13.3B) uses fans or blowers to force air flow. During combustion in a forced draft furnace, positive pressure is created by the fan or blower. This positive pressure forces air into the burner air registers through a header or plenum. Rising combustion gases inside the furnace create draft or negative pressure.

A **balanced draft furnace** (shown in Figure 13.3C) uses both forced and induced draft fans to facilitate air flow. One fan forces air flow into the burner registers, while the other fan pulls (induces) combustion gases out of the stack.

13.2 Furnace Sections and Components

Furnaces are divided into two sections: the radiant section and the convection section. Both of these sections are identified in Figure 13.4.

The **radiant section** is the furnace section located at the bottom of the unit, closest to the heat source. The **convection section** is the upper area of a furnace. The following section describes the flow of heat from the bottom of the radiant section through the top of the convection section.

Radiant Section

In the radiant section the primary method of heat transfer is **radiation** (the transfer of heat by electromagnetic waves). This is the hottest section of the furnace, with the tubes being directly exposed to the heat source. Located at or near this section are fuel lines, fuel valves, burners, draft gauges, the purge system, the bridgewall, radiant tubes, and refractory lining.

FUEL LINES AND BURNERS The **fuel lines**, found at or near the radiant section, provide the fuel required to operate the burners. Common furnace fuels are natural gas, fuel oil, and fuel gas (waste gases from other processes). The flow of the fuel through these lines is controlled by fuel valves.

Induced draft furnace a furnace that uses fans, located between the convection section and the stack, to pull air up through the furnace and induce air flow by creating a lower pressure in the firebox.

Forced draft furnace a furnace that uses fans or blowers to force the air required for combustion into the burner's air registers.

Balanced draft furnace a furnace that uses fans to facilitate air flow; fans induce flow out of the firebox through the stack (induced draft) and provide positive pressure to the burners (forced draft).

Radiant section the lower portion of a furnace where heat transfer is primarily through radiation.

Convection section the upper and cooler section of a furnace where heat is transferred and convection tubes absorb heat, primarily through the method of convection; the part of the furnace where process feed and/or steam enters.

Radiation the transfer of heat by electromagnetic waves.

Fuel lines supply lines that provide the fuel required to operate the burners.

Figure 13.4 Cabin furnace with components identified.

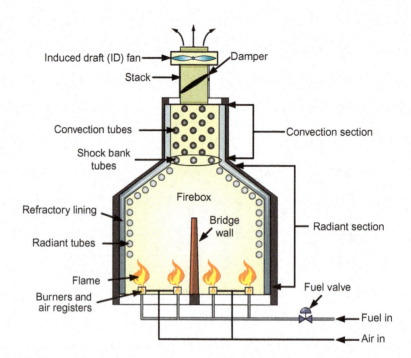

Pilot   an initiating device used to ignite the burner fuel.

Burners   devices with open flames used to provide heat to the firebox; number of burners needed depends on the furnace's size and capacity.

Air register   also called the burner damper, a device located on a burner that is used to adjust the primary and secondary air flow to the burner; they provide the main source of air entry to the furnace.

Firebox   the combustion area and hottest section of the furnace where burners are located and radiant heat transfer occurs.

Radiant tubes   tubes located in the firebox that receive heat primarily through radiant heat transfer.

Refractory lining   a form of insulation used inside high temperature boilers, incinerators, heaters, reactors, and furnaces; a common type is called fire brick.

As the fuel leaves the fuel line, it passes into the burners where a **pilot** ignites the burner fuel. The **burners**, like the burners on a kitchen stove, are mechanical devices where fuel is burned in a controlled manner to produce heat. As the fuel is transferred to the burners, air is introduced through **air registers**. This air facilitates the combustion process. In some burners, the air is mixed with the fuel before it ignites at the burner tip. This type of burner is called a premix burner.

During the combustion process, draft gauges, calibrated in inches of water, are used to measure the differential pressure between the outside of the furnace and the flue gas inside of the furnace.

The heat that is produced by the burners is radiated into the main body of the furnace called the **firebox**. Because of the intense heat generated, the firebox must be lined with a special refractory lining. This refractory serves two purposes: to protect the furnace shell from the heat inside the firebox and to reflect the heat back into the firebox, preventing its escape from the furnace.

PURGE SYSTEM   Before lighting a furnace, it is important to remove any combustible materials or unburned fuel from the firebox. Failure to do so can result in an explosion. The system that is used to remove these materials from the firebox is called a purge system.

In forced, induced, or balanced draft furnaces, fans are used to purge material from the firebox using air. Steam is used in natural draft furnaces. Steam purge systems also can be used to extinguish uncontrolled fires in the firebox. This type of steam is referred to as *snuffing steam*.

BRIDGEWALL   Also contained in the firebox of some furnaces is a bridgewall, a wall or vertical partition in the fire box or radiant section of the furnace that is used to deflect heat. The bridgewall compartmentalizes the radiant section and helps redirect the heat toward the tubes located along the outer wall. This redirected heat allows for radiant heating of the process fluids passing through the tubes. The term bridgewall is also used to describe the area of a furnace between the radiant and convection sections.

RADIANT TUBES   The tubes located in the radiant section are called **radiant tubes** (see Figure 13.4) because they receive radiant heat from the furnace burners. Radiant tubes can be mounted either vertically or horizontally, and can be placed in different locations or arrangements depending on the type of furnace.

REFRACTORY LINING   A **refractory lining** (Figure 13.5) is a castable or bricklike form of insulation used inside high-temperature furnaces. The refractory lining helps maintain a

Figure 13.5 Example of insulating firebricks.

CREDIT: RomanKorytov/Shutterstock.

uniform temperature on tube walls, minimizes heat loss, and keeps the outside wall from becoming too hot. (Excessive heat on the exterior walls could cause personal injury or damage to the firebox).

Refractory linings are typically composed of specialized heat-resistant materials such as firebrick, castable refractories, insulating brick, and ceramic fiber batts. Burners are lined with *burner tiles* (specially shaped firebrick) which can handle the high temperatures and abrasive properties of the flame. Ceramic fiber insulation batts and firebricks are used on the ceiling and walls of the furnace, while the furnace floor usually is composed of castable refractories (stronger materials that can be molded) because it must be hard enough to walk on during maintenance. Each of these insulation types is held together with heat-resistant mortar and is attached to the furnace wall by studs welded to the outside steel structure wall.

Convection Section

As the heat moves up through the furnace, it eventually leaves the radiant section and moves into the convection section (see Figure 13.4). The convection section is located at the top of the furnace firebox and is farther away from the heat source. The primary method of heat transfer in this section is convection (heat transfer through circulation or movement of a fluid). Feed and steam coils are located in this section; this is the entry and preheat point for the feedstock. In most furnaces the tubes physically exit the furnace and re-enter into the radiant section. The point at which tubes exit and re-enter the furnace is referred to as the crossover piping. Boiler feedwater is preheated here and steam is superheated. It is considered the cooler section of the furnace. Within this section are convection tubes, often finned or studded, to increase surface area and heat transfer rate. Above this section are the stack and stack **damper**.

SHOCK BANK   The convection section has relatively few components. At the base of the convection section is a row of shock bank tubes which receive both radiant and convective heat. Shock bank tubes protect the convection section from direct exposure to the radiant heat of the firebox and are usually made of materials that have greater heat resistance than the tubes in the convection section. The **shock bank** is sometimes called the *shield section*.

CONVECTION TUBES   Above the shock bank tubes is a series of tubes called convection tubes. **Convection tubes** receive most of their heat through convection because they are protected from direct exposure to radiant heat by the shock bank. Convection tubes can be made of the same material as the radiant tubes and can be finned or studded to increase heat transfer.

STACK AND STACK DAMPER   At the top of the furnace is the **stack**, a cylindrical outlet that removes flue gas from the furnace. Within the stack is a valve or movable plate called the **stack damper**, which regulates the flow of flue gases leaving the furnace.

Damper   a device that regulates the flow of air into a furnace and flue gases leaving a furnace.

Shock bank (shock tubes)   also called the *shield section*, a row of tubes located directly above the firebox (radiant section) in a furnace that receives both radiant and convective heat; protects the convection section from exposure to the radiant heat of the firebox.

Convection tubes   furnace tubes, located above the shock bank tubes, that receive heat through convection.

Stack   a cylindrical outlet, located at the top of a furnace, through which flue (combustion) gas is removed from the furnace.

Stack damper   a device that regulates the flow of flue gases leaving the furnace.

13.3 Furnace Operating Principles

Furnaces burn fuel inside a containment area (firebox) to produce heat. Process fluids are pumped through tubes located in this heated firebox. The heat from the firebox is transferred through the tube walls by radiation and through the process fluid by convection. The products of combustion (flue gases) then flow from the furnace, up through and out the stack where they are released to the atmosphere.

Fuel Supply

The fuel supply flow is determined by temperatures inside the furnace system. In many process furnaces, the temperature of the product exiting the furnace tubes indicates how much fuel is needed.

Fuel flow is typically controlled by adjusting the fuel pressure with some type of regulating valve. The regulator can be adjusted manually or automatically. When heavier liquid fuels are used (e.g., No. 6 fuel oil), heat must be applied to reduce viscosity. This is accomplished by raising the temperature of the fuel oil to ensure proper atomization (creation of a fine mist) at the burner tip. The temperature control of the liquid fuel is usually accomplished by attaching electrical or steam-jacketed heat tracing to the fuel supply line. At the burner tip, the fuel oil is mixed with steam, which atomizes it, turning it into a fine mist which is more easily ignited. Figure 13.6 shows an example of fuel and steam lines leading into a set of furnaces.

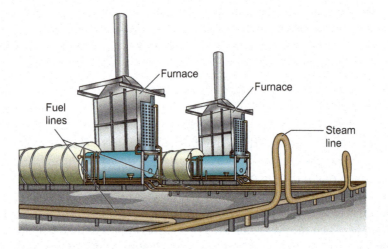

Figure 13.6 Fuel lines entering and steam line exiting a set of furnaces.

Combustion Air

All furnaces require the right mixture of air and fuel for complete combustion to occur. Without sufficient air, the furnace could become too fuel-rich, which could result in inefficient burning, smoke production, high carbon monoxide (CO) emissions, or furnace explosion. With too much air, however, energy is wasted and the amount of hot flue gas exiting the furnace increases.

To maintain proper ratios and prevent hazardous situations, online analyzers are used to monitor the excess oxygen (O_2) content of the flue gas. If O_2 levels are too low, then CO emissions increase. If O_2 levels are too high, then energy efficiency decreases and nitrogen oxide (NO_x) emissions increase. **Nitrogen oxides (NO_x)** are a form of undesirable emission produced from reactions of nitrogen and oxygen. NO_x is produced more often by too high of a flame temperature, not by excess O_2.

Nitrogen oxides (NO_x) undesirable air pollution produced from reactions of nitrogen and oxygen. The primary NO_x species in process heaters are nitric oxide (NO) and nitrogen dioxide (NO_2).

Flame Temperature

Autoignition is the term used to describe the temperature at which a mixture of fuel and air ignites spontaneously. The autoignition temperature of most fuels used in the process

Did You Know?

Carbon monoxide (CO) is extremely poisonous. It is a colorless, odorless gas produced during incomplete combustion of carbon-based fuels (e.g., gasoline, wood, or oil). When carbon monoxide enters the body, it binds with the hemoglobin in the blood, where oxygen normally binds. This prevents the blood from carrying oxygen to the cells. Loss of consciousness occurs as the brain is deprived of oxygen, and the result can be serious injury or death.

Because of the danger of CO poisoning, process technicians must be careful when working in or around combustion equipment. Personal gas monitors are worn by employees to ensure sufficient oxygen levels and detect the presence of harmful gases such as CO or H_2S.

CREDIT: zulkamalober/Shutterstock.

industries is in the range of 800 to 1,300 degrees Fahrenheit (427–704 degrees C). For the fuel to ignite automatically, the furnace's burner region must operate at a temperature exceeding the autoignition temperature of the fuel mixture.

Because burner flames can exceed 3,000 degrees Fahrenheit (1,649 degrees C), peak flame temperature must be controlled by maintaining the correct fuel/air ratio to the burners. Lowering the flame temperature will also significantly decrease the production of nitrogen oxides (NO_x), an undesirable byproduct of furnace operation.

Furnace Pressure Control

Industrial furnaces are operated at negative pressure. The furnace pressures fluctuate with the burner firing rate (e.g., the pressure is lowest at the lowest firing rate and highest at the highest firing rate). To compensate for this constantly changing condition, furnaces are equipped with draft control systems. These control systems consist of a stack damper that is controlled to maintain the desired pressure in the combustion chamber. As the burner firing rates decrease, the damper throttles the flow of flue gas out of the stack to hold the pressure constant. It is important that a sufficient amount of air enters the furnace to ensure good combustion of the fuel. This requirement must be achieved while at the same time, allowing enough flue gas to exit the stack to create draft and maintain negative pressure in the firebox.

Draft control systems regulate and stabilize the pressure in the combustion chamber. A pressure gauge in the furnace chamber or duct gives a visual indication of the air flow needed to maintain the target pressure. Pressure controls can be manual or automatic.

Interlock Systems

Interlock control systems are safety devices used to protect the furnace from developing dangerous operating conditions. Instrument control systems use sensors and logic control to shut off the supply of furnace fuel or process feed and to shut down the furnace under certain conditions. Some examples of items that can trip an interlock and shut down a furnace are:

- Loss of flame at the burners
- High coil outlet temperature
- Low or high burner fuel pressure/flow
- Low speed on the forced or induced draft fan, or loss of a fan
- Loss of process flow through the tubes
- High firebox pressure
- High stack temperature

- Excessive NO_x emissions
- Excessive CO emissions
- Excessive stack smoking

13.4 Potential Problems

When operating furnaces, process technicians must be aware of certain situations that can result in system shutdown or equipment damage. Some of these problems are listed below.

- Burner flames can impinge on tubes, causing coking (the deposit of carbon) inside the process tubes which may lead to localized tube metal overheating and failure.
- A change in the flow rate or composition of the process fluid flowing through the tubes can create or increase **coke** deposits in the tubes, cause high temperatures on the tube walls, produce an extreme pressure drop in the tubes, and ultimately cause tube failure.
- Poor flame distribution can cause overheating and tube failure.
- Liquid entrained in the fuel gas can cause flame failure or temperature shock in the furnace.
- Improper warmup or cool-down of furnaces can result in failure of the refractory materials and hot spots on the furnace walls.
- Poor atomization of liquid fuels can result in pooling of liquids inside the firebox, which can ignite.
- Ignition of excess unburned fuel inside the firebox can cause furnace explosions.
- Loss of furnace draft or improper draft control can cause burner flame instability.

Coke carbon deposits inside process tubes (*coking*), which can result in poor heat transfer to process fluid and overheating and failure of the tube metal.

Equipment Age and Design

Over time, equipment ages and without proper preventative maintenance, will fail. Rotating equipment, such as a forced or induced draft fan, is subject to failure because of normal wear over time. Burner air dampers and linkages can wear out or become inoperable as a result of inadequate lubrication or dust and grit build up, or they might fail because of corrosion.

Freezing/thawing and earth movements can crack concrete equipment foundations and cause external corrosion of structural components. Tube supports and spring hangers that have not been properly adjusted and maintained or have been subjected to flame impingement can create stress in furnace tubes and allow sagging or contact with other tubes. Damage to insulating refractory materials can result in hot spots on the skin of the furnace. Burner fuel lines need periodic cleaning and the burner tiles often need repair.

Instrument Problems

Instrument malfunctions can contribute to a number of problems, including an improper fuel/air ratio or a complete trip of the furnace system. Faulty sensors can activate furnace interlocks, causing unnecessary shutdowns.

Fouling

Some furnace problems are caused by fuel or process-related fouling. Fuels can contain impurities that produce corrosive materials and cause fouling. In some cases, the lower limit of the flue gas temperature in the furnace stack is set by the dew point of these acidic materials which are carried in the flue gas. Operating below the dew point temperature corrodes the tubes and refractory lining in the upper sections of the furnace. Other fuels contain trace metal impurities such as vanadium pentoxide (V_2O_5). Vanadium ash collects on upper furnace tubes, causing oxidation failures and attacking refractory insulation.

Process feedstocks into the furnace can decompose from the high furnace temperatures. This decomposition causes the formation of coke deposits on the inside surface of furnace tubes. Some furnaces, such as ethylene crackers, operate for long periods of time by controlling coke deposits through the use of proper temperatures, upstream purification, and addition of steam in the feed coils with the feedstock. Other furnaces are designed for periodic decoking (offline removal of coke deposits with steam and air), essentially burning off the coke deposits.

Tube Life and Temperature

Several factors determine the length of time that a tube can be in service before it must be repaired or replaced. These factors include the uniformity of burner firing, internal fouling, condition of tube support hangers, tube design, and tube manufacturing processes such as *carburization* (a process by which carbon is introduced into a metal in order to make the surface harder and more abrasion-resistant). Overheating can damage tubes. This can happen when operating the furnace with too little process fluid flow in the process tubes without adjusting the temperature. Tubes can also be damaged by direct flame impingement. If the fuel contains sulfur and water, heat from the flame causes the formation of sulfuric acid, which can damage the tubes.

External Factors

Other factors that contribute to furnace problems include the loss of steam supply, electrical failure, cold weather and upstream process upsets. Loss of steam supply can affect steam tracing on liquid fuel lines and atomizing steam needed for atomizing the fuel. Electrical failure impacts instrumentation and electric heat tracing. Cold weather can freeze the water in a process or utility, as well as in instrument lines. Upsets in the upstream processing units can cause off-specification feed to be sent to the furnace.

Insufficient Maintenance

Furnaces are shut down for periodic maintenance. This maintenance includes the inspection and repair of the insulating refractory materials, tubes, and burner tiles. Failure to properly maintain the furnace and burners can lead to potentially serious incidents. For example, burner tips must be clean and in good condition to ensure proper combustion and a good burner flame pattern. If the tips become plugged, they could cause the flame to go out temporarily, which can lead to an explosion when the fuel re-ignites on a hot surface in the furnace.

To achieve the highest efficiency and lowest emissions, the air and fuel need to mix and burn at the burner tip. Improperly sealed furnaces allow air to enter the furnace through leaks. Analyzers monitor the excess oxygen (O_2) exiting the stack but cannot detect the amount of combustion air which entered through leaks rather than the burners. Since the air that enters a furnace through leaks will show up in the stack excess O_2 reading, air leaks can hide the fact that there is insufficient combustion air entering the furnace at the burners. This could lead to excess CO and NO_x emissions as well as reduced thermal efficiency.

Figure 13.7 shows an example of a gas burner requiring maintenance attention.

Figure 13.7 Gas burner in need of maintenance; plugged fuel tips causing the flames to be nonuniform.

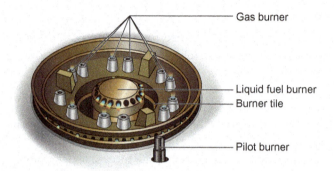

Gas burner

Liquid fuel burner

Burner tile

Pilot burner

13.5 Safety and Environmental Hazards

Hazards are associated with both normal and abnormal furnace operations. Hazards can affect the personal safety of the process technician, the equipment, facility operations, and the environment. Process technicians must always follow proper safety precautions when working around furnaces. Table 13.1 lists some of the hazards associated with improper furnace operations.

Table 13.1 Hazards Associated with Improper Furnace Operations

Improper Operation	Possible Effects			
	Individual	Equipment	Production	Environment
Failure to follow furnace safety procedure (e.g., lighting off burners without purging the firebox)	Burns, injuries, or death, loss of employment	Explosion in the firebox or flashbacks	Lost production because of downtime for repairs	Exceeding Environmental Protection Agency (EPA) limits for opacity
Poor control of excess air and draft	Burns, injuries, or death	Flame impingement and/or tube rupture and explosion	Lost production because of downtime for repairs; loss of efficiency	Exceeding EPA CO, NOx, and opacity limits
Bypassing safety interlocks	Burns, injuries, or death, loss of employment	Explosion in the firebox	Lost production because of downtime for repairs	Fines from operating above environmental limits
Opening furnace inspection ports when the firebox pressure is greater than atmospheric pressure	Burns, injuries, or death caused when hot flames or combustion gases are forced out of the inspection port	Sudden drop in box pressure	Cooling of the furnace	Hot combustion gases released into the environment
Failure to wear proper protective equipment (e.g., gloves and face shield) when opening furnace inspection ports	Burns, injuries, or death if the pressure in the firebox is greater than atmospheric pressure			

Fires, Spills, and Explosions

Because process furnaces usually contain combustible gases and liquids under pressure, there is always the potential for uncontrolled fires, spills, or explosions. For example, tube failures can cause large and sudden fires. Furnace leaks that allow flammable product to escape from the furnace also create a fire hazard.

To prevent dangerous conditions from occurring, it is critical that all instrument and interlock problems be corrected immediately. Proper maintenance of equipment is necessary to prevent incidents caused by things like plugged burner tips and flame impingement Interlocks must never be bypassed.

An accumulation of flammable liquids in the furnace firebox is an explosion hazard, and startup operations are especially hazardous. Each type of furnace has its own specific startup and emergency procedures that must be followed closely. To prevent flashbacks or explosions when preparing to light off the burners, make sure that the firebox has been purged properly and that adequate draft is provided.

RESPONDING TO UNCONTROLLED FIRES Every process technician must know how to extinguish an uncontrolled fire. In most cases, the proper response is to shut off all fuels, close the damper, and flood the firebox with snuffing steam. The technician should always be familiar with plant-specific procedures before attempting to extinguish any type of fire. The plant safety/emergency response organization is usually engaged during any plant fire.

Hazardous Operating Conditions

Hazardous conditions are present during both normal and abnormal furnace operations. For example, contact with hot pipelines or equipment can cause burns. Incomplete combustion creates a potential ignition or overpressure hazard in the firebox. Failure to control the balance of the fuel and air supplies can cause symptoms such as puffing (a sound like the chugging of a locomotive), and flame or smoke emission from the furnace stack.

If electrical power is lost, the furnace is usually shut down automatically by the furnace's safety instrumented system (SIS). When this occurs, the process technician must make sure the fuel supply is completely shut down and then check all the interlock actions for proper activation.

Personal Safety with Furnaces

In addition to wearing standard personal protective equipment, process technicians should observe the following precautions when performing routine preventive maintenance on furnace equipment:

- Always wear a face shield when inspecting a furnace at any open port or inspection door, or lighting the burners.
- Stand to the side when opening the inspection door, not in front of the door. Always minimize the amount of time that the inspection door is open, and never assume that the firebox is at negative pressure. During upsets, the possibility of positive pressure in the firebox must be assumed, because that situation could cause very hot flue gases to flow out the inspection port and severely burn an operator.
- Always stay in contact with the control room operator.
- Be aware of hot surfaces and hot flue gases (e.g., at the entrance and exit points around the tubes).
- Always wear long sleeves and gloves to prevent contact burns.
- Wear flame-retardant clothing when required.

Environmental Impact

It is important for furnaces to operate properly because excessive smoke, SO_x (sulphur oxide), CO (carbon monoxide), or NO_x (nitrogen oxide) emissions from furnace operations can have a negative impact on the environment and can result in citations and/or fines.

NITROGEN OXIDE (NO_x) EMISSIONS   Nitrogen oxides (NO_x) are byproducts of the combustion of fuel at high temperatures or with fuel containing nitrogen (N_2) compounds. High combustion temperatures oxidize atmospheric nitrogen and produce nitric oxide (NO), which can be further oxidized to nitrogen dioxide (NO_2). Both compounds are atmospheric pollutants and are collectively referred to as NO_x. In recent years, many industries have been required to comply with stricter standards for NO_x emissions.

The amount of NO_x formation depends on combustion characteristics such as temperature, residence time, and the concentrations of oxygen and nitrogen in the flame zone. Of these, peak flame temperature is the most important parameter in determining potential NO_x formation. Atmospheric pollution by nitrogen oxides can be controlled by reducing the amount of source emissions and by treating the exhaust gas.

Low-NO_x burners modify the combustion process by staging the air or fuel (i.e., having the correct air/fuel ratio), or by providing premixed fuel and recycling flue gas into the burner throat. These measures are designed to reduce peak flame temperature by distributing the flame over a larger space than is provided by traditional burners. Some companies have developed ultra-low-NO_x burners that combine premixing and staging to achieve lower NO_x emissions. Although beneficial to the environment, these burners are more difficult to operate and require more operator attention.

To keep NO_x emissions under control, the burner air registers and fuel valves must always be operated properly. The process technician must monitor burner flame appearance and adjust burner conditions as needed.

CARBON MONOXIDE EMISSIONS   Carbon monoxide (CO) typically forms from the incomplete combustion of a fuel containing carbon. Besides environmental impacts, CO is an asphyxiant that can cause death in high enough concentrations. It is a regulated pollutant that can be controlled by ensuring there is adequate air available for combustion. This is why some amount of excess air should be supplied to account for imperfect mixing of the fuel and combustion air. Excessive CO emissions are an indication of reduced thermal efficiency, because CO is a fuel that becomes CO_2 (carbon dioxide) when it has been combusted.

SMOKE EMISSIONS   In addition to creating a hazard in the firebox, smoke creates an emissions (opacity) violation. Smoking is caused by incomplete combustion of fuels or insufficient excess O2 levels. Understanding and controlling burner operations can provide the right balance for complete combustion and control of smoke emissions.

13.6 Process Technician's Role in Furnace Operations and Maintenance

Furnace operations have a direct and immediate impact on the operation of other process equipment units such as reactors or separation systems. Deviating from normal furnace operation can result in loss of production, off-spec product, the high cost of repairing or replacing failed or damaged components, and fines for emissions outside the air permit standards.

To achieve the highest efficiency and lowest level of hazardous emissions, the process technician balances the fuel to the furnace with the process flow rates. Then, using the stack O_2 analyzer data, the technician adjusts air registers to match the fuel combustion and the damper to achieve the proper firebox pressure.

The process technician is responsible for monitoring for problem conditions associated with furnaces and taking immediate action if problems are found, in order to ensure personal and process safety. A process technician's inspection of the furnace should include examining the exterior metal skin for hot spots, the burner flames for uniform color and size, and the process tubes for uniformity in color and/or leaks. The process technician must take immediate action if problems are found. Some companies also have special groups of people assigned to do this monitoring. The mechanical maintenance or inspection group often completes the inspection of the furnace's internal components when the furnace is periodically taken out of service.

13.7 Typical Procedures

Furnace procedures are specific to each process unit. Process technicians are required to have a clear understanding of site and unit-specific procedures, and are responsible for knowing how to implement the procedures in the event of an emergency.

Furnace Startup and Shutdown

Furnaces must be heated and cooled gradually to prevent damage to the furnace firebox insulation and refractory materials because these components are susceptible to damage from thermal shock. Specific time and temperature guidelines in furnace startup and shutdown procedures should always be followed because startups have the greatest risk for spills and fires. When using combination burners, the firebox temperature must be near the operating range before the fuel oil is introduced. The protection of instrument safety interlocks is also essential during startup operations and during normal operations.

Furnace procedures for the tube cleaning operation also should be followed, with specific attention to the type of furnace and its metallurgy. The decoking procedure introduces a combination of air and steam into the tubes to heat them and burn off deposits. The process effluent lines are diverted to the atmosphere or to a vessel specially designed for the decoking process.

Summary

A furnace is a piece of equipment in which heat is discharged by burning fuel and transferred directly or indirectly to a fluid mass for the purpose of increasing the temperature of a fluid. The most common types of industrial furnaces are box, vertical cylindrical, and cabin.

Box furnaces can be used to heat process fluids or generate steam. They also can be used for heat treating applications such as tempering, hardening, or firing.

Vertical cylindrical furnaces are similar to box furnaces, but they are cylindrical in shape and stand upright, so they require less space. Tubes in the radiant section are laid out in a vertical configuration along the walls of the cylinder.

Cabin furnaces are used in high temperature processes and are shaped like a log cabin with a chimney.

Air flow inside furnaces can be provided by natural draft or assisted draft. Natural draft furnaces use natural air currents instead of fans to create air flow. Assisted draft furnaces use electric motor or steam turbine-driven rotary fans or blowers. The three main types of air flow in assisted draft furnaces are induced draft, forced draft, and balanced draft.

Furnaces are divided into two sections: the radiant section and the convection section. The radiant section is at the bottom of the furnace and is closest to the heat source. Located at or near this section are fuel lines, fuel valves, burners, draft gauges, the purge system, the bridgewall, radiant tubes, and the refractory lining.

The convection section is at the top of the furnace and is farther away from the heat source. Within this section are convection tubes and the shock bank or shield section. The stack and stack damper are located above the convection section.

Furnaces burn fuel inside a firebox to produce heat. Process fluids are pumped through this heated firebox in a series of tubes. The heat from the firebox is transferred through the walls of the tubes to the process fluid via radiation and convection. The flue gas flows from the firebox up through a stack and into the atmosphere.

Furnaces are designed to operate at a capacity that maximizes the operating unit's product output, but maintains safety and ensures that environmental emission limits are met. Burner fuel and air flow are regulated to help maintain these specifications. Environmental regulations limit excess releases of known harmful gases into the atmosphere, the most common of these being nitrous oxides (NOx).

Interlock controls help protect the furnace system from dangerous operating conditions by shutting off the supply of furnace fuel or process feed and shutting down the furnace before a dangerous situation exists. Improperly operating a furnace can lead to safety and process problems, including injury, explosion or tube rupture, lost production, and environmental exposure to harmful gases.

Process technicians should always monitor furnaces for puffing (excessive smoke), flames, pressure drops, temperature drops, or other abnormal conditions and take immediate action when something is not right. They should conduct preventive maintenance as needed.

Checking Your Knowledge

1. Define the following terms:

   a. Pilot

   b. Stack

   c. Burner

   d. Firebox

   e. Induced draft furnace

   f. Natural draft furnace

   g. Air register

   h. Vertical cylindrical furnace

   i. Shock bank

   j. Radiant tube

   k. Cabin furnace

   l. Radiant section

   m. Refractory lining

2. Furnaces are used to:

   a. increase the temperature of a fluid

   b. release gases to the environment

   c. decrease temperature and pressure

   d. increase the viscosity of a fluid

3. Which of the following is NOT a common type of furnace?

   a. Cabin

   b. Spherical

   c. Box

   d. Vertical cylindrical

4. Select the components that are associated with a furnace. (Select all that apply.)

   a. Radiant tubes

   b. Bridgewall

   c. Fuel valve

   d. Tube sheet

5. List three possible consequences for a furnace when there is insufficient air and the furnace becomes too fuel-rich.

6. (True or False) Industrial furnaces are operated at positive pressure.

7. Periodic cleaning is required on the _____.

   a. stack

   b. burner fuel lines

   c. bridgewall

   d. shock bank

8. Which of the following hazards can cause cooling of the furnace? (Select all that apply.)

   a. Poor control of excess air and draft

   b. Bypassing safety interlocks

   c. Opening furnace inspection ports when the firebox pressure is greater than the atmospheric pressure

   d. Increasing the air into the furnace

9. (True or False) CO is an asphyxiant that can cause death in high enough concentrations.

10. Which of the following is NOT a potential problem in furnace operation?

    a. Quick startup

    b. Plugging of the process tubes

    c. Positive pressure in the firebox

    d. Uniform flame height and uniform process flame color

11. Match the furnace adjustment to its reason to achieve maximum efficiency and lowest emissions.

I. Adjust the fuel to the furnace to achieve the correct	**a.** firebox pressure
II. Adjust the stack damper to achieve the proper	**b.** O_2 analyzer reading
III. Adjust the burner air registers to get the correct	**c.** balance with the process flow rates

12. (True or False) A process technician completes the inspection of the furnace's internal components.

13. (True or False) When using combination burners, firebox temperature must be near the operating range before the fuel oil is introduced.

14. What is required for the decoking procedure? (Select all that apply.)

   a. Air

   b. Water

   c. Steam

   d. Bleach

NOTE: Answers to Checking Your Knowledge questions are in the Appendix.

Student Activities

1. Given a diagram or a cutaway of a furnace, identify the various components and explain the purpose of each.

2. Describe the importance of operating a furnace properly. Be sure to include emissions concerns, environmental impact, and fines or citations.

3. Complete a one-page report about one of the following gases. Include its impact on the environment, its harm to the atmosphere, and its harm to an individual.

   Nitrogen oxide (NOx)

   Carbon monoxide (CO)

   Carbon dioxide (CO_2)

Chapter 14
Boilers

Objectives

After completing this chapter, you will be able to:

14.1 Identify the components of boilers and the purpose of boilers in the process industries. (NAPTA Boilers 1, 3-5*) p. 292

14.2 Identify the common types of boilers and their applications. (NAPTA Boilers 2) p. 296

14.3 Explain the operating principles of boilers. (NAPTA Boilers 6) p. 298

14.4 Identify potential problems associated with boilers. (NAPTA Boilers 11) p. 302

14.5 Describe safety and environmental hazards associated with boilers. (NAPTA Boilers 7, 10) p. 303

14.6 Describe the process technician's role in safe boiler operation, maintenance, and operator qualification. (NAPTA Boilers 8, 9) p. 304

14.7 Identify typical procedures associated with boilers. (NAPTA Boilers 7) p. 304

* North American Process Technology Alliance (NAPTA) developed curriculum to ensure that Process Technology courses will produce knowledgeable graduates to become entry level employees in process technology. Objectives from that curriculum are named here in abbreviated form. For example, "(NAPTA Boilers 2)" means that this chapter's objective relates to objective 2 of the NAPTA curriculum about boilers).

Key Terms

Air registers—devices that control the flow of air to the burners to maintain the correct fuel-to-air ratio and to reduce smoke, soot, or NO$_x$ (nitrogen oxide) and CO (carbon monoxide) formation, **p. 297.**

Blowdown—the process of taking water out of a boiler to reduce the concentration level of impurities. There are two types of blowdown, continuous and intermittent, **p. 293.**

Burners—devices that introduce, distribute, mix, and burn a fuel (e.g., natural gas, fuel oil, or coal) for heat, **p. 297.**

Coagulation—a method for concentrating and removing suspended solids in boiler feedwater by adding chemicals to the water, which causes the impurities to cling together, **p. 301.**

Condensate—condensed steam, which often is recycled back to the boiler, **p. 294.**

Damper—a movable plate that regulates the flow of air or flue gases in boilers, **p. 297.**

Deaeration—removal of air or other gases from boiler feedwater by increasing the temperature using steam and stripping out the gases, **p. 301.**

Demineralization—a process that uses ion exchange to remove mineral salts; also known as *deionization*. The water produced is referred to as deionized water, **p. 301.**

Desuperheated steam—superheated steam from which some heat has been removed by the reintroduction of water. It is used in processes that cannot tolerate the higher steam temperatures, **p. 300.**

Desuperheater—a system that controls the temperature of steam leaving a boiler by using water injection through a control valve, **p. 297.**

Downcomers—tubes that transfer water from the steam drum to the mud drum, **p. 297.**

Draft fan—a fan used to control draft in a boiler, **p. 297.**

Economizer—the section of a boiler used to preheat feedwater before it enters the main boiler system, **p. 293.**

Filtration—the process of removing particles from water or some other fluid by passing it through porous media, **p. 301.**

Firebox—the area of a boiler where the burners are located and where radiant heat transfer occurs, **p. 296.**

Fire tube boiler—a device that passes hot combustion gases through the tubes to heat water on the shell side, **p. 298.**

Igniter—a device (similar to a spark plug) that automatically ignites the flammable air and fuel mixture at the tip of the burner, **p. 292.**

Impeller—a fixed, vaned device that causes the air/fuel mixture to swirl above the burner; different from an impeller in a turbine or pump, **p. 292.**

Knockout pots—devices designed to remove liquids and condensate from the fuel gas before it is sent to the burners, **p. 292.**

Mud drum—the lower drum in a boiler; also called the water drum; serves as a settling point for solids in the boiler feedwater, **p. 293.**

Pilot—an initiating device used to ignite the burner fuel, **p. 297.**

Premix burner—a device that mixes fuel gas with air before either enters the burner tip, **p. 297.**

Radiant tubes—tubes containing boiler feedwater that are heated by radiant heat from the burners and boiled to form steam that is returned to the steam drum, **p. 297.**

Raw gas burner—a burner in which gas has not been premixed with air, **p. 297.**

Refractory lining—a bricklike form of insulation used to reflect heat back into the box and protect the structural steel in the boiler, **p. 296.**

Reverse osmosis—method for processing water by forcing it through a membrane through which salts and impurities cannot pass (purified bottled water is produced this way), **p. 301.**

Riser tubes—tubes that allow water or steam from the lower drum to move to the upper drum, **p. 297.**

Saturated steam—steam in equilibrium with water (e.g., steam that holds all of the moisture it can without condensation occurring), **p. 300.**

Softening—the treatment of water that removes dissolved mineral salts such as calcium and magnesium, known as hardness in boiler feedwater, **p. 301.**

Spiders—devices with a spiderlike shape that are used to inject fuel into a boiler, **p. 292.**

Spuds—devices used to inject fuel into a boiler, **p. 292.**

Stack—an opening at the top of the boiler that is used to remove flue gas, **p. 297.**

Steam drum—the top drum of a boiler where all of the generated steam is collected before entering the distribution system, **p. 293.**

Steam trap—a device used to remove condensate or liquid from steam systems, **p. 294.**

Superheated steam—steam that has been heated to a very high temperature so that a majority of the moisture content has been removed (also called *dry steam*), **p. 300.**

Superheater—tubes located near the boiler outlet that increase (superheat) the temperature of the steam, **p. 297.**

Waste heat boiler—a device that uses waste heat from a process to produce steam, **p. 297.**

Water tube boiler—a type of boiler that contains water-filled tubes that allow water to circulate through a heated firebox, **p. 296.**

14.1 Introduction

Steam has many applications and a long history in the process industries. Steam provides efficient heat transfer and contains a high amount of latent heat. It is used to heat and cool process fluids, power and purge equipment, fight fires, facilitate distillation, and induce other physical and chemical reactions.

Boilers are an important source of energy in the process industries because they supply steam to operate process equipment and produce the steam used throughout the process facility. Examples of process equipment that uses steam includes turbines, reactors, distillation columns, stripper columns, and heat exchangers.

General Components of Boilers

Boilers are devices in which water is boiled and converted into steam under controlled conditions. Boiler components can vary, but the most common components include a firebox, burners, drums, tubes, an economizer, a steam distribution system, and a boiler feedwater system.

Firebox

Like other process furnace and direct-fired heater fireboxes, boiler fireboxes have a refractory lining, burners, a convection-type section, a radiant section, fans, air flow control, a stack, and dampers. The boiler firebox is insulated to reduce the loss of heat and enhance the heat energy being transferred to the boiler's internal components.

Burners

Burners inject air and fuel through a distribution system that mixes them in proper concentrations so combustion can occur. Most boilers use natural gas, fuel oil, or coal burners to provide heat to the boiler.

Knockout pots devices designed to remove liquids and condensate from the fuel gas before it is sent to the burners.

Impeller a fixed, vaned device that causes the air/fuel mixture to swirl above the burner; different from an impeller in a turbine or pump.

Spiders devices with a spiderlike shape that are used to inject fuel into a boiler.

Igniter a device (similar to a spark plug) that automatically ignites the flammable air and fuel mixture at the tip of the burner.

Spuds devices used to inject fuel into a boiler.

The key components of natural gas burners include pilots, impellers, spuds, spiders, and igniters. **Knockout pots** remove liquids and condensate from the fuel gas before it is sent to the burners. Dampers regulate the flow of air to the burner. Impellers in boilers are not the same as impellers in turbines or pumps. **Impellers** are fixed, vaned devices that cause the air/fuel mixture to swirl above the burner. Spuds or **spiders** (devices with a spiderlike shape) are used to inject fuel into the boiler (Figure 14.1). **Igniters** automatically ignite the flammable air and fuel mixture at the tip of the burner.

Did You Know?

Low NO$_x$ formation burners use nozzles, called **spuds**, to inject fuel into a boiler.

Spuds promote better combustion than traditional burners.

Figure 14.1 Fuel injection devices. **A.** Spud. **B.** Spider.

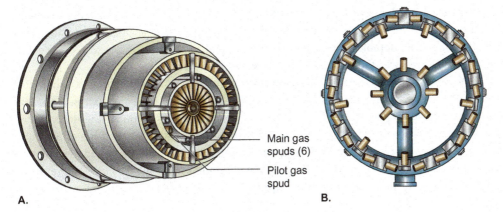

Main gas
spuds (6)

Pilot gas
spud

A.

B.

Drums

The drums that comprise a water tube boiler resemble a large water distribution header connected by a complex network of tubes. The **mud drum** is the lower drum in a boiler. The **steam drum** is the upper drum of a boiler where all of the generated steam is collected. The mud drum and water tubes are filled completely with water, while the steam drum is only partially full. Maintaining this vapor space in the upper drum allows the saturated steam to collect and pass out of the header.

Feedwater to the boiler is treated to achieve the required chemical composition. Water lost in the boiler is replaced through a makeup water line. Sediment accumulates in the bottom of the mud drum and is removed through blowdown. **Blowdown** is the process of removing small amounts of water from the boiler to reduce the concentration of impurities. Blowdown can be either continuous or intermittent. Continuous blowdown is the constant removal of a small quantity of boiler water from the steam drum to remove suspended solids and salts that could concentrate in the steam drum. Intermittent (or *bottoms*) blowdown is the occasional opening of a valve on the bottom of the mud drum to remove solids that have settled.

Mud drum the lower drum in a boiler; serves as a settling point for solids in the boiler feedwater.

Steam drum the top drum of a boiler where all of the generated steam is collected before entering the distribution system.

Blowdown the process of taking water out of a boiler to reduce the concentration level of impurities. There are two types of blowdown, continuous and intermittent.

Economizer

The **economizer** is the section of a boiler used to preheat feedwater before it enters the main boiler system. Preheating the water increases boiler system efficiency. This heat exchanger transfers heat from the stack gases to the incoming feedwater. The economizer is usually located close to the stack gas outlet of the boiler. Economizers can be supported from overhead or from the ground. The feedwater line that serves the boiler is piped into and travels through the economizer. No additional feedwater control valves or stack gas dampers are required.

An economizer is similar to the convection section in a direct-fired heater. Both operate under the energy-saving concept of recovering some of the heat from the hot flue gases before they are lost out of the stack. The typical improvement in efficiency of a boiler with an economizer is 2 to 4 percent.

Economizer the section of a boiler used to preheat feedwater before it enters the main boiler system.

Steam Distribution System

The steam distribution system consists of valves, fittings, piping, and connections suitable for the pressure of the steam being transported. Steam exits the boiler at sufficient pressure required for the process unit or for electrical generation. For example, when steam is used to drive steam turbine generators to produce electricity, the steam must be produced at a much higher pressure than that required for process steam. The steam pressure can then be reduced for the turbines that drive process pumps and compressors that require lower pressure steam.

Most steam used in a process facility is ultimately condensed to water. **Condensate** is typically returned through condensate return systems and reused as boiler feedwater. This saves the plant considerable money because the condensate, which has already been treated with chemicals, reduces the cost of boiler feedwater treatment.

A facility's steam system is usually composed of two or three piping systems with different pressure steam headers. There are different methods of maintaining several steam headers operating at different pressures. One method is through a letdown valve or a reducing station in which a pressure control valve is used. In this type of arrangement, a pressure drop across the valve equalizes with the pressure contained in the lower pressure steam header. In many instances, the steam header itself has another letdown valve that repeats the same function, supplying steam to yet another, lower pressure steam header.

Another method of supplying steam to headers of different pressures is by piping the exhaust of steam turbines into headers. For example, a steam turbine that operates at 1,500 PSIG pressure might exhaust into a 550 PSIG steam header. The turbine that uses the 550 PSIG steam then might exhaust into a 50 PSIG header. A letdown valve or reducing station (see Figure 14.2) offers the advantage of maintaining a steam header at an interim pressure that might not be suitable for all steam turbines.

Condensate condensed steam, which often is recycled back to the boiler.

Figure 14.2 A. Diagram of a letdown station. **B.** Letdown station with pressure-reducing valve, safety valves, and separator.

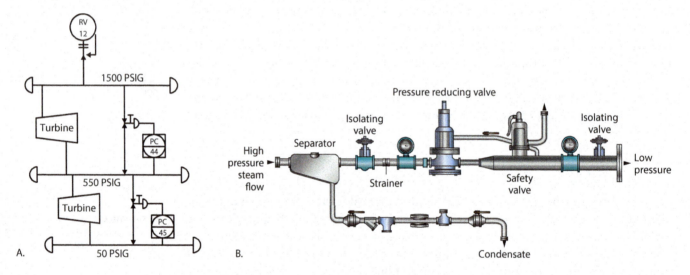

Some process plants are supported by a cogeneration plant, in which electricity and steam are generated in the same unit. Steam is brought to a very high temperature and pressure. This high-pressure steam is used to turn a turbine that turns an electrical generator. The exhaust from the turbine goes into a lower pressure steam header. In this way, a cogeneration plant supplies the process facility with both electricity and steam.

An important component used in steam systems is a steam trap. **Steam traps** are used to remove condensate or liquid from steam systems. There are several steam trap designs. The most common, however, are the mechanical trap (Figure 14.3) and the thermostatic trap (Figure 14.4). See also Figures 4.18 and 4.19.

Mechanical traps operate based on the density difference between steam and condensate. Internal floats attached to mechanical linkages are found inside most mechanical traps. As the condensate levels rise, the linkage causes the valve to open. As the levels drop, the valve closes. Inverted bucket and float traps are both common examples of mechanical traps.

Steam trap a device used to remove condensate or liquid from steam systems.

Figure 14.3 Examples of different designs for mechanical steam traps. **A.** Inverted bucket trap. **B.** Float trap.

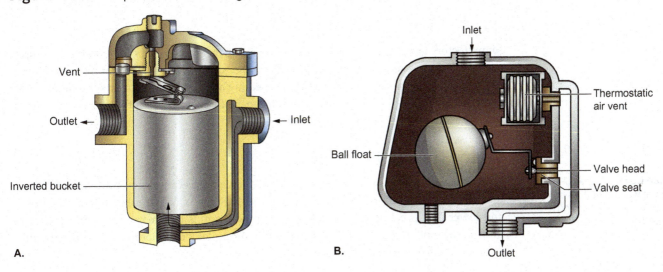

A.

B.

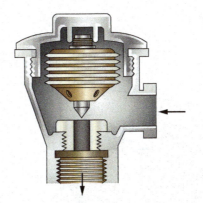

Figure 14.4 Example of a type of Bellows trap.

Thermostatic traps operate on the principle of temperature change (steam is hotter than condensate). These traps contain valves that are opened or closed by thermal expansion and contraction. Bimetallic and bellows traps are examples of temperature-operated traps. Condensate will cause the bellows to collapse or bimetallic element to move, releasing the condensate. When the condensate is removed and steam is present, the bellows expands or the bimetallic element moves the opposite way, and the steam trap closes.

Boiler Feedwater System

The boiler feedwater supply is a critical part of steam generation. There must always be as many pounds of water entering the system as there are pounds of steam leaving.

The water used in steam generation must be free of contaminants such as minerals and dissolved impurities that can damage the system or affect its operation. Suspended materials such as silt and oil create scale and sludge, and must be filtered out. Dissolved gases such as carbon dioxide and oxygen cause boiler corrosion and must be removed by deaeration and other methods of treatment. Because dissolved minerals cause scale, corrosion, and turbine blade deposits, boiler feedwater must be treated with lime or soda ash to precipitate these minerals from the water. Recirculated condensate must be deaerated to remove dissolved gases.

Depending on the individual characteristics of the raw water, boiler feedwater can be treated by clarification, sedimentation, filtration, ion exchange, deaeration, membrane processes, or a combination of these methods. Boiler feedwater treatment is discussed in greater detail in the following section.

14.2 Water Tube, Waste Heat, and Fire Tube Boilers

Many factors are considered when selecting a boiler. These factors include the pressures and temperatures required, total capacity, number of generating tubes, number of drums, type of circulation (natural or forced), superheating and desuperheating requirements, tube configuration, and cost. The most common types of boilers used in the process industries are water tube, waste heat, and fire tube boilers.

Water Tube Boilers

Water tube boiler   a type of boiler that contains water-filled tubes that allow water to circulate through a heated firebox.

Water tube boilers (shown in Figure 14.5) are so-called because they contain water-filled tubes that allow water to circulate through a heated firebox. Water tube boilers have upper and lower drums connected by tubes. The upper drum is the steam drum, and the lower drum is the mud drum. Chemicals are added to the boiler feedwater that enters these drums in order to prevent fouling and corrosion.

Figure 14.5 A. Diagram of water tube boiler components. **B.** Water tube boiler.
CREDIT: B. melnikofd/Fotolia.

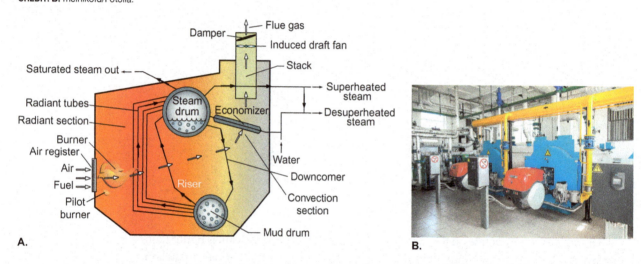

Heat is generated in the boiler through a direct-fired heater with a natural gas, oil, or combination burner. The heat from the burner is transferred to the water tubes. As water flows through the tubes, the combustion gases heat the water and produce steam. This steam is collected in the upper drum, while combustion gases exit the boiler stack as flue gas.

Several types of tubes are found inside the boiler. Generating tubes are attached to the upper and lower drums. Water flows through the tubes from the upper (steam) drum down to the lower (mud) drum and back up to the steam drum. The *downcomer tube* is the cold water line between the upper and lower drums. The *riser tube* is the hot water/steam line between the lower and upper drums. *Superheater tubes* are tubes where steam is removed from the steam drum and heated to remove moisture content without an increase in pressure.

Firebox   the area of a boiler where the burners are located and where radiant heat transfer occurs.

The water level in a water tube boiler is controlled in the steam drum. The water level in this drum must be maintained for safety reasons and for compliance with standard operating procedures. Loss of water level can damage boiler equipment. Excessively high water levels can result in carryover, which can cause steam to become saturated with water containing chemicals, which in turn can result in fouling in the steam system. An even greater risk is that water carryover will result in wet steam, which can cause operational upsets.

Refractory lining   a bricklike form of insulation used to reflect heat back into the box and protect the structural steel in the boiler.

WATER TUBE BOILER COMPONENTS   The heating portion of a boiler is similar to that of a furnace. Like furnaces, boilers contain a **firebox** where the burners are located and radiant heat transfer occurs. A special **refractory lining** (a bricklike form of insulation) is used to reflect heat back into the box and protect the structural steel in the boiler.

Located in the firebox area of the boiler are radiant tubes. **Radiant tubes** and riser tubes both contain boiler feedwater that is heated by radiant heat from the burners and boiled to form steam that is returned to the steam drum.

Burners are devices that introduce, distribute, mix, and burn a fuel (e.g., natural gas, fuel oil, or coal) for heat. **Pilots** are used to light burners. A burner can be a **premix burner** (fuel gas and air mixed before either enters the burner tip), a **raw gas burner** (gas not pre-mixed with air), or a combination.

Air registers control the flow of air to the burners to maintain the correct fuel-to-air ratio and to reduce smoke, soot, or NO_x (nitrogen oxide) and CO (carbon monoxide) formation.

A **draft fan** is used to control draft in a boiler. Depending on the design of the boiler, the draft fan either forces the air through the boiler (*forced draft*) or pulls the air through the boiler (*induced draft*). In some boilers, a combination of the two is used (*balanced draft*).

Water tube boilers have a **stack** at the top of the boiler to remove flue gas. Contained within the stack is a **damper** for regulating the flow of flue gases.

A steam drum at the top of the boiler is where all generated steam gathers before exiting the boiler. Depending on the boiler design, water enters the steam drum or the mud drum from the economizer.

At the bottom of the boiler is the mud drum, where sediment accumulates. This sediment is removed by intermittent blowdown, which is done manually.

Downcomers are tubes that transfer water from the steam drum to the mud drum. As cooler water descends from the steam drum and flows through the downcomers, it picks up heat from the firebox and replenishes the water supply to the mud drum. **Riser tubes** allow water or steam from the lower drum to move to the upper drum.

The **superheater** is a set of tubes located toward the boiler outlet that increases (super-heats) the temperature of the steam flow. The steam drum usually is connected to the super-heater through a coil or pipe.

A **desuperheater** is a temperature control point at the outlet of the boiler steam flow that maintains a specific steam temperature by using boiler feedwater injection through a control valve. The purpose of the desuperheater is to lower the temperature of the steam.

Waste Heat Boilers

Waste heat boilers use excess or waste heat from a process to produce steam. Waste heat boilers have two functions: to produce steam and to provide cooling for a process in order for it to proceed or to recover heat that would otherwise be released to the atmosphere, losing a tremendous amount of usable energy. Figure 14.6 shows a waste heat boiler that can be used to recover waste heat energy and cool the flue gas stream from a turbine exhaust.

Radiant tubes tubes containing boiler feedwater that are heated by radiant heat from the burners and boiled to form steam that is returned to the steam drum.

Burners devices that introduce, distribute, mix, and burn a fuel (e.g., natural gas, fuel oil, or coal) for heat.

Pilot an initiating device used to ignite the burner fuel.

Premix burner a device that mixes fuel gas with air before either enters the burner tip.

Raw gas burner a burner in which gas has not been premixed with air.

Air registers devices that control the flow of air to the burners to maintain the correct fuel-to-air ratio and to reduce smoke, soot, or NO_x (nitrogen oxide) and CO (carbon monoxide) formation.

Draft fan a fan used to control draft in a boiler.

Stack an opening at the top of the boiler that is used to remove flue gas.

Damper a movable plate that regulates the flow of air or flue gases in boilers.

Downcomers tubes that transfer water from the steam drum to the mud drum.

Riser tubes tubes that allow water or steam from the lower drum to move to the upper drum.

Superheater tubes located near the boiler outlet that increase (superheat) the temperature of the steam.

Desuperheater a system that controls the temperature of steam leaving a boiler by using water injection through a control valve.

Waste heat boiler a device that uses waste heat from a process to produce steam.

Figure 14.6 Diagram of a waste heat boiler.

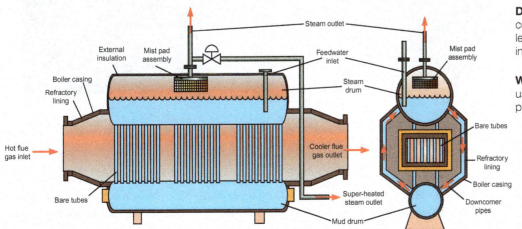

Did You Know?

Steam locomotives and home water heaters are both examples of fire tube boilers.

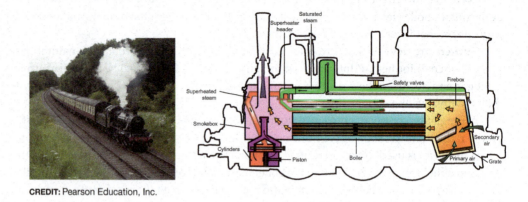

CREDIT: Pearson Education, Inc.

Waste heat boilers improve efficiency and save money by allowing steam to be produced through the use of waste gases. Use of waste gases as a heat source reduces the amount of money spent on burner fuels and reduces environmental impact.

Because of the duty required of waste heat boilers, construction is usually thick-walled and designed to withstand high pressures and temperatures. Waste heat boilers usually are single-pass, floating-head type heat exchangers that experience a considerable amount of expansion and contraction of the tubesheet.

In many furnaces, the waste heat boiler is on the outlet of the furnace. This design recovers heat by generating steam and thus cooling the flue gas stream exiting the furnace stack. Because of this, waste heat boilers are sometimes referred to as *steam generators*.

Fire Tube Boilers

Fire tube boiler a device that passes hot combustion gases through the tubes to heat water on the shell side.

Fire tube boilers pass hot combustion gases through the tubes to heat water on the shell side of the boiler. In this type of boiler, combustion gases are directed through the tubes while water is directed through the shell. As the water begins to boil, steam is formed. This steam is directed out of the boiler to other parts of the process, and makeup water is added to compensate for the fluid loss. In this type of system, the water level within the shell must always be maintained so that the tubes are covered. Otherwise, the tubes could overheat and become damaged. Figure 14.7 shows examples of fire tube boilers.

Figure 14.7 Fire tube boilers.

CREDIT: **A.** Jay Petersen/Shutterstock. **B.** By Milen Mkv/Shutterstock.

A.

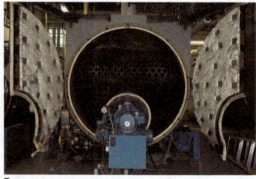

B.

14.3 Principles of Operation of Boilers

Boilers use a combination of radiation, convection, and conduction to convert heat energy into steam energy. Proper boiler operation depends on controlling many variables, including boiler feedwater quality, water flow and level in the boiler, furnace temperatures and pressures, burner efficiency, and air flow.

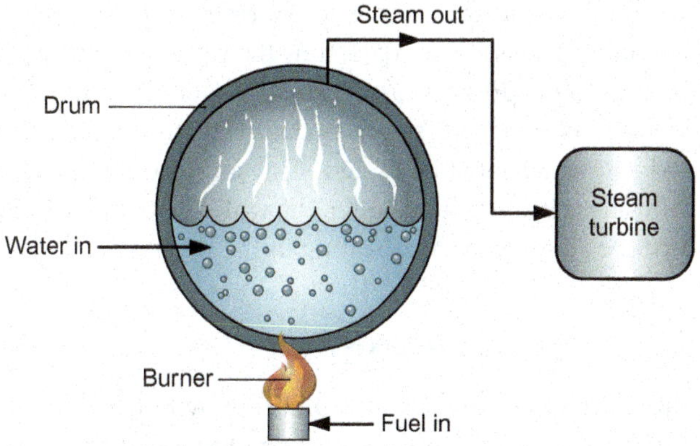

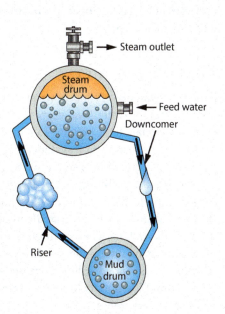

Figure 14.8 Simple boiler.

To illustrate how boilers work, consider the simple boiler shown in Figure 14.8. Simple boilers consist of a heat source, a water drum, a water inlet, and a steam outlet. In this type of boiler, the water drum is partially filled with water and then heat is applied. Steam forms after the water is heated sufficiently. As the steam leaves the vessel, it is captured and sent to other parts of the process (e.g., used to turn a steam turbine, or sent to a heat exchanger to heat a process fluid). Makeup water is then added to the drum to compensate for the liquid lost as steam.

Boilers use the principle of differential density when it comes to fluid circulation. For boilers to work properly, they must have adequate amounts of heat and water flow. Factors that affect boiler operation include pressure, temperature, water level, and differences in water density. As fluid is heated, the molecules expand and the fluid becomes less dense. When cooler, denser water is added to hot water, convective currents are created that facilitate water circulation and mixing.

Water Circulation

The circulation of boiler water (shown in Figure 14.9) is based on the principle of convection. A fluid that is heated expands and becomes less dense, moving upward through heavier, denser fluid. Convection and conduction transfer heat through pipe walls and water currents, resulting in unequal densities. Cold water flows through the downcomer to the bottom of the mud drum and then flows upward through the riser (water wall tubes) as it is heated.

In a water tube boiler, circulation occurs because the temperature of the fluid in the downcomer is always lower than the temperature in the boiler and generating (riser) tubes. Steam bubbles are formed as the liquid temperature continues to increase. These bubbles increase

Figure 14.9 Water circulation.

the circulation as they move up the riser tubes. The pressure builds as the water vapor collects in the upper drum. Each time the water passes through the tubes, it picks up more heat energy. As the pressure increases, the boiling point of the water increases. When the target pressure is achieved, steam is delivered to the steam header. To maintain this pressure, makeup water must be added, heat must be continually applied, and circulation must be controlled. In a fire tube boiler, the water level in the boiler shell must be maintained above the tubes to prevent overheating of the tubes.

Superheated Steam

Saturated steam is steam in equilibrium with water (e.g., steam that holds all of the moisture it can hold and still remain a vapor). Saturated steam can be used to purge process equipment or perform other functions, or it can be superheated.

As long as the steam and water are in contact with each other, the steam is in a saturated condition. Saturated steam cannot absorb additional water vapor, but the boiler can continue to add heat energy to it. Steam that continues to take on heat energy or get hotter is known as superheated steam.

Superheated steam, which is produced downstream of the steam drum (typically in the firebox), is steam that has been heated to a temperature above its saturated temperature. Superheated steam is typically 200 to 300 degrees F (93 to 149 degrees C) hotter than saturated steam. Typical uses for superheated steam include:

- Driving turbines
- Catalytic cracking
- Product stripping
- Maintenance of steam pressures and temperatures over long distances
- Producing steam for systems that require dry, moisture-free steam.

Desuperheated Steam

Superheated steam might not be the best choice for heat transfer in some heat exchangers because the amount of energy given up by superheated steam is relatively small compared to the energy given up by saturated steam. Also, some facility processes cannot tolerate the high temperatures of superheated steam. The process of cooling the superheated steam is called desuperheating. **Desuperheated steam** is superheated steam from which some heat has been removed by the reintroduction of boiler feedwater. Typically, desuperheating does not occur at the boiler but at specific points in the process where boiler feedwater is injected into superheated steam.

Boiler Feedwater

Boiler feedwater levels and flows are critical to proper boiler operation. If feedwater flow is reduced and the water level decreases to the point where the boiler runs dry, the tubes will overheat and fail. If the boiler water level becomes too high, excess water will be carried over into the steam distribution system. This negatively affects process facility steam consumers and can damage turbines and other equipment.

During the boiling process, most suspended solids stay in the water section of the drum while steam is sent to the distribution system. Suspended solids are removed from the steam drum by sending a small amount of the feedwater, called continuous blowdown, to a blowdown tank. This continuous blowdown is usually released to a waste water treatment processing unit. Boilers utilize both continuous and intermittent blowdown systems to remove suspended solids from the steam drum and solids that have settled from the mud drum. Blowdowns limit the scale buildup that can negatively affect turbine blades and superheater tubes.

Feedwater must be free of contaminants that could affect boiler operation. As a general rule, the higher the steam pressure, the stricter the feedwater quality requirements will be. Important feedwater parameters include pH (alkalinity or acidity of the water), hardness (amount of mineral content in the water), oxygen and carbon dioxide concentration, presence of silicates, dissolved or suspended solids, and concentration of organics. Water treatment techniques include reverse osmosis, ion exchange, deaeration, membrane contactors, and electrode ionization or demineralization.

Did You Know?

When a volume of water is converted to steam, the volume increases more than 1,600 times its original volume. This is why steam generation is a dangerous part of industrial processes.

Water drop   Same water drop vaporized

Water Treatment Methods

Raw water can come from a variety of sources, such as lakes, rivers, or wells. Each water source has its own components and treatment requirements. In general, however, the water chemistry required for steam production must meet standards. The water needs to be filtered and have minerals and oxygen removed. Raw water goes through the following steps to become boiler feedwater.

1. Clean the water. This step removes suspended solids. Depending on water source this could include:

   a. Coagulation/sedimentation

   b. Filtration

2. Remove minerals. This step is done to the clean water (from step 1) to remove minerals that could build up on steam turbines or other process equipment. Depending on the water source, this step could be one or more of these processes:

   a. Softening

   b. Demineralization (ion exchange)

   c. Reverse osmosis (membrane)

3. Remove the oxygen. Dissolved oxygen and other gases (primarily CO_2) in boiler feedwater are major causes of boiler system corrosion. While oxygen results in localized corrosion (pitting), CO_2 forms carbonic acid and damages condensate piping. This step could include:

   a. Deaeration

   b. Oxygen scavenging

   Coagulation adds chemicals to reduce coarse suspended solids, silt, turbidity, and colloids through the use of a clarifier. The impurities gather together into larger particles and settle out of the chemical/water solution (*sedimentation*).

   Filtration removes coarse suspended matter and sludge from coagulation or from water softening systems. Gravel beds and anthracite coal are common materials used for filter beds.

   Softening is the treatment of water to remove dissolved mineral salts such as calcium and magnesium, known as hardness, in boiler feedwater. Softening methods include the addition of calcium carbonate (lime soda), phosphate, and/or zeolites (crystalline mineral compounds).

   Demineralization is the removal of ionized mineral salts by ion exchange. The process is also called *deionization*, and the water produced is called deionized water.

   Reverse osmosis uses pressure to remove dissolved solids from boiler feedwater by forcing the water from a more concentrated solution through a semipermeable membrane to a less concentrated solution.

   Deaeration removes oxygen or other gases from boiler feedwater by increasing the temperature, using steam, to strip out the dissolved gases.

Coagulation a method for concentrating and removing suspended solids in boiler feedwater by adding chemicals to the water, which causes the impurities to cling together.

Filtration the process of removing particles from water or some other fluid by passing it through porous media.

Softening the treatment of water that removes dissolved mineral salts such as calcium and magnesium, known as hardness in boiler feedwater.

Demineralization a process that uses ion exchange to remove mineral salts; also known as *deionization*. The water produced is referred to as deionized water.

Reverse osmosis method for processing water by forcing it through a membrane through which salts and impurities cannot pass (purified bottled water is produced this way).

Deaeration removal of air or other gases from boiler feedwater by increasing the temperature and stripping out the gases.

Specific terms describe the water as it moves through these steps. Water starts as raw water and becomes demineralized water, then deaerated water, and finally boiler feedwater.

Steam that has been condensed (*condensate*) is already clean and can be fed back into the system at the deaerator. This saves the cost of treating more raw water, making it practical to recycle any used steam condensate.

Burner Fuel

Boilers use a single fuel or a combination of fuels, including refinery gas, natural gas, fuel oil, and powdered coal. In some complexes, scrubbed off-gases are collected from process units and combined with natural gas or liquefied petroleum gas in a fuelgas balance drum. The balance drum establishes a constant system pressure and fairly stable BTU (British thermal unit) content. It also provides for separation of suspended liquids in the gas vapors to prevent large slugs of liquid from being carried over into the fuel distribution system. As the scrubbed gases enter the balance drum, heavier liquids fall to the bottom along with any gases that have condensed into liquid. The lighter gas leaves the top of the balance drum and goes to the fuel distribution system. The fuel oil system delivers fuel to the boiler at the required temperatures and pressures. The fuel oil is heated to pumping temperature, sent through a coarse suction strainer, pumped to a temperature-control heater, and then pumped through a fine mesh strainer before being burned.

14.4 Potential Problems

Boiler system operations have a direct and immediate impact on the operation of other process equipment, such as distillation columns and turbines. The boiler must be operating properly to produce steam energy at steady pressures. Failure to maintain normal boiler operation can result in loss of production, compromised product quality, and high costs to repair or replace failed or damaged components. Table 14.1 lists problems associated with boilers.

Table 14.1 Problems Associated with Boilers

Condition	Problem
Burner flame impingement or poor flame distribution	Overheating and failure of the tubes
A change in the fuel flow	Creates or increases soot deposits on the tubes, causing excessively high temperature on the tube walls
A change in the composition of the fuel	Creates or increases soot deposits on the tubes, causing excessively high temperature on the tube walls
Loss of boiler feedwater flow	Allows the steam drum to run dry, causing the tubes to overheat and fail and leading to catastrophic equipment failure and potential injuries
A high water level in the steam drum	Causes water to carry over into the steam distribution system, which can damage equipment downstream
Poor control of feedwater treatment	Creates formation of scale in the tubes, resulting in a loss of boiler efficiency and can cause tube failure
Power outage	Loss of pumps and draft fans

Contributing Factors

Several factors can contribute to problems in boiler operation, including the design and age of the equipment, instrument problems, fouling, tube temperatures, and external factors.

EQUIPMENT AGE OR DESIGN   The types of boiler equipment that are most subject to failure from aging include boiler feedwater pumps, blowdown valves, and system piping. In addition, linkages on stack dampers can wear out from inadequate lubrication or from the collection of dust and grit. Freezing and earth movements can crack concrete equipment foundations and cause external corrosion of structural components. Worn or damaged tube supports can create stress problems for the boiler tubes.

WATER AND OTHER CONTAMINANTS Water and contaminants can contribute to instrument malfunctions and other problems that can activate boiler interlocks and trip a boiler. For example, water can condense in the pressure sensing leads in the firebox and cause faulty pressure readings, automatically shutting down the boiler.

Tube life in a boiler can also be significantly affected by the quality of the feedwater. For example, improper feedwater treatment can create salt deposits inside the tubes. These salt deposits cause tube corrosion and/or hot spots, leading to premature tube failure.

EXTERNAL FACTORS Other factors that contribute to boiler problems include interrupted water supply, electrical failure, cold weather, and downstream process upsets. For example, if a large steam user suddenly stops taking steam, steam header pressure can increase, causing abrupt pressure swings and instrument sensing problems faster than the boiler firing controls can respond. Another factor is cold weather, which can cause water in the process and utility lines to freeze and plug the line. Upsets in the upstream processing units can cause off-spec fuel to be sent to the boiler.

14.5 Safety and Environmental Hazards

Hazards associated with both normal and abnormal boiler operation can negatively affect personal safety, equipment, and the environment. Process technicians must always take proper safety precautions when working around boilers. Table 14.2 lists some of the hazards associated with improper operation of boilers and possible effects.

Table 14.2 Hazards Associated with Boiler Operation

Improper Operation	Possible Effects			
	Individual	Equipment	Production	Environment
Opening header drains, vents, and peepholes	Burns and eye injuries			
Failing to purge the firebox (startup)	Burns or injuries	Explosion in the firebox; damage to the internal components of the boiler	Process facility upset; lost steam production; downtime for repairs	Exceeding regulatory limits for opacity (the amount of light blocked by a medium)
Poor control of excess air and draft control		Flame impingement	Reduced boiler efficiency	Exceeding opacity limits
Loss of boiler feedwater	Burns or injury	Tube rupture; loss of downstream equipment use	Lost production because of downtime for repairs	
Loss of fuel gas or oil		Loss of downstream equipment use	Lost steam production	

NO$_X$ and Smoke Emissions

Excessive smoke or nitrogen oxide (NO$_x$) emissions from boiler operation have a negative impact on the environment. Both are considered air pollutants and can lead to emissions violations. To keep NO$_x$ emissions under control, the burner air registers and fuel valves must always function properly. Understanding and controlling oil burner operations allows the right balance for complete combustion and control of smoke emissions. Process technicians should also be cautious when handling chemicals to reduce environmental impact.

Fire Protection and Prevention

The most potentially hazardous operation in steam generation is boiler startup. During startup, a flammable mixture of gas and air can build up as a result of flame loss at the burner during *light-off* (when the pilot flame is ignited). Each type of boiler requires specific startup and emergency procedures, including purging before light-off and after misfire or loss of burner flame.

Hazardous Operating Conditions

Hazards exist during both normal and abnormal boiler operation. For example, there is a potential for exposure to feedwater chemicals, steam, hot water, radiant heat, and noise. Alternate fuel sources might be available onsite in the event fuel gas is lost because of a process unit shutdown or emergency.

Personal Safety

Process technicians must always follow safe work practices and wear appropriate personal protective equipment when working around boilers or performing process sampling, inspection, maintenance, or turnaround activities. Process technicians should also avoid skin contact with boiler blowdown because it can contain hazardous chemicals.

14.6 Process Technician's Role in Operation and Maintenance

Process technicians are responsible for performing specific procedures to safely operate and maintain boiler system equipment. Starting up a boiler, for example, requires filling the drum with water, lighting the burner, bringing the boiler up to pressure, and then placing the boiler online. Each of these steps requires that the process technician perform a number of tasks and follow specific procedures, which vary according to the site and boiler type. When monitoring and maintaining boilers, process technicians must always remember to look, listen, and check for all the factors listed in Table 14.3. Failure to perform proper maintenance and monitoring could affect personal safety as well as the process and could result in equipment damage.

Table 14.3 Process Technician's Role in Operation and Maintenance

Look	Listen	Check
■ Look at firebox for flame impingement on tubes ■ Observe burner flame color and pattern ■ Observe for wall hotspots (external and internal) ■ Look at draft balance (pressures) ■ Look at temperature gradient ■ Monitor burner balance (fuel and air) ■ Inspect controlling instruments (water level, fuel flow, feedwater flow, pressure, steam pressure, and temperature) ■ Look at the boiler stack for smoke ■ Watch for proper steam flow ■ Monitor the air flow and oxygen level, and adjust draft as needed ■ Ensure that the blowdown on the mud drum is operating efficiently	■ Listen for abnormal noise (e.g., fans, burners, water leaks, steam leaks, or external alarms) ■ Listen for huffing or puffing, either of which can indicate improper draft operation	■ Check for excessive vibration (fans and burners) ■ Check burner atomization for uniformity (if not uniform, burner tips on the atomization gun or the burner will require replacement) ■ Check firing efficiency (CO_2, CO, and O_2 in the stack) ■ Collect boiler feedwater samples and ensure proper chemical treatment ■ Check for burner wear, which can cause uneven flame or hotspots

14.7 Typical Procedures

Process technicians must be familiar with boiler procedures, including startup, shutdown, lockout/tagout, and emergency procedures. The procedures that follow are generic in nature. Process technicians should always refer to site-specific procedures before performing any work on their unit.

Startup

1. Inform control room personnel that the boiler is going to be put into service.
2. Use the peepholes to inspect the inside of the boiler and verify that it is free of debris (e.g., scaffold boards, rain suits, and tools) and that the refractory lining, tubes, and tube supports are all intact.
3. Inspect the outside of the boiler for loose flanges around inlet and outlet piping. Ensure that all blinds have been removed, valves are in their proper startup positions, and the pressure relief valve is properly lined up.
4. Inspect the condition of the draft fan, the damper, and the fuel gas system to ensure all are in satisfactory operating condition.
5. While in close communication with control room personnel, open the damper and start up the draft fan per standard operating procedures.
6. Fill the boiler to its normal water level to satisfy the interlock (control safety systems that have to be verified before boiler startup) for proper water level.
7. After the draft in the boiler is stable and has been purged for the required length of time, open the fuel gas supply to the burner and the pilot gas block valves.
8. Inform control room personnel that the pilot is being ignited.
9. After all of the pilots are lit and burning (confirmed through visual observation), open the primary and secondary air registers as required. Slowly open the main burner block valves one at a time until all burners are burning.
10. Remain in the boiler area to inspect the flames for proper flame patterns and general operation (e.g., draft fan and fuel supply).
11. Use the peepholes to inspect the inside of the boiler and look for uniform color of the tubes.

Note: Many of these steps are included in a boiler automated control system.

Shutdown

With some exceptions, boiler shutdown is the reverse of the startup procedure.

1. Gradually reduce load and firing rate.
2. Turn off fuel.
3. Shut steam header valve.
4. Shut down feedwater pump.
5. Open vents and drains.
6. Leave fans on to help cool boiler.

The control room operator should monitor conditions and inform the outside operator when it is safe to block in all the burners, shut down the fan, and close the damper.

Emergency

In an emergency, one fuel gas block valve (often referred to as the fireman) usually is designated as the main shutoff valve. Generally, the draft fan continues to run and the damper remains open. The only thing the outside operator does is block in the burners, block in the fuel, and stop all pumps.

Lockout/Tagout for Maintenance

Each company has its own lockout/tagout procedures. Process technicians must be familiar with these procedures before performing any maintenance.

Summary

Boilers convert water into steam, which is supplied to steam consumers throughout a process facility. The steam produced by boilers provides heat to other process equipment and supplies the steam energy to drive turbines and compressors.

Boilers use a combination of radiant, conductive, and convective methods to transfer heat. These devices consist of a number of tubes that carry the water-steam mixture through the boiler for maximum heat transfer.

The process industries use three types of boilers: water tube, waste tube, and fire tube. Water tube boilers are the most commonly used boilers in process operations. In water tube boilers, the water is circulated through tubes that run between the upper steam drum and the lower water collection drum (mud drum). Water circulation is created using the principles of differential density. The downcomer is the cooler water line that goes from the upper drum to the lower drum. The riser is the hotter water line that goes from the lower drum to the upper drum.

A fire tube boiler is similar to a shell and tube exchanger. A combustion tube equipped with a burner transfers heat through the tubes and out of the boiler. The tubes in a fire tube boiler are submerged in water and designed to transfer heat energy to the liquid through conduction and convection. Sediment accumulated in the bottom of the mud drum is removed by water blowdown. Steam from the upper drum can be superheated before entering the steam distribution system. Desuperheaters are used to lower the steam temperature for processes that cannot tolerate the higher steam temperatures.

Steam used in a process facility is usually condensed to water. This condensate can be reused as boiler feedwater.

Oil- or gas-fired burners provide the boiler heat. A damper is used to regulate combustion air flow. Combustion gases exit the boiler stack as flue gas.

The boiler feedwater supply is a critical part of steam generation. Feedwater is treated to remove contaminants that can damage system equipment or compromise its efficient operation. An economizer is used to preheat the feedwater as it enters the boiler. Makeup water is added to maintain proper water levels in the system.

Proper boiler operation depends on controlling feedwater quality, maintaining a specific water level in the system, controlling boiler temperatures and pressures, and maintaining burner efficiency. Process technicians are responsible for monitoring boiler system operation and performing preventive maintenance on boiler system equipment. Typical tasks include checking the boiler stack for smoke; inspecting burners and flame patterns; checking burner fuel pressure, temperature, and air flow; and maintaining proper steam flows, temperatures, and pressures.

Problems associated with boiler operation include the impingement of burner flames on the tubes, loss of boiler feedwater, inadequate treatment of feedwater, or a high water level in the steam drum. A change in the flow or composition of the fuel can increase soot deposits and cause excessively high temperature on the tubes.

Some problems are a consequence of equipment age and design. Other problems are the result of instrument malfunctions, which can be caused by rain, dust, or moisture. Performing the required preventive maintenance, such as making sure that the damper linkages are properly lubricated, can prevent many problem conditions.

Boiler safety is built into procedures for operation. As is the case with all furnaces or direct-fired heaters, there is always the potential for fire or explosion in the boiler firebox. This is of greatest concern during boiler startup. Because of this, the boiler firebox must be purged of combustibles before light-off and in the event of misfire or loss of burner flame. Process technicians must always follow the specific startup and emergency procedures for the particular boiler in order to ensure personal safety as well as the safety of peers, the site, and the community.

Other potential hazards associated with boiler operation include the potential for exposure to feedwater chemicals, steam, hot water, radiant heat, and noise. Process technicians should wear appropriate personal protective gear and avoid skin contact with blowdown materials, which can contain hazardous chemicals.

Checking Your Knowledge

1. Define the following terms:
   a. Superheated steam
   b. Saturated steam
   c. Blowdown
   d. Fire tube boiler
   e. Water tube boiler
   f. Waste heat boiler

2. (True or False) Boilers deliver steam at a desired pressure and temperature that is used throughout the process facility.

3. List three examples of process equipment that use steam energy.

4. The common component(s) of a boiler is (are) (select all that apply):
   a. radiant tubes
   b. burners
   c. a bridgewall
   d. a firebox

5. Burners inject air and fuel through a distribution system. The key components of a natural gas burner include (select all that apply):
   a. dampers
   b. impellers
   c. spuds or spiders
   d. drums

6. (True or False) The economizer section is a fixed, vaned device that causes the air/fuel mixture to swirl above the burner.

7. Dissolved minerals in the boiler feedwater system include metallic salts and calcium carbonates that cause scale, corrosion, and turbine blade deposits. Boiler feedwater is treated with _____ to precipitate these minerals from the water.
   a. lime
   b. salt
   c. lemon
   d. ethanol

8. A water treatment method that uses ion exchange to remove all ionized mineral salts is:
   a. deaeration
   b. coagulation
   c. reverse osmosis
   d. demineralization

9. (True or False) Burner flame impingement or poor flame distribution can cause overheating and failure of the tubes.

10. When should a boiler be purged? (Select all that apply.)
    a. Before light-off
    b. After misfire
    c. At the loss of burner flame
    d. When flame impingement is present

11. (True or False) If atomization is not uniform, the burner tips on the atomization gun or burner require replacement.

12. What can be caused by burner wear? (Select all that apply.)
    a. Uneven flame
    b. Flame impingement
    c. Hot spots
    d. Loss of boiler feedwater flow

13. The _____ is a common name for the one fuel gas block valve that is designated as the main shutoff valve.

NOTE: Answers to Checking Your Knowledge questions are in the Appendix.

Student Activities

1. Write a one-page paper about the three boiler types and their functions. Include drawings of the boilers and label their components.

2. Research boiler accidents. Choose one and write a one-page paper about the causes of the accident and how this type of accident can be avoided in the future.

3. Within a group, select one type of water treatment method used in boiler feedwater. Research this treatment method and present what you learned to the class. Be sure to include drawings and explain how the water treatment can enhance productivity for the process unit.

4. With a partner, review the typical startup procedure mentioned in the chapter. After reviewing the procedure, decide what steps can be added to make the procedure safer for the individual, the equipment, the production goals, and the environment. Be prepared to present your information to the class.

Chapter 15
Vessels

Objectives

After completing this chapter, you will be able to:

15.1   Identify the purpose, types, and applications of major vessels. (NAPTA Vessels 1, 2*) p. 311

15.2   Identify the components of vessels and their purpose. (NAPTA Vessels 3, 4) p. 320

15.3   Explain the operating principles of vessels. (NAPTA Vessels 5) p. 330

15.4   Identify potential problems associated with vessels. (NAPTA Vessels 9) p. 331

15.5   Describe safety and environmental hazards associated with vessels. (NAPTA Vessels 6) p. 332

15.6   Describe the process technician's role in vessel operation and maintenance. (NAPTA Vessels 8) p. 333

15.7   Identify typical procedures associated with vessel operation and maintenance. (NAPTA Vessels 7) p. 334

* North American Process Technology Alliance (NAPTA) developed curriculum to ensure that Process Technology courses will produce knowledgeable graduates to become entry-level employees in process technology. Objectives from that curriculum are named here in abbreviated form. For example, "(NAPTA Vessels 1, 2)" means that this chapter's objective 1 relates to objectives 1 and 2 of the NAPTA curriculum about vessels).

Key Terms

Agitator/mixer—a device used to mix the contents inside a vessel, **p. 320.**

Articulated drain—a hinged drain, attached to the roof of an external floating roof tank, that moves up and down with the roof as the fluid level rises and falls, **p. 313.**

Atmospheric storage tanks—vessels in which atmospheric or low pressure is maintained, **p. 312.**

Baffle—a metal plate, placed inside a vessel, that is used to alter the flow of chemicals, facilitating mixing, **p. 320.**

Barge—a flat-bottomed boat used to transport fluids (e.g., oil or liquefied petroleum gas) and solids (e.g., grain or coal) across shallow bodies of water such as rivers and canals, **p. 319.**

Blanketing—the process of putting an inert gas, usually nitrogen, into the vapor space above the liquid in a vessel to prevent air leakage into it (often called a *nitrogen blanket*), **p. 326.**

Boot—a section in the lowest area of a process vessel where water or other liquid is collected and removed, **p. 321.**

Bullet vessel—cylindrically shaped container used to store contents at moderate to high pressures, **p. 314.**

Containment wall—an earthen berm or constructed wall used to protect the environment and people against tank failures, fires, runoff, and spills; also called *bund wall, bunding, dike,* or *firewall,* **p. 328.**

Cylinders—vessels that can hold extremely volatile or high-pressure materials, **p. 318.**

Distillation—the process of separating two or more liquids by their boiling points, **p. 314.**

Drum—a specialized type of storage vessel or intermediary process vessel, **p. 316.**

Dryer—a vessel in which moisture is removed from process streams, **p. 316.**

Fixed roof tanks—storage tanks that have a roof permanently attached to the top. The roof can be dome or cone-shaped or flat, **p. 312.**

Floating roof—a type of vessel covering, used on storage tanks, that floats upon the surface of the stored liquid; it is used to decrease vapor space and reduce the potential for evaporation, **p. 312.**

Foam chamber—a fixture or piping installed on liquid storage tanks which contains fire-extinguishing chemical foam, **p. 322.**

Gauge hatch—an opening on the roof of a vessel that is used to check levels and obtain samples of the contents, **p. 322.**

Grounding—the process of using the earth as a return conductor in an electrical circuit, **p. 327.**

Hemispheroid vessel—a pressurized container used to store material with a vapor pressure slightly greater than atmospheric pressure (e.g., 0.5 PSI to 15 PSI). The walls of a hemispheroid tank are cylindrically shaped and the top is rounded, **p. 314.**

Hopper car—a type of railcar designed to transport solids such as plastic pellets and grain, **p. 319.**

Hoppers/bins—vessels that typically hold dry solids, **p. 318.**

Level indicator—a gauge placed on a vessel to denote the height of the liquid level within the vessel, **p. 322.**

Lining—a coating applied to the interior wall of a vessel to prevent corrosion or product contamination, **p. 323.**

Manway—an opening in a vessel or tank that permits entry for inspection or repair, **p. 323.**

Mist eliminator—a device in the top of a vessel, composed of mesh, vanes, or fibers, that collects droplets of mist (moisture) from gas to prevent it from leaving the vessel and moving forward with the gas flow (also known as a demister), **p. 323.**

Platform—a strategically located structure designed to provide access to instrumentation and to allow the performance of maintenance and operational tasks. Platforms usually are accessed by stairs or ladders, **p. 323.**

Pressure relief device—a component that discharges or relieves pressure in a vessel or piping system in order to prevent overpressurization (e.g., pressure relief valve, vacuum breaker, or rupture disc), **p. 325.**

Pressure/vacuum relief valve—valve that maintains the pressure in the tank at a specific level by allowing air flow into and out of the tank. Also known as *breather valve, conservation vent,* and *tank breather vent,* **p. 324.**

Pressurized tanks—enclosed vessels in which a greater-than-atmospheric pressure is maintained, **p. 314.**

Reactor—vessel in which a controlled chemical reaction is initiated and takes place either continuously or as a batch operation, **p. 316.**

Scale—dissolved solids deposited on the inside surfaces of equipment, **p. 331.**

Ships—seagoing vessels used to transport cargo across oceans and other large bodies of water, **p. 319.**

Skirt—support structure attached to the bottom of freestanding vessels, **p. 329.**

Spheres—pressurized storage tanks that are used to store volatile or highly pressurized material; their spherical shape allows stress to be uniformly distributed over the entire vessel; also called Hortonspheres, **p. 314.**

Sump—an area of temporary storage located in the bottom of a vessel from which undesirable material is removed, **p. 321.**

Tank car—a type of railcar designed to transport liquids, **p. 319.**

Tank trucks—a vehicle with a container designed to transport fluids in bulk over roadways, **p. 319.**

Underground storage tank—a container in which process fluids are stored underground, **p. 313.**

Vapor recovery system—process used to capture and reclaim vapors, **p. 329.**

Vessel—a container in which materials are processed, treated, or stored, **p. 311.**

Vortex—cyclone-like rotation of a fluid, **p. 329.**

Vortex breaker—a metal plate, or similar device, that prevents a vortex from being created as liquid is drawn out of a vessel, **p. 329.**

Weir—a flat or notched dam or barrier to liquid flow that is usually used either for the measurement of fluid flow or to maintain a given depth of fluid, as on a tray of a distillation column, in a separator, or in another vessel, **p. 330.**

15.1 Introduction

Vessels are containers in which materials are processed, treated, or stored and are a vital part of the process industries. Without this type of equipment, process industries would be unable to process, manufacture, and store large quantities of feedstocks and products. By using vessels, companies can improve both cost effectiveness and efficiency.

In addition to long-term storage, vessels can be used to provide intermediate storage between processing steps and to provide residence time so reactions can be completed or materials can settle. Vessels also provide a container in which the separation of products can occur, using temperature (heat or cold), pressure, or other (mechanical) means.

Vessels vary greatly in design (e.g., size and shape) based on the requirements of the process. Factors that affect vessel design and construction (wall thickness and materials used) include pressure and temperature requirements, product type (liquid, gas, or solid), chemical properties, corrosion factor, and volume. Other factors such as ambient conditions and geographic location (e.g., earthquake or flood zone) of the vessel need to be considered as well.

Vessels are used to carry out process operations such as distillation, drying, filtration, stripping, and reaction. These operations usually involve many different types of vessels, ranging from large towers to small additive and waste collection drums.

Vessel a container in which materials are processed, treated, or stored.

Types of Vessels

The most common process vessel types are tanks, towers and columns, reactors, drums, dryers, filters, cylinders, hoppers, and bins. Additionally, mobile vessels such as tank cars, tank trucks, and hopper cars are used to transport materials across land, while barges, tankers, and container ships move material over waterways.

Vessels operating between 15 and 3,000 PSI are considered to be pressure vessels. *Code vessel* is the term applied to this type of vessel, which indicates that it was designed and built according to the American Society of Mechanical Engineers (ASME) code. The ASME code developed and maintains rules regarding construction materials, methodology, and fabrication practices used with pressure-containing vessels.

Atmospheric Storage Tanks

Atmospheric storage tanks are vessels in which atmospheric or low pressure is maintained. They are not appropriate for substances that contain toxic vapors or high vapor pressure. This tank type can be an above-ground storage tank (AST) or an underground storage tank (UST), as is the case with most gasoline storage tanks at service stations. Most above-ground tanks are cylindrical in shape and equipped with either a fixed or floating roof. Atmospheric storage tanks range in size from small tanks holding a few hundred barrels to mammoth tanks capable of storing more than 1 million barrels and measuring more than 400 feet (122 meters) in diameter.

Design and construction of atmospheric tanks is governed by standards set forth by the American Petroleum Institute (API), while the operation of tanks is regulated by the EPA and local environmental agencies. Large modern tanks usually are made of steel plates that are welded together in large sections, but in early steel tank construction, sections were bolted or riveted together. The bolted tank design (Figure 15.1) is still used today for tanks intended to store nonhazardous materials such as potable water, liquid and solid food, and grain products. Some tanks are constructed using alternative materials such as fiberglass, polyethylene, or concrete.

Figure 15.1 Example of a bolted storage tank.

CREDIT: pedrosala/Shutterstock.

Atmospheric storage tanks are only appropriate for substances that do not contain toxic vapors or high vapor pressure liquids. Types of material that can be stored in an atmospheric storage tank include raw materials (unit feedstocks), additives, intermediate products (not yet finished), final products (process outputs), and wastes (recoverable or nonrecoverable off-spec products and byproducts).

Fixed roof tanks have a roof permanently attached to the top. The roof can be dome or cone-shaped, or flat. This type of tank is used for storing material with a high flash point such as fuel oil, diesel oil, and heating oil.

A **floating roof**, such as the one shown in Figure 15.2A, floats on the surface of a stored liquid to minimize vapor space. This decreased vapor space makes floating roof tanks ideal for storing more volatile material than can be stored in a fixed roof tank. The use of a floating roof tank can reduce product loss because of evaporation by 80–90 percent. Floating roofs can be external (exposed to the environment), or internal (enclosed with a standard fixed cone roof above the floating roof, and protected from the elements). Flexible seals around the roof's edge prevent vapor loss and water intrusion on an external floating roof tank.

External floating roof tanks are suitable for storage of liquids such as crude oil, kerosene, and jet fuel, while materials with very low flash points (gasoline, ethanol) most often are stored in internal floating roof tanks.

In addition to their floating roofs, these tanks also have a stairway from the outside top edge of the roof to the top of the floating roof (Figure 15.2B). The stairway is essential for inspection of the roof and periodic maintenance activities. Specific site safety procedures for accessing a floating roof must be followed.

Figure 15.2 **A.** Example of an external floating roof with an articulated drain. **B.** Floating roof ladder.

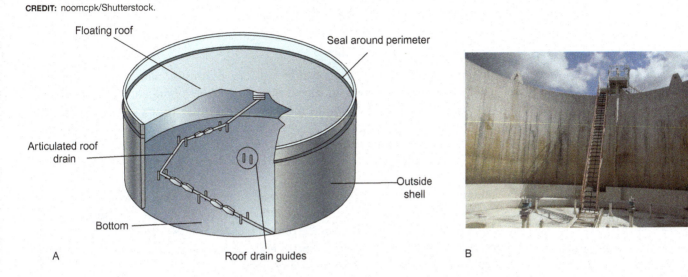

An **articulated drain** is a hinged drain that is attached to a floating roof. Its purpose is to allow drainage of rain water and melting snow from the roof. See Figure 15.2A for an example of an external floating roof tank with an articulated drain.

An **underground storage tank** is a container and its connected piping system in which a minimum of 10 percent of the combined process fluids is stored underground. Because an underground storage tank is installed in the ground, this type of vessel must be built to structurally withstand the pressure exerted by the soil around the tank. Integral ribs often are included in the tank design to provide this additional strength.

Underground storage tanks, such as the one shown in Figure 15.3, must be watertight to prevent groundwater from contaminating the tank contents and to prevent tank contents from contaminating the soil and groundwater surrounding the tank. All underground storage tanks must have protection against spills, overfill, and corrosion incorporated into their design. Underground tanks that store certain hazardous materials are regulated under

Articulated drain a hinged drain, attached to the roof of an external floating roof tank, that moves up and down with the roof as the fluid level rises and falls.

Underground storage tank a container in which process fluids are stored underground.

Figure 15.3 Underground storage tank system.

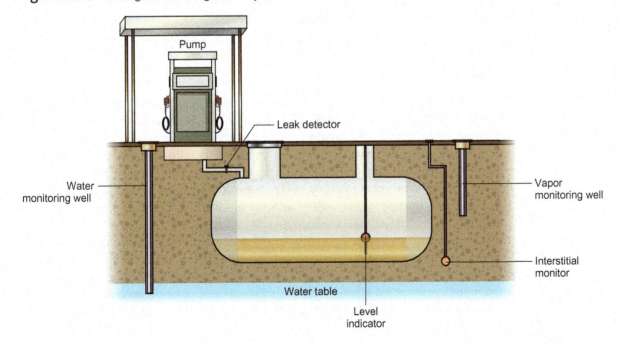

the Comprehensive Environmental Response, Compensation, and Liability Act (CERCLA). These tanks must have secondary containment systems with interstitial monitoring (devices that indicate the presence of a leak in the space between the first and the second wall).

Pressure Storage Tanks

Pressurized tanks are enclosed containers in which the pressure is maintained at a level that is higher than atmospheric pressure. The most common types of pressure storage tanks are spheres and spheroids, bullets, and hemispheroids. Figure 15.4 shows examples of these vessels. The design and construction of pressurized tanks are governed by codes and standards developed and maintained by the American Society of Mechanical Engineers (ASME). Many of these standards have been adopted by government agencies as rules of construction.

Spheres are spherically shaped vessels designed to distribute the pressure evenly over every square inch of the container. Because of their shape, spheres are very strong vessels, having no weak points. Spherical vessels often are used to store volatile or materials pressurized at 15 pounds per square inch (PSI) or higher. Spheres are completely round vessels, while spheroids have a slightly flattened appearance.

A **bullet vessel** (bullet) is a horizontal cylindrically shaped vessel used to store contents at moderate to high pressures. The rounded ends of a bullet vessel, called heads, help distribute pressure more evenly than in a non-rounded vessel. Bullets are often used as intermediary storage for light ends such as propane and butane.

A **hemispheroid vessel** is used to store material with a vapor pressure slightly greater than atmospheric pressure (e.g., 0.5 PSI to 15 PSI). The walls of hemispheroid vessels are cylindrically shaped and the top is rounded.

Pressurized tanks enclosed vessels in which a greater-than-atmospheric pressure is maintained.

Spheres pressurized storage tanks that are used to store volatile or highly pressurized material.

Bullet vessel cylindrically shaped container used to store contents at moderate to high pressures.

Hemispheroid vessel a pressurized container used to store material with a vapor pressure slightly greater than atmospheric pressure (e.g., 0.5 PSI to 15 PSI). The walls of a hemispheroid tank are cylindrically shaped and the top is rounded.

Figure 15.4 Spherical, cylindrical (bullet), and hemispheroid vessels.

CREDIT: **A.** Arcady/Fotolia. **B.** supakitmod/Fotolia. **C.** Aunging/Shutterstock.

A. Spherical   **B.** Cylindrical (Bullet)   **C.** Hemispheroid

Towers

Towers are the central pieces of equipment in many process units. They can be classified by the function they perform, which can be **distillation** (fractionation), stripping, absorbing, adsorbing, or extracting. The height and diameter of a tower varies depending on the requirements of the process: liquid and vapor rates, required product purity, spacing required for tower components, etc. Towers are usually some of the tallest pieces of equipment in a plant, some higher than 300 feet (91 meters). Towers can operate at vacuum pressure (less than atmospheric) as well as at hundreds of pounds higher than atmospheric pressure.

Towers are vertical vessels, usually cylindrical with heads at both top and bottom. Openings in the shell that allow for liquid flow into and out of the tower are called *nozzles*. Nozzles are also provided for instrumentation connections, drains, and other operational requirements. Tower internals include components such as trays (Figure 15.5) or packing (Figure 15.6), which facilitate vaporization and condensation in a distillation tower.

Distillation the process of separating two or more liquids by their boiling points.

Figure 15.5 Examples of distillation tower tray types: **A**. Sieve tray; **B**. Valve tray; **C**. Bubble cap tray.

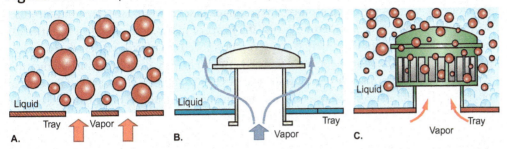

A.

B.

C.

Figure 15.6 Examples of structured and random packing used in distillation towers. **A**. Structured packing. **B**. Raschig rings (random packing). **C**. Berl/Intalox saddles (random packing). **D**. Pall ring (random packing). **E**. Assorted random packing materials.

CREDIT: **E**. magnetix/Shutterstock.

A.

B.

C.

D.

E.

Tower tray parts include weirs and downcomers (Figures 15.7 and 15.8). Weirs maintain a liquid level on each tower tray. Downcomers direct liquid from a tray to the tray below it.

Figure 15.8 Illustration showing the operation of a distillation tower—bubble cap trays, weirs, and downcomers.

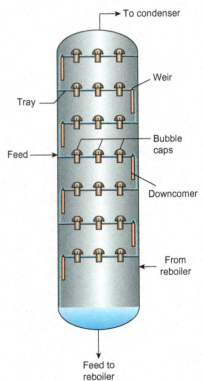

Figure 15.7 Sieve tray showing weirs and two alternating downcomers.

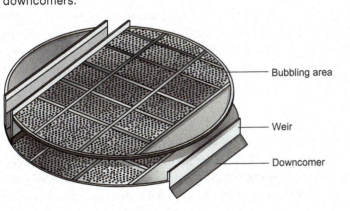

Reactors

Reactor a vessel in which a controlled chemical reaction is initiated and takes place either continuously or as a batch operation.

Reactors are vessels in which a process takes place that converts feedstock into products with different properties. **Reactor** vessels, the heart of many petrochemical facilities, can be horizontally or vertically situated. Most reactions occur in reactor vessels that are either spherical or cylindrical in shape. Reactors can operate at very high pressures (higher than 2,000 PSI) and temperatures exceeding 800 degrees F (427 degrees C). Because of the extreme operating conditions of some reactors, the design and construction of reactor vessels vary greatly. Reactors are discussed in depth in Chapter 16.

Drums

Drum a specialized type of storage tank or intermediary process vessel.

A **drum** is a process vessel used for intermediary storage on process units, for auxiliary and utility storage, or for additive storage. Drums can function as receivers, separators, surge vessels, or mixing vessels. Horizontal drums frequently are used to separate oil and water and can have a boot for water collection. Vertical drums, often used for gas/liquid separation, are sometime called knockout drums and are seen in compressor suction lines and in fuel gas lines before furnaces. Figure 15.9 shows some drum types used in the process industries.

Figure 15.9 A. Liquid knock-out drum. **B.** Example of a horizontally oriented drum.

CREDIT: B. aor/Shutterstock.

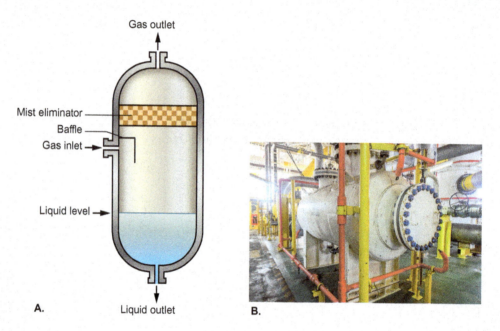

Dryers

Dryer a vessel in which moisture is removed from process streams.

A **dryer** is used to remove liquid from a solid or gas using an adsorbent or by heat transfer (Figure 15.10). The different types of dryers include rotary drum dryers, tray dryers, column dryers, paddle dryers, double cone dryers, and fixed bed dryers. Applications for dryers in the process industries include their use in plant air systems, petrochemical manufacturing, and pharmaceutical manufacturing.

Filters

Filters are used to remove impurities such as scale, dirt, or any type of particulate matter from a liquid or gas stream. Filtration can be accomplished by the use of a vessel in which sand or anthracite coal is the filtering medium used. This type of filter is used extensively in wastewater treatment and boiler feedwater treatment. In other

Figure 15.10 A. A diagram of a rotary dryer used in process industries. **B.** A rotary drum dryer used to dry fertilizer.

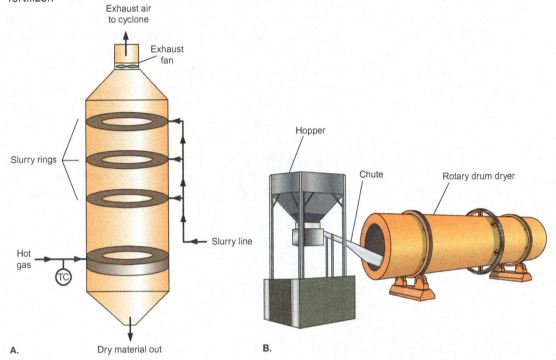

A.

B.

applications, cartridges can be used in the filtration vessel (Figure 15.11), especially in services such as a lube oil filter, furnace fuel gas filter, or a compressor suction line filter. A filter can be indistinguishable from other types of vessels, without knowledge of the specific process.

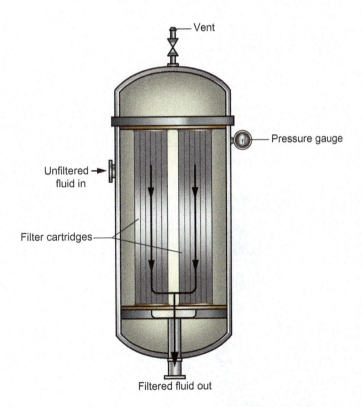

Figure 15.11 Cartridge filter, an example of a filter used in process industries.

High Pressure Cylinders

Cylinders are vessels used to hold extremely volatile or high pressure compressed gaseous material. Material stored in cylinders can be in liquid form when stored in the cylinder at normal temperatures but gaseous at atmospheric pressure. Examples of this type of material include propane, chlorine, and anhydrous ammonia. Some gases (such as nitrogen and oxygen) do not become liquid even at the higher pressure inside a cylinder.

Materials used in the construction of cylinders include steel, aluminum, and composite materials. Material flow out of the cylinder is controlled by a valve and pressure regulator. Valve caps are installed to prevent damage by covering the valve when the cylinder is not in use or is being transported.

The International Organization for Standardization (ISO), OSHA, and the US Department of Transportation all have standards in place that govern the construction, inspection, storage, and maintenance of cylinders. It is important that cylinders be stored properly, in a dry, cool, well-ventilated area, with caps installed to protect the valves. They must also be held securely in place to prevent tipping. Full and empty cylinders should be stored separately. Figure 15.12 shows some high-pressure cylinders.

Cylinders vessels that can hold extremely volatile or high-pressure materials.

Figure 15.12 High pressure gas cylinders.
CREDIT: VikOl/Shutterstock.

Hoppers and Bins

Hoppers and **bins** are vessels used for bulk storage or transport of dry materials, including powders and pelletized materials. They also are used to handle surge capacities in material processing operations. Hoppers (Figure 15.13) allow for material received in bags or sacks to be introduced into a process in a controlled and measured way.

Hoppers/bins vessels that typically hold dry solids.

Figure 15.13 Hopper.
CREDIT: Richard Thornton/Shutterstock.

Bins and hoppers vary greatly in their size and shape, as well as in wall thickness and construction material. They are used in many processes, including those in the chemical, oil and gas, power, and pulp and paper industries.

Mobile Containers

Process industry products are transported from manufacturing sites to customers in a variety of mobile tanks and vessels. For example, tank trucks and tank cars move products on land, while barges and ships transport products across water.

A **tank car** (Figure 15.14A) is a type of railcar designed to transport liquids in bulk. Tank cars can be insulated to maintain their loads at specific temperatures or pressurized to maintain their loads at set pressures. Tank cars also can be divided into compartments in order to carry several different products in one car.

A **hopper car** (see Figure 15.14B) is a railcar designed to transport solids such as plastics and grain.

Tank trucks (see Figure 15.14C) are vehicles with containers designed to transport fluids in bulk over roadways. Tank trucks are similar to tank cars because they can be insulated, pressurized, and compartmentalized.

Did You Know?

Spherical vessels are much stronger than nonspherical vessels. That is why they are used for extremely high-pressure applications. Spherical vessels are double walled and insulated when used in cryogenic applications for materials such as hydrogen, nitrogen, oxygen, ethylene, and LNG.

CREDIT: Muratart/Fotolia.

Tank car   a type of railcar designed to transport liquids.

Hopper car   a type of railcar designed to transport solids such as plastic pellets and grain.

Tank trucks   vehicles with containers designed to transport fluids in bulk over roadways.

Figure 15.14 A. Tank car used to transport liquid cargo by rail. **B.** Hopper car used to transport solids by rail. **C.** Tank truck used to transport fluid cargo over roadways.

CREDIT: **A.** Richard Thornton/Shutterstock. **B.** 3DMI/Shutterstock. **C.** DK Arts/Shutterstock.

A.

B.

C.

Ships are seagoing vessels used to transport cargo across oceans and other large bodies of water. The most common types of ships used by the process industries are container ships, tankers, and supertankers. Container ships transport truck-size storage containers loaded with products. Tanker ships transport bulk liquids such as oil and chemicals. Figure 15.15A shows an example of a container ship, and Figure 15.15B. shows an example of a tanker. A **barge** (see Figure 15.15C) is a flat-bottomed boat used to transport fluids (e.g., oil or liquefied petroleum gas) and solids (e.g., grain or coal) across shallow bodies of water such as rivers and canals. Barges can be pushed or pulled by tugboats.

Both barges and tankers are compartmentalized to stabilize the load. Each compartment must be filled and unloaded in a prescribed sequence to keep the vessel level while it is loaded or unloaded.

Ships   seagoing vessels used to transport cargo across oceans and other large bodies of water.

Barge   a flat-bottomed boat used to transport fluids (e.g., oil or liquefied petroleum gas) and solids (e.g., grain or coal) across shallow bodies of water such as rivers and canals.

Figure 15.15 A. Example of a container ship. **B.** Example of a tanker. **C.** Barge used to transport heavy goods.

CREDIT: **A** and **B.** EvrenKalinbacak/Shutterstock. **C.** DoublePHOTO studio/Shutterstock.

A.

B.

C.

15.2 Common Components of Vessels

Vessels, including tanks and drums, have many components with which process technicians should be familiar.

A shell is the primary component of a vessel, making up the main part of its body. A shell can be cylindrical, spherical, or conical and it can be constructed of different materials, although steel is the most common.

The *heads* of a vessel are located on both ends of the shell and provide closure for the shell. They are usually welded to the shell and constructed of the same material as the shell. Vessel heads can be flat, but most are of a curved design, which provides a greater amount of strength to the vessel. The three designs of curved heads are hemispherical, elliptical, and torispherical. See Figure 15.16 for examples.

Figure 15.16 Examples of pressure heads of vessels.

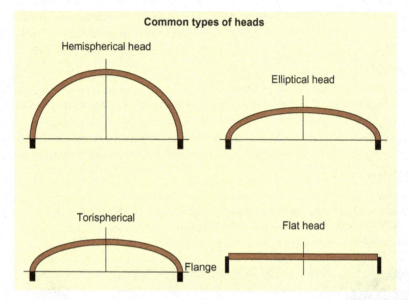

Common types of heads

Hemispherical head

Elliptical head

Torispherical

Flat head

Flange

Vessel nozzles are connections that are attached to openings in a vessel shell. They allow for the necessary attachments for a vessel to function. Nozzles are indicated for such things as piping to allow material to enter and exit the vessel, connections for instrumentation, sight glasses, manways, and vents. Figure 15.17 shows a vessel with a variety of nozzles.

Figure 15.17 Vessel nozzles.

CREDIT: ChiccoDodiFC/Shutterstock.

Agitator/mixer   a device used to mix the contents inside a vessel.

Baffle   a metal plate, placed inside a tank or other vessel, that is used to alter the flow of chemicals, facilitating mixing.

An **agitator/mixer** is a device used to combine the contents inside a vessel. Mixers are operated using motors located outside the vessel. Pumps can be used to mix the liquid by recirculating the contents of the vessel. Figure 15.18 shows vessels equipped with agitators.

A **baffle** is a metal plate placed inside a vessel. It is used to alter the flow of a vessel's contents, facilitate mixing, or cause turbulent flow. Baffles can be vertical or horizontal. An *impingement baffle* absorbs the kinetic energy in a fluid entering a vessel, redirects the high velocity inlet flow downward, and prevents it from splashing across the vessel to the opposite wall. In gas/liquid separators, the inlet baffle provides the initial separation of the gas and liquid components. Some vessel inlet baffles are perforated to facilitate distribution of

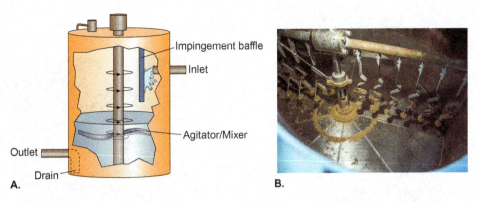

A.
Impingement baffle
Inlet
Agitator/Mixer
Outlet
Drain

B.

Figure 15.18 A. Vessel with agitator. **B.** View of a beer tank where hops, malt, and water are being mixed.

CREDIT: B. Henny van Roomen/Shutterstock.

the material entering the vessel. In vessels with mixers, horizontal baffles along the walls are used to break the vortex that a mixer creates and improve mixing. Figures 15.18A and 15.19 show examples of impingement baffles.

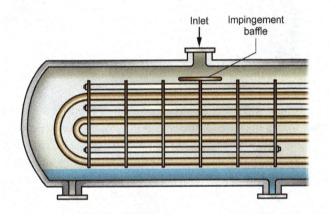

Inlet Impingement baffle

Figure 15.19 Impingement baffle (also called an impingement plate).

A **sump** (also called a **boot**) is the lowest section at the base of a vessel. In a tank, the sump allows material to be removed from the bottom of the tank and the tank to be completely emptied. In a drum, a boot allows water to settle out of the process fluid and be removed. Figure 15.20 shows examples of a sump and a draw-off boot.

Sump an area of temporary storage located in the bottom of a vessel from which undesirable material is removed.

Boot a section in the lowest area of a process vessel where water or other liquid is collected and removed.

Figure 15.20 A. Example of a vessel with a sump. **B.** Vessel draw-off boot.

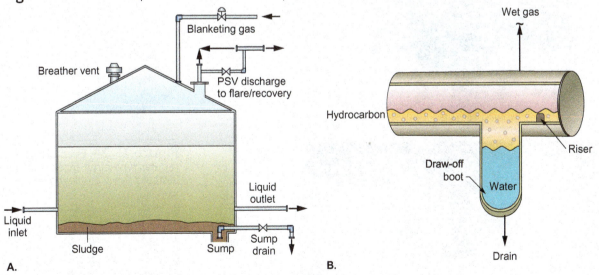

A heating/cooling jacket or coil can be attached to the external shell of a vessel. In some cases, the coil is inside the vessel. When both heating and cooling are required at different times in the vessel's operation, the coil can be designed to circulate either a heating or a cooling medium. Figure 15.21 shows heating/cooling coils and jackets.

Figure 15.21 A. External heating/cooling coil. **B.** Heating/cooling jacket. **C.** Internal heating/cooling coil.

External heating/cooling coil

Heating/cooling jacket

C.

Foam chamber a fixture or piping installed on liquid storage tanks which discharges fire-extinguishing chemical foam.

Gauge hatch an opening on the roof of a vessel that is used to check tank levels and obtain samples of the contents.

Level indicator a gauge placed on a vessel to indicate the height of the liquid level within the vessel.

A **foam chamber** (see Figure 15.22) is a fixture attached near the roof joint of a storage tank that is designed to discharge chemical foam onto the surface of a flammable or combustible liquid. The foam can be used to extinguish fires within the tank or on the tank roof.

A **gauge hatch** (see Figure 15.23) is an opening on the roof of a tank that is used to check tank levels and obtain samples of the product or chemical.

A sight glass (or gauge glass) **level indicator** (see Figure 15.24) is a gauge placed on a vessel to indicate the height of the liquid level within the vessel.

Figure 15.22 A. Foam chamber structure. **B.** Foam extinguishing a fire.

CREDIT: **B.** akiyoko/Shutterstock.

A.

B.

Figure 15.23 Gauge hatch.

Figure 15.24 Example of a tubular type sight glass level indicator on a small vessel.

A **lining** is a coating applied to the interior wall of a vessel to prevent corrosion or product contamination. Linings usually are made of polymers, elastomers, or glass. Figure 15.25 shows an example of a vessel lining.

Lining a coating applied to the interior wall of a vessel to prevent corrosion or product contamination.

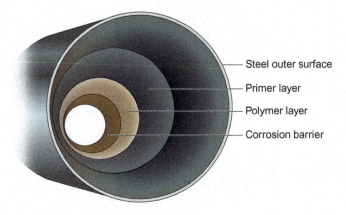

Steel outer surface
Primer layer
Polymer layer
Corrosion barrier

Figure 15.25 Cross section of a vessel lining.

A **manway** is an opening in a vessel that permits entry for inspection or repair. In towers and drums, the opening is typically a 24-inch diameter circle, but the opening in tanks can be a larger rectangular manway to allow access by equipment used to clean the tank. The number of manways, the size of the opening, and manway styles vary from vessel to vessel. Most manways are circular flanged openings on the side of the vessel for entry. Figure 15.26 shows examples of vessel manways.

Manway an opening in a vessel or tank that permits entry for inspection or repair.

Figure 15.26 Examples of a manway. **A.** Manway with bolts and nuts in place. **B.** Manway with bolts and nuts removed in preparation for entry. **C.** Photo showing manway location.
CREDIT: **A.** Mr. Nirut Nuamkerd/Shutterstock. **B.** noomcpk/Shutterstock. **C.** wittybear/Fotolia.

A. **B.** **C.**

A **mist eliminator** is a device in the top of a vessel designed to collect droplets of mist from gas before it exits the vessel. It is composed of mesh, vanes, or fibers. The droplets coalesce (merge) to form larger droplets until they are too large and heavy to remain on the mesh; at this point, they drop back into the vessel. Figure 15.27 shows an example of a mist eliminator.

A **platform** is a strategically located structure on a vessel designed to provide access to instrumentation and to allow performance of maintenance and operational tasks. Platforms are generally accessed by stairs or ladders. Their use and construction are regulated by the Occupational Safety and Health Administration (OSHA). Figure 15.28 shows examples of platforms installed on a process vessel and the side of a storage tank.

Mist eliminator a device in the top of a vessel, composed of mesh, vanes, or fibers, that collects droplets of mist (moisture) from gas to prevent it from leaving the vessel and moving forward with the gas flow (also known as a demister).

Platform a strategically located structure designed to provide access to instrumentation and to allow the performance of maintenance and operational tasks. Platforms are usually accessed by stairs or ladders.

Figure 15.27 Mist eliminator.

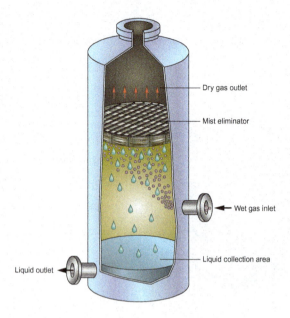

Dry gas outlet

Mist eliminator

Wet gas inlet

Liquid outlet

Liquid collection area

Figure 15.28 A. Platform on a horizontal process vessel. **B.** Platform attached to a storage tank.

CREDIT: A. SergBob/Shutterstock.
B. Thiradech/Shutterstock.

A.

B.

Pressure/Vacuum Relief Valves

Atmospheric storage tanks are built to hold large volumes of liquid, but they cannot withstand pressure much higher than atmospheric pressure or a vacuum. Unfortunately, many catastrophic tank farm fires and oil spills have been caused by overpressurizing or collapsing a tank. Because of this, devices have been developed to prevent tank damage from overpressurization and to prevent associated hazards from occurring.

Because of changing pressure/temperature of tank contents during day/night and filling/emptying operations, fixed roof atmospheric tanks must have a mechanism for relieving excess pressure and preventing vacuum conditions. This protection is provided by a **pressure/vacuum relief valve** (also referred to as a breather valve, tank breather vent or conservation vent). These valves maintain the pressure in the tank at a specific level by allowing air flow in and out of the tank. Figure 15.29 shows an example of a tank pressure/vacuum relief valve.

Tank breather vents are installed on the outer roof of storage tanks that have no internal floating roof. These relief devices conserve tank vapors while protecting against overpressurization and vacuum. Inside a breather vent is a *pallet* (a guided disk that opens and closes in response to a pressure differential). This disk permits the intake of air or the escape of vapors as needed to keep the tank within safe working pressures. Pallets are weight-loaded to correspond to set pressures for each tank. Tank breather valves have the pressure and vacuum pallets mounted side by side on the tank.

Pressure/vacuum relief valve   valve that maintains the pressure in the tank at a specific level by allowing air flow into and out of the tank. Also known as *breather valve*, *conservation vent*, and *tank breather vent*.

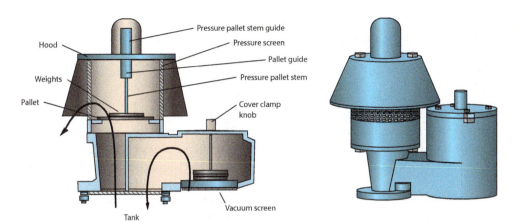

Figure 15.29 Pressure/vacuum relief valve.

Pressure Relief Devices

A **pressure relief device** (see Figure 15.30) is a component that discharges or relieves pressure in the event of overpressurization. It is considered to be the last line of defense to protect life, the environment, and equipment. A pressure relief device should activate only if all other levels of vessel/system protection have failed. A key design feature of pressure relief devices is that they are activated by the material they protect, relying on no other energy source (electric, pneumatic pressure, etc.) for operation. Pressure relief devices include relief valves, safety valves, and rupture discs. These components are designed to open when a set pressure is reached, in order to protect a vessel or system from exceeding its design pressure and to prevent bursting and rupture of a vessel and its associated piping. The design, installation, and maintenance of pressure relief valves is governed by the code enacted by the American Society of Mechanical Engineers (ASME) and guidelines set forth by the American Petroleum Institute (API) for the oil and gas industry.

Pressure relief device   a component that discharges or relieves pressure in a vessel or piping system in order to prevent overpressurization (e.g., pressure relief valve, vacuum breaker, or rupture disc).

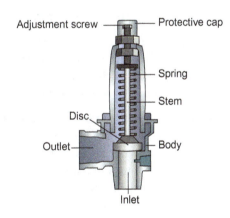

Figure 15.30 Pressure relief device (valve) used to prevent overpressurization.

CREDIT: Dale Stagg/Shutterstock.

RELIEF VALVES   Relief valves are used in liquid (incompressible fluid) service to limit the pressure in a vessel or system. Increased pressure can be the result of a process upset condition, instrument or equipment failure, or a fire in the surrounding area that heats up the vessel or piping. In the event that pressure exceeds the relief valve setpoint, the valve opens, creating a path for the pressure to exit the vessel or system. Relief valves can be spring or weight loaded. After the pressure decreases to a point below the relief valve setpoint and safe operating conditions have been restored, the relief valve closes (reseats).

SAFETY VALVES   Safety valves are used in compressible fluid service (gas, steam) and are spring-loaded devices. There are three distinct differences between safety valves and relief valves.

- Safety valves are intended for use in gas service; relief valves are used in liquid service
- Safety valves have a larger outlet than relief valves, because compressible gases require that a greater amount be released to reduce pressure.

- Safety valves open fully and quickly when set pressure is exceeded, but relief valves open more slowly and in proportion to the pressure increase.

SAFETY RELIEF VALVES Safety relief valves are spring-loaded pressure relief devices that may be used in either liquid or gas service.

RUPTURE DISCS A rupture disc is another type of pressure relief device. It differs from relief and safety valves in that it does not reseat after vessel pressure returns to a safe level. It is a one-time use device that remains open after operation and must be replaced following each use. For this reason, rupture discs are not usually the primary protection mechanism used.

If the differential pressure between the upstream and downstream sides of a rupture disc reaches the designated pressure, the disc bursts, allowing free flow of the process fluid and reducing pressure in the equipment or piping.

Rupture discs (Figure 15.31) frequently are used in conjunction with relief valves or safety valves to protect the internal components of those valves from corrosion (or from plugging if the system contains solids). Rupture discs also ensure that leakage does not occur through the relief or safety valve, especially important in toxic service.

Figure 15.31 Examples of rupture discs, new and used.

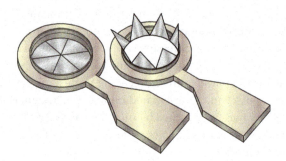

Auxiliary Equipment for Vessels

In addition to the components previously mentioned, vessels contain additional auxiliary equipment. This equipment includes components associated with blanketing, corrosion monitoring, fire protection, heating and cooling, insulation and tracing, secondary containment, and vapor recovery systems.

BLANKETING SYSTEMS **Blanketing** is the process of introducing an inert gas (usually nitrogen) into the vapor space above the liquid in a storage tank to prevent air leakage into the tank. This layer of gas is referred to as a *nitrogen blanket*. Blanketing a tank reduces the amount of oxygen present and decreases the risk of fire and explosion by preventing air (oxygen) from reacting with tank contents. Nitrogen blankets also reduce the risk of tank implosion (collapse) by preventing a vacuum from being created when a tank is emptied. Figure 15.32 shows an example of a tank blanketed with an inert gas.

Blanketing the process of putting an inert gas, usually nitrogen, into the vapor space above the liquid in a vessel to prevent air leakage into it (often called a *nitrogen blanket*).

Figure 15.32 Tank blanketed with inert gas.

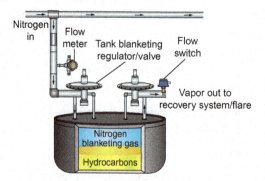

CORROSION MONITORING SYSTEMS   Corrosion in tanks can cause serious safety, operational, and environmental problems, so it must be controlled to the extent possible. Since corrosion most often occurs around a tank bottom, it can lead to leakage from the tank. Corrosion control is often handled by corrosion monitoring systems, such as that seen in Figure 15.33.

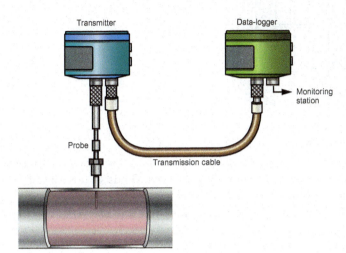

Figure 15.33 Corrosion monitoring system.

Typical corrosion monitoring systems consist of corrosion probes, transmitters, communication lines, and receivers. The probes are attached to the vessel or piping and are connected to the transmitter. The transmitter sends corrosion data to a receiver via a communication cable. The data then can be downloaded for analysis by authorized testing personnel (individuals who are qualified to test for metal thickness and symptoms of corrosion).

FIRE PROTECTION SYSTEMS   Fire protection can be provided through water sprinkler systems, which keep vessel supports and shells cool, or by chemical foam systems, which extinguish fire by covering the product with a foam blanket to separate the flame source from the product surface and smothering it, removing air.

GROUNDING SYSTEMS   **Grounding** involves connecting an object to the earth using copper wire and a grounding rod to provide a path for electricity to dissipate harmlessly into the ground. Grounding is installed on vessels in order to redirect any static electricity that might build up on their surfaces. This prevents the ignition of flammable vapors by sparks that are generated when static electricity is discharged. Grounding also is used on electrical equipment to prevent electrical shock.

Grounding   the process of using the earth as a return conductor in a circuit.

HEATING AND COOLING   Heating and cooling can be applied to vessels by a variety of methods. For example, water can be used to heat or cool the contents of a vessel. This can be accomplished by direct injection of steam or refrigerated water into the vessel. Additionally, heating media (such as steam or hot oil) and cooling media (such as refrigerated water or ammonia) can be circulated through internal or external coils or jackets attached to the vessel to adjust the temperature of the contents.

INSULATION AND TRACING   Insulation and tracing (either steam, hot oil, or electrical) are provided to prevent vessel contents from freezing or to keep lines flowing. Insulation also is used as a safety measure to keep people from coming in contact with hot lines. Figure 15.34 shows an example of heat tracing on a storage tank.

Did You Know?

Static electricity can ignite gasoline vapors. Static electricity can be generated by:

- Getting in and out of a car while the car's gas tank is being filled.
- Filling a portable gas can in the bed of a pickup truck. The safe way to fill a gas can is with the gas can placed on the ground.

CREDIT: D and D Photo Sudbury/Shutterstock.

Figure 15.34 Steam tracing system on the side of a storage tank.

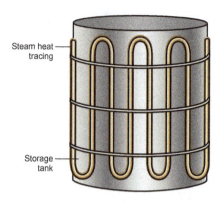

Steam heat tracing

Storage tank

Some pipes contain fluids that will freeze or set up if they are not kept warm. This might not have anything to do with the temperature outside, but might be caused by the chemical properties (viscosity) of the fluid inside the pipe. Some fluids set up and stop flowing at very high temperatures (e.g., some tar sets up if its temperature goes below 200 degrees F [93 degrees C]). Without adequate heat tracing, these fluids would be unable to pass properly through a piping system.

Secondary Containment Systems

A secondary containment system involves the equipment, including valves, piping, and walls built around vessels to contain the vessel's contents in the event of a leak, spill, or vessel failure. OSHA requires all above-ground tanks to have a containment system built around them for personal and environmental safety. A containment wall, usually made from earth or concrete, also can be referred to as a dike or a firewall. A **containment wall** (see Figure 15.35) is a wall built around a piece of equipment that protects people and the environment against tank failures, fires, runoff, and spills. In some cases, a group of vessels can be protected within a common containment system.

Containment wall   an earthen berm or constructed wall used to protect the environment and people against tank failures, fires, runoff, and spills; also called *bund wall, bunding, dike,* or *firewall.*

Figure 15.35 Containment wall surrounding a tank to protect against tank failures, fires, runoff, and spills.

CREDIT: travelview/Shutterstock.

Containment wall

The main purposes of containment systems are to:

- Minimize safety risks by containing chemical spills in a small area and thus trapping contaminants before they can spread to other areas
- Protect the soil and the environment from contaminants
- Protect humans from potential hazards (e.g., fire or chemical release)
- Contain wastewater and contaminated rainwater until it can be drained into a proper sewage line
- Facilitate cleanup after a release of hazardous material
- Protect the environment and people from product release caused by tank failures.

In the event of rainfall accumulation within the containment wall, a process technician must open a valve and drain the water to prevent the tank from lifting off its foundation (floating). Water inside the containment wall, however, must be tested for possible pollutants before it can be drained into the sewer system.

VAPOR RECOVERY SYSTEM A **vapor recovery system** is the process used to capture and reclaim vapors. Vapors are captured by methods such as chilling or scrubbing. The vapors are then purified and returned to the process, stored, or incinerated. See Figure 15.36.

Vapor recovery system process used to capture and reclaim vapors.

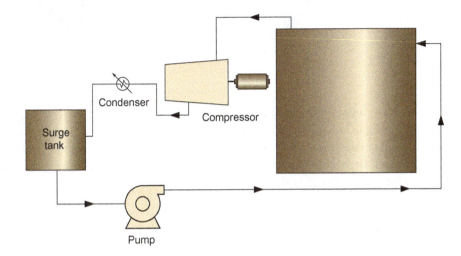

Figure 15.36 Vapor recovery system.

OTHER COMPONENTS Cylindrical pressure vessels are typically supported by vessel skirts. A support **skirt** is a cylindrical steel section that is welded to the lower vessel shell or the bottom head. Figure 15.37 shows an example of a vessel skirt on a vessel being delivered on a barge to a process facility.

Skirt support structure attached to the bottom of freestanding vessels.

Figure 15.37 Skirt surrounds the bottom of a pressure vessel.

CREDIT: shinobi/Shutterstock.

A **vortex breaker** is a metal plate or similar device that prevents a **vortex** (the cyclonelike rotation of a fluid) from being created as liquid is drawn out of the vessel. This prevents gas or vapor from escaping with the liquid and creating cavitation in a downstream pump, which can damage it over time. Examples of various vortex breaker designs are shown in Figure 15.38.

Vortex breaker a metal plate, or similar device, that prevents a vortex from being created as liquid is drawn out of the vessel.

Vortex cyclone-like rotation of a fluid.

Figure 15.38
Vortex breaker designs.

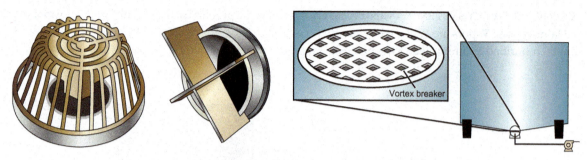

Weir a flat or notched dam or barrier to liquid flow that usually is used either for the measurement of fluid flows or to maintain a given depth of fluid as on a tray of a distillation column, in a separator, or in another vessel.

A **weir** is a flat or notched dam that functions as a barrier to flow. A weir also can be used to separate two substances (e.g., oil and water) in a mixture. Figure 15.39 shows an example of a vessel (decanter) with a weir. In this example, anything that is above the weir flows over it. Because oil is lighter than water, oil is the substance that goes over the weir and is separated out of the mixture.

Figure 15.39 Oil flow over the weir in a three-phase separator.

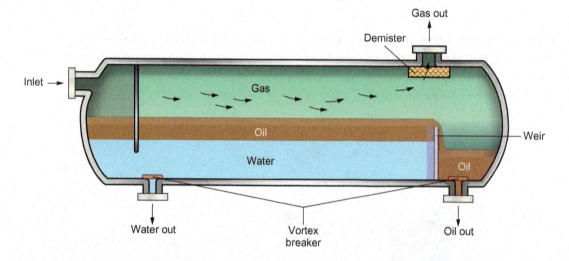

15.3 Operating Principles of Vessels

When operating a process vessel, tank, tower, column, or drum, proper control of the pressure, temperature, flow, level, and composition of the contents must be maintained and all procedures for operating each vessel must be followed.

Each vessel is designed to work within a design range of pressure, and both overpressurization and undepressurization must be avoided. When putting a vessel into service, the pressure should be increased gradually to operating pressure because rapid pressure changes can damage internal components. Vessel pressure relief devices must always be operational when the vessel is in service.

There is a design range of temperature for each unique vessel. Care must be taken not to allow the temperature to exceed or drop below the temperature range for the specific vessel. Like pressure, temperature is increased gradually when putting a vessel into service. Rapid temperature changes can crack welds and damage vessels.

Each vessel is designed to operate within a design flow range. The flow is monitored to ensure that neither high flows nor low flows occur. Rapid flow changes can damage internal components, such as valve trays, sieve trays, and packing.

A liquid level range is provided for each vessel that is designed. Overfilling or underfilling the vessel can result in damage to equipment, individuals, and the environment. Liquid levels should be raised and lowered gradually to prevent damage to internal vessel components. Liquid flows in and out of vessels are monitored to ensure that operations are maintained in the proper range.

Each vessel is designed to hold certain types of material. The process composition must be appropriate to the vessel type. Valve and pump alignment is essential to prevent cross-contamination of material within a vessel.

15.4 Potential Problems with Vessels

As with any piece of process equipment, various problems can arise with vessels. Potential problems include corrosion, scale buildup, over- and underpressurization, operating upsets, shutdowns, and malfunctions during startup.

Corrosion

Corrosion occurs when metal is deteriorated through a chemical reaction (e.g., iron rusting). Corrosion can occur for a variety of reasons, whether from direct contact with chemicals (e.g., acids, bases, and salts), temperature extremes, oxidation, or water exposure. Corrosion can have a catastrophic effect on the integrity of pipelines, vessels, and equipment.

Corrosion causes thinning, pitting, or cracking (e.g., chloride stress corrosion) of vessel walls, which can lead to leaks or cause the integrity of the vessel to be compromised. Compromised vessel integrity can result in tank rupture or collapse. Where corrosion is a possibility, an additional metal thickness (corrosion allowance) is calculated into the vessel design.

Metal corrosion must be checked at regular intervals. Vessel wall thickness checks can be made while the vessel is in operation, but more accurate thickness measurements are made when a vessel is empty and clean.

Scale Buildup

Scale consists of dissolved solids deposited on the inside surfaces of equipment. Scale usually does not affect vessel integrity, but it can affect the quality or purity of the vessel's contents. Scale also can cause blockage of vessel components and instrumentation.

Scale dissolved solids deposited on the inside surfaces of equipment.

Over- and Underpressurization

Serious consequences can result if vessels are subjected to overpressurization or underpressurization. Vessels can rupture as a result of overpressurization; they can collapse inward (*implode*) if underpressurized. Overpressurization can occur if too much material is pumped into a vessel or if the rate of filling a tank is too high and the pressure relief valve fails to work properly. Instrumentation failure, change in feed composition, loss of cooling, or water in the feed stream are other conditions that can lead to a pressure upset in a vessel. Underpressurization can occur if the tank vent is plugged while a tank is being emptied or if nighttime temperatures decrease significantly. This creates a vacuum within the vessel and can result in implosion. Figure 15.40 shows an implosion that resulted from vacuum pressure.

Operating Upsets

Upsets within a process can allow corrosive chemicals to enter equipment that was not designed to handle such chemicals. This can lead to rapid corrosion and vessel damage. Changes or deviations (*excursions*) in operating temperatures and extreme levels also can accelerate corrosion rates and damage vessels by overheating.

Did You Know?

Special attention should be given when "steaming out" or applying steam to low-pressure tanks because the steam pressure can overpressurize the vessel. Also, rapid cooling from a rain shower can cause the steam to collapse, resulting in a drop of internal pressure to vacuum (below atmospheric pressure) and tank collapse.

Figure 15.40 Imploded vessel.

Courtesy of Jeffrey Laube.

Shutdowns and Startups

Prior to startup after a shutdown for maintenance, vessels should be checked for any foreign materials that might have been left behind during vessel maintenance. During vessel closure, gaskets must be installed properly and bolts tightened properly. On vessels that are operated at high temperatures, flange bolts can become loose as the vessel heats up. These bolts must be retightened while the vessel is heating up.

15.5 Hazards Associated with Improper Operation

The improper operation of a vessel can create unsafe or hazardous situations. Table 15.1 lists some of these scenarios.

Table 15.1 Hazards Associated with Improper Vessel Operation

Improper Operation	Possible Effects			
	Individual	Equipment	Production	Environment
Overfilling	Exposure to hazardous chemicals; injury	Damage to vessel and equipment, especially floating roof tanks	Cleanup cost; lost product	Spill, fire, vapor release
Putting wrong or off-spec material in a vessel	Exposure to hazardous chemicals; injury	Damage to vessel and equipment because of undesirable chemical reactions from incompatible chemical mixtures	Vessel cleaning costs; rerun material; reduced storage; cleanup cost	Spill, fire, vapor release
Misalignment of blanket system	Exposure to hazardous chemicals; injury	Loss of blanket; damage to vessel and equipment; collapse of tank because of vacuum	Cleanup cost; lost product; reduced storage	Spill, fire, vapor release
Misalignment of pump systems	Exposure to hazardous chemicals; injury	Damage to vessel and equipment because of undesirable chemical reactions from incompatible chemical mixtures; damaged pump	Vessel cleaning costs; rerun material; reduced storage; cleanup cost; contamination of other equipment	Spill, fire, vapor release
Pulling a vacuum on a vessel while emptying	Exposure to hazardous chemicals; injury	Collapse of vessel because of vacuum; damage to vessel and equipment	Cleanup cost; lost product; reduced storage	Spill, fire, vapor release
Overpressurization	Exposure to hazardous chemicals; injury	Rupture of vessel; damage to vessel and equipment	Cleanup cost; lost product; reduced storage	Spill, fire, vapor release

15.6 Process Technician's Role in Operation and Maintenance

Monitoring and Maintenance Activities

Process technicians are often responsible for the routine monitoring of vessels. These activities are usually performed during scheduled rounds when a technician is specifically performing vessel equipment checks. Many of the activities listed below, however, should be part of a technician's regular routine when traveling throughout the facility to perform other tasks. Technicians must always remember to look, listen, check, and smell for the items listed in Table 15.2. Failure to conduct proper monitoring can affect the process and result in injuries, equipment damage, or an environmental incident.

Table 15.2 Process Technician's Role in Operation and Maintenance

Look	Listen	Check	Smell
■ Monitor levels ■ Look carefully at firewalls, sumps, and drains ■ Inspect auxiliary equipment associated with the vessel ■ Visually inspect for leaks (especially if associated with abnormal odor) ■ Use level gauges and sight glasses to monitor level ■ Monitor level and pressure ■ Inspect for corrosion	■ Listen for abnormal noise	■ Check for abnormal heat on vessels and piping ■ Check for excessive vibration on pumps and mixers	■ Be aware of abnormal odors that could indicate leakage ■ Use gas-testing equipment to detect gas leaks and vapors

Special Procedures

Vessels must be taken out of service to prepare for turnarounds or large-scale maintenance projects. To prepare a vessel for maintenance, the vessel must be isolated from all types of hazardous energy, including electrical, mechanical, hydraulic, pneumatic, chemical, and thermal. Process technicians are often involved in the isolation of vessels and use their company's lockout/tagout procedures to perform this isolation.

After vessels are removed from service and isolated, vessel entry is often necessary to perform various maintenance tasks or repairs. Process technicians should follow their company's confined-space entry procedures to enter and work inside vessels safely. Process technicians must understand the hazards of confined-space entry. For example, looking inside a manway on a vessel might not appear to be very hazardous, but it could be deadly if it is in a low-oxygen environment.

VESSEL SYMBOLS   To locate vessels accurately on a Piping and Instrumentation Diagram (P&ID), process technicians should be familiar with the symbols that represent different types of vessels.

Figure 15.41 shows some examples of vessel symbols. It is important to note, however, that the appearance of these symbols can vary from facility to facility.

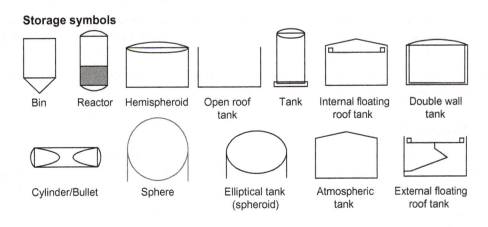

Storage symbols

Bin   Reactor   Hemispheroid   Open roof tank   Tank   Internal floating roof tank   Double wall tank

Cylinder/Bullet   Sphere   Elliptical tank (spheroid)   Atmospheric tank   External floating roof tank

Figure 15.41 Vessel symbols.

15.7 Typical Procedures

Startup, shutdown, lockout/tagout (a procedure used in industry to isolate energy sources from a piece of equipment), and emergency procedures vary, depending on the specific process facility. Each process technician is required to be familiar with his or her site-specific procedures. Discussed below, however, are a few typical procedure examples.

Startup

The following are common steps that must be followed when returning a storage tank to service. They can vary from facility to facility, however, so process technicians should always follow procedures that are specific for their unit and/or facility.

- Inspect the interior of the tank for cleanliness.
- Inspect for leaks by hydrotesting the tank.
- Check the gauging system.
- Connect swing line cables (if applicable).
- Check to see that all manways are in place.
- Remove blinds and check to see that flanges and valves are bolted properly.
- Close and seal water draw valves (if applicable).
- Ensure that all necessary walkways are in place and free of debris.
- Open roof drains on external floating roof tanks.
- Check to see that mixers are in place (if applicable).
- Ensure that valves associated with venting and vacuum protection devices are properly aligned.

The following startup activities are initiated after the checklist above is completed.

1. Slowly begin product flow into the tank by gradually opening the appropriate tank valve.
2. Observe the tank at the prescribed times listed on the Returning Tank to Service Form to ensure that no leaks are present.

Shutdown

Six common steps must be performed during shutdown.

1. Confirm that all of the liquid has been removed from the vessel.
2. Refer to site-specific procedures if personnel will not be entering the vessel.
3. Remove all of the harmful vapors from the vessel if personnel will be entering it (refer to site-specific procedures as required).
4. Open the manway hatches and place an electrically grounded air mover in one of the hatches to remove harmful vapors if a person is going to enter the vessel.
5. Monitor the lower explosive limit (LEL), the minimum percentage of fuel required for combustion. Do not let any personnel inside the vessel until an acceptable LEL reading is obtained.
6. Do not allow anyone to enter the vessel until all the above have been completed and a trained and certified attendant is present.

Emergency

All process facilities have specific emergency procedures. Process technicians are responsible for knowing and being able to execute emergency procedures for their area of responsibility.

Summary

Vessels are used to store raw materials and additives, intermediate products, final products, and waste materials. Vessels include low pressure tanks, drums, bins, and hoppers; high pressure columns, towers, reactors, and cylinders. Underground storage tanks are used by the process industries to store materials beneath the surface of the soil. Tank trucks, tank cars, and hopper cars move products on land, while barges and ships transport products across water.

Process technicians must recognize and understand vessel components, including floating roofs, articulated drains, foam chambers, sumps, mixers, manways, vapor recovery systems, vortex breakers, baffles, weirs, boots, and mist eliminators.

Process technicians must operate vessels within their design ranges for pressure, temperature, flow, level, and material composition. Operating outside the normal range for any variable could result in injury to personnel, damage to the vessel, loss of production, and damage to the environment.

Proper operations can avoid spills, fires, environmental contamination, and injury.

Most vessels have a variety of auxiliary systems (e.g., fire protection and secondary containment) incorporated into their designs to enhance safe operation and prevent environmental incidents such as leaks and spills. Pressure relief devices must be properly aligned when a vessel is in service. A common problem with vessels is pulling a vacuum when draining the vessel. Vacuum can cause a vessel to collapse.

Process technicians should follow all vessel operating procedures. Most vessels require a gradual increase in pressure, temperature, and flow to avoid damaging the vessel or the vessel internals. Vessels must be inspected on a regular basis to ensure their integrity. Process technicians must always look, listen, check, and smell when monitoring and maintaining vessels and take appropriate action when something does not seem right. They also must be aware of the hazards of improper vessel operations.

Checking Your Knowledge

1. Define the following terms:

a. Articulated drain

b. Baffle

c. Blanketing

d. Boot

e. Containment wall

f. Floating roof

g. Gauge hatch

h. Manway

i. Mist eliminator

j. Sump

k. Vortex breaker

l. Weir

2. Name four factors that have an impact on decisions about vessel design and construction.

3. Which of the following vessels would be the most appropriate choice for storing volatile substances under high pressure?

a. Spherical

b. Bullet

c. Atmospheric

d. Hemispheroid

4. What is used for bulk storage or transport of dry materials?

a. Weirs

b. Dryers

c. Hoppers

d. Downcomers

5. Which of the following types of storage containers is used to transport liquid across roadways?

a. Tank car

b. Tank truck

c. Container ship

d. Tanker

6. A _____ is a metal plate or similar device placed inside a vessel to prevent a cyclone-like rotation from being created as the fluid is drawn out of the vessel.

a. sump

b. manway

c. mist eliminator

d. vortex breaker

7. List three differences between safety valves and relief valves.

8. (True or False) When putting a vessel into service, the pressure should be increased quickly to operating pressure.

9. Which of the following can be a direct result of rapid temperature changes? (Select all that apply.)

   a. Cracked welds

   b. Cross-contamination

   c. Damage to vessels

   d. Damage to internal components

10. (True or False) Accurate vessel wall thickness measurements can be made while the vessel is empty and clean.

11. Which of the following accelerates corrosion rates and damages vessels by overheating?

   a. Overpressurization

   b. Underpressurization

   c. A vacuum within the vessel

   d. Changes in operating temperatures

12. (True or False) Overpressurization can cause a rupture of the vessel.

13. List three details that a process technician should look for during normal vessel monitoring and maintenance.

14. What type of vessel does the following symbol represent?

   a. Open roof tank

   b. Double wall tank

   c. Internal floating tank

   d. External floating tank

15. Match the following steps with the correct procedure. Insert either "a" for startup or "b" for shutdown.

I.	Inspect for leaks by hydrotesting the tank.	
II.	Close and seal water draw valves.	
III.	Confirm that all of the liquid has been removed from the vessel.	
IV.	Open roof drains on external floating roof tanks.	

NOTE: Answers to Checking Your Knowledge questions are in the Appendix.

Student Activities

1. On the cross-section of the storage tank below, identify the labeled components.

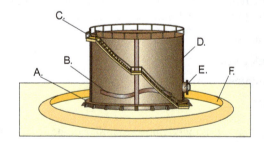

2. Research and write a one- to two-page paper about a tank implosion or explosion.

3. On a piping and instrumentation diagram (P&ID), identify all of the vessels.

Chapter 16
Reactors

Objectives

After completing this chapter, you will be able to:

16.1 Identify the purpose, common types, and applications of reactors. (NAPTA Reactors 1, 2*) p. 338

16.2 Identify the components of reactors and the purpose of each. (NAPTA Reactors 3, 4). p. 344

16.3 Identify auxiliary equipment associated with reactors. p. 345

16.4 Explain the operating principles of reactors. (NAPTA Reactors 5) p. 346

16.5 Identify potential problems associated with reactors. (NAPTA Reactors 9) p. 347

16.6 Describe safety and environmental hazards associated with reactors. (NAPTA Reactors 6) p. 350

16.7 Describe the process technician's role in reactor operation and maintenance. (NAPTA Reactors 7, 8) p. 351

*North American Process Technology Alliance (NAPTA) developed curriculum to ensure that Process Technology courses will produce knowledgeable graduates to become entry-level employees in process technology. Objectives from that curriculum are named here in abbreviated form. For example, "(NAPTA Reactors 1, 2)" means that this chapter's objective relates to objectives 1 and 2 of the NAPTA curriculum named Vessels – Part II – Reactors).

Key Terms

Activation energy—the minimum amount of energy that is required for a reaction to take place, **p. 347.**

Autoclave reactor—a vessel in which the reaction occurs at high pressure and temperature, **p. 340.**

Batch reaction—a carefully measured and controlled process in which raw materials (reactants) are added together to create a reaction that makes a single quantity (batch) of final product, **p. 338.**

Catalyst—a substance used to facilitate the rate of a chemical reaction without being consumed in the reaction, **p. 340.**

Continuous reaction—a chemical process in which raw materials (reactants) are continuously fed into a reactor vessel and products are continuously being formed and removed from the reactor vessel, **p. 338.**

Endothermic—a chemical reaction that requires the addition or absorption of energy, **p. 346.**

Exothermic—a chemical reaction that releases energy in the form of heat, **p. 346.**

Fixed bed reactor—a reactor vessel in which the catalyst bed is stationary and the reactants are passed over it. In this type of reactor, the catalyst occupies a fixed position and is not designed to leave the reactor with the process, **p. 341.**

Fluidized bed reactor—a reactor that uses high-velocity fluid to suspend or fluidize solid catalyst particles, **p. 342.**

Inhibitor—a substance used to slow or stop a chemical reaction, **p. 349.**

Mixing system—a system consisting of an agitator, circulating pump, gas spargers, and a series of baffles to provide proper mixing of reactants and a catalyst, **p. 345.**

Relief system—a system designed to prevent overpressurization or a vacuum, **p. 345.**

Runaway reaction—an uncontrolled reaction that is accelerated by the heat generated from the reaction. As the reaction rate increases, additional heat is released; the release of heat causes the reaction rate to increase and release more heat, **p. 348.**

Sparger—a pipe with holes drilled along its length; holes allow for the introduction and distribution of a pressurized gas into a liquid, **p. 342.**

Stirred tank reactor—a reactor vessel that contains a mixer or agitator to improve the mixing of reactants, **p. 339.**

Temperature control system—a system that allows the temperature of a reaction to be adjusted and maintained, **p. 345.**

Tubular reactor—a continuous flow vessel in which reactants are converted in relation to their position within the reactor tubes, not influenced by residence time in the reactor, **p. 340.**

Vessel heads—components on the top and bottom of a reactor shell that enclose it, **p. 344.**

16.1 Introduction

Reactors are vessels in which controlled chemical reactions are initiated and take place. In these types of vessels, raw materials enter the reactor as feedstock and are converted by means of a chemical reaction to form the desired chemical or chemical mixture. This reactor effluent, which has properties different from those of the feed streams, can be either a final product or an intermediate stream that requires additional separation or processing to recover the desired products. Reactors rarely generate a product that is free of initial reactants and/or solvents. Therefore, a downstream separation process usually follows a reaction process. Reactors are used in many process industries, including food and beverage, chemical, and pharmaceutical manufacturing to create a wide variety of products, such as gasoline, plastics and beer and wine.

In the process industries, chemical reactions are used to generate new products with more desirable properties. These reactions occur in a reactor and can be categorized as **batch reactions** or **continuous reactions**.

Batch reaction   a carefully measured and controlled process in which raw materials (reactants) are added together to create a reaction that makes a single quantity (batch) of final product.

Continuous reaction   a chemical process in which raw materials (reactants) are continuously fed into a reactor vessel and products are continuously being formed and removed from the reactor vessel.

Did You Know?

Two-part epoxy is a batch reaction process. In this type of reaction, two liquid components (the epoxy and the hardener) are mixed together in predetermined amounts and then allowed to react to form a strong, solid bonding material.

CREDIT: Ajt/Shutterstock.

Types of Reactors

Reactors come in many shapes and sizes, most commonly a spherical or cylindrical vessel, which can be vertically or horizontally situated. The most common types of reactors are shown in Figure 16.1.

Some reactors operate at very high pressures (exceeding 2,000 PSI) and temperatures (more than 800 degrees Fahrenheit/427 degrees Celsius). Because of these extreme operating conditions, the design, construction materials, and construction methods of reactor vessels vary greatly.

Stirred Tank Reactor

A **stirred tank reactor** contains one or more mixers or agitators mounted on a shaft inside the tank. Depending on the process and the design of the reactor, this type of reactor can be heated or cooled by an internal or external coil, an external jacket, an external fired heater, or an external heat exchanger. It also can use external vapor phase condensation for cooling. Stirred tank reactors can be used for both batch and continuous reaction processes. When

Stirred tank reactor a reactor vessel that contains a mixer or agitator to improve the mixing of reactants.

Figure 16.1 **A.** Continuous stirred tank reactor. **B.** Tubular reactor. **C.** Fixed bed reactor. **D.** Fluidized bed reactor.

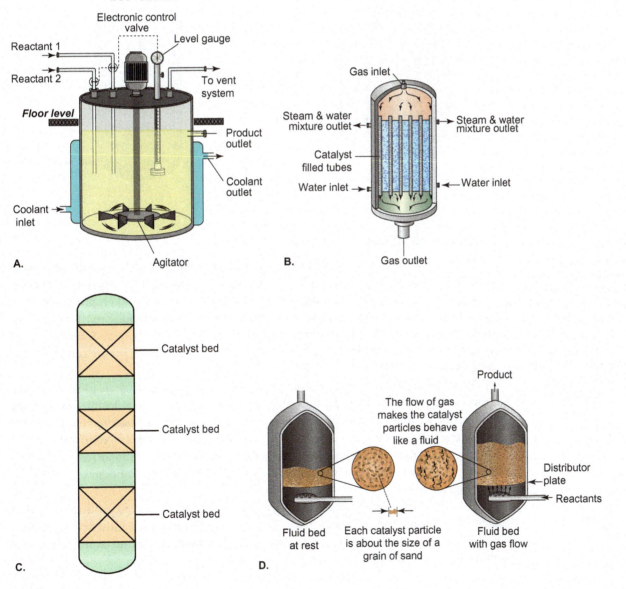

they operate at higher pressures, they are sometimes called **autoclave reactors**. Figure 16.2 shows examples of stirred tank reactors.

Figure 16.2 Stirred tank reactor with: **A.** External heat exchanger. **B.** External coil. **C.** External jacket. **D.** Internal coil.

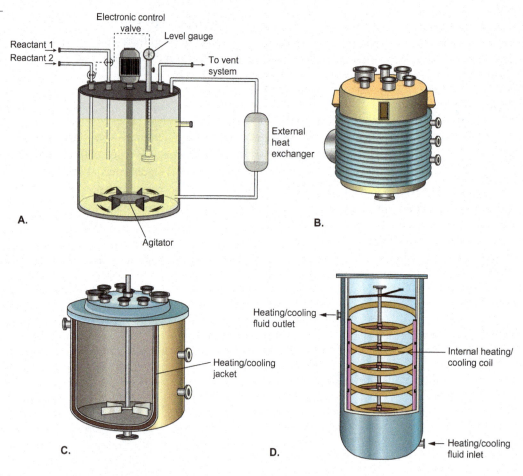

Tubular Reactor

A **tubular reactor** is composed of one or more pipes or tubes in which reactants flow and a chemical reaction takes place. Tubular reactors may be required if the reaction must occur quickly. Figure 16.3 shows an example of a loop tubular reactor. Based on process requirements (e.g., volume, surface area, and pressure), the design of a tubular reactor can range from a simple jacketed tube to a multipass shell and tube arrangement. **Catalyst** can be packed into the tubes to facilitate the reaction process. Tubular reactions are categorized as continuous reactions.

Flow through some tubular reactors is described as *plug flow*, an ideal condition meaning that all of the material flows through the reactor at high velocities and has the same residence time (time in reactor). This ensures that all product leaving the reactor has undergone the same reaction conditions. Plug flow reduces the chance of unwanted side reactions and improves product yield. Because of the design of a tubular reactor, a larger amount of surface area per unit volume is available for heat transfer than is available in other reactor types.

Tubular reactors are widely used by chemical manufacturers to produce materials such as polyester and ammonia. They are also used by the food industry in production of high fructose corn syrup, in wastewater treatment, and in the manufacturing of pharmaceutical products.

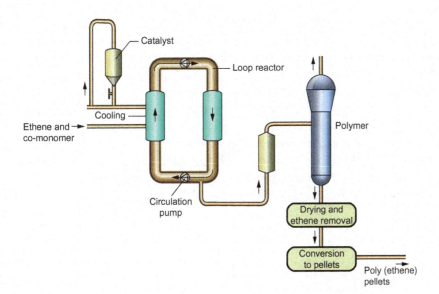

Figure 16.3 Loop-type tubular reactor.

Another type of tubular reactor is simply tubes inside a furnace. The tubes might or might not contain catalyst. Examples of these reactors are steam reformers, used to produce hydrogen and cracking furnaces.

Fixed Bed Reactor

A **fixed bed reactor** is a reactor in which the catalyst bed is stationary. A chemical reaction occurs as the reactants are passed through the catalyst. In this type of reactor, the catalyst occupies a fixed position and is not designed to leave the reactor with the process. Figure 16.4 shows an example of a fixed bed reactor.

Fixed bed reactors contain one or more catalyst beds, a catalyst support system, and inlet and outlet nozzles. The catalyst support system keeps the catalyst in place during operation. This system contains a screen or plates with holes that allow liquid or gas to flow through the reactor.

Fixed bed reactor   a reactor vessel in which the catalyst bed is stationary and the reactants are passed over it. In this type of reactor, the catalyst occupies a fixed position and is not designed to leave the reactor with the process.

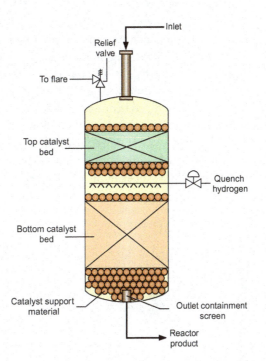

Figure 16.4 Fixed bed reactor with multiple beds.

A fixed bed that does not contain catalyst is called a *packing bed*. Its purpose is to create the turbulence necessary for good mixing of the reactants and to provide additional surface area, improving heat transfer, and increasing reaction rate. In a fixed bed reactor, products are formed when the reactant flows into the reactor and over the catalyst. The product then flows out of the reactor as finished product or as an intermediate stream that requires further processing. Proper contact between the reactants and the catalyst is critical. Uniform distribution of the reactants across the catalyst bed is accomplished by a distribution device such as a tray, spray nozzle, **sparger**, or gas diffuser.

Figure 16.5A shows the top side of a tubular distribution device, with the inset showing the underside. Figure 16.5B shows how the underside spray nozzles of a distribution device function to achieve even distribution of material.

Sparger   a pipe with holes drilled along its length; holes allow for the introduction and distribution of a pressurized gas into a liquid.

Figure 16.5 A. Tubular distribution device with outlet holes on the underside. **B.** Spray nozzles for the distribution of liquid.

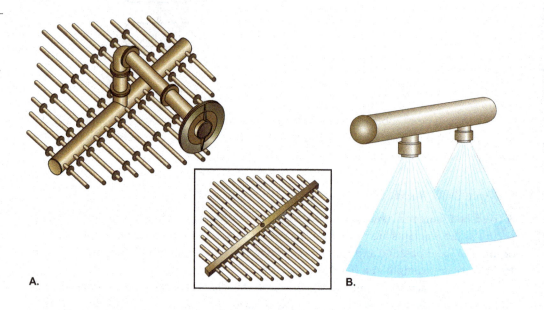

A.

B.

Fluidized Bed Reactor

A **fluidized bed reactor** (see Figure 16.1D) is a reactor that uses high-velocity fluid (gas or liquid) to suspend or fluidize solid catalyst particles. After the catalyst has been used, it is sent to a regenerator to be regenerated (cleaned) and returned to the reactor. Equipment abrasion, particle agglomeration, particle fracturing, particle size distribution, particle density, gas velocity, gas density, and pressure drop are all issues that must be addressed when operating a fluidized bed. Most catalysts foul over time or when a system is operated improperly. Fluidized bed reactors allow for easy regeneration of the catalyst because it can be taken out, cleaned, and returned to the reactor.

An example of a fluidized bed reactor is the fluid catalytic cracking unit (FCCU), which is used to "crack" (break down) heavy gas oils into gasoline and diesel fuel in a refinery. Figure 16.6 shows an example of a fluidized catalytic cracking unit.

Fluidized bed reactor   a reactor that uses high-velocity fluid to suspend or fluidize solid catalyst particles.

Did You Know?

The catalytic converter on a car is a fixed bed reactor. Catalytic converters are used to remove harmful pollutants from engine exhaust. A car's catalytic converter uses metals, like platinum or rhodium in the catalyst. Many industrial processes use these same metals on the active catalyst sites.

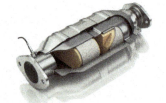

CREDIT: Slavoljub Pantelic/ Shutterstock.

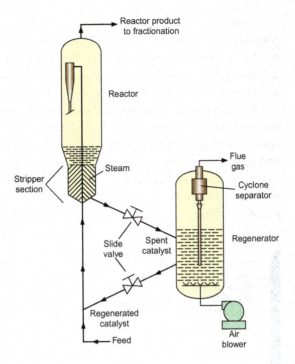

Figure 16.6 Fluid catalytic cracking unit (FCCU).

Nuclear Reactor

Nuclear reactors are high-temperature and high-pressure reactors used to generate the steam necessary to produce electricity. Instead of a chemical reaction (e.g., burning coal), a nuclear reaction (splitting uranium atoms through fission) is used to produce heat. The reactor, its components, and the water pumped through the reactor control the nuclear chain reaction and carry away the heat generated. Figure 16.7 is an example of a nuclear reactor.

Did You Know?

In 2016, the United States had 99 commercial nuclear reactors operating at 61 nuclear power plants in 30 states. These plants generated close to 20 percent of the total US electricity demand

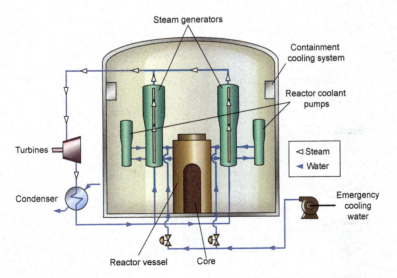

Figure 16.7 Example of a nuclear reactor.

When uranium atoms are split, they release a tremendous amount of heat. Water surrounding the fuel rods absorbs this heat. For this reason, the water is referred to as the reactor coolant. Commercial nuclear reactors in the United States can be divided into two groups: boiling water reactors and pressurized water reactors. The classification of the reactor depends on whether the reactor coolant is used directly or indirectly.

In boiling water reactors, the heat from the nuclear fission is used to heat the reactor coolant water and boil it into steam so it can be used to drive a steam turbine. In pressurized water reactors, the water in the reactor is kept at a higher pressure so it remains a liquid and does not boil. The water then is pumped through heat exchangers, where it is used to boil water. Because they produce steam, these heat exchangers are often referred to as *steam generators*. This secondary system fluid (the water boiled into steam) then is used to drive a steam turbine.

Did You Know?

Nuclear power plant employees typically receive less radiation exposure in their jobs than the amount people receive from everyday sources, such as cosmic rays from the sun, medical X-rays, and radon.

CREDIT: Kaspri/ Shutterstock.

The fuel rods in a nuclear reactor typically last from 18 to 24 months. After that, about one-third of the fuel rods are replaced with fresh fuel rods and the remaining rods are rearranged so that they are consumed more evenly. Spent nuclear fuel is very radioactive, so it must be shielded at all times.

16.2 Components of Reactors

While different reactor types consist of various components, some of the main components include a shell, vessel heads, agitator, baffles, tank, mixing system, heating and cooling system, and a relief system. Figure 16.8 shows an example of a reactor and some of its components.

The shell is the outer surface, casing, or external covering of a vessel, consisting of side walls and vessel heads. The shell consists of various types of materials that protect it from chemical effects, depending on the chemical reaction taking place inside the equipment. The side walls of a reactor vessel contain the pressure inside the reactor. The **vessel heads** are the

Vessel heads components on the top and bottom of a reactor shell that enclose it.

Figure 16.8 Reactor components.

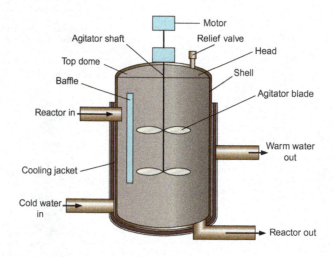

components at the top and bottom of the shell that enclose it. Nozzles welded onto the shell and heads are used for a variety of purposes, such as attaching piping for flow into and out of the reactor, providing a location for instrumentation connections, and connecting auxiliary equipment (pressure/vacuum relief devices, manways, and agitators).

The agitator of a stirred tank reactor typically consists of a motor, shaft, and blades. The agitator shaft is a cylindrical rod that holds the agitator blades. These blades stir the chemicals in the reactor. Agitator blades are similar to those in a kitchen blender. They act on the reactor fluid by generating turbulence, which promotes mixing. In some reactors, two or more sets of agitator blades are used to increase the amount of mixing. Because turbulence is required for proper mixing, baffles are installed. A baffle is a stationary plate, mounted on the wall of the tank or vessel, which is used to alter the flow of a reactor's contents and facilitate mixing. Baffles are typically vertical or spiral strips or horizontal rings fixed to the reactor walls.

Table 16.1 contains an overview of common reactor types and their components.

Table 16.1 Reactor Types and Their Common Components

Reactor Type	Example	Common Components
Stirred tank reactor	Sulfuric acid alkylation reactor	Vessel, agitator, internal baffles, heating/cooling system, catalyst mix tank, and relief system
Tubular reactor	Steam methane reformer reactor	Relief system, heating/cooling jacket, gas quench system
Fixed bed reactor	Hydrotreating reactor	Inlet distributor, catalyst support material, gas quench system, catalyst support tray, multiple beds, inner screen and outer screen, and relief system
Fluidized bed reactor	Fluid catalytic cracking reactor	Relief system, baffles, cyclone separators, risers, standpipes, reaction chamber, plenum chamber, slide valves, and stripping section

Reactor Systems

The **mixing system** includes components such as an agitator, circulating pump, gas spargers, and a series of baffles to provide proper mixing of reactants and catalyst. Because mixing is critical in most reactions, instrumentation (e.g., motion indicators and process interlocks) can be incorporated to ensure proper operation.

A **temperature control system** allows the temperature of a reaction to be adjusted and maintained. Simple systems can contain a cooling or heating jacket or coil. More complex systems contain a circulation system with heat exchangers. Depending on the requirements, some systems might contain both jackets/coils and circulation systems. The temperature control system is especially critical in exothermic reactions. Because of this, instrumentation (e.g., process interlocks) is usually included to ensure proper operation.

A **relief system** is designed to prevent overpressurization. It includes a piping system that collects the discharge of the pressure relief valve and routes it to the atmosphere, to a flare, or to another containment system.

Many reactor systems also require filtration or decanting systems downstream of the reactor to recover catalyst or product or to remove unwanted reaction byproducts.

Mixing system a system consisting of an agitator, circulating pump, gas spargers, and a series of baffles to provide proper mixing of reactants and a catalyst.

Temperature control system a system that allows the temperature of a reaction to be adjusted and maintained.

Relief system a system designed to prevent overpressurization or a vacuum.

16.3 Auxiliary Equipment Associated with Reactors

Auxiliary equipment associated with reactors can include internal heating and cooling coils, heating and cooling jackets, shell and tube heat exchangers, agitators, various types of internal lining, static mixer elements, continuous circulating water systems, pressure/vacuum relief systems, fire and explosion suppression systems, programmable logic and digital control systems, and temperature control systems. Table 16.2 lists examples of various types of auxiliary equipment and their benefits.

Table 16.2 Auxiliary Equipment Associated with Reactors and Their Related Benefits

Auxiliary Equipment	Benefits
Internal heating and cooling coils or external circulating heating and cooling systems	Used to heat and cool the material in a reactor
Reactor shell jacket	Provides a shell through which heating or cooling media passes
Corrosion-resistant liners and construction materials (e.g., glass, ceramics, polymers, precious metals, or alloys)	Prevent the corrosive properties of the reactants and products from damaging the reactor shell
Static mixer elements in tubular or inline reactors	Increase mixing and turbulence. Without adequate mixing, complete reaction might not occur. Also, without adequate mixing, "hot spots" can occur, leading to product deterioration and/or poor yields, and they could initiate a runaway reaction
Continuously circulating water, hot oil, or other fluid systems in jackets and internal and external heat exchangers	Allow for heating and cooling of the reaction materials
Temperature control systems	Control the temperature in the reactor by regulating the flow rate of the heating/cooling fluid to the reactor heating/cooling system
Pressure or vacuum control and relief systems (including rupture discs)	Prevent overpressurization or a vacuum that could result in the catastrophic failure of the reactor
Fire and explosion suppression systems (e.g., sprinklers, explosion barriers, reaction quench systems, inert gas, or foam blanketing)	Protect personnel and equipment and control runaway reactions or other abnormal conditions such as solvent or reactant fires
Programmable logic and digital control systems	Manage reaction steps and ensure correct conditions for continuous reactions

16.4 Operating Principles of Reactors

Reactor systems are the heart of many process operations. A chemical reaction is a process in which a *chemical change*, rather than a physical one, occurs to the substances fed into a reactor, called reactants. This transformation is accomplished by the breaking apart of chemical bonds between the atoms in molecules of the reactants and the creation of new bonds to form products with properties different from those of the reactants.

Reactors provide the conditions necessary for controlled reactions to occur efficiently and safely. Reactions can require agitation for mixing or solids suspension, heat addition or removal to control temperature, gas addition or removal to control pressure, or analyzers to provide information on composition. The reaction rate (the rate at which reactants are converted to products) generally increases as the temperature increases. To achieve the desired conversion of reactants to products, sufficient residence time must also be provided.

Chemical reactions can either be endothermic or exothermic. **Endothermic** refers to reactions that absorb or require heat. **Exothermic** describes reactions that release energy in the form of heat.

Endothermic and exothermic reactions both require energy, usually in the form of heat, for a reaction to begin. In order for new products to be created, existing chemical bonds must be broken so that new bonds can be established. These bonds are broken when collisions

Endothermic a chemical reaction that requires the addition or absorption of energy.

Exothermic a chemical reaction that releases energy in the form of heat.

Did You Know?

A simple way to remember the difference between endothermic and exothermic reactions is to think of **en**dothermic reactions as having heat **en**tering the reactor. **Ex**othermic reactions must have heat **ex**iting the reactor.

- Endo = Enter
- Exo = Exit

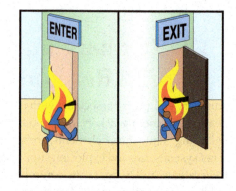

occur between molecules. In order to break the existing bonds, energy must be added for the atoms to overcome the attachments holding them together. This heat energy is called **activation energy**, the minimum amount of energy that is required for a reaction to take place.

Activation energy can be compared to riding a bicycle over a hill (Figure 16.9). Energy must be added by the rider for the bicycle to go uphill. If the rider doesn't pedal fast enough, then the bicycle will not be able to reach the top of the hill and will roll backward to the starting point. When the rider has pedaled fast enough for the bicycle to reach the highest point on the hill and begin the descent, then the energy the rider has stored is used to propel the bicycle down the hill.

Activation energy   the minimum amount of energy that is required for a reaction to take place.

Energy IN

Energy OUT

Figure 16.9 An "energy hill" illustrating the use of activation energy.

CREDIT: udaix/Shutterstock.

One way that activation energy in a chemical reaction can be produced is by adding heat. Increasing the temperature increases the amount of energy in the system, which in turn increases the reaction rate. Some reaction rates are very sensitive to temperature. As little as an 18-degree Fahrenheit (10-degree Celsius) change in reaction temperature can double the reaction rate. There can be disadvantages in increasing the reaction temperature to increase the reaction rate to overcome the "energy hill." For example, the temperature could become excessively high and pose a safety risk. As the temperature increases, the system pressure increases. The rate of unwanted side reactions (a chemical reaction that occurs in addition to the main reaction) could increase, resulting in increased production of waste products. Higher temperatures also can accelerate equipment corrosion and sometimes require the use of more expensive construction materials.

Another way to get over the "energy hill" is to make it less steep and easier to climb. The addition of a catalyst achieves this goal by lowering the amount of activation energy required and allowing reactions to occur more easily. The catalyst increases the reaction rate without increasing the temperature. Catalysts can also be designed to lower the activation energy required for the desired reaction and thus prevent unwanted side reactions.

A third way that the rate of reaction can be increased is by increasing the number of collisions between molecules. The more often these molecules collide, the faster the reaction. In a gas phase reaction, the collision rate can be increased by raising the pressure. In a liquid phase reaction, mixing the reactor contents will increase the collision rate.

In summary, the rate of reaction is a key variable in reactor operation. Process technicians must understand the variables (e.g., temperature, using a catalyst, pressure, concentrations, residence time, and mixing) that are critical for controlling the rate of reaction for the process they are operating.

16.5 Potential Problems

When working with reactors, process technicians should always be aware of potential problems with the reactor and its associated equipment, such as runaway reactions, loss of cooling/heating systems, pressure variations, leaks, loss of mixing system, cyclone separator or sparger failure, and issues with instrumentation systems. Potential problems must be addressed in order to keep equipment functioning properly, ensure personal safety, ensure safety of coworkers and the community, and prevent environmental incidents.

Runaway reaction an uncontrolled reaction that is accelerated by the heat generated from the reaction. As the reaction rate increases, additional heat is released; the release of heat causes the reaction rate to increase and release more heat.

Runaway Reaction

One of the highest risks in the operation of reactors is a **runaway reaction**. A runaway reaction is an uncontrolled reaction in which the reaction rate increases and releases heat, the heat causes an increased reaction rate, and the increased reaction rate releases even more heat. A horrific example of this was the runaway reaction in 1984 at a pesticide plant in Bhopal, India, where more than 40 tons of methyl isocyanate (cyanide gas) was released into the atmosphere. The gas immediately killed more than 3,800 people, and many thousands more died from the effects of the release during the following years.

Loss of Cooling/Heating System

Temperature in a reactor has a direct effect on the reaction rate. It is estimated that for many reactions that occur at room temperature, the reaction rate will double for every 18-degree Fahrenheit (10-degree Celsius) increase in temperature. Excessive heating or failure to remove heat can produce off-specification products that must be reworked, sold at a loss, or sent to waste. In addition, a loss of cooling could result in a rapid temperature increase and might lead to a runaway reaction, which could cause serious injury to personnel and damage to the reactor equipment.

A loss of a reactor's heating system will lead to a decrease in reaction rate and off-spec product. In an endothermic reaction process, excessive cooling or loss of a heating system can cause problems such as crystallization in liquids, higher viscosities, or freezing of polymers.

Pressure Variations

High pressure in a reactor can affect the rate of reaction, especially in reactions involving gases. Increased pressure will compress the gas, leading to an increase in the number and intensity of collisions between atoms, which will increase the rate of reaction. High pressure will lead to activation of the pressure relief system equipment and could result in equipment damage or vessel failure. In reactors where it is possible to achieve a vacuum condition, a vacuum relief system also is required to prevent damage to the reactor, its associated piping, and auxiliary equipment.

Leaks

Leaks can lead to exposure to harmful or hazardous substances, fires, injury to personnel, or equipment damage. Depending on the compatibility of reactor contents and equipment, corrosion and equipment degradation can result.

Loss of Mixing System

Loss of a mixing system component will have a negative effect on reactor product quality and yield. Failure of one of the components of the reactor agitator (coupling, motor, blades, or shaft) or a problem with a catalyst mixing system can result in poorly mixed reactor contents and off-spec product or reduced yield.

Cyclone Separator Failure

In a fluidized bed reactor system, cyclone separators in the reactor are used to separate spent catalyst from cracked hydrocarbon vapor before the vapor goes to the fractionator. In a regenerator, cyclone separators are used to remove entrained catalyst from the exiting flue gas. A failure of a cyclone separator could lead to a loss of catalyst, resulting in higher catalyst replacement cost, and can result in a unit shutdown and great economic impact. Another result of regenerator cyclone failure could be environmental impact or exceeding environmental limits on stack emissions.

Sparger Failure

Spargers are used to introduce and distribute a gas stream into a liquid in a reactor. Sparger failure can lead to an uneven distribution of material in the reactor, and channeling can occur, resulting in localized temperature increases. Off-spec product and reactor hot spots can result and lead to reactor failure.

Instrumentation Issues

Because reactor systems operate at critical flow rates, temperatures, and pressures, the consequences associated with instrumentation failures can have a significant impact on both the quality and quantity of the product, as well as equipment reliability and plant operating costs. The rate of reactants and catalyst entering the reactor must be controlled in order for the desired reaction to occur. Without reliable temperature control, a reactor could exceed design conditions, leading to equipment failure. Most critical instruments are redundant, which means there is a backup instrument that can be checked to verify accuracy of the primary reading or to provide information should the primary instrument fail.

Although there are a variety of different types of problems that could be encountered with the operation of reactor equipment, Table 16.3 lists some of the potential problems

Inhibitor a substance used to slow or stop a chemical reaction.

Table 16.3 Potential Problems Associated with Reactor Equipment Operation

Problem	Cause	Solution
Runaway reaction	■ Inadequate heat removal (failure of cooling system), inadequate agitation, excessive amount of catalyst, inadequate amount of **inhibitors** ■ Excessive gas generation resulting from above ■ Contaminated or wrong reactants	■ Maximize heat removal ■ Reduce pressure to a minimum ■ Stop all feed streams ■ Add inhibitors if possible
Agitator or agitation failure	■ Motor trip out ■ Coupling failure ■ Shaft failure ■ Agitator blade failure ■ Excessive reaction mass viscosity	■ Reset breaker on motor starter ■ Repair broken coupling ■ Address reaction problem ■ Replace agitator
Leaks	■ Corrosion ■ Overpressure	■ Reduce pressure ■ Repair leak
Lining failure	■ Corrosion ■ Poor installation ■ Glass liner breakage	■ Repair or replace liner
Incomplete reactions, product deterioration, or undesirable byproducts	■ Thermocouple or temperature-sensor failure ■ Improper reaction conditions ■ Sparger not functioning properly	■ Repair or replace thermocouple or sensor ■ Repair sparger ■ Return conditions to normal operating status ■ Confirm proper instrument operation and repair or recalibrate if required
Product yield low/off-spec product	■ Inadequate mixing ■ Catalyst addition rate low ■ Deterioration of catalyst ■ Sparger not functioning properly	■ Correct agitation ■ Increase catalyst addition rate ■ Replace catalyst ■ Correct sparger operation
Temperature increasing or decreasing abnormally	■ Inadequate heat addition/removal (heating/cooling system malfunctioning) ■ Improper agitation ■ Temperature instrumentation failure	■ Correct problem with heat addition/removal system ■ Correct agitation ■ Repair instrumentation
Competing reactions	■ Temperature, pressure, and reactant concentrations are incorrect ■ Jacket or coil heating/cooling temperatures are incorrect ■ Feed, catalyst, or solvent (reaction media) pump failure ■ Feed ratio controller system failure	■ Bring temperature and pressure back into control ■ Repair cooling/heating system ■ Restart pump ■ Restart ratio control system ■ Re-establish proper quantities

associated with reactors, as well as possible causes and solutions for each. It is important to note that with each potential scenario, process technicians need to know the standard operating procedures for their process units.

16.6 Safety and Environmental Hazards

Significant hazards, which can affect personnel safety, equipment, and the environment, are associated with both normal and abnormal reactor operations. Reactor operation under normal conditions can be hazardous because of the presence of high-pressure, high-temperature, and/or hazardous materials. During process upsets and emergencies, additional hazards are present, such as an increased possibility of contact between the chemicals involved in the process and the skin and eyes of nearby workers.

Feed and product streams can be flammable or toxic substances, which can present a risk of fire because of the flammability or reactivity of these materials. Certain types of catalyst can cause irritation or burns if skin contact occurs. In some processes, a powdered form of catalyst must be added to a reactor. Care should be taken to avoid inhalation of the dust associated with this task. Some catalyst also presents the potential for dust explosions caused by static electricity created when materials are loaded into a reactor.

Some reactors use a nitrogen blanket to control pressure. This adds a number of potential hazards that the process technician must know and understand, such as oxygen deficiency. Nitrogen also can be used in reactors as an inhibitor, to control temperature by deactivating catalyst and slowing the rate of a reaction.

When some reactors are shut down for catalyst change-out, they are maintained under a nitrogen blanket to prevent fires that can occur as a result of pyrophoric materials deposited on the surface of the catalyst. When exposed to air, pyrophoric materials can ignite spontaneously and create a fire hazard as well as an environmental hazard because of the release of sulfur dioxide. The pyrophoric materials also can create enough heat to ignite any remaining hydrocarbons in the reactor. Process technicians must be aware of these potential hazards. They should observe safety zones established around such "inerted" vessels to prevent exposure to an oxygen-deficient environment.

Personal protective equipment (PPE) should be worn during normal and abnormal reactor operations. In addition to basic PPE (gloves, safety shoes and glasses, hardhat, hearing protection, and flame-resistant clothing), additional PPE can be required, including:

- Goggles when sampling or working with catalyst
- Breathing protection if toxic gases or a nitrogen-rich environment might be present or when working with certain catalyst types.

Environmental effects are also important hazards to keep in mind. Many chemicals are toxic and harmful to the environment if leaked to the atmosphere. Fires can release harmful compounds such as sulfur dioxide into the atmosphere. Because of this, regulatory standards limit the amount of specific hazardous materials that can be released. Exceeding the regulated amounts requires reporting, and fines are possible if reporting and regulations are not followed.

Table 16.4 lists some of the potential hazards to personnel safety, equipment, and the environment associated with normal and abnormal operation of equipment in reactor systems.

Table 16.4 Hazards Associated with Reactors

Personnel Safety Hazards	Equipment	Environment
Exposure to hazardous/toxic feed/product material	Leaks because of cycling pressures and temperatures (startup/shutdown)	Discharges to ground caused by leaks or spills
Rotating equipment hazards with agitators, feed hoppers, etc.	Fires associated with flammability or reactivity of reactor material	Particulate matter (catalyst) released into air
Skin contact hazards with some catalyst types (acids)	Dust explosion hazard because of static electricity created when materials are loaded	Release of hazardous/toxic vapors into air
Burns associated with hot equipment	Erosion of equipment caused by catalyst abrasion	Spill of liquid catalyst
Slips and falls resulting from leaks or spills	Hot equipment surfaces	Release of toxic substances into air in the event of fire
Inhalation hazard associated with some catalyst types (powdered)		
High pressure hazards		

16.7 Process Technician's Role in Operation and Maintenance

Process technicians play an important role in the safe operation and maintenance of reactors (see Table 16.5). It is especially important that operators have a thorough understanding of the materials with which they are working, the chemistry of the reactions taking place, and the reactivity potential and hazards associated with both feed and catalysts.

Through frequent monitoring, process technicians can ensure that the equipment is operating safely and properly. During this monitoring, technicians must be alert to unusual sounds that can signal reactor problems. Process technicians should check for leaks from reaction system equipment on a frequent basis.

Process technicians also must monitor outside instrumentation on the reactor on a regular basis and as requested by the process board operator. Process board technicians also must monitor all controllers to ensure that the process variables are being held at the desired set points and that all alarm conditions have been properly addressed.

Process technicians should be familiar with the appropriate procedures for monitoring and maintaining reactor components and conditions. The typical procedures (including startup, shutdown, emergency, and lockout/tagout) vary from site to site and for each type of reactor. Because of this, process technicians need to follow the standard operating procedures (SOPs) for their assigned unit(s). These procedures involve multiple steps that must be carried out in the specific order they are listed.

Table 16.5 Process Technician's Role in Reactor Operation and Maintenance

Look	Listen	Check
■ Confirm that agitator is operating properly ■ Monitor proper operating temperatures, pressures, flows, and levels ■ Confirm proper operation of heating/cooling system equipment ■ Ensure that safety systems are in good operating condition ■ Look for hot spots on reactor shell	■ Listen for abnormal noises	■ Check for excessive equipment vibration ■ Check for excessive heat on motorized equipment ■ Examine equipment to make sure there are no leaks ■ Check quality of reactor products

Summary

Reactors are specialized vessels that are used to contain a controlled chemical reaction and convert raw materials into finished products, which have characteristics different from the raw materials. Reactor designs vary widely because of the wide range of operating conditions and reactions that take place.

A stirred tank reactor contains a mixer or agitator mounted inside the tank that is used to mix products. This type of reactor also is referred to as an autoclave reactor if it operates at high temperatures and pressures.

A tubular reactor has a shape similar to a tubular heat exchanger. Many tubular reactors contain a section of pipe designed with a shell, baffles, jackets, and inlet and outlet ports to mix two fluids together to create a chemical reaction.

A fixed bed reactor contains a stationary bed of catalyst over which liquid or gas flows to create the desired product. A fluidized bed reactor is a reactor in which a fine catalyst is fluidized and contacted with the reactants.

Nuclear reactors are used within the power industry and can be divided into boiling water reactors and pressurized water reactors. In this type of reactor, water is pumped around fuel rods to remove the heat of the nuclear reaction and ultimately form steam. In both cases, steam turbines drive large electrical generators, which produce electricity for many consumers.

Chemical reactions are either endothermic or exothermic. Endothermic reactions require heat to occur, and exothermic reactions generate heat. Reactors are designed to accommodate the conditions needed for either of these reaction types.

Reactor components include equipment such as the reactor vessel (shell and heads), agitator, and baffles. Important systems that support the operation of a reactor include the mixing system, relief system, catalyst recovery system, and heating/cooling system.

Potential problems associated with reactor equipment include runaway reactions, malfunction of cooling/heating systems, overpressurization, leaks, failure of mixing system components, cyclone separator or sparger issues, and instrumentation failures.

Some of the safety and environmental hazards that must be considered with reactors are the potential hazardous nature of the feed and product streams, inhalation and skin contact hazards associated with certain catalyst types, risk of fires because of the flammability or reactivity of the materials involved in the reaction process, catalyst and hazardous vapor releases to the atmosphere, the presence of a nitrogen-rich environment, and the potential for dust explosions caused by static electricity created when materials are loaded into a reactor.

Process technicians need to have a good understanding of the materials with which they are working, including chemistry and reactivity information. Technicians are responsible for knowing and following standard operating procedures associated with the reactors on their units, including startup, shutdown, and emergency procedures. Technicians also must understand the importance of making regular unit rounds to inspect equipment and ensure that it is operating within acceptable operating limits.

Checking Your Knowledge

1. Define the following terms:
 a. Catalyst
 b. Endothermic
 c. Exothermic
 d. Inhibitor
 e. Runaway reaction

2. _____ are installed in a reactor to provide turbulence for proper mixing.

3. What reactor type is also called an autoclave reactor?
 a. Stirred tank reactor
 b. Fixed bed reactor
 c. Tubular reactor
 d. Fluidized bed reactor

4. Which of these problems creates the greatest risk in the operation of reactors?
 a. Leaks
 b. Runaway reactions
 c. Loss of cooling system
 d. Loss of heating system

5. What type of reactors are widely used by chemical manufacturers to produce materials like polyester and ammonia?

   a. Stirred tank reactors

   b. Fixed bed reactors

   c. Tubular reactors

   d. Fluidized bed reactors

6. What are the types of nuclear reactors generating electricity in the United States? (Select all that apply.)

   a. Boiling water reactors

   b. Fixed bed reactors

   c. Pressurized water reactors

   d. Tubular reactors

7. What effect will a failure of the agitator have on the operation of a reactor? (Select all that apply.)

   a. Reduced yield

   b. Off-spec product

   c. Reduced reactor contents

   d. Low temperatures in the reactor

8. Cyclone separators are used in _____ reactors to separate catalyst from gas streams.

   a. tubular

   b. fixed bed

   c. fluidized bed

   d. stirred tank

9. What system is designed to prevent overpressurization?

   a. Mixing system

   b. Vessel system

   c. Relief system

   d. Temperature control system

10. List two ways that activation energy in a chemical reaction can be produced.

11. What is the benefit of static mixer elements in tubular or in-line reactors?

    a. They prevents corrosion.

    b. They increase mixing and turbulence.

    c. They prevent overpressurization.

    d. They protect personnel and equipment.

12. (True or False) Reactor operation under normal conditions can be hazardous due to the presence of high-pressure, high-temperature, and/or hazardous materials.

13. Which of the following is a personnel safety hazard associated with reactors? (Select all that apply.)

    a. Burns

    b. Skin contact with acid

    c. Inhalation hazard

    d. Fire hazard

14. (True or False) Through frequent training, process technicians can ensure that the equipment is operating safely and properly.

NOTE: Answers to Checking Your Knowledge questions are in the Appendix.

Student Activities

1. Write a paper describing stirred tank reactors, tubular reactors, fixed bed reactors, and fluidized bed reactors. In the paper, discuss how each reactor works, its components, the type of process, and examples of when it is used.

2. As a group, research how catalytic cracking systems ("cat crackers") work to produce gasoline.

3. Draw a schematic of a reactor system with associated auxiliary systems attached, and include a description of the process. Present your work to the class.

4. Research additional uses for reactors. Summarize and report your findings to the class.

Chapter 17
Filters and Dryers

Objectives

After completing this chapter, you will be able to:

17.1 Identify the purpose, common types, components, and applications of filters and dryers. (NAPTA Filters 1, 2 and NAPTA Dryers 1, 2, 4-6*) p. 355

17.2 Identify potential problems associated with filters and dryers. (NAPTA Filters 10 and NAPTA Dryers 10) p. 363

17.3 Describe safety and environmental hazards associated with filters and dryers (NAPTA Filters 7 and NAPTA Dryers 7) p. 364

17.4 Describe the process technician's role in filter and dryer operation and maintenance (NAPTA Filters 9 and NAPTA Dryers 9) p. 365

17.5 Identify typical procedures associated with filters and dryer (NAPTA Filters 8 and NAPTA Dryers 8) p. 365

*North American Process Technology Alliance (NAPTA) has developed curriculum to ensure that Process Technology courses will produce knowledgeable graduates to become entry level-employees for the career field of process technology. Objectives from that curriculum are named here in abbreviated form (e.g., NAPTA Filters 10 addresses objective 10 of the NAPTA curriculum about Filters).

Key Terms

Bag filter—a tube-shaped filtration device that uses a porous bag to capture and retain solid particles, **p. 357.**

Breakthrough—the condition of any adsorbent bed or filter medium when the component being adsorbed or filtered starts to appear in the outlet stream, **p. 359.**

Cartridge filter—a filtration unit that uses a fine mesh that is tightly folded or pleated to remove suspended contaminants from a liquid, **p. 356.**

Channeling—formation of a path through a dryer or filter that allows feed to flow through without proper contact with the filtering medium or drying agent, **p. 364.**

Cyclone separator—a mechanical device that uses a swirling (cyclonic) action to separate heavier and lighter components of a gas or liquid stream, **p. 363.**

Desiccant—a specialized substance contained in a dryer that removes hydrates (moisture) from a process stream, **p. 359.**

Filter aid—a substance applied as a slurry to a filtering medium, which coats it and aids in the filter's efficiency, **p. 357.**

Filter cake—layers of solids that are deposited on the surface of a filter medium, **p. 357.**

Filter medium—the material that actually does the filtering in the filter unit, **p. 355.**

Fixed bed dryer—a device that uses a desiccant to remove moisture from a process stream, **p. 359.**

Fixed bed filter—a filter in which the fluid passes over a stationary bed of filtering medium, **p. 358.**

Flash dryer—a device that forces hot gas (usually air or nitrogen) through a vertical or horizontal flash tube causing the moist solids to become suspended, **p. 360.**

Fluid bed dryer—a device that uses hot gases to fluidize solids and promote the contact between dry gases and solids, **p. 360.**

Leaf filter—a type of filter that uses a precoated screen, located inside the filter vessel, to hold the filtered material, **p. 357.**

Micron—a unit of measure equal to 0.000001 meters. Filters are rated based on the size of the particles they are capable of filtering, **p. 356.**

Plate and frame filter—a type of filter used to remove liquids from a slurry feed by using a deep filtration process that includes clarification and prefiltration, **p. 358.**

Pleated cartridge filter—a type of cartridge filter that uses a filter medium in a pleated form to provide more surface area, **p. 356.**

Rotary drum filter—a type of filter that contains a cloth screen on a rotating drum to hold solids being filtered, **p. 357.**

Rotary dryer—a device composed of large, rotating, cylindrical tubes; reduces the moisture content of a material by bringing it into direct contact with a heated gas, **p. 361.**

Slurry—a liquid solution that contains a high concentration of suspended solids, **p. 362.**

Slurry dryer—a device designed to convert wet process slurry into dry powder, **p. 362.**

Spray dryer—a device that sprays a moist feed stream into a vertical drying chamber using a nozzle or rotary wheel, **p. 361.**

17.1 Introduction

Filters are used to remove solid particles, and dryers are used to remove moisture. Many process operations depend on the use of filters and dryers for efficient unit operation and/or to deliver a high-quality product by removing undesirable components from process streams.

A filter is a device that removes solid particles from a process fluid and contains them, allowing the clean product to pass through the filter. Many factors are considered when selecting an appropriate filter, including the type of material being filtered, compatibility of the feed stream with the filtering medium, the number and size of the particles to be removed from the feed stream, the degree of filtering required, and cost.

Dryers are devices used to remove moisture from a process stream, usually a gas or solid. They are essential to the production of quality, on-spec final products.

Types of Filters and Dryers

Filters are classified by the way they achieve their task. Three common classifications of filters are mechanical, gravity, and vacuum. Mechanical filters use a mechanical means, such as a strainer or cartridge to remove particles from the feed stream. Gravity filters make use of simple gravitational forces by having the feed enter at the top of the filter, cascade down though a filtering medium, and exit at the bottom of the filter. Fixed bed filters are an example of this type of filter. A vacuum filtration system uses a vacuum pump to pull feed through a **filter medium**, trapping the contaminants on the surface of the medium and allowing the clean fluid stream to pass through. A rotary drum filter operates in this way. The most common types of filters are cartridge, bag, leaf, rotary drum, fixed bed, plate and frame, and strainers.

Filter medium   the material that actually does the filtering in the filter unit.

The most common types of dryers used in the process industries are fixed bed, fluid bed, flash, spray, and rotary dryers.

Filters

The main component of a filter system is the filter unit itself. All filter systems have a means of accessing the inside of the filter to allow for cleaning or replacing the filtering medium. Filters are rated on their ability to remove particles of various sizes from a fluid. Filters have a size rating that refers to the smallest size of particle that can be expected to be trapped by the filter and prevented from flowing through it. For example, a filter with a 10-micron rating can stop a particle that is 10 microns or larger in size. Particles smaller than 10 microns would not be removed from the fluid by the filter. A **micron** is a unit of measure (1 micron = 0.001 millimeters or 0.000001 meters).

Filters are usually given an absolute rating or a nominal rating. With an absolute rating, the filter is capable of removing 99.9 percent of the particles larger than a specified micron rating. With a nominal rating, the majority (85 percent to 95 percent) of an approximate particle size is removed and does not pass through the filter.

CARTRIDGE FILTER A **cartridge filter**, the most common type of filter used in the process industries, uses a fine mesh that is tightly folded or pleated to remove suspended contaminants from a liquid. Cartridge filters consist of a housing and one or more filter tubes that fit inside it. The filter tube consists of a central core that is wrapped or surrounded with a filter medium (the material that performs the filtering action). This filter medium, which can be made of various materials, including stainless mesh, cotton, fiberglass, polypropylene, and other natural or synthetic fabrics, works like a strainer. As the fluid moves through the filter medium, solid particles become trapped by the cartridge. Figure 17.1 shows examples of cartridge filters.

Within cartridge filters, the process fluid flows from the outside of the cartridge to the inside. Over time, the cartridge filter fills with particles, requiring it to be replaced or cleaned. One indication that the cartridge filter is full is when the process flow becomes restricted and the differential pressure (ΔP) across the cartridge increases.

In a **pleated cartridge filter** (shown in Figure 17.1B), the actual filter medium is in a pleated form to provide more surface area for filtration. This results in more pounds of filtered material per cubic foot of available space (volume). The pleated material is supported

Micron a unit of measure equal to 0.000001 meters. Filters are rated based on the size of the particles they are capable of filtering.

Cartridge filter a filtration unit that uses a fine mesh that is tightly folded or pleated to remove suspended contaminants from a liquid.

Pleated cartridge filter a type of cartridge filter that uses a filter medium in a pleated form to provide more surface area.

Figure 17.1 Cartridge filters. **A.** Most common cartridge filters use fine mesh. **B.** Pleated cartridge filter.

CREDIT: **B.** Roongzaa/Shutterstock.

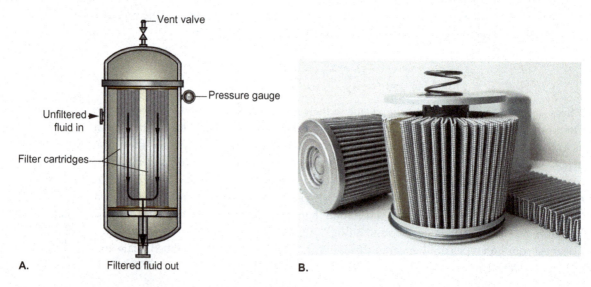

A.

B.

by a coarse, perforated core to minimize pressure drop and provide good strength to the cartridge. A common example of a pleated cartridge filter is a swimming pool filter or the filter on a wet/dry shop vacuum.

BAG FILTER A **bag filter** (shown in Figure 17.2) is a tube-shaped filtration device that uses a porous bag to capture and retain solid particles. Depending on the filter design, process fluids can flow from the inside of the bag to the outside or from the outside to the inside. In some industrial applications, bags are shaken in order to release the particles into a chamber so they can be returned to the process.

Common examples of bag filters include the bags used in household vacuum cleaners and those attached to automatic swimming pool cleaners. An industrial example of a bag filter is a baghouse system. Baghouse systems are pollution control devices used to remove particulates from an air stream prior to discharge into the atmosphere. Figure 17.3 is a diagram of a baghouse system.

Bag filter a tube-shaped filtration device that uses a porous bag to capture and retain solid particles.

Figure 17.2 Examples of filter bags for baghouse.

Figure 17.3 A baghouse system.

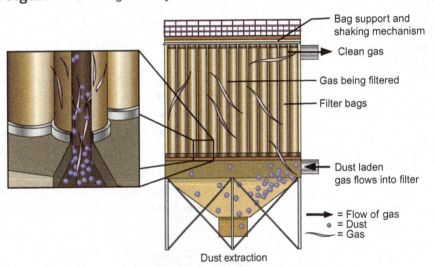

Bag support and shaking mechanism

Clean gas

Gas being filtered

Filter bags

Dust laden gas flows into filter

= Flow of gas
= Dust
= Gas

Dust extraction

LEAF FILTER A **leaf filter** uses a **filter aid** on a fine mesh screen inside the filter vessel to hold the filtered material. A filter precoat slurry is prepared in the mix tank, circulated by a pump, and deposited on the filter. During operation, the filter precoat becomes plugged with the granular filtered material being separated from the process fluid. When the filtering medium is expended (clogged), the filter is backwashed to flush the precoat out of the filter. It is discarded, and a new precoat is applied. If left unattended, this plugging eventually would restrict the flow to the point of zero flow. When this occurs, the filter must be disassembled, cleaned, and recoated with precoat material. Filter precoat systems consist of a mix tank, an agitator, and a pump. An example of a leaf filter is a water filter that uses retaining screens, such as a pool filter that is found in many backyard pools. Figure 17.4 shows an example of a leaf filter.

ROTARY DRUM FILTER A **rotary drum filter** (Figure 17.5), also called a cake filter, contains a cloth screen on a drum that retains the solids being filtered. As the partially submerged drum rotates in a vat of slurry, a vacuum draws the liquid through the filter medium (cloth) on the drum surface and leaves the solids on the outside of the drum so they can be scraped off. The accumulated layers of solids are called **filter cake**. The vacuum pulls air (or gas) through the cake and continues to remove liquid as the drum rotates. If required, the cake can be washed prior to final drying and discharge.

Leaf filter a type of filter that uses a precoated screen, located inside the filter vessel, to hold the filtered material.

Filter aid a substance applied as a slurry to a filtering medium, which coats it and aids in the filter's efficiency.

Rotary drum filter a type of filter that contains a cloth screen on a rotating drum to hold solids being filtered.

Filter cake layers of solids that are deposited on the surface of a filter medium.

Figure 17.4 Leaf filter.

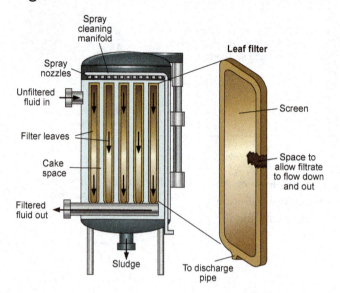

Spray cleaning manifold

Spray nozzles

Unfiltered fluid in

Filter leaves

Cake space

Filtered fluid out

Sludge

Leaf filter

Screen

Space to allow filtrate to flow down and out

To discharge pipe

Figure 17.5 Rotary drum filter showing rotation through the drum.

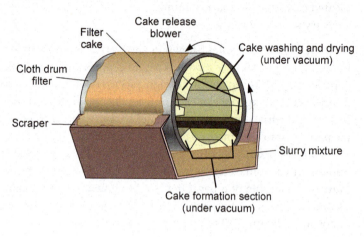

Filter cake

Cake release blower

Cloth drum filter

Cake washing and drying (under vacuum)

Scraper

Slurry mixture

Cake formation section (under vacuum)

Fixed bed filter   a filter in which the fluid passes over a stationary bed of filtering medium.

Plate and frame filter   a type of filter used to remove liquids from a slurry feed by using a deep filtration process that includes clarification and prefiltration.

FIXED BED FILTER   In a **fixed bed filter**, the filtering medium is stationary in the filter unit and held in place by a support structure. The process fluid passes over and through the filtering medium, which traps the contaminant particles. When the filtering medium becomes saturated with the contaminant, it must be regenerated or discarded and replaced with fresh material to restore filtering capability. Commonly used material for filtering medium includes sand, clay, and activated carbon. Applications for fixed bed filters include swimming pool filters, jet fuel, boiler feedwater, and wastewater treatment processes. Figure 17.6 shows a diagram of a fixed bed filter.

PLATE AND FRAME FILTER   A **plate and frame filter** (shown in Figure 17.7) is used to remove liquids from a slurry feed by using a deep filtration process that includes clarification and prefiltration. In a plate and frame filter, feed enters the filter and travels into chambers

Figure 17.6 Fixed bed filter.

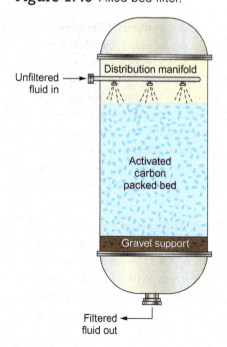

Unfiltered fluid in

Distribution manifold

Activated carbon packed bed

Gravel support

Filtered fluid out

Figure 17.7 Plate and frame filter.

CREDIT: Yps71/Shutterstock.

between the plates. Liquid passes through filter cloth into the plate and is removed from the filter, while solids collect on the filter cloth. Once the chambers between the plates are full of solid material, plate shifters move the plates to release and discharge the solids. Plate and frame filters are used in coatings, resins, and ink processes in small and large-scale applications.

STRAINER A strainer is a metal screen installed in a line to trap and remove impurities from a process fluid stream. In the petrochemical industry, strainers are routinely placed in pump suction lines to capture particles of rust, coke, or other impurities to prevent damage to the pump. Strainers are also common in steam trap installations, because very small particles of dirt or rust could lodge in the steam trap internal parts, causing it to fail in the open position and allow live steam to blow through.

Like filters, strainer size is measured by the size of the openings in the screen (mesh), which determines the size of particles that will be captured. For example, a 60 mesh screen (60 openings per inch) is the same as a 250 micron screen and is capable of filtering particles larger than a grain of fine sand.

Common types of strainers are "Y" type, "T" type, and basket strainers. The "Y" type usually is used when a smaller amount of solids is present, since the "Y" holding capacity is much less than that of "T" type and basket strainers. See Figure 17.8 for examples of strainers.

Figure 17.8 Types of strainers. **A.** "Y" type strainer. **B.** "T" type strainers. **C.** Basket strainer.

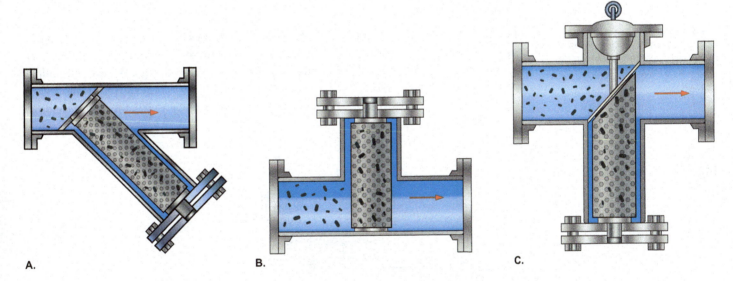

A.   B.   C.

Dryers

The most common types of dryers are fixed bed, fluid bed, flash, spray, and rotary dryers.

FIXED BED DRYER **Fixed bed dryers** are devices that use **desiccant** (substances that attract and adsorb moisture) to remove moisture from a process stream. In this type of dryer, the desiccant is distributed evenly in a bed and remains there throughout the life of the bed. Desiccant selection is based on its strong affinity for moisture and on its capacity to contain the moisture being removed.

Over time, the desiccant becomes full, loses its capacity for retaining additional moisture and must be replaced or regenerated. **Breakthrough** is the condition of any adsorbent bed when the component being adsorbed starts to appear in the outlet stream. Analyzers located on the dryer outlet continuously monitor moisture content and detect breakthrough, indicating the need for desiccant regeneration or replacement.

In applications where the moisture content or the dryness requirement is especially high, fixed bed dryers are sometimes used in series.

Fixed bed dryer a device that uses a desiccant to remove moisture from a process stream.

Desiccant a specialized substance contained in a dryer that removes hydrates (moisture) from a process stream.

Breakthrough the condition of any adsorbent bed or filter medium when the component being adsorbed or filtered starts to appear in the outlet stream.

Dual Bed Air Dryer   The dual-bed air dryer (Figure 17.9) is an example of a fixed bed dryer typically used on instrument air systems. This type of dryer uses a desiccant such as a silica gel installed in two beds. One bed removes water from the air while the other bed is being regenerated. As the desiccant becomes saturated with water, the beds are switched so that the air flows through the fresh bed. The saturated bed is heated and purged with hot gas (usually air) to strip the water from the desiccant so that it can be reused. Air dryers are used to dry the air for instrumentation and for process use where moisture in the air cannot be tolerated (e.g., unloading a liquid chlorine car).

Figure 17.9 Dual bed air dryer, a type of fixed bed dryer.

CREDIT: Mohd Nasri Bin Mohd Zain/ Shutterstock.

Fluid bed dryer   a device that uses hot gases to fluidize solids and promote the contact between dry gases and solids.

FLUID BED DRYERS   A **fluid bed dryer** (Figure 17.10) uses hot gases (usually air) passed upward through wet solids at a velocity high enough to lift the solids and cause them to act like a fluid. This fluidization promotes contact between the dry gases and wet solids, resulting in very high rates of heat transfer. This type of dryer is best suited for the drying of crystals, granules, short fibers, and some powders, and is often used in the pharmaceutical industry.

Figure 17.10 Fluid bed dryer.

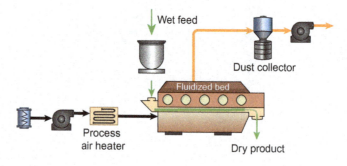

Flash dryer   a device that forces hot gas (usually air or nitrogen) through a vertical or horizontal flash tube causing the moist solids to become suspended.

FLASH DRYERS   A **flash dryer** is a device that forces hot gas (usually air or nitrogen) through a vertical or horizontal flash tube, causing the moist solids to become suspended. Because the feed enters the hot gas stream near the gas inlet, drying takes place within 1–3 seconds. The dry product then is carried to the collection system, which in most cases is a cyclone or bag collector. Figure 17.11 shows an example of a flash dryer system.

In this type of system, the dryer must achieve good gas dispersion in order to increase the surface area of the wet solid exposed to the hot gases. To accomplish this, the flash tube incorporates a Venturi tube (a tube with a constriction used to control fluid flow), which allows high-velocity gas to aid in product dispersion. Flash dryers are the most economical choice for drying nonsticky, moist, powdery, granular, and crystallized materials. They are simple and take up less space than other drying systems such as fluid bed dryers. Because of the relatively short residence time, they are also good for heat-sensitive materials. They are not good, however, for materials that require greater dryer residence time.

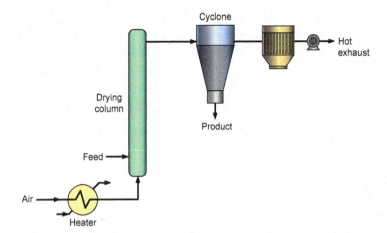

Figure 17.11 Flash dryer system.

SPRAY DRYER A **spray dryer** is a device that sprays a moist feed stream into a vertical drying chamber using a nozzle or rotary wheel. The nozzle or rotary wheel atomizes (makes into a fine spray) the feed stream where it comes into contact with hot gas. The moisture evaporates quickly, leaving small, dry particles. The wet gas and dry particles exit from the bottom of the drying chamber and flow to a cyclone, where the dry particles are separated from the wet gas. A bag filter is used downstream of the cyclone to capture very fine particles (called *fines*). The wet gas then exits the bag filter and is vented to the atmosphere, recycled, or sent to a recovery system. Figure 17.12 shows an example of a typical spray dryer.

Spray dryer a device that sprays a moist feed stream into a vertical drying chamber using a nozzle or rotary wheel.

Figure 17.12 Typical spray dryer.

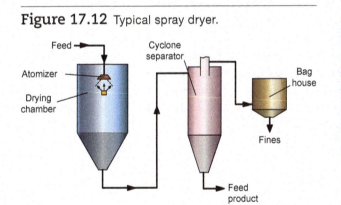

Did You Know?

Food and beverage manufacturers use freeze dryers and spray dryers to create products such as crystals of instant coffee.

CREDIT: Boris Bulychev/Shutterstock.

ROTARY DRYER A **rotary dryer** is a device composed of large, rotating, cylindrical tube that reduces the moisture content of a material by bringing it into direct contact with a heated gas. Rotary dryers employ a slightly inclined rotating cylinder in which the product discharge end is lower than the inlet end where moist feed enters the dryer (Figure 17.13). Depending on the application, hot gases can be introduced into either end of the cylinder.

Rotary dryer a device composed of large, rotating, cylindrical tubes; reduces the moisture content of a material by bringing it into direct contact with a heated gas.

Figure 17.13 Outside view of a rotary dryer.

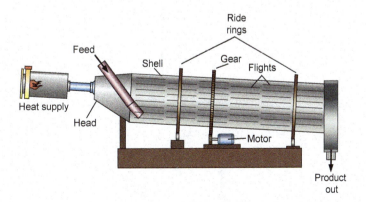

Figure 17.14 Cross-section of a rotary dryer.

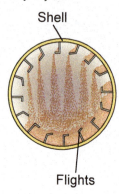

Shell

Flights

In this type of dryer, the feed is lifted by a series of fins known as *lifting flights*, which are attached to the cylinder wall in a helical or spiral configuration (see Figure 17.14). The moist feed collects on the flights, and as the cylinder rotates, the feed is lifted to the top of the dryer. The feed then falls by gravity from the top to the bottom of the cylinder. The product continues to cascade down the length of the dryer, again aided by gravity. This process is known as *rolling forward* or *downhill*. Finally, the dry product and wet exhaust exit the cylinder at the lowest end of the dryer.

Did You Know?

The clothes dryer in your home is a type of rotary dryer, that always includes a filter for lint.

CREDIT: studiovin/ Shutterstock.

Slurry dryer a device designed to convert wet process slurry into dry powder.

Slurry a liquid solution that contains a high concentration of suspended solids.

SLURRY DRYER A **slurry dryer** is a device designed to convert wet process **slurry** (a liquid solution that contains a high concentration of suspended solids) into dry powder. In one type of slurry system, wet slurry is pumped to a set of rings in the tower and is atomized by a set of nozzles. This atomized slurry then passes through heated air in the tower and is dried. The nozzles in the tower are designed to atomize the slurry in a consistent pattern so the tower can dry the particles into powder form.

Figure 17.15 is a diagram of a standard countercurrent tower dryer where slurry enters the tower through three nozzle rings and a heated gas (usually air) stream enters at the bottom. The furnace heats the air before it enters the tower. This heated air flow dries the slurry to a powder. The dried solid material falls to the bottom of the tower, while the moist air exits the top of the tower into a cyclone separator.

Figure 17.15 Diagram of a standard counter-current tower dryer used to dry slurry.

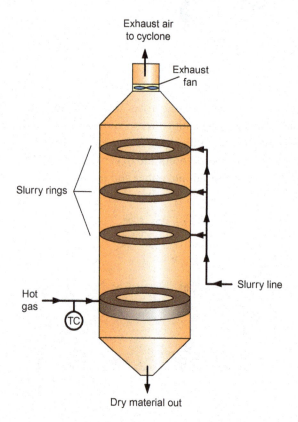

Exhaust air to cyclone

Exhaust fan

Slurry rings

Hot gas

TC

Slurry line

Dry material out

A **cyclone separator**, or simply a cyclone, is a mechanical device that uses a swirling (cyclonic) action to separate heavier and lighter components of a gas or liquid stream. Because of their design, cyclone separators are essentially free of moving parts.

Inside a cyclone, air and entrained solids enter the side of the cylinder tangentially through a slot. This creates a rotation from the top to the bottom of the cylinder, similar to the vortex created by a tornado. As the gas and solids rotate toward the bottom of the cylinder, the heavier material moves by centrifugal force to the cyclone wall and drops to the conical bottom, where it collects. The lighter particles and gas move to the center of the cyclone and are drawn up as exhaust out of the top (see Figure 17.16).

The majority of cyclones have a conical bottom section. This section spins the solids even faster toward the bottom outlet, allowing for disengagement (separation) from the gas stream. At the top, there is a small outlet pipe protruding from the cyclone through which the cleaned gas stream, along with a small amount of fines, exits. This protrusion into the center of the cyclone ensures that few solids leave with the gas stream.

Cyclones are used in saw mills to remove sawdust from extracted air. In the petrochemical field, cyclones are used in catalytic crackers to remove catalyst from the vapor.

17.2 Potential Problems with Filters and Dryers

Improper operation of filters and dryers can have a serious negative impact on downstream processes and final product quality. The severity of the impact is relative to the consequences of contamination. For example, the presence of contaminant material in the final product can result in costly product rejection and rework. Even worse, if contaminated product reaches the customer, it could be rejected or shipped back to the manufacturer, which can result in monetary penalties, diminished company reputation, and loss of business. The following are some additional examples of negative consequences associated with improper filter and dryer operation.

- In a polymer process facility, moisture breakthrough in a feedstock guard bed dryer can cause major loss of polymer catalyst activity and the production of off-specification product.

- A dryer failure in moisture-containing process streams containing highly corrosive compounds, such as hydrochloric acid (HCl), can lead to massive mechanical failure of the downstream equipment because of corrosion.

- A filter that releases solids can damage pumps and plug downstream equipment, requiring operations to be shut down for equipment repair and limiting process facility production rates.

- Traces of water can poison a catalyst bed, leading to costly downtime and catalyst replacement.

- Moisture in the instrument air system can eventually lead to broad failure of control system components. For example, water in an instrumentation air line could freeze during cold weather, thus rendering instrumentation inoperable.

- Failure of a mud filter to properly remove sulfurous compounds in a lime kiln can lead to elevated environmental emissions that are higher than permissible limits.

- The manufacture of chlorine dioxide requires filtered water and air. Failure of the filtering systems can lead to chlorine dioxide decomposition and process shutdown.

- Failure to filter solids from a fluid process stream can plug or damage downstream equipment such as pumps. This reduces facility production rates and can lead to environmental violations because of releases to the atmosphere and/or spills to the ground.

This section describes some of the typical problems associated with filter and dryer system operation and how they can be prevented or corrected.

Cyclone separator a mechanical device that uses a swirling (cyclonic) action to separate heavier and lighter components of a gas or liquid stream.

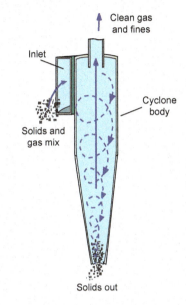

Figure 17.16 Cyclone showing flow in and solids out.

Breakthrough

Moisture breakthrough occurs in a dryer when the output contains excessive moisture. Causes of breakthrough include poor bed regeneration because of insufficient heating and poor contact of the material to be dried with the hot drying gases. In a filter system, breakthrough can occur as a result of a hole in the filtering medium, a loss of filter aid, or a failure of an individual element in the filter. In dryer systems using heat rather than a desiccant, a partial or total loss of the heat source can result in breakthrough. Breakthrough can also result if the outlet analyzers, used to monitor for moisture, fail and dryer beds do not switch and regenerate properly.

Compressor systems use filters to capture oil entrained in the compressed air prior to entry into the air dryer. Filter failure allows oil to coat the desiccant, rendering it useless and leading to premature breakthrough. Breakthrough can also result from an instrument malfunction, resulting in incomplete bed regeneration. To prevent breakthrough, the oil filter element and the prefilter system should be inspected and/or replaced at regularly scheduled intervals.

Fouling

Fouling occurs in filters when there is a buildup of solids or if the filter medium becomes coated with sticky materials. Excessive fouling leads to a reduction in flow rate and an increase in differential pressure across the filter. Because of this, differential pressure should be monitored continuously to determine when the filter needs to be taken off line for cleaning or replacement.

Channeling

Channeling formation of a path through a dryer or filter that allows feed to flow through without proper contact with the filtering medium or drying agent.

Channeling occurs when flow is not evenly distributed across a dryer or filter and a path is formed that allows feed to flow through without properly contacting the filter medium or drying agent. If channeling is suspected, the bed should be regenerated to try and disrupt the channel. Failure of an orifice or valve setting can allow too much flow through the unit, which can also lead to channeling. Ultimately, physical inspection of the vessel's internal components can be required to ensure that internal baffling, supports, and other components are in place and functioning properly.

The following steps will help to prevent channeling.

- Stay within pressure and flow limits of the system; do not overpressurize.
- Operate within the design limits (e.g., temperatures and flows).
- Ensure that the correct drying agent or precoat is installed.

17.3 Safety and Environmental Hazards

Improper operation of equipment in filter and dryer systems can affect personnel safety, equipment, production, and the environment. Table 17.1 lists some of the hazards associated with the normal and abnormal operation of equipment in filter and dryer systems.

Table 17.1 Hazards Associated with Filter/Dryer Systems

Improper Operation	Possible Effects			
	Personnel Safety	**Equipment**	**Production**	**Environment**
Equipment leakage	■ Inhalation hazard ■ Skin contact hazards ■ Burn hazard ■ Slips and falls resulting from leaks or spills	■ Fires associated with flammability ■ Dust explosion hazard ■ Corrosion caused by exposure to moisture or corrosive feed materials	■ Filtering medium can be considered waste ■ Lost production	■ Release of hazardous/toxic vapor into air ■ Discharges to ground caused by leaks or spills

Improper Operation	Possible Effects			
	Personnel Safety	Equipment	Production	Environment
Rotating equipment guard missing	Injury or death from contact with moving equipment			
Breakthrough		Damage to equipment from unfiltered or undried process fluid	Lost production	
Fouling		Damage to equipment from unfiltered or undried process fluid	Lost production	
Channeling		Damage to equipment from unfiltered or undried process fluid	Lost production	
Sampling without proper PPE	■ Inhalation hazard ■ Skin contact hazards ■ Burn hazard			

17.4 Process Technician's Role in Operation and Maintenance

The role of the process technician is to ensure the safe, proper operation of filters and dryers and to be aware of problems that can affect the system. Table 17.2 lists some of the activities that should be carried out as part of a process technician's regular unit rounds when working with filters and dryers.

Table 17.2 Process Technician's Role in Operation and Maintenance

Look	Listen	Check
■ Observe for desiccant on the floor or ground under dryers ■ Look for pools of liquid on the floor or ground around equipment ■ Observe instrumentation regularly ■ Look for excessive pressure drop across filters ■ Look for leaks around the filter closure ■ Monitor and clean/replace filter elements as necessary ■ Ensure that a proper micron rating filter is used	■ Listen for unusual noises	■ Check for high temperatures ■ Check for excessive vibrations ■ Check for excessive heat on motorized equipment ■ Check that dryer heating system is functioning correctly ■ Examine the filtered/dried product for evidence of contamination/breakthrough ■ Check strainer baskets and clean as necessary

17.5 Typical Procedures

Process technicians are responsible for monitoring filter and dryer system equipment for proper operation and for detecting and correcting small problems before they become large problems. When performing routine and preventive maintenance, technicians must comply with all clearing, isolation, and lockout/tagout requirements.

All process technicians must be familiar with normal startup and shutdown procedures and emergency shutdown procedures. They must know how to respond to emergencies and any regulatory issues concerning the process involved.

Monitoring

Process control systems and log sheets are used to document equipment and process operations. Information found on log sheets and in control system history databases provides trends and operating conditions that allow process technicians to troubleshoot and problem-solve issues as they arise. In many regulated systems, these log sheets and databases are required as proof of proper operation for the systems.

Lockout/Tagout

Lockout/tagout procedures ensure that the equipment is in a zero-energy state while maintenance or inspection is performed. Process technicians are required to follow the lockout/tagout procedure for their process unit. A generic summary of lockout/tagout procedures follows.

1. The equipment is isolated from the process; that is, pump breakers are de-energized, equipment is depressurized and drained, and the equipment is cleaned (purged) of process fluids or gases.
2. All forms of possible energy sources (for example, pneumatic, hydraulic, electrical, mechanical, and nuclear [radioactive]) are removed and prevented from being reactivated.

Startup and Commissioning

Strainers often are used during commissioning when larger particles can be expected. During the startup of a new system, large-mesh strainers are gradually replaced with smaller mesh until all particles are removed.

Housekeeping

Housekeeping is a routine part of daily operations. Spills and leaks must be cleaned up promptly to avoid the possibility of slips and falls. During routine or preventive maintenance, proper housekeeping should be maintained to help reduce the risk of injury.

Sampling

Process technicians are responsible for keeping a process facility operating safely and efficiently. Process technicians also routinely sample and test process fluids and solids at various stages of the production process. These tasks are necessary to help ensure that a high level of product quality is maintained.

Samples are taken in process systems to detect problems early so that equipment is not damaged and products are not wasted. If testing indicates that a material is off specification, actions must be taken immediately to correct the problem and minimize profit loss. Samples are also taken to verify that any material discharged into the environment is in compliance with company and government regulations.

When sampling or handling materials, process technicians must take all applicable precautions to prevent exposure. Any excess or unused portions of samples should be disposed of properly according to company procedures.

Another important point to consider when taking samples is to make sure that the samples are not contaminated with other materials. To safeguard against this, the process technician taking the sample must ensure that the sample container is clean and free from contaminants. After the sample container is filled, the process technician should immediately seal it from the atmosphere by tightly closing the cover or cap to prevent contamination while en route to the laboratory.

Summary

Filters and dryers are important pieces of equipment in many process operations. Filters are used to remove solid particles from process fluids, and dryers are used to remove moisture from a solid or a gas. Proper operation of these systems prevents potentially serious problems from developing in downstream equipment and processes.

Cartridge filters force process fluids through a filter medium that covers a filter tube, while trapping solid particles. Pleated cartridge filters employ a filter medium in a pleated form to provide more surface area for filtration. Bag filters contain a bag through which process fluids flow from either the inside out or the outside in and particles are collected. In a leaf filter, a precoated screen is mounted inside the filter vessel to hold filtered material. Rotary drum filters use a cloth covered rotating drum to retain solids. Leaf filters and drum filters use a precoated screen to catch and hold the filtered material. Plate and frame filters are used to remove liquids from slurry and are used in coatings, resins, and ink processes. Strainers are metal screens installed in lines to capture solid impurities from a fluid stream.

Filter ratings refer to the smallest size of particle that can be expected to be trapped by the filter and prevented from flowing through it.

A fixed bed dryer uses a desiccant to adsorb moisture. Many fixed bed dryers have two beds: one in service, and the other in regeneration or in standby. Some fixed bed dryers use two beds in series so that any contaminant material that escapes the first bed is caught in the second bed.

Fluid bed dryers use hot gases to fluidize a solid desiccant or process product to promote contact between the dry gases and the solids. Gas and liquid phases are separated, and the desiccant usually is rejuvenated and recycled to the dryer.

Flash dryers are used to remove moisture from hot process fluids. A sudden pressure drop across an orifice causes the moisture to flash, converting it from a liquid to a vapor so it can be routed away from the process and removed.

Improper operation of filters and dryers can cause serious problems in downstream equipment and affect the quality of the finished product. For example, breakthrough can diminish drying and filtering performance, causing the product to be off specification. Solids that are not trapped can damage or plug downstream equipment, creating a need for shutdown and limiting production rates.

Process technicians are responsible for the routine monitoring of filter and dryer equipment and for the early detection of problems. Process technicians are also responsible for performing routine housekeeping tasks, taking samples for testing, and performing routine and preventive maintenance according to established procedures while complying with all lockout and tagout requirements. Technicians must also know how to perform normal startups and shutdowns and emergency shutdowns, and how to respond to emergencies.

Problems associated with filter and dryer systems include breakthrough, fouling, and channeling. Close monitoring of operating conditions can prevent some problems from occurring and minimizing their impact if they do occur.

Checking Your Knowledge

1. Define the following terms:

a. Breakthrough

b. Desiccant

c. Filter aid

d. Micron

2. One purpose of filters is to:

a. add air or other gases to a process stream

b. remove moisture from a process stream

c. remove solid particles from a process stream

d. none of the above

3. List the three common classifications of filters.

4. One purpose of dryers is to:

a. remove trace chemical components of a stream

b. remove moisture

c. clean the water in process streams and products

d. remove solid particles from a process stream

5. A _____ is a piece of equipment that uses swirling action to separate heavier and lighter components of a gas or liquid stream.

a. cyclone

b. conveyer

c. dryer

d. decanter

6. In the case of a dryer, _____ occurs when the dryer output contains excessive moisture.

   a. breakthrough

   b. saturation limit

   c. contamination replacement

   d. channeling

7. List two steps to help prevent channeling.

8. Which of the following operational hazards can cause damage to the equipment from unfiltered or undried process fluid? (Select all that apply.)

   a. Fouling

   b. Channeling

   c. Breakthrough

   d. Equipment leakage

9. (True or False) Process technicians may routinely take samples for lab testing.

10. _____ are often used during commissioning when larger particles of impurities can be expected.

NOTE: Answers to Checking Your Knowledge questions are in the Appendix.

Student Activities

1. Write a two-page paper about the types of filters and dryers used in the process industries. Be sure to explain the purpose of each type.

2. With a classmate, prepare a presentation about the differences among fluid bed dryers, flash dryers, and fixed bed dryers.

3. Discuss safety and environmental hazards associated with filters and dryers.

4. Research breakthrough with a classmate and list possible corrective actions and steps for prevention.

Chapter 18
Solids Handling Equipment

Objectives

After completing this chapter, you will be able to:

18.1 Identify the common types, components, and applications of solids handling equipment. (NAPTA Equipment Tools 7, 8*) p. 370

18.2 Describe potential problems and hazards associated with solids handling equipment. (NAPTA Equipment Tools 9) p. 383

18.3 Describe the process technician's role in solids handling equipment operation and maintenance. (NAPTA Equipment Tools 11, 12) p. 384

*North American Process Technology Alliance (NAPTA) has developed curriculum to ensure that Process Technology courses will produce knowledgeable graduates to become entry-level employees for the career field of process technology. Objectives from that curriculum are named here in abbreviated form (e.g., NAPTA Equipment Tools 9 addresses objective 9 of the NAPTA curriculum about equipment such as solids handling equipment).

Key Terms

Bin—a vessel that typically holds dry solids, **p. 377.**

Blower—a mechanical device, either centrifugal or positive displacement in design, that has a lower ratio of pressure change from suction to discharge and is not considered to be a compressor. The resistance to flow is downstream of the blower, **p. 375.**

Bucket elevator—a continuous line of buckets attached by pins to two endless chains running over tracks and driven by sprockets, **p. 376.**

Conveyor—a mechanical device used to move solid material from one place to another, **p. 373.**

Die—a metal plate with specifically shaped perforations designed to give materials a specific form as they pass through an extruder, **p. 372.**

Extruder—a device that forces materials through the opening in a die so it can be formed into a particular shape, **p. 371.**

Feeder—a piece of equipment that conveys material to a processing device (e.g., an extruder) at a controllable and changeable rate, **p. 370.**

Gravimetric feeder—a device designed to convey material at a controlled rate by measuring the weight lost over time, **p. 371.**

Hopper—a funnel-shaped, temporary storage container that is usually filled from the top and emptied from the bottom, **p. 377.**

Interlock—a system for connecting mutually independent equipment that can stop upstream equipment when downstream equipment is shut down, **p. 383.**

Live bottom—a device at the bottom of a bin or silo designed to facilitate the smooth discharge of material, **p. 377.**

Pneumatic conveyor— a device used to move powdered or granular material using pressurized gas flow from one point to another, **p. 375.**

Rotary airlock valve—a valve that uses rotating pockets to meter a fixed volume of solids at a fixed rate; also called an *airlock, rotary valve,* or *rotary feeder,* **p. 380.**

Saltation velocity—the minimum gas velocity required to maintain particle entrainment; if gas velocity slows below the saltation velocity, particles will fall from suspension in a pipe, **p. 375.**

Screener—a piece of equipment that uses sieves or screens to separate dry materials according to particle size, **p. 372.**

Silo—a tower-shaped container for storage of materials; typically used to store product destined for distribution, **p. 376.**

Solids handling equipment—equipment used to process and transfer solid materials from one location to another in a process facility; can also provide storage for those materials, **p. 370.**

Trickle valve—a valve used to continuously transfer a fixed weight of solids between two different pressure zones at a constant rate, **p. 380.**

Volumetric feeder—a device designed to convey materials at a controlled rate by running at a set motor speed, **p. 371.**

Weighing system—a system used to weigh material before shipping, **p. 381.**

18.1 Introduction

The process industries use many types of equipment to process, transfer, store, and ship solids. When solid material is required as a feedstock, process technicians must often unload solids from their containers, such as railcars, into larger, long-term storage equipment. This solids movement occurs through both mechanical and pneumatic means.

Solids handling equipment is designed so process solids can behave more like liquids. For example, a conveyor moves solids from one place to another in a continuous stream, much like liquid flows through a pipe. Bins hold a large volume of solid material just as a chemical storage tank holds a large volume of liquids. Compressed air is often used to push material from a bin to a hopper car or truck, the way a pump is used to push liquids from a tank to a tank car or truck. An inert gas such as nitrogen can be used to push material if the potential for static electricity in solids handling operations creates risk for an explosive atmosphere. Nitrogen or other inert gases also can be used to blanket bins and hoppers to prevent ignition or an explosion, such as a dust explosion.

Types of Solids Handling Equipment

Solids handling equipment
equipment used to process and transfer solid materials from one location to another in a process facility; also can provide storage for those materials.

Solids handling equipment is used to process and transfer solid materials from one location to another in a process facility. It also can provide storage for those materials. Some of the most common types of solids handling equipment include feeders, extruders, screening systems, conveyors, elevators, storage containers, valves, and weighing systems.

Feeders

Feeder a piece of equipment that conveys material to a processing device (e.g., an extruder) at a controllable and changeable rate.

A **feeder** is a piece of equipment that carries (conveys) material to a processing device (e.g., an extruder) at a controllable and changeable rate. This ability to change feed rate distinguishes feeders from conveyors.

Feeders can operate *gravimetrically* (based on weight) or *volumetrically* (based on volume). The rate of the feeder can be set based on material rate (e.g., pounds or kilograms per hour) for gravimetric feeders, or on motor speed (e.g., revolutions per minute) for volumetric feeders (shown in Figure 18.1).

Figure 18.1 A. Example of a gravimetric feeder. **B.** Example of a dry solids volumetric feeder.

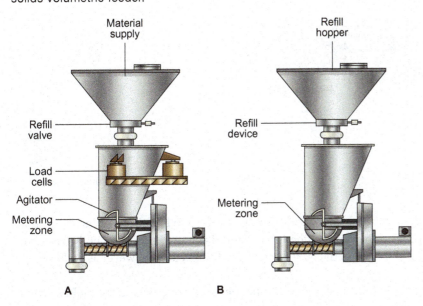

A **gravimetric feeder** (Figure 18.1A) is a device designed to convey material at a controlled rate by measuring the weight lost over time. While in operation, gravimetric feeders take an internal weight measurement for the material loaded into the feeder, record that weight, feed material for a specific amount of time, and then take another internal weight measurement. The feeder compares the weight lost over time (actual rate) with the target rate and makes automatic adjustments if needed. This type of feeder is usually referred to as a *loss in weight feeder*.

A **volumetric feeder** (Figure 18.1B) is a device designed to convey material at a controlled rate by running at a set motor speed. Volumetric feeders must have a predetermined maximum feed rate that is calculated by the feeder's maximum motor speed and material density. For example, if a feeder is capable of moving 400 pounds per hour (181 kg/hr) at its maximum motor speed (100 percent), then the controller would be set to 50 percent output to run the feeder at 200 pounds per hour (91 kg/hr).

Gravimetric feeder a device designed to convey material at a controlled rate by measuring the weight lost over time.

Volumetric feeder a device designed to convey materials at a controlled rate by running at a set motor speed.

Extruders

An **extruder** is a device that forces materials through the opening in a die so it can be formed into a particular shape. Extrusion is a continuous process used for making products such as pasta, chewing gum, certain candies, plastic films, sheets, tubes, plastic pipes, and many other items. Extruders are also used to make the plastic pellets used to manufacture products like beverage containers and plastic outdoor furniture.

In the plastics extruding process, plastic material (e.g., granules, pellets, or powder) from a hopper is fed into a long, cylindrical, heated chamber where the material moves by the action of a continuously rotating screw. This cylindrical chamber is known as an extruder and is composed of three zones: the feed zone, the compression zone, and the metering zone (Figure 18.2).

Material enters the extruder through the feed zone (no melting takes place in this zone) and then moves to the compression zone. At the start of the compression zone, friction

Extruder a device that forces materials through the opening in a die so it can be formed into a particular shape.

Figure 18.2 Example of a continuous extrusion feeder showing the feed zone, compression, zone, metering zone, and extrusion.

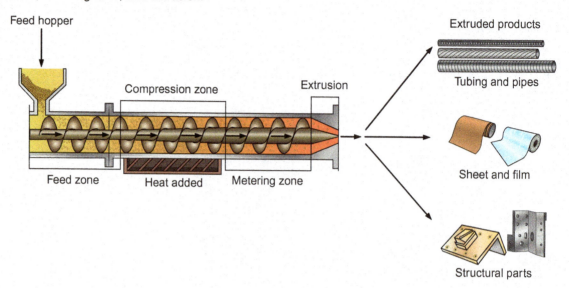

and compression-induced heat begin the melting process. While the machine is running, most of the energy required for melting is mechanical energy resulting from the compression and shearing of feed material by the screw. Additional heat, however, is added to the extruder wall.

At the end of the compression zone is the metering zone, where the molten plastic is forced out through a small opening called a **die** (a metal plate with a precisely shaped perforation or opening, designed to give the extruded material a specific form). The die is responsible for giving the final product its shape. After the plastic passes through the die, the extruded plastic is cut by a rotating knife and fed into a water bath for cooling.

Figure 18.3 shows examples of dies used to shape final products.

Die a metal plate with specifically shaped perforations designed to give materials a specific form as they pass through an extruder.

Figure 18.3 Dies are used to shape extruded products.

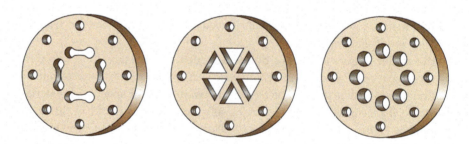

Screening Systems

Screening systems (e.g., sieves, sifters, and screens) are used to separate particles by size. Particles that are too small (called *fines*) are reprocessed or recycled, and particles that are too large (called *overs*) are remilled or recycled.

A **screener** is a piece of equipment that uses sieves or screens to separate dry materials according to particle size. In a screening system, materials enter the top of the screener and are distributed across the surface of the screen. Vibration is then used to further distribute the material across the screen and convey it toward the discharge end.

Some screening systems contain multiple levels; the top tier contains the largest screen openings and the bottom tier contains the smallest (Figure 18.4). Some screening systems also contain balls confined in pocket areas beneath the screen surface. The bouncing action of these balls dislodges particles by direct contact, thereby cleaning the screens. These resilient balls also provide agitation to separate particles that might stick together.

Screener a piece of equipment that uses sieves or screens to separate dry materials according to particle size.

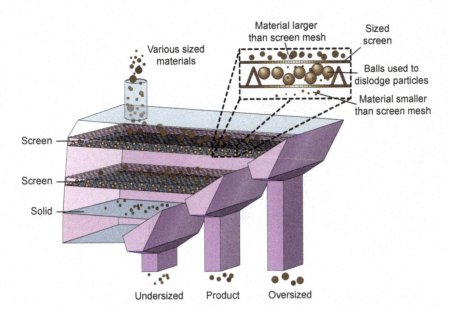

Figure 18.4 Separating screens from largest to smallest.

Conveyors

A **conveyor** is a mechanical device used to move material from one place to another. Types of conveyors include belt, screw, moving floor, drag chain, vibratory, and powered roller.

A belt conveyor consists of a belt supported by pulleys that runs in a continuous loop. Belt conveyors can be used to move items such as coal, petroleum coke, or grain to a storage or packaging area. Conveyor belts are not always horizontal. In fact, they often move material on an incline, either up or down. Examples of commonly used belt conveyors are the ones found in grocery store checkout lines. These conveyors help move the groceries toward the cashier and speed up the checkout process. Figure 18.5 shows examples of belt conveyors.

Conveyor   a mechanical device used to move solid material from one place to another.

Figure 18.5 A. Belt conveyor diagram. **B.** Ore conveyor.

CREDIT: B. Courtesy of Martha McKinley.

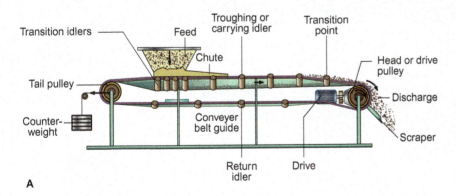

A

B.

Screw conveyors are conveyors that use a helical screw, rotating within a stationary tube or trough, to convey material. This rotating screw propels liquids or fluidized solids along the tube through the pushing action of the screw blades. Screw conveyors can move liquids or fluidized solids either horizontally or upward at an incline. Figure 18.6 shows an example of a screw conveyor.

Figure 18.6 Inside view of a screw (helical) conveyor.

CREDIT: WindVector/Shutterstock.

Did You Know?

An escalator in an office building or department store is considered a moving floor conveyor.

CREDIT: Tetra Images/Alamy Stock Photo.

Moving floor conveyor systems are ground-based conveyer systems designed to transport both light and heavy loads, such as pallets into a trailer. Figure 18.7 shows an example of a moving floor conveyor.

Drag chain conveyors consist of two or more parallel chains moving in the same direction along a conveying duct. A gripping mechanism made from a rubberlike material is attached to the chain links. These gripping mechanisms grab and push the conveyed material along the duct in the direction of the moving chains. Figure 18.8 shows an example of a drag chain conveyor.

Figure 18.7 Example of a moving floor conveyor.

CREDIT: Media Whalestock/Shutterstock.

FIGURE 18.8 Example of a drag chain conveyor.

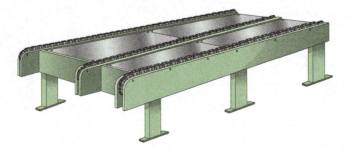

Vibratory conveyors typically include a conveyor trough or deck that is shaken by a vibratory driver to produce motion in the material carried in the trough. Mechanical vibratory drivers typically include a number of eccentrically mounted weights that, when rotated, create vibration in the trough or deck, conveying the materials along the deck.

Roller systems are conveyor systems that use a series of rotating cylinders (rollers) to move materials. Roller movement in these types of systems can be either gravity-driven or mechanically powered. In a gravity-driven system, the cylinders are mounted in parallel and suspended in a frame by bearings so they turn freely as materials (e.g., bags or boxes) move across the rollers. Powered roller systems use chains or belts to turn the rollers and move the load. Because they are mechanically driven, powered roller systems can be used to move materials horizontally, at an upward incline, or in a downward direction. Gravity-driven rollers, on the other hand, are designed to move only horizontally and in a downward direction. Figure 18.9 shows an example of a powered roller system.

Figure 18.9 Example of mechanically powered roller system.

CREDIT: Cultura Creative/Alamy.

PNEUMATIC CONVEYORS   A **pneumatic conveyor** is a device used to move powdered or granular material using pressurized gas flow from one point to another. There are two types of pneumatic conveying systems: dilute phase and dense phase. Both systems use gas (usually air) to transfer suspended solids both horizontally and vertically through pipelines. These systems can convey a wide range of solids, from fine powders to pellets.

Dilute phase systems use a conveying gas volume and velocity sufficient to keep the solids that are being transferred in suspension. These systems are used to convey solids continuously so that the material does not accumulate in the bottom of the conveying line at any point. There are three types of dilute phase conveying systems: pressure-driven operation, vacuum-driven operation, and pressure/vacuum operation. A pressure-driven dilute phase system (Figure 18.10) uses a **blower** to blow (push) gas (air) through a line, where it picks up solids dropped into the line from a hopper/silo rotary valve. The solids entrained in the flowing gas are transferred to a bin where they drop and are then stored. The solids-free gas is then vented from the bin.

Pneumatic conveyor   a device used to move powdered or granular material using pressurized gas flow from one point to another.

Blower   a mechanical device, either centrifugal or positive displacement in design, that has a lower ratio of pressure change from suction to discharge and is not considered to be a compressor. The resistance to flow is downstream of the blower.

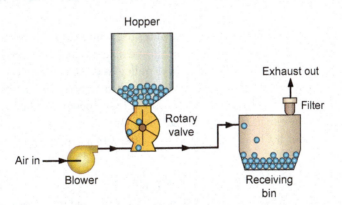

Figure 18.10 Dilute phase pressure driven pneumatic conveyor.

Dense phase systems operate with product velocities below the **saltation velocity** (the velocity at which particles fall from suspension in the pipe); thus, the flow of material is not continuous but moves in waves or plugs. Typical dense phase conveying systems use a small volume of high-pressure gas to push plugs or waves of solids through a conveying line. In moving bed flow, wave movement occurs as air flows over the material that has collected at

Saltation velocity   the minimum gas velocity required to maintain particle entrainment; if gas velocity slows below the saltation velocity, particles will fall from suspension in a pipe.

Figure 18.11 Comparison of dilute and dense phase conveying systems.

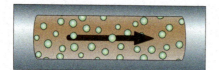

Dilute phase conveying system
(constant suspension)

Dense phase conveying system
(plugs or waves)

the bottom of the pipe, and material moves in dune-like fashion. In plug type flow, zones of gas separate plugs as they move through the transfer line. Figure 18.11 shows a comparison of flow through dilute phase and dense phase conveying systems.

The advantage of a dense phase conveying system is lower gas requirements. This translates to lower energy requirements and lower solids velocities. Lower velocities mean that abrasive and friable (easily crumbled) materials can be conveyed without significant transfer line erosion or product degradation.

Process Elevators

A **bucket elevator** (Figure 18.12) is a continuous line of buckets that are attached with pins to two endless chains running over tracks and driven by sprockets. Bucket elevators are designed to move bulk material vertically. Materials carried by bucket elevators can be light or heavy, and they can range in size from large to fine (e.g., large pieces of gravel to small pieces of grain).

Bucket elevator a continuous line of buckets attached by pins to two endless chains running over tracks and driven by sprockets.

Figure 18.12 Example of a bucket elevator.

CREDIT: fosgen/Shutterstock.

Most modern bucket elevators use rounded rubber buckets mounted on a rubber belt that is rotated by very large pulleys (several feet in diameter), both at the top and bottom of the elevator. The top pulley (the drive pulley) is driven by an electric motor. After the load is picked up from a pit or a pile of material, it then is carried to an alternate location and discharged.

True vertical conveyors run at high speeds, so they discharge the material at the top of the elevator by centrifugal force . Incline conveyors run at a slower speed and discharge under the head pulley, and they often use a trip or wiper device to empty the bucket.

Silos, Bins, and Hoppers

Silos, bins, and hoppers (Figure 18.13) are used in process industries to store materials (usually dry) that are to be used later in a process or sold to a client. Bins, silos, and hoppers can be made of many materials (e.g., concrete, wood, and steel) and usually have a valve (slide or rotary) fitted at the bottom.

A **silo** (shown in Figure 18.13A) is a large, tower-shaped container. Silos are typically used to store products destined for distribution.

Silo a tower-shaped container for storage of materials; typically used to store product destined for distribution.

Figure 18.13 A. Example of a silo. **B.** Example of a steel grain bin. **C.** Example of a hopper.

CREDIT: **A.** Leigh Trail/Shutterstock. **B.** stocksolutions/Shutterstock. **C.** Richard Thornton/Shutterstock.

A.

B.

C.

A **bin** (Figure 18.13B) is a container used to store solid products or materials. It is typically designed to hold a finished product destined for distribution.

A **hopper** (Figure 18.13C) is a funnel-shaped, temporary storage container typically filled from the top and emptied from the bottom. Hoppers are smaller than bins or silos and are usually mounted to other types of equipment that are fed by them. An example of a hopper is a storage vessel that directly feeds an extruder.

Bins, silos, and hoppers are designed to receive and store material from a process until it is eventually transferred to bagging operations, hopper cars, or hopper trucks. To facilitate the transfer of materials, many of these containers are fitted with a gyrating or rotating spiral bin discharger called a live bottom.

LIVE BOTTOM A **live bottom** is a device used on a silo, bin, or hopper to facilitate the smooth, uninterrupted discharge of material. A live bottom may be a vibratory device attached to the bottom of a bin or silo, or a screw, which keeps material with poor flow characteristics moving through the hopper or silo. Live bottoms contain a flow-inducing device that:

- Keeps the weight of the product inside from compacting onto the equipment below and jamming the exit area of the cone
- Changes the powder draw from ratholing (center feed) to a more uniform feed.
- Prevents bridging, where material arches over the exit area of the cone.

Figure 18.14A shows a live bottom attachment. Figure 18.14B, Figure 18.14C, and Figure 18.14D show how products flow with and without a live bottom, respectively.

Bin a vessel that typically holds dry solids.

Hopper a funnel-shaped, temporary storage container that usually is filled from the top and emptied from the bottom.

Live bottom a device at the bottom of a bin or silo designed to facilitate the smooth discharge of material.

Figure 18.14 A. Live bottom attachment. **B.** How product flows with a live bottom. **C.** Product without a live bottom may flow out a steep hole in the center (called "ratholing"). **D.** Without a live bottom, product flow may be blocked (called "bridging").

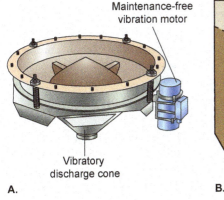

Maintenance-free vibration motor

Vibratory discharge cone

A.

B.

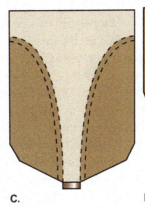

C.

D.

Figure 18.15 Piston vibrator attached to a hopper car.

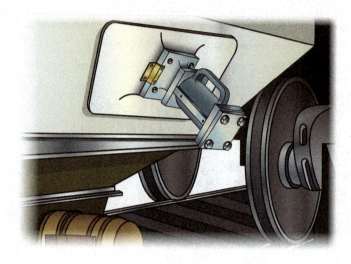

A device called a piston vibrator (Figure 18.15) serves the same purpose in a hopper car.

HOPPER CARS AND HOPPER TRUCKS   Hopper cars and hopper trucks are designed to carry solid materials by rail or over roads. The descriptions included in this text are specific to hopper cars, although hopper trucks employ the same technology.

Hopper cars are used in process industries to transport material in large quantities from one facility to another. Some hopper cars are a single bin, while others have multiple bins or chambers.

Figure 18.16A shows an open hopper car for transporting solids. Figure 18.16B shows the exterior of a multibin covered hopper car. Figure 18.16C provides a diagram of the internal configuration of a multibin hopper car.

Figure 18.16 A. Open hopper car used to transport solids. **B.** External picture of a multibin covered hopper car. **C.** Illustration of the internal configuration of a multibin hopper car.

CREDIT: **A.** safephoto/Shutterstock. **B.** RyzhkozSergey/Shutterstock.

A.   B.   C.

Hopper cars may be covered (having a roof) or open (without a roof). Covered cars are required for the transport of materials like sugar, grains, or other dry bulk products, which must be protected from the elements. Open hopper cars can be used for materials unaffected by the elements, like coal. There are two types of access covers on the top of most covered hopper cars. One is a standard access cover used to gain entry to a product bin. The other is a cover with a breather vent designed to allow air to come in and out of the bins while preventing rain or other contaminants from entering.

Figure 18.17 shows different types of access covers on the top of hopper cars.

In the case of some dry material, a hopper car can be offloaded by gravity, when steel plates on the bottom of each compartment are opened. Other hopper cars can be offloaded using a transfer hose attached to a piping connection on the bottom of the hopper car. In

Figure 18.17 Top loading area of hopper cars.

Figure 18.18 Latching mechanism of access doors on top of a hopper car.

this case, the material will be transferred to storage using a vacuum system. A hopper car may also be pressurized in order to push the product from the hopper car to an atmospheric bin for storage or further processing. The access doors at the top of these hopper cars are the same as regular solid access doors (i.e., simple open/close doors). They have different latching mechanisms and sizes, however, and they are designed to maintain a pressurized environment inside the car. Since these hopper cars are pressurized during offloading, the access doors are thick and have multiple latches to ensure their safety under pressure. This type of hopper car has one product bin divided into multiple sections, and each bin section is usually offloaded individually. Figure 18.18 shows the top side of a hopper car with latching systems.

Another method used to unload hopper cars is to drop the material from the car into a pit below the tracks or into a tanker's hold for transport. The interior of this type of bin is very simple. There are no working parts other than the slide gate, which operates from the outside. Individual hopper car bins of this type must be lined up directly over the top of the pit or tanker hold to begin offloading. Figure 18.19 shows a hopper car unloading into a tanker's hold for transport.

Figure 18.19 Hopper car dropping product into a tanker's hold for transport.

Figure 18.20 Hopper car with air pressurized piping system for offloading.

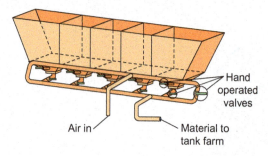

Hand operated valves

Air in

Material to tank farm

Powder railcars, used to transport products such as coal dust or lime, can be designed as a single bin with multiple unloading ports. Although this type of railcar has multiple unloading ports, there are no divider walls between the sections.

Figure 18.20 shows a hopper car with an air pressurized piping system for offloading.

Valves

A **trickle valve** is used to transfer a fixed weight of solids between two different pressure zones at a constant rate. Trickle valves operate under a vacuum or negative pressure on the upstream (or inlet) side of the valve flap, and the pressure differential holds the flap against the bottom of the valve housing. This free-hanging flap between the two different pressure environments maintains a seal. This type of valve is often used at the bottom of a cyclone. Product from the cyclone collects on the upstream or negative pressure side of the channel. When the pressure from the accumulated weight of the product is greater than that on the positive pressure side of the channel, the product will begin to trickle through the flap. Figure 18.21A shows how the trickle valve flap works in a negative pressure environment within a chute.

A **rotary airlock valve** (also called a rotary airlock, rotary valve, or rotary feeder) uses rotating pockets to meter a fixed volume of solids at a fixed rate. Rotary airlock valves are designed to operate in much the same way as the trickle valve, separating areas of low and high pressure. The term "airlock" refers to the fact that the air is restricted from moving between inlet and outlet sides of the valve. Generally, the low-pressure environment is located above, or upstream from, the valve, and the higher-pressure environment is located downstream from, or below, the valve. This type of valve uses pockets separated by vanes which are attached to a rotor that rotates inside a cylindrical housing. The valve is open to the process on the upstream side or top of the valve, where material flows into one of the pockets. The vanes rotate through the sealed chamber, and the material drops from the valve pocket when it reaches the downstream side or bottom of the valve, maintaining the differential pressure between the upstream and downstream pressure environments. Figure 18.21B shows a diagram of a rotary airlock valve.

Trickle valve a valve used to continuously transfer a fixed weight of solids between two different pressure zones at a constant rate.

Rotary airlock valve a valve that uses rotating pockets to meter a fixed volume of solids at a fixed rate; also called an *airlock*, *rotary valve*, or *rotary feeder*.

Figure 18.21 A. Trickle valve flap works in a negative pressure environment within a chute. **B.** Diagram of a rotary valve (airlock).

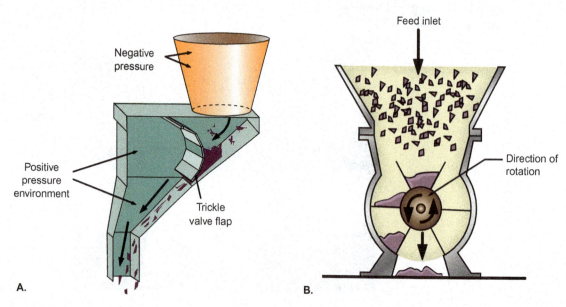

Negative pressure

Positive pressure environment

Trickle valve flap

Feed inlet

Direction of rotation

A.

B.

Weighing Systems

A **weighing system** is used to weigh material before shipping. Weighing systems are similar to the scales we use in our homes. Weighing systems are important because of weight requirements or restrictions. They are also used to establish the weight shipped to a customer.

Weighing system a system used to weigh material before shipping.

An example of a weighing system is a series of load cells used on surge bins to show the current amount of material in storage. The load cell contains a strain gauge. A strain gauge load cell equates an amount of bin weight to a specific amount of electrical resistance. Electrical resistance occurs when the bin weight increases and deforms the strain gauge in the load cell, which is attached to each leg of the bin. This amount of electrical resistance is converted into a weight measurement. Most strain gauges are of a double-switch type. This allows one switch to fail without the system shutting down. Because strain gauges are often mounted under bins and hoppers, replacing one usually requires considerable downtime and expense. Figure 18.22 shows an example of a strain gauge.

Multiple strain gauges (load cells) are typically used to measure one bin. The use of three gauges is preferred because it allows a more stable system than a two-point system, where overloading on one side or the other could occur.

Weighing systems for hopper cars and hopper trucks use a system of load cells mounted to a steel frame under the rails or loading platform to provide the weight measurement. Modern railroad weigh scales allow the automated weighing of entire trains as they cross the scale at speeds up to 6 mph (10 km/hour).

Figure 18.22 Train scale using a strain gauge system.

Compression increases the amount of resistance

Resistance is measured between two points

Bulk Bag Station

A bulk bag (also referred to as a *supersack*) station is a piece of equipment designed to fill bulk bags to a predetermined weight. Bag filling can be manual or automated. Solid material flows into the bulk bag from a storage location. Bag weight is constantly measured using a load cell, and material flow is terminated when the desired weight is achieved. Automated systems are designed to have an automatic feed system that fills, weighs, and seals the bag and then removes and transfers it to a storage location for shipment. When filled, bulk bags typically weigh between 1,000 and 3,000 pounds (454 to 1361 kg). Figure 18.23 shows finished bags being moved to a bulk bag stacking area and bags awaiting shipment.

Figure 18.23 A. Bulk bags filled and moving to bulk bag stacking. **B.** Bulk bags being stacked by forklift before shipment.

CREDIT: **A. and B.** Mr. Amarin Jitnathum/Shutterstock.

A.   **B.**

Large bags can be desirable because they contain a large quantity of product in one container. Large bulk bag handling equipment can be costly, however, so smaller bags (typically 50 pounds; 23 kg) are often used.

Bagging Operations

Some customers require finished products to be packaged in smaller quantities. In these instances, bags often are the preferred container. Bags come in all sizes, such as 50-pound (23-kg) bags of corn, 25-pound (11-kg) bags of dog food, and 80-pound (36-kg) bags of concrete. In general, these bags use the same basic packing equipment. Packaging operations may be fully automated or totally manual.

After the finished product is classified (e.g., run through screening equipment so it can be sorted to the desired size), the material is pneumatically transferred to a hopper that is part of the bagging operation. Depending on the type of bag to be filled, modern bagging equipment will pick up an empty bag, place it under a filling nozzle, and meter the product flow into the bag, which sits on load cells until the bag reaches the target weight, when the flow of product stops. In a valve bag (Figure 18.24A), the bag is placed over and encloses the filling spout, which minimizes product spillage. After being filled, the bag is automatically sealed, or a sealer is used to seal the bag. In an open mouth bag (Figure 18.24B), the bag is filled, then transferred immediately to a sealing mechanism that glues or sews the bag closed. After being filled and sealed, the bag is conveyed to a palletizer.

Figure 18.24 A. Valve bag filling. **B.** Open mouth bag filling and sealing.

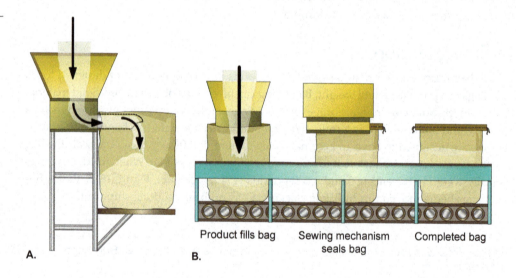

Product fills bag   Sewing mechanism   Completed bag
                       seals bag

A.   **B.**

A palletizer (Figure 18.25A) is a piece of equipment designed to stack the filled bags in a specific pattern on a pallet. The pattern (shown in Figure 18.25B) is determined by the size of the bag and the number of layers or tiers required to achieve the maximum allowed height or weight and still maintain a stable stack for shipping. The last step in the bagging process is automated stretch wrapping. A stretch wrapper (shown in Figure 18.25C) wraps the pallet with a wide film of polyethylene wrap (36 to 48 inches/ 91 to 122 cm wide) to prevent the bags from shifting. The number of revolutions around the pallet depends on the thickness of the stretch wrap needed to provide a secure and stable bag stack on the pallet. The pallet is now ready for shipping to customers.

The shipping container determines the final pallet height and weight. Pallet design must take into consideration the available room in the truck trailer, sea container, or railcar. The availability of the equipment needed to move the pallet from the packaging floor to the shipping container and the ultimate weight of the container filled with pallets must also

Figure 18.25 **A.** Robotic palletizer. **B.** Pallet stacking pattern. **C.** Stretch wrapping equipment.

CREDIT: **A.** Chesky/Shutterstock. **B.** Slavoljub Pantelic/Shutterstock. **C.** Baloncici/Shutterstock.

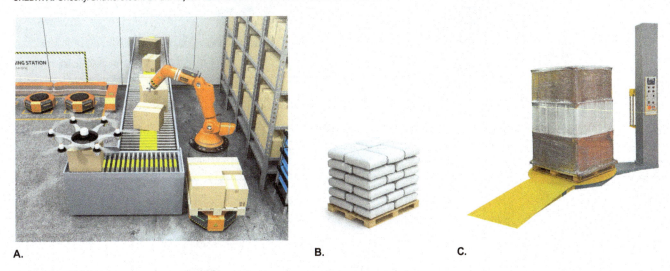

A.   B.   C.

be considered. If a tractor trailer is over a legal weight limit, the trucking company can be fined when the truck is weighed at an interstate weigh station. All of these variables must be considered when designing the packaging system for a specific product.

18.2 Potential Problems and Hazards

Potential Problems

When working with solids handling equipment, process technicians should always be aware of potential problems, such as equipment breakdown, equipment plugging, electrical power surges, or failure, pinch points, slips, and falls. Equipment malfunction can create hazardous conditions and result in product waste or rework, impacting profitability. If equipment breaks down, it can cause production interruptions throughout the process facility. Plugged equipment can cause facility downtime and require process technicians to have to enter equipment to clear the pluggage, which can result in the potential for exposure to dangerous materials and result in product spillage. Spills, which can cause slipping hazards and lost production, also are potential problems that face process facilities. Some equipment is protected by a process **interlock** (a system for connecting mutually independent equipment that can stop upstream equipment when downstream equipment is shut down). A process interlock is a logic constraint within a control system to prevent damage to equipment. It also provides a safety function to prevent the risk of injury to plant personnel.

Interlock a system for connecting mutually independent equipment that can stop upstream equipment when downstream equipment is shut down.

  Human error can result in equipment downtime for clearing pluggage or for unplanned cleaning. It can also result in equipment failure. The most important reason to eliminate human error, however, is that it can result in serious injury to personnel. Because of this, it is very important that process technicians follow all standard operating procedures to ensure that equipment operates properly and safely.

Safety and Environmental Hazards

Hazards associated with normal and abnormal operation of solids handling equipment can affect personal safety, equipment, production, and the environment. They can lead to injury to personnel or an environmental incident. Product spills and dust explosions can endanger personnel and damage the surrounding area. Table 18.1 lists several potential hazards and their effects.

Table 18.1 Hazards Associated with Solids Handling Equipment

Improper Operation	Possible Effects			
	Individual	**Equipment**	**Production**	**Environment**
Product spills	Injury from slipping or chemical exposure	Damage and repair Shutdown and cleaning before restart	Loss of production caused by downtime	Ground or water contamination
Excessive dust accumulation	Explosion Dust exposure	Damage and repair Shutdown and cleaning before restart	Loss of production caused by downtime	Atmospheric dust pollution
Solids plugging	Confined space entry required to clear pluggage Chemical exposure	Shutdown and cleaning before restart	Loss of production from equipment shutdown	Ground or water contamination
Equipment failure	Personal injury Chemical exposure	Repair or replacement	Loss of production caused by downtime	

18.3 Process Technician's Role in Operation and Maintenance

Process technicians should be familiar with the appropriate procedures for safely monitoring and controlling solids handling equipment components and conditions. The typical procedures, which include startup, shutdown, emergency, and lockout/tagout, vary from site to site and for each type of solids handing equipment. Process technicians must follow the standard operating procedures (SOPs) for their assigned unit(s) in order to ensure personal and process safety and to prevent spills or explosive events.

A process technician's role in the safe operation and maintenance of solids handling equipment is critical and is the same as for all processing equipment (i.e., following standard operating procedures for the facility and the unit). Process technicians are required to report all abnormal events, such as spills and injuries, to their supervisor. Process technicians must understand operational procedures relating to each piece of equipment in order to maintain safe operating conditions, report any violation to their supervisor, and be able to respond in the event of an emergency.

Process technicians should routinely monitor solids handling equipment to ensure that it is operating properly. When monitoring and controlling this equipment, process technicians should always remember to look, listen, and check for the items indicated in Table 18.2.

Table 18.2 Process Technician's Role in Operation and Maintenance

Look	Listen	Check
■ Observe for leaks or the accumulation of solids or dust as these can lead to blockage or explosion ■ Be alert for systems operating at higher or lower than normal speeds ■ Look for system pulling higher amps than normal ■ Monitor for gas or solids leaking into the atmosphere	■ Abnormal noises such as banging or hammering ■ Normal noises with abnormal pitches	■ Excessive heat in equipment and production areas ■ Surging through the lines (e.g., normal flow with instances of excessively high or low flow)

Summary

Solids handling equipment is used to process, transfer, store, package, and ship solids. Solids can be moved or transferred in a variety of ways.

A conveyor is a mechanical device used to move solid material from one place to another. Types of conveyors include belt, screw, moving floor, vibratory, powered rollers, and both dilute phase and dense phase pneumatic. Although each conveyor is designed differently, their basic use is the same: to move material from one place to another.

A feeder is a piece of equipment that conveys material to a processing device. Feeders operate gravimetrically or volumetrically. Gravimetric feeders convey material at a controlled rate, measuring the weight lost over time. Volumetric feeders convey material at a controlled rate by running at a set motor speed.

Extruders create shaped products (e.g., pellets, strings, or sheets) by forcing malleable solids through a perforated plate, called a die.

Bucket elevators use a series of rotating buckets attached to a belt to move material from a pit or pile to an elevation above the pit or pile.

Screening systems are used for particle-size distribution. Screeners classify (separate) dry materials according to particle size. Material enters at one end and is conveyed across the screen. Smaller particles fall through the screen, while larger particles remain above the screen surface and exit the screener at the lower end.

Silos, bins, and hoppers are storage containers used to hold material (usually dry). A silo is a large container used for on-site storage. Silos are typically vertical storage containers that can reach up to 275 feet (84 m) high. Bins are smaller containers that are designed to hold a finished product destined for distribution. Hoppers are the smallest of the three and are usually mounted directly to solids processing equipment. Bins and silos can have mechanisms at the bottom (e.g., a live bottom) that allow material to be removed easily.

A trickle valve uses negative air pressure to contain a quantity of product until that product reaches a specific weight. A free-hanging flap is used between two pressure environments to maintain a seal. After the weight of the product reaches a set limit, the flap opens and the product passes through.

A rotary valve uses a set of rotating pockets to meter a fixed volume of solids at a fixed rate from a low-pressure environment to a higher-pressure environment.

Hopper cars and hopper trucks are designed to carry solid materials by rail or over the road. There are single-bin hoppers and multiple-bin hoppers. Hoppers discharge their load by either dropping directly from the bottom of the hopper into a pit or tanker, or by pneumatic transfer of the material through hoses and piping systems. Whether the solid material is stored in bins or hopper cars or trucks, load cells are used to determine the weight of the stored material.

Packaging systems transfer product into containers such as bags, supersacks, or boxes. Bulk bags typically contain quantities ranging from 1,000 to 3,000 pounds. Bagging equipment fills smaller bags manually or automatically. The bagging equipment also transfers the bags or boxes to pallets and secures them with plastic wrap called stretch wrapping.

Process technicians must be aware of potential problems that can exist with solids handling equipment. Technicians must also monitor equipment for safety and environmental hazards and follow proper operating procedures at all times.

Checking Your Knowledge

1. Define the following terms:

   a. Extruder

   b. Screener

   c. Bin

   d. Conveyor

   e. Feeder

   f. Hopper

2. (True or False) A conveyor is a piece of equipment used to mix raw product to form a new product (usually dry material).

3. (True or False) A gravimetric feeder conveys material at a controlled rate by running at a set motor speed.

4. (True or False) A volumetric feeder must have a predetermined maximum feed rate that is calculated by the feeder's maximum motor speed and material density.

5. Which of the following is a small storage container usually attached to another piece of equipment that it feeds?

   a. Silo
   b. Bin
   c. Hopper
   d. Cyclone

6. (True or False) A trickle valve is a free-hanging flap between two different pressure environments that maintains an air seal inside a flow area.

7. Which of the following is not a piece of bag packaging equipment?

   a. Stretch wrapper
   b. Fluid bed dryer
   c. Valve bag filler
   d. Palletizer

8. Which of the following is the most common effect of product spills?

   a. Dust pollution
   b. Dust exposure
   c. Loss of production
   d. Chemical exposure

9. Which of the following solids handling equipment can be weighed using strain gauge load cells? (Select all that apply.)

   a. Palletizer
   b. Volumetric feeder
   c. Surge bin
   d. Bulk bag

NOTE: Answers to Checking Your Knowledge questions are in the Appendix.

Student Activities

1. Write a one-page paper about how a bucket elevator system works. Describe the pieces of equipment. Give an example of when a bucket elevator system is or can be used in industry. Draw a diagram of the system.

2. Describe how a pneumatic conveyor in a grain processing facility works from beginning to end. Explain how you would know how much material is moving through the system.

3. Use the Internet to research three (3) different industries that use bins or silos. Create a one-paragraph explanation about each industry and share with the class.

Chapter 19
Environmental Control Equipment

*North American Process Technology Alliance (NAPTA) has developed curriculum to ensure that Process Technology courses will produce knowledgeable graduates to become entry-level employees for the career field of process technology. Objectives from that curriculum are named here in abbreviated form (e.g., NAPTA Intro to Tools and Equipment 7, 8 means that this chapter's objective addresses objectives 7 and 8 of the NAPTA curriculum about tools and equipment).

Key Terms

Baghouse—air pollution control equipment that contains fabric or bag filter tubes, envelopes, or cartridges designed to capture, separate, or filter particulate matter, **p. 389.**

Dike—a wall (earthen, metal or concrete) built around a piece of equipment to contain any liquids should the equipment rupture or leak, **p. 396.**

Electrostatic precipitator—a device that contains positively charged collecting plates and a series of negatively charged wires, **p. 389.**

Flare system—a process safety system in which unwanted process gases are burned to prevent their release into the atmosphere, **p. 390.**

Incinerator—a device that uses high temperatures to destroy solid, liquid, or gaseous wastes, **p. 392.**

Landfill—a human-made or natural pit, typically lined with an impermeable, flexible substance (e.g. rubber) or clay, which is used to store residential or industrial waste materials, **p. 398.**

Recuperator—a heat exchanger that preheats the waste gas feed stream with the flue gas exiting the thermal oxidizer, **p. 393.**

Scrubber—a device used to remove contaminants from a gas stream by washing the stream with water or some other neutralizing agent, **p. 391.**

19.1 Introduction

Several factors have spurred the process industries to maximize the use of environmental controls in processes and operating equipment. These factors include the ever-increasing need to protect Earth's natural resources, the advent of emerging technologies, and the enactment of stricter environmental regulations. Environmental disasters such as the release of heavy metals into the drinking water of Flint, Michigan, and the Deepwater Horizon oil spill in the Gulf of Mexico have reinforced the need for stricter environmental controls and procedures when dealing with both hazardous and nonhazardous chemicals.

The environment suffers when emission of impurities, by-products, or other materials into the air, land, or water reaches unacceptable levels. When the environment suffers, people and other living things suffer. Laws are in place at federal, state, and local levels to ensure that the environment is protected at all times. Industrial companies nationwide have stepped forward to enhance operations of their facilities in order to meet or exceed those standards.

The purpose of designing and implementing environmental controls is to maintain the operation of process facility equipment within the prescribed standards allowable by law. The process of environmental control is accomplished through monitoring, sampling, and using control devices. Control devices can be categorized by the natural resource being preserved and by the type of operating equipment that is regulated by each control mechanism.

Types of Environmental Control Equipment

Air Pollution Control

Many industrial processes (e.g., catalytic cracking, sulfur recovery, vessel loading) and equipment components (e.g., furnaces, boilers, incinerators, dryers, and gas turbines) vent to the atmosphere and require the combustion of fossil fuels in order to operate. As hydrocarbon-based fossil fuels are burned, nitrogen oxides (NO_x), carbon monoxide (CO), and sulfur dioxide (SO_2) are produced. NO_x, CO, and SO_2 emissions are extremely harmful to living organisms. Carbon dioxide (CO_2) and methane (CH_4), which are also produced, are not toxic at normal levels but are greenhouse gases, affecting the environment by contributing to global warming. The release of particulates by some process equipment has been linked to adverse health effects such as asthma and cancer. These pollutants contribute to acid rain and join with hydrocarbons to create ground-level ozone. Because of this, these emissions must be carefully monitored and maintained according to the terms of the Clean Air Act.

The Clean Air Act, which was enacted by Congress in 1977 and amended in 1990, requires industry to reduce and control hazardous emissions that affect ambient air (atmospheric air that we breathe). Title V, one of several programs in the 1990 amendment to the Clean Air Act, requires local and state air quality agencies to issue operating permits for facilities that produce significant amounts of air pollution.

In light of these rules and regulations, many technologies and operating strategies have been developed to reduce NO_x emissions. Processes such as selective catalytic reduction (SCR) and selective noncatalytic reduction (SNCR) use a substance, such as ammonia, to facilitate a chemical reaction which yields nitrogen and water to reduce the amount of NO_x

emissions generated in combustion equipment. Other measures, such as switching to a fuel with a lower nitrogen content, flue gas recirculation (FGR), or the use of low NO_x burners reduce the amount of NO_x formed in combustion processes. These types of emission control processes maintain high NO_x reduction rates while minimally affecting the performance of the process equipment.

BAGHOUSES AND PRECIPITATORS (ELECTROSTATIC)   Boilers and power generation facilities that burn fossil fuel, such as coal, use baghouses and electrostatic precipitators as pollution control devices. These devices prevent combustion residue from escaping the stacks and passing into the atmosphere. Combustion residue is commonly particulate matter, such as unreacted carbon particles.

Baghouses consist of fabric or bag filter tubes, envelopes, or cartridges designed to capture, separate, or filter particulates as the effluent gas stream passes through them. **Electrostatic precipitators** contain positively charged collecting plates and a series of negatively charged wires or electrodes. As particulates in the gas pass over the wires, they become negatively charged and are attracted to the positively charged collecting plates. Once these particles have been captured, the clean effluent gas is discharged to the atmosphere and the particulates are collected for disposal or incorporation into products like cinder blocks (blocks made of ash and cement that are used in construction). Figure 19.1A shows a baghouse; Figure 19.1B shows how an electrostatic precipitator processes particles.

Baghouse   air pollution control equipment that contains fabric or bag filter tubes, envelopes, or cartridges designed to capture, separate, or filter particulate matter.

Electrostatic precipitator   a device that contains positively charged collecting plates and a series of negatively charged wires.

Figure 19.1 **A.** Baghouse. **B.** Electrostatic precipitator.

CREDIT: **B.** Artwork studio BKK/ Shutterstock.

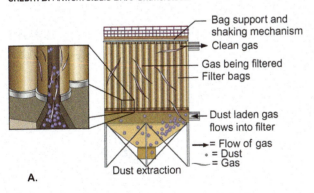

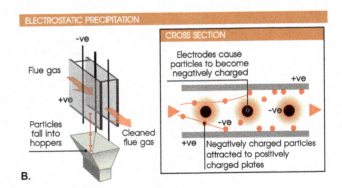

COAL GASIFICATION   Another environmental control technology being implemented today is coal gasification. In this process, coal is converted from a solid to a gaseous state before being burned in power plants to produce electricity. Gasification helps minimize carbon monoxide (CO), nitrogen oxides (NO_x), and sulfur oxides (SO_x). It also reduces the amount of carbon dioxide (CO_2) emitted to the atmosphere. CO_2 contributes to the *greenhouse effect* (trapping of CO_2 in the earth's atmosphere), which is commonly thought to promote global warming.

VAPOR AND GAS EMISSION CONTROLS   Several other types of vapor and gas emission equipment are used to capture and/or recycle process materials rather than emit them to the atmosphere. For example, some tanks and vessels incorporate vapor recovery systems that take the vapor emitted by these vessels, cool it to a liquid phase, and return the liquid back to the vessel. Figure 19.2A shows a flare gas recovery system. Figure 19.2B is a schematic of a flare gas recovery system.

Figure 19.2 A. Photo of flare gas recovery system. **B.** Schematic of flare gas recovery system.
CREDIT: A. Courtesy of John Zink Company, LLC.

A.

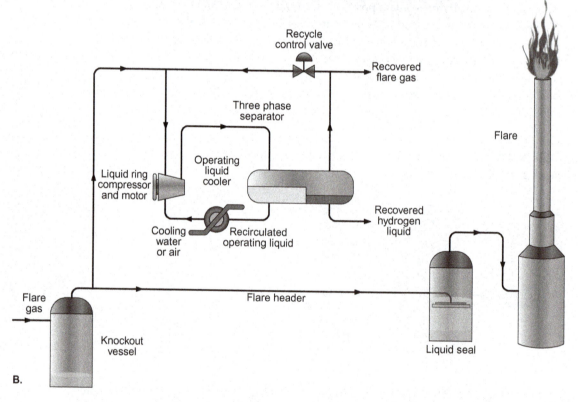

B.

FLARE SYSTEMS   **Flare systems** are important process safety devices. They are used to safely burn excess process gases, which cannot be recovered, before they are released into the atmosphere. Flare systems are most often used during emergency situations and unit startups and shutdowns to rapidly dispose of gases. They can be vertical or ground mounted. Figure 19.3 shows an example of a vertically mounted flare stack.

Figure 19.3 Vertical flare stack.

CREDIT: Nightman1965/ Shutterstock.

Most flares are equipped with steam injection systems to aid in the combustion process and to protect the tip of the flare from damage caused by heat and flame. When operating properly, a flare reduces the gases disposed through it to carbon dioxide and water vapor. In order to maintain a flare's capability to receive and burn gas at any time, these units continuously burn a small amount of pilot gas to ensure there is an ignition source present.

Flare systems are considered the last resort for venting process gases, and every effort is made to minimize flaring. They can be noisy and may emit smoke if they are not properly controlled. The improper flaring of process gases creates emissions such as NO_x, SO_2 and CO_2. The white "smoke" leaving a flare stack is steam. Flare systems are highly regulated by the Environmental Protection Agency (EPA) as well as state agencies in the states where these facilities are located.

GAS TREATMENT (CATALYTIC AND NONCATALYTIC)   Different forms of catalytic processes are used to remove SO_2 (sulfur dioxide), H_2S (hydrogen sulfide), and NO_x (nitrogen oxides) from the effluent gas streams of chemical manufacturing, natural gas production, and oil and gas refining facilities. These gas streams are also commonly referred to as *flue gas* or *off gas*. In one process, called the Claus process, H_2S and SO_2 in refinery off gas is converted into elemental sulfur and water. The sulfur is then used to manufacture products like cosmetics, fertilizer, and pharmaceuticals. In another catalytic gas treatment process, called selective catalytic reduction (SCR), NH_3 (ammonia) is added to a flue gas stream, which then flows over a catalyst bed that converts the NO_x to nitrogen gas (N_2) and water—elements that are more compatible with the environment.

A common method of noncatalytic treatment of gas streams is to inject NH_3 (ammonia) into the gas stream, a process called *selective noncatalytic reduction* or SNCR. This redox reaction (an oxidation-reduction reaction) converts the NO_x to water and nitrogen, elements that are more environmentally friendly. Another material that can be used for this process is urea (an organic compound of carbon, nitrogen, oxygen, and hydrogen that is produced from the reaction of synthetic ammonia and carbon dioxide). When injected into flue gas, urea converts the NO_x to nitrogen, carbon dioxide (CO_2), and water (H_2O).

SCRUBBERS **Scrubbers** are devices that remove or neutralize unwanted gases or particulates from a gas stream by spraying the stream with liquid (water or reagents), sometimes mixed with crushed limestone, which acts as an absorbent. In a scrubber, the gas is passed through a liquid that is moving in the opposite direction (counterflow). The liquid absorbs the contaminants from the gas before it is released to the atmosphere or sent for further treatment elsewhere. Figure 19.4 shows an example of a scrubber system.

Another type of contaminant removal system is a fixed bed adsorption system, typically using an activated carbon adsorber. In an activated carbon adsorber, gas passes through a fixed bed of activated carbon. As the gas flows through the system, contaminants are attracted to the pores in the carbon granules, attach to the carbon, and are removed while the clean gas is released to the atmosphere or sent for further treatment (Figure 19.5).

Scrubber a device used to remove contaminants from a gas stream by washing the stream with water or some other neutralizing agent.

Figure 19.4 Example of a scrubber system.

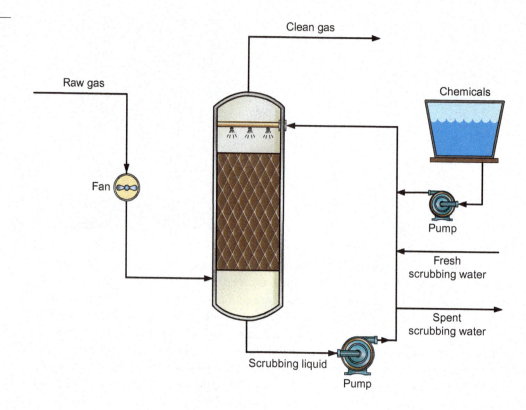

Figure 19.5 Example of a carbon adsorption system.

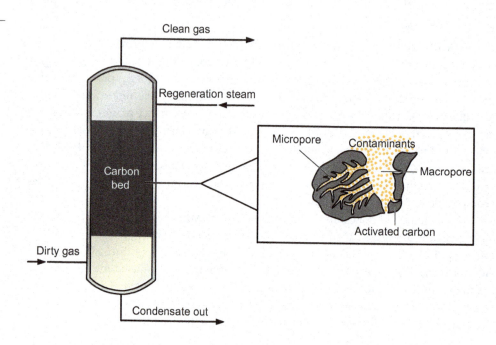

Incinerator a device that uses high temperatures to destroy solid, liquid, or gaseous wastes.

INCINERATORS (THERMAL OXIDIZERS) **Incinerators** are devices that use high temperatures to destroy solid, liquid, or gaseous wastes. In the process industries, these wastes are typically hydrocarbon-based gases and volatile organic compounds (VOCs). In an incineration system, a small amount of the waste stream remains after incineration. For this reason, these systems are strictly regulated by the EPA and state and local environmental agencies, and must meet the quality mandates for waste disposal (usually at least 95% destruction of the products of the incineration process) before vapors can be introduced into the atmosphere.

Hazardous waste incinerators utilize extremely high temperatures, usually well above 1000 degrees Fahrenheit (538 degrees Celsius), to thermally destroy waste material. The resulting flue gas stream is sometimes cooled before being returned to the atmosphere in order to conserve energy and to minimize thermal pollution.

During incineration, there is often enough heat energy left in the flue gas stream to generate steam in a waste heat boiler. This secondary steam production reduces the operating costs associated with incinerator operation. Figure 19.6 displays a schematic for a thermal oxidizer with a **recuperator** (a heat exchanger that preheats the waste gas feed stream with the flue gas exiting the thermal oxidizer) and a waste heat boiler.

Recuperator   a heat exchanger that preheats the waste gas feed stream with the flue gas exiting the thermal oxidizer.

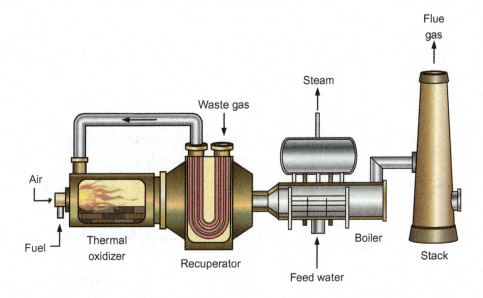

Figure 19.6 Schematic of thermal oxidizer with recuperator and waste heat boiler.

Water and Soil Pollution Control

Other areas of environmental concern are Earth's waterways and soil. Dumping into the soil or pollution of waterways can lead to the destruction of ecosystems and the leaching of contaminants into the drinking water supply.

To prevent contamination of the country's waterways, the Federal Water Pollution Control Act was enacted by the U.S. government in 1948. It was revised in 1972 and renamed the Clean Water Act. To address soil contamination, Congress passed the Resource Conservation and Recovery Act (RCRA) in 1976. This law governs the disposal of both nonhazardous and hazardous solid waste and establishes requirements for underground storage tanks.

The Clean Water Act regulates the discharge of pollutants into waterways in the United States and prohibits their discharge without a permit. A facility's National Pollutant Discharge Elimination System (NPDES) permit also establishes specific monitoring and reporting requirements that the facility must uphold. The act authorized the EPA to set wastewater standards for industry.

Water in and around process facilities must be constantly monitored for contaminants. Some of these contaminants may come from process leaks, overburdened process sewer flows, or surface water or rain that has been contaminated by contact with a piece of operating equipment. Other pollutants are produced as a direct result of a process unit's operation. For example, condensate produced from the heating or cooling of a process, water used for equipment cleaning, and wastewater are all potential sources of pollutants that require a disposal plan.

Wastewater is one of the main disposal concerns in process facilities. Most of this water comes from process sewers or from storm drain overflow during periods of heavy rainfall. Before it can be introduced into the environment, this water must be cleaned of all hydrocarbon material, and the pH must be brought within an acceptable range. It must be thermally adjusted (heated or cooled), and it must be properly oxygenated to meet all federal, state, and local environmental regulations.

PRIMARY WASTEWATER TREATMENT PROCESS   The wastewater cleanup process is accomplished in several steps. First, the water is processed through a primary treatment step, where any oil or solids are separated from the wastewater and the pH is adjusted to ensure optimum performance of the activated sludge process.

Separators and Interceptors   In an attempt to recover and recycle oil from a process wastewater or storm water stream, facilities use various devices to skim the oil from the water's surface. These units are called American Petroleum Institute (API) separators, parallel plate interceptors (PPIs), or corrugated plate interceptors (CPIs). Each of these devices works in a similar manner. For example, each unit must include an upstream sediment bay, skimmer, weir, separator basins, grit chambers, bar screens, and sludge hoppers. Figure 19.7 shows an example of an API separator.

Figure 19.7 Diagram of an API separator.

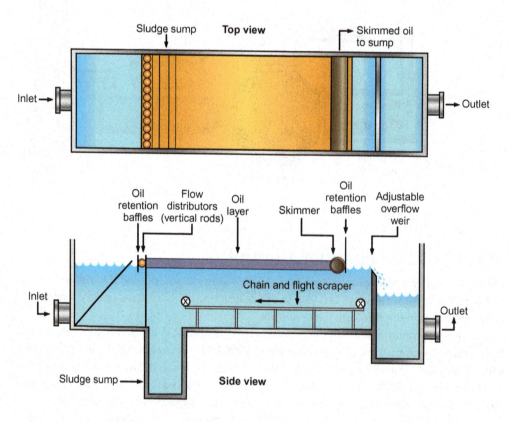

The design of the API separator is based on specific gravity differences between the oil and the wastewater. Because of their different densities, a solids layer settles to the bottom of the separator, the oil rises to the top, and the wastewater becomes the middle layer between the oil and the solids.

Parallel plate interceptors (PPIs) perform the same function as API separators, but with a parallel plate assembly. PPIs, like API separators, rely on the specific gravity difference between the oil and the water for separation. One advantage of this type of separator is that its design takes up less space than the standard API separator.

Corrugated plate interceptors were a design improvement of the parallel plate interceptor. Rather than plates being positioned horizontally, they use a series of corrugated plate packs at an inclined angle in a vessel to separate oil, water, and solids. Corrugated plate interceptors (CPIs) are more efficient in separating oil and solids from the wastewater. The plates are made from fiberglass, rather than metal, to prevent corrosion. Figure 19.8 shows an example of a corrugated plate interceptor.

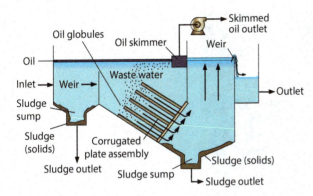

Figure 19.8 Diagram of a corrugated plate interceptor (CPI).

SECONDARY WASTEWATER TREATMENT PROCESS

Activated Sludge Process After undergoing primary treatment, the water is then mixed with bacteria as it enters the secondary treatment step, where it is aerated and the bacteria (often referred to as bugs) interacts with the wastewater. These microorganisms (bacteria) are attracted to and ingest (consume) hydrocarbons and other pollutants. The activated sludge process is an aerobic process, meaning that oxygen is required for the process to be effective. Figure 19.9 shows a secondary treatment process.

Figure 19.9 Secondary wastewater treatment process.
CREDIT: Kekyalyaynen/Shutterstock.

Did You Know?

The United States spends more money on environmental control than any other country in the world.

The EPA's Annual Performance Plan and Budget for fiscal year 2016 was $8.6 billion.

CREDIT: Mongkolchon Akesin/Shutterstock

CLARIFIER After going through the activated sludge process, the wastewater enters a secondary clarifier. In this step of the process, the bacteria settle to the bottom, are removed as waste sludge, or are recycled for reuse in the aeration basins.

Figure 19.10 shows an example of a *clarifier*. The clarifier feeds wastewater up through a pipe in its center and distributes it into the "settling zone" where the bacteria settles out of the water and falls to the bottom of the clarifier.

If removed as waste, the bacterial sludge is then dried into a cake and burned to ash in an incinerator.

Exhaust gases from this incineration process can be passed over water tubes in a waste heat boiler, where steam is generated for facility use. The exhaust gases are cooled before being discharged into the atmosphere. The effluent from this incinerator is scrubbed or washed with water and a caustic spray mixture to ensure that the ash or combustion products do not reach the atmosphere.

Figure 19.10 Clarifier.

CREDIT: Kekyalyaynen/Shutterstock.

Water leaving the clarifier enters a third treatment stage, which may include carbon filtration or other treatment methods that remove volatile organic compounds (VOCs) such as benzene or toluene. Water is sampled for both VOCs and oxygen content to be sure it complies with the discharge permit requirements before it is returned to the waterway.

Activated carbon is utilized in most of these treatment facilities to provide an adsorbing (adhering) surface. The organics and microorganisms in the stream will attach to this surface. Figure 19.11 is an example of a typical process facility wastewater treatment plant.

The activated sludge plant is the most common type of wastewater treatment plant used in industry, but there are other EPA-approved methods of treating wastewater.

Figure 19.11 Diagram of typical process facility wastewater treatment plant. Wastewater enters at the API separator.

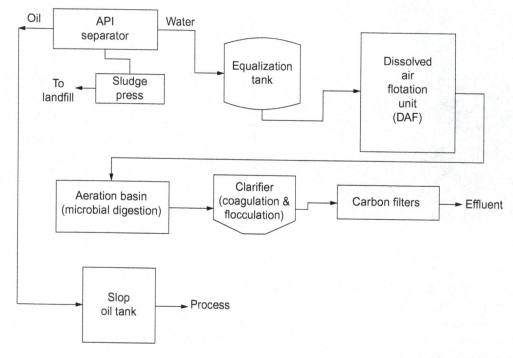

Dike a wall (earthen, metal, or concrete) built around a piece of equipment to contain any liquids should the equipment rupture or leak.

DIKES Dikes are another environmental control measure. **Dikes** are containment devices (earthen berms, metal or concrete walls) built around a piece of equipment to contain any liquid should the equipment rupture or leak. These barriers serve as dams designed to contain overflows or spills from process equipment. They are designed to hold at least 10% more liquid than the equipment they surround. They also hold surface water such as rain in the event that water becomes contaminated. The water trapped inside these dikes must be tested for chemical compounds and treated, if necessary, before it is allowed to return to a waterway. If a tank overflows, the contaminated earth must often be removed and replaced with clean fill.

Dike failure can be catastrophic. Tailings dams are built to contain mining wastes and are some of the largest engineered structures in the world. A failure of a tailings dam can have a significant impact on nearby communities and the environment. For example, a 2015 accident in Brazil killed 19 people and contaminated the drinking water supply for thousands when a tailings dam experienced a catastrophic failure and collapsed. A tailings dam failure in British Columbia, Canada, dumped millions of cubic yards of gold and copper mine waste into a neighboring lake, polluting the area's watershed. Dike failures lead to property destruction, ruined fisheries, contaminated drinking water, and polluted water sources.

SETTLING PONDS   Wastewater is also treated in settling ponds or basins before being returned to a natural waterway. These settling ponds may be constructed of concrete or simply earthen pits dug out and lined with a heavy rubber liner. During the treatment process, water is diverted into these ponds, where it is mechanically aerated to increase the oxygen content and remove chemical gases. These ponds allow sufficient residence time for solids to settle to the bottom. The effluent water is then removed, filtered, and chemically treated (neutralized), and its temperature adjusted before being returned to waterways.

Retention time in settling ponds can range from less than one day to several days. These ponds are sometimes operated in series so that each body of water becomes cleaner than the one before it. Figure 19.12A is an example of a settling pond with a liner.

For settling ponds to be effective, they must be aerated (have oxygen added). Figure 19.12B shows an example of aeration inside a settling pond.

Figure 19.12 A. Settling pond with liner. **B.** Aeration device in a wastewater settling pond.

CREDIT: **A.** Linda Armstrong/Shutterstock. **B.** JL Jahn/Shutterstock.

A.

B.

LANDFILLS   Another type of environmental control used for solid waste is a landfill. All landfills come under the regulation of the RCRA Subtitle D for solid waste and Subtitle C for hazardous waste. Examples include the following:

- Municipal solid waste landfills
- Bioreactor landfills
- Industrial waste landfills
  - Construction and demolition debris
  - Coal combustion residual

Hazardous wastes are regulated by the Toxic Substances Control Act, Subtitle C, and include the following:

- Hazardous waste, not including solid waste
- Polychlorinated biphenal (PCB) waste

Landfill a human-made or natural pit, typically lined with an impermeable, flexible substance (e.g. rubber) or clay, which is used to store residential or industrial waste materials.

Landfills are human-made or natural pits, typically lined with an impermeable, flexible substance (e.g., rubber) or clay. They are used to dispose of residential or industrial waste materials. Landfills are classified as sanitary landfills if they contain the type of waste disposed of in homes and restaurants. Hazardous landfills contain industrial materials and byproducts that can be harmful to the environment if not contained and treated properly.

Did You Know?

The decomposition of organic materials in landfills produces methane gas. This gas can be harnessed and used to power homes and businesses, reducing energy costs while also reducing the amount of greenhouse gas emissions released into the atmosphere.

CREDIT: Kamil Macniak/Shutterstock.

Because the contents of landfills pose potential environmental risks, their construction is complex. Landfills must first be lined with a layer of impermeable material (e.g., rubber). A layer of absorbent clay is placed on top of this liner, and then another impermeable layer is placed on top before the waste can be placed in the facility. In addition to the liners, systems to detect leakage into the surrounding soil are included in the design. There may also be deep water wells on the perimeter of the landfill that extend into aquifers. These wells can be tested periodically to detect leakage into the watershed. Figure 19.13 is a detailed diagram of a landfill's components.

Other disposal methods may be required for different types of waste. For example, environmentally harmful liquids must be absorbed by sand or sawdust and placed in drums before disposal. Due to the limited amount of space in a landfill and the strict regulations that apply to these types of facilities, incinerators are used as a means of waste disposal when possible. Asbestos must be marked plainly in heavy plastic bags and disposed of in an area designated for hazardous wastes, before being taken to a landfill which must be permitted to accept asbestos.

Figure 19.13 Landfill diagram.

CREDIT: Courtesy of Eastman Chemical.

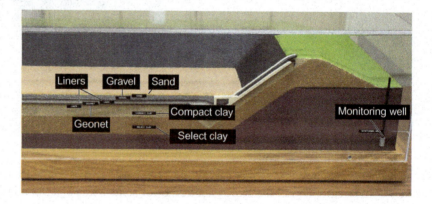

19.2 Environmental Rules and Regulations

Process technicians must be familiar with the rules and regulations associated with the environmental control equipment in their process areas. Environmental rules and regulations vary depending on the facility, the process, and federal and state requirements.

Federal Regulations

The Clean Air Act required the EPA to establish the National Ambient Air Quality Standards (NAAQS) for pollutants found in outdoor air that are harmful to people and the environment. These standards have been set for six common materials: ground level ozone, carbon monoxide (CO), sulfur dioxide (SO$_2$), particulate matter, lead, and nitrogen dioxide (NO$_2$). Each state is responsible for formulating a plan to address the pollutants and maintain the standards set by the EPA. State environmental agencies hold companies accountable for the release of these and other air pollutants. Through the Clean Water Act, a system of permits is used for eliminating the discharge of known pollutants into waterways. This system is called the National Pollutant Discharge Elimination System (NPDES). The Resource Conservation and Recovery Act (RCRA), passed by Congress in 1976, establishes a program to manage both hazardous and nonhazardous solid wastes, as well as underground storage tanks.

19.3 Potential Problems

Potential safety and health problems associated with environmental control equipment are based on several different issues. For example, nature itself can overburden a wastewater treatment facility when rains are torrential and sewers become overloaded. If inlet flows are too high, contaminated water or water that does not meet EPA standards may be released to rivers, lakes, or streams, causing environmental damage and resulting in fines for a company.

Equipment malfunction can also result in releases into the air, land, or water if the problem results in the inability of the process unit to meet EPA standards. One example would be a flare system that emits smoke over a long period of time, indicating that combustion is incomplete and contaminants are being released into the atmosphere.

Landfills that are not lined properly or that have leaking liners allow contaminants to leach into the soil and eventually into the ground water. Another problem is placing inappropriate materials into a landfill. For example, asbestos waste is prohibited from landfills unless the landfill meets additional requirements under the Clean Air Act.

Human error is another factor that may cause problems in an environmental control system. An employee who reads process instrumentation incorrectly or makes an incorrect adjustment can create a situation that allows hazardous material to be released to the environment. This can result in negative health effects for site employees and members of the general public living near the facility, as well as damage to the environment and equipment.

Noncompliance with federal regulations can lead to significant penalties and even jailtime for the responsible individual. Any emission that is federally regulated needs to be carefully monitored so that annual limits are not exceeded. Errors like leaving a vapor recovery unit offline can cause several issues (e.g., violating a government regulation or exceeding a permit limit). Preventing spills is very important, especially with operations in environmentally sensitive areas onshore or offshore.

19.4 Process Technician's Role in Operation, Maintenance, and Compliance

Due to the strict regulations and laws that govern the process industries, the process technician's role is critical to the success of a company's waste management and environmental control program. The EPA, along with other federal, state, and local government agencies, places regulations on and requires facilities to obtain permits to operate specific types of waste management processes and equipment. These permits specify policies and procedures for the safe operation of these types of equipment.

Process technicians must receive training which is mandated by the Occupational Safety and Health Administration (OSHA). This training is conducted through the company to ensure that permits are followed explicitly. Process technicians must also understand operational procedures relating to each piece of environmental equipment, understand the reporting requirements associated with the equipment, and be prepared to respond to any type of emergency situation that may arise.

Process technicians must be familiar with appropriate procedures for monitoring and maintaining environmental control equipment. Typical procedures such as startup, shutdown, emergency, and lockout/tagout vary at each individual site and for each type of environmental control equipment, so process technicians must always follow the standard operating procedures (SOPs) for their assigned unit.

Summary

The process industries use a wide variety of environmental control devices to protect the air, land, and waterways surrounding process facilities. The burning of fossil fuels and their resultant byproducts, the heat produced by process units, and the chemicals used to complete the production of goods must be managed to minimize their impact. This protects our health, safety, and the environment.

Equipment such as baghouses, precipitators, flares, wastewater treatment facilities, and scrubbers work together to eliminate emissions into the air, land, and water. These components are enhancements to specific processes and are strictly regulated by the EPA. Laws such as the Clean Air Act, the Clean Water Act, the Resource Conservation and Recovery Act (RCRA), and many others dictate the guidelines by which these equipment components can operate in a process facility.

Environmental controls have raised the level of awareness in the process industries regarding the need to protect the Earth's vital resources. That need has driven industry to modify, redesign, and create new technologies for the production of goods and to maintain their environmental control systems to better protect people and the environment.

Checking Your Knowledge

1. Define the following terms:

   a. Baghouse

   b. Dike

   c. Flare system

   d. Incinerator

   e. Landfill

2. List three measures, other than treating the NO_x generated by those processes, that can be taken to reduce the amount of NO_x produced in combustion processes.

3. List the three harmful compounds produced by the burning of carbon-based fossil fuels.

4. (True or False) Flare systems are used most often during emergency situations and unit startups and shutdowns to dispose of gases rapidly.

5. In a gas treatment process, a common method of non-catalytic treatment of gas streams is to scrub or wash the gas with

   a. sulfur monoxide

   b. hydrogen sulfide

   c. ammonia

   d. sulfur dioxide

6. What purpose does a facility's NPDES permit serve? (Select all that apply.)

   a. It allows the facility to set wastewater standards.

   b. It establishes specific monitoring and reporting requirements.

   c. It regulates the discharge of pollutants into waterways.

   d. It regulates the discharge of pollutants into the atmosphere.

7. List four common materials that the National Ambient Air Quality Standards regulate in the outdoor air.

8. (True or False) Landfills allow contaminants to leach into the soil to complete their proper decomposition.

9. ____ mandates training in environmental control rules and regulations.

   a. OSHA

   b. NOAA

   c. EPA

   d. HHS

NOTE: Answers to Checking Your Knowledge questions are in the Appendix.

Student Activities

1. Research one of the following acts and write a one-page report on it:

   a. Clean Water Act

   b. Clean Air Act

   c. Resource Conservation and Recovery Act (RCRA)

2. Work in teams to discuss and brainstorm a procedure for one of the following:

   a. disposal of salty wastewater

   b. disposal of used lithium batteries

   c. disposal of used light bulbs

3. Research one of the environmental incidents mentioned in the text (Flint, Michigan drinking water contamination or Deepwater Horizon oil spill) or another that is of interest. Present a report on it to the class.

Chapter 20
Auxiliary Equipment

 Objectives

Objectives

After completing this chapter, you will be able to:

20.1 Describe the purpose, types, components, and operations of common auxiliary equipment: agitator, mixer, agglomerator, eductor, ejector, centrifuge, hydrocyclone, and demister. (NAPTA Vessels 3 and 4, Reactors 3 and 4, Turbines 3 and 4, Pumps 4 and 5*) p. 403.

20.2 Identify hazards associated with auxiliary equipment. (NAPTA Vessels 8, Reactors 8) p. 412.

20.3 Describe the process technician's role in the use of auxiliary equipment. (NAPTA Vessels 8, Reactors 8) p. 412.

*North American Process Technology Alliance (NAPTA) developed curriculum to ensure that Process Technology courses will produce knowledgeable graduates to become entry level employees in process technology. Objectives from that curriculum are named here in abbreviated form. For example, "(NAPTA Reactors 3 and 4)" means that this chapter's objective relates to objectives 3 and 4 of NAPTA's course content on reactors.

Key Terms

Agglomerator—a mixer that feeds dry and liquid materials together to produce an agglomerated (clustered together) product like powdered laundry detergent, **p. 405.**

Agitator—a device that acts on a fluid by generating turbulence, which promotes mixing, **p. 403.**

Bernoulli principle—a principle stating that, as the speed of a fluid increases, the pressure of the fluid decreases, **p. 406.**

Centrifuge—a device that uses centrifugal force to separate solids from liquids or to separate fluids based on their density. **p. 409.**

Demister—a device that promotes separation of liquids from gases. Also called a mist eliminator, **p. 411.**

Drum mixer—a screw-type device used to stir, agitate, mix, or blend the contents of a drum into a consistent composition, **p. 406.**

Dynamic mixer—a piece of equipment that rotates to mix products rapidly, **p. 405.**

Eductor—a device that uses a fluid as a motive force to move another fluid using a Venturi design; also called a *jet pump eductor* or *aspirator*, **p. 407.**

Ejector—a device that is used to maintain vacuum by using a Venturi design powered by a motive gas (usually steam or air) flow. The ejector helps evacuate inert gases from the vacuum atmosphere; sometimes called a *vacuum ejector, injector,* or *steam ejector,* **p. 407.**

Hydrocyclone—a mechanical device that promotes separation of heavy and light liquids by centripetal force, a force that acts on a body moving in a circular path and is directed toward the center around which the body is moving (e.g., like a cyclone), **p. 410.**

Inline mixer—a mixer that has no moving parts, it operates by creating a complex flow path that results in contact between fluid streams; also called a *static mixer*, **p. 405.**

Jet pump eductor—a device that uses a high pressure liquid to create a Venturi effect to transfer fluid without the need for an external power source, **p. 409.**

Mixer—a device used to combine chemicals or other substances mechanically using a rotary motion, **p. 404.**

Venturi effect—the drop in pressure that occurs when liquid flows through a constricted section of pipe; as the speed increases, the pressure decreases, **p. 406.**

20.1 Introduction

The process industries use many different types of auxiliary equipment. The term auxiliary refers to items that act in a supporting capacity. In other words, auxiliary equipment is equipment that is secondary to the main processing unit but is necessary for the process to be completed safely.

Auxiliary equipment can be used for a variety of activities, such as creating a vacuum or mixing and transferring materials in order to produce a desired product. While the list of auxiliary equipment is quite extensive, this chapter covers only a few of the items commonly used in the process industries.

Types of Auxiliary Equipment

There are many different types of auxiliary equipment. Some of the most common types, however, include agitators, mixers, eductors, centrifuges, and hydrocyclones.

Agitators

Agitators and mixers are used throughout the process industries for mixing and blending of feedstocks and products. An **agitator** is a device that generates turbulence, which promotes mixing. Process technicians operate agitators using motors located outside the vessels. Agitators and mixers come in many different shapes and sizes.

Agitators are used in a wide range of mixing duties (e.g., solids suspension, liquid blending, and destratification), and can be used in many different industries and operations

Agitator a device that acts on a fluid by generating turbulence, which promotes mixing.

Figure 20.1 **A.** Example of a tank with an agitator inside. **B.** Agitator with components labeled.

CREDIT: A. Courtesy of Ecrecon, Inc.

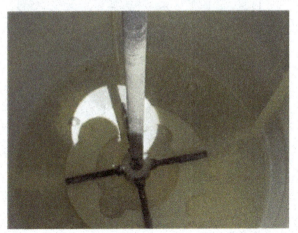

A.

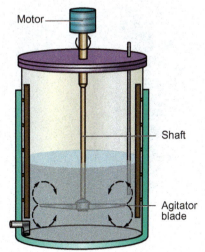

B.

Did You Know?

Many home washing machines contain an agitator. As the agitator rotates back and forth, it moves the clothes through the water/detergent mixture.

Once the cleaning process is complete, the washer drum spins rapidly to produce the centrifugal motion necessary to force the excess water out of the clothes.

CREDIT: Paul Velgos/Shutterstock.

(e.g., food and beverage manufacturing, waste-water treatment, and chemical production). Figure 20.1A shows an example of a tank with an agitator inside.

The components of an agitator are relatively simple. Agitators consist of a driver, an agitator shaft and agitator blades. Figure 20.1B shows an agitator with the components labeled.

The agitator shaft is a cylindrical rod that holds the agitator blades. Agitator blades act directly on the fluid. Agitator blades come in many different designs and provide various degrees of mixing within a specific tank or reactor. In some tanks or reactors, two sets of agitators, or a single agitator shaft with multiple blades, are used to increase the mixing process. Baffles (shown in Figure 20.2) are also used with agitators to facilitate mixing.

Agitators function by stirring the contents of a vessel and creating turbulence. This keeps any suspended solids from settling to the floor of the vessel and forming a layer of solids.

An example in the petrochemical industry is an agitator in a crude oil tank. Crude oil from an oil well contains sediment and other oil-wetted solids. If the crude oil were allowed to sit undisturbed in a tank, these solids would settle to the bottom. Over time, the solid layer would build up and require that the tank be removed from service for cleaning (a costly and time-consuming process). By employing an agitator, it is possible to keep the solids in suspension, resulting in less buildup of solids and less tank maintenance. The agitator also prevents the crude oil from stratifying (settling into different layers).

Mixers

Mixer a device used to combine chemicals or other substances mechanically using a rotary motion.

A **mixer** is a piece of equipment that is designed to combine products. In the process industries, mixers can be operated continuously or for a set period of time. Mixers can be used to mix wet or dry material during a process or to produce a final product. There are many

Figure 20.2 Tank with baffles. **A.** Top view. **B.** Side view.

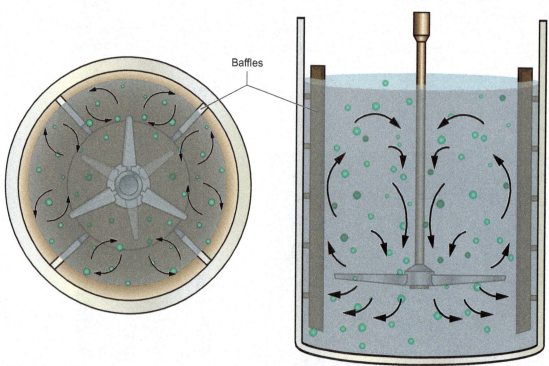

A. Top view

B. Side view

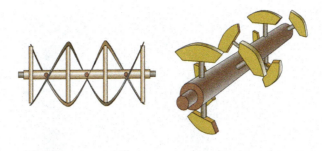

Figure 20.3 Examples of different types of mixer blades.

Spiral mixer   Paddle mixer

different kinds of mixers and mixer blades, including dynamic mixers, inline mixers, agglomerators, and drum mixers. Figure 20.3 shows examples of various types of mixer blades.

Each type of mixer can introduce liquids through a series of nozzles or blend various powdered materials. For example, in the oil and gas industries, mixers are used to blend gasoline into different grades. In the pharmaceutical and cosmetics industries, mixers are used to make powders, creams, lipsticks, and ointments.

DYNAMIC MIXERS   A **dynamic mixer** (shown in Figure 20.4) is a piece of equipment that rotates to combine products. The components of a dynamic mixer include a shaft, blade, coupling, bearing housing, and driver.

Dynamic mixer   a piece of equipment that rotates to mix products rapidly.

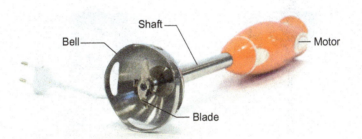

Bell   Shaft   Motor

Blade

Figure 20.4 Small, handheld, dynamic mixer used in food preparation.

CREDIT: Golik Alexander/Shutterstock.

INLINE MIXERS   An **inline mixer** (or *static mixer*) has no moving parts and operates by creating a complex flow path that results in an intermingled contact between fluid streams. The components of a static or inline mixer include a design element inside the pipe where the mixing occurs. Static or inline mixers are usually limited to the mixing of liquid streams. Figure 20.5 shows an example of an inline mixer.

Inline mixer   a mixer that has no moving parts; it operates by creating a complex flow path that results in contact between fluid streams; also called a *static mixer*.

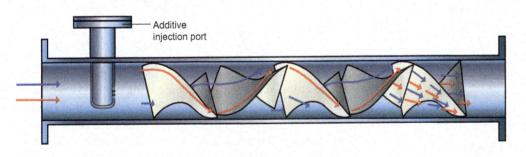

Additive injection port

Figure 20.5 Inline mixer. The motion of fluids flowing over the blades creates a gradual mixing of the different fluids as they move from entry point to discharge point.

AGGLOMERATORS   An **agglomerator** is a mixer that feeds dry and liquid materials together to produce an agglomerated (clustered) product like powdered laundry detergent. Figure 20.6 displays an image of agglomerators used in the pharmaceutical industry.

Some agglomerators are vertical mixers that rotate. During rotation, liquid material is introduced into the powder through a series of nozzles. Once the product has been hydrated (moisture added to it) and agglomerated (formed into clusters), it is then sent to a material dryer. The proper flow of powder and liquid, alignment of the nozzle air caps, and cleanliness of the nozzle assemblies is extremely important when trying to achieve correct density in agglomerated products.

Agglomerator   a mixer that feeds dry and liquid materials together to produce an agglomerated (clustered together) product like powdered laundry detergent.

Figure 20.6 Agglomerators used in the pharmaceutical industry.

CREDIT: NavinTar/Shutterstock.

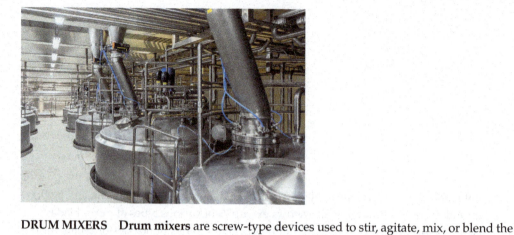

Drum mixer a screw-type device used to stir, agitate, mix, or blend the contents of a drum into a consistent composition.

DRUM MIXERS **Drum mixers** are screw-type devices used to stir, agitate, mix, or blend the contents of a drum into a consistent composition. Drum mixers can be handheld or stand-alone devices. Common components of a drum mixer are a driver (usually electric or air) and an agitator. The size and shape of a drum mixer can vary based on the design need. For example, a hand-held drum mixer (shown in Figure 20.7A) resembles a drill with a long, screw-like drill bit that is inserted into a mixing drum.

Figure 20.7 A. Handheld drum mixer. **B.** Rotating drum mixer with baffles.

CREDIT: B. Dario Sabljak/Shuttertock.

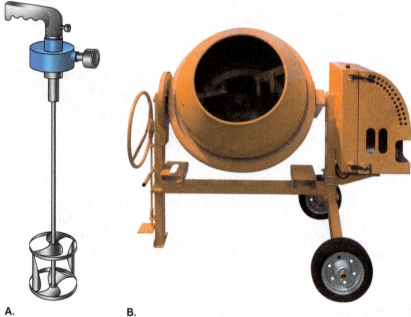

A.   B.

Did You Know?

A common kitchen mixer is a type of vertical (dynamic) mixer.

Like an industrial mixer, this vertical (dynamic) mixer can combine wet or dry materials to form a desired mixture.

CREDIT: Doomu/Shutterstock.

Rotating drum mixers are barrel-shaped devices that contain baffles or screws (see Figure 20.7B) that facilitate mixing as the barrel rotates along its axis. Drum mixers can also be used for material that needs to be rotated slowly for a constant mixture (e.g., cement in a cement truck). Figure 20.8 shows a concrete truck drum mixer. When the truck's drum rotates in one direction, the screw inside the drum mixes the concrete. When the drum turns in the other direction, the screw forces the cement out of the truck and onto a chute.

Venturi effect the drop in pressure that occurs when liquid flows through a constricted section of pipe; as the speed increases, the pressure decreases.

Bernoulli principle principle stating that, as the speed of a fluid increases, the pressure of the fluid decreases.

Eductors and Ejectors

Eductors and ejectors are similar pieces of equipment, both of which operate using the **Venturi effect**. The Venturi effect is a fluid flow principle which states that the velocity of a fluid will increase through a constricted flow area, in order to maintain the same flow rate. Based on **Bernoulli's principle**, we know that when the velocity of a fluid increases, there will be a decrease in pressure at that point.

Figure 20.8 Concrete truck drum mixer.

CREDIT: Mattanin Nonchang/Shutterstock.

A Venturi tube is a device based on the Venturi principle which is designed to create a constriction in a piping section. It can be used for measuring flow rate, as well as in equipment like eductors and ejectors. It consists of a converging section, a throat, and a diverging section. Within a Venturi tube, the velocity of a gas or liquid increases as a result of the decreasing internal diameter in the pipe or duct. Figure 20.9 shows an example of a Venturi tube.

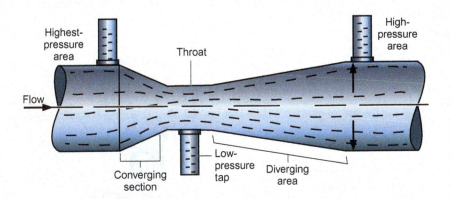

Figure 20.9 Diagram of a Venturi tube.

Eductor   a device that uses a fluid as a motive force to move another fluid using the Venturi effect; also called a *jet pump eductor* or *aspirator*.

An **eductor** is a device that uses a fluid as a motive force to move another fluid using a Venturi design. The internal components of an eductor are a nozzle and the Venturi section (converging section, throat, diverging section). Eductors are reliable pieces of equipment because they contain no moving parts. They are also very resistant to corrosion and erosion. They can be made from almost any material. An eductor's motive fluid can be a liquid or gas. The eductor is used to move the second fluid, which can also be a liquid or a gas. Eductors can be used in a variety of applications. One common use of eductors is for the application of firefighting foam used to extinguish flammable liquid fires.

Ejector   a device that is used to maintain vacuum by using a Venturi design powered by a motive gas (usually steam or air) flow. The ejector helps evacuate inert gases from the vacuum atmosphere; sometimes called a *vacuum ejector*, *injector*, or *steam ejector*.

The eductor is connected to the fire hose and the stream of firewater is started. As the firewater flows through the eductor, it pulls the foam out of its container, mixing it with the stream of water and discharging it onto the fire.

An **ejector** is a device that creates and maintains a vacuum system. Similar to eductors, the operation of ejectors is based on the Venturi effect. Figure 20.10 shows ejectors and eductors. A steam ejector (shown in Figure 20.10A) uses a flow of steam to produce a Venturi effect. The components of an ejector system

Did You Know?

A garden hose chemical sprayer is a type of eductor. As the water flows from the hose into the eductor, the chemicals in the container are pulled up through a tube into the water stream. The chemicals and water are mixed and then discharged through the spray nozzle.

CREDIT: John E Heintz Jr/ Shutterstock.

include the Venturi (diffuser throat), ejector housing, nozzle, and a downstream steam condenser. In an ejector, the nozzle inside the ejector increases the velocity of the steam, thereby creating an area of low pressure at the low-pressure suction inlet. This low-pressure area allows lower-pressure gases to be drawn into the stream that is moving through the ejector. Ejectors are commonly used with condensing turbines to remove inert gases from the vapor space and maintain vacuum in the condenser section of the condensing steam turbine. They are also widely used to maintain vacuum on vacuum distillation columns. This lowers the boiling point and causes the feed mixture components to separate at a lower temperature, thus saving energy costs and improving product quality. Ejectors have no moving parts. Steam is usually the motive fluid for an ejector. Gas leaving through the ejector is either vented to the atmosphere or is sent to a condenser. The condenser is a type of heat exchanger where the vapor is cooled and condensed into liquid, occupying less space. This is often an economical way to improve the efficiency of the vacuum.

The advantage of ejectors is that they are easy to install and maintain. They can also be used in various types of industries because they can be made of a variety of materials (e.g., carbon steel, bronze, stainless steel, and Teflon) in order to meet the needs of the process.

The disadvantages of ejectors are that a considerable amount of energy (e.g., high-pressure steam) is consumed, and the performance of the ejector is very sensitive to steam supply conditions (e.g., steam pressure and steam temperature). Figure 20.10B is a diagram of a foam eductor used for fire suppression. Figure 20.10C is an in-line foam eductor.

Figure 20.10 A. Ejector used to maintain vacuum on a distillation column, showing flow of vessel gases to be exhausted. **B.** Diagram of a foam eductor. **C.** Photo of in-line foam eductor with components labeled. **D.** A jet pump eductor raising a liquid to a higher elevation.

CREDIT: C. Jonathan Lingel/Shuttertock.

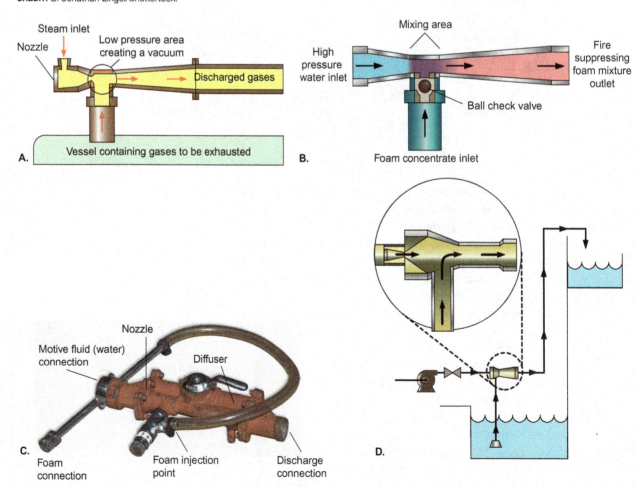

JET PUMP EDUCTOR   A **jet pump eductor** also functions like an aspirator. It is a device that uses the movement of high-pressure fluid to create suction (Venturi effect) to transfer material without the need for an external power source. A jet pump eductor requires a high velocity motive liquid, which is sometimes provide by a centrifugal pump, as shown in Figure 20.10D.

Jet pump eductors are effective at lifting fluids from wells, removing combustible fluids (which could ignite if exposed to a spark from an electric pump), and removing excessive amounts of debris (materials that would damage the screws or blades in a conventional pump). When a centrifugal pump is used, it does not come in contact with the material being pumped, so it will not be damaged by debris. Jet pump eductors can be used to vacuum liquid from sumps and cellars. The advantages of jet pump eductors are that they have no moving parts and they are simple and lightweight.

Jet pump eductor   a device that uses a high pressure liquid to create a Venturi effect to transfer fluid without the need for an external power source.

Centrifuges

A **centrifuge** is a device that uses centrifugal force, rather than gravity, to separate solids from liquids or to separate liquids based on their density difference. In a centrifuge, a spinning drum separates lighter liquid from heavier liquids or solids, causing the lighter liquid to exit from the center while the heavier liquids or solids are spun to the outside edge of the drum. As the drum spins, centrifugal force (the force that causes something to move outward from the center of rotation) causes the components to separate. The centrifugation process can be a single-stage or multistage operation depending on the nature of the material being separated. Figure 20.11A shows a diagram of a type of batch centrifuge called a chamber bowl centrifuge. Feed (liquid containing solids) enters the center of the centrifuge at the top. As the bowl spins, material moves outward, depositing solids on the outer wall in each chamber. Clarified liquid exits from the outermost chamber. Centrifuges can be used for such tasks as separating whole milk into skim milk and cream. They are also used extensively in the wastewater treatment process to dewater sludge prior to disposal.

A centrifuge designed to handle liquid containing a higher concentration of solids is the decanter centrifuge. The decanter centrifuge shown in Figure 20.11B is a continuous scroll centrifuge. Operating horizontally, it consists of a bowl with a helical screw. While the bowl spins at a very high speed, the discharge screw operates at a different speed, moving the solids that have collected on the walls toward the sediment outlet.

The main benefit of centrifugal extractors over extraction columns or a mixer-settler system is that they take up less space. The main disadvantage of a centrifugal extractor is that they have gearboxes and motors, which are more expensive and require additional maintenance.

Centrifuge   a device that uses centrifugal force to separate solids from liquids or to separate fluids based on their density.

CREDIT: Romaset/Shutterstock.

Figure 20.11 Diagrams of
centrifuges. **A.** Batch centrifuge.
B. Continuous scroll (decanter)
centrifuge.

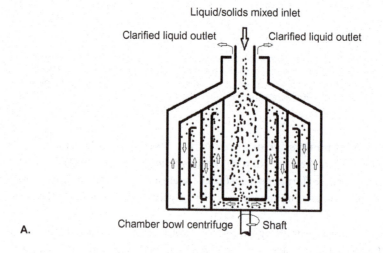

A.

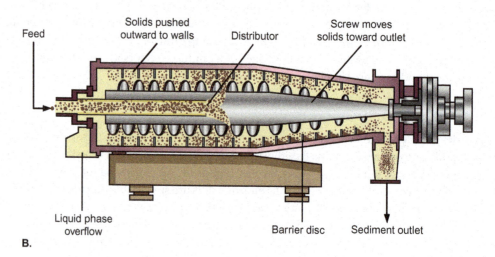

B.

Hydrocyclones and Cyclone Separators

Hydrocyclone a mechanical device
that promotes separation of heavy
and light liquids by centripetal force,
a force that acts on a body moving in
a circular path and is directed toward
the center around which the body is
moving (e.g., like a cyclone).

Hydrocyclones are mechanical devices that promote separation of heavy and light components of liquids by centripetal force (Figure 20.12). In a hydrocyclone, incoming liquid enters near the top of the vessel. The liquid then begins to rotate (like a cyclone) as a result of natural centrifugal force caused by the orientation of the inlet flow pipe. Centrifugal force causes the heavy liquid to move toward the outer walls of the vessel, where it flows down the walls of the cylinder and out the bottom. The lighter liquid remains in the center of the vessel, and is forced out the top of the vessel. Unlike a centrifuge, the hydrocyclone is a static device and does not spin. The flow of fluid entering the hydrocyclone must be in the operational-design range. If it is too low, the cyclone rotation will no longer promote separation.

Hydrocyclones differ from centrifuges in that they have no moving parts. One advantage of hydrocyclones is that they are less expensive than centrifuges. A disadvantage is that they are less effective than centrifuges.

In the process industries, hydrocyclones are used in a variety of applications. For example, hydrocyclones are used to remove solids and other impurities from seal flush systems on pumps.

The operation of cyclone separators (see Figure 20.12B) is much like that of a hydrocyclone. However, the cyclone separator is used to remove particulates from a flue gas stream. In the cyclone separator, heavy particulates are flung to the outside and drop down out of the bottom of the cyclone while clean flue gas exits through the top.

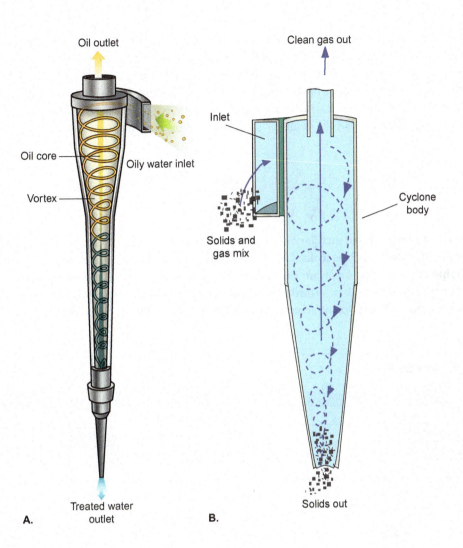

Demisters

A **demister** (also called a mist eliminator) is a mechanical device that promotes separation of liquids from gases. Demisters make vapors change direction many times. This allows gases to pass through the device while trapping liquids. These liquids coalesce (collect) and fall back into the vessel. Figure 20.13 shows a diagram of a demister. Demister designs include:

- A porous woven pad or a steel wool pad (also called *demister pads*)
- A stack of screenlike material
- A zigzag vane(chevron) design.

Demister a device that promotes separation of liquids from gases. Also called a mist eliminator.

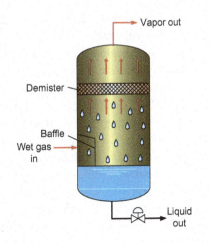

Figure 20.13 Diagram of a demister.

Each type causes the gas/liquid mixture to change direction many times while moving through the demister. The droplets of liquid are captured by the demister and prevented from reaching the outlet. They coalesce until they become a droplet large enough to fall back into the vessel. For proper operation, the gas flow rate must be maintained in the design operational range. Too low a flow may result in wet gas passing through a demister, and too high a rate will overload the demister's liquid handling capacity.

20.2 Hazards Associated with Auxiliary Equipment

When working around auxiliary equipment, process technicians must always be aware of potential hazards such as leaks, failure associated with wear and tear, exposure to rotating equipment, vibration on steam jets and nozzles, and steam conditions not being at design settings. These and other hazards can affect personal safety, equipment, and the environment. Table 20.1 lists potential hazards associated with auxiliary equipment.

Table 20.1 Hazards Associated with Auxiliary Equipment

Personal Safety	Equipment	Environment
Rotating equipment hazards	Rotating equipment hazards	Overfilling, causing spills
Hazardous/flammable materials	Equipment failure due to vibration	Underfilling, causing splashing and discharge to ground
Liquid splashing	Overheating/over pressurizing	Unanticipated reactions, causing release to environment
Slips/trips due to leaks	Unexpected equipment restarts	
Hot surfaces	Static electricity during loading or mixing	
Inhalation hazards	Unbalanced load, causing excessive vibration	
High operating speeds		

20.3 Process Technician's Role in Operation and Maintenance

The process technician's role in the safe operation and maintenance of auxiliary equipment includes following standard operating procedures for the facility and the unit. Process technicians are also required to report readings, conduct field checks, and troubleshoot equipment problems. They should be familiar with common symbols that may be used for the equipment (Figure 20.14).

Process technicians must understand operational procedures related to each piece of equipment in order to maintain safe operating conditions, to report any abnormal equipment behavior to their supervisor, and to respond appropriately in the event of an emergency.

Potential safety hazards when operating rotating auxiliary equipment such as the agitator, mixer, agglomerator, and centrifuge include entangled hair or clothing and contact with body parts. Rotating equipment guards can help minimize accidental contact; make sure all guards are in place for all operational equipment. For rotating equipment to perform properly, the RPMs (rotations per minute) of that equipment must remain within its design operating range. If equipment is rotating too slowly, the process task may not be performed successfully, resulting in a loss of product quality. Rotation that is too fast can exceed equipment design parameters and result in equipment failure.

When operating nonrotating auxiliary equipment such as the eductor, ejector, hydrocyclone, and demister, the rate of flow is critical for proper operation. This type of equipment has a flow range within which it must operate. Too little flow may not allow the desired process task to occur. Flow that is too high can result in equipment damage.

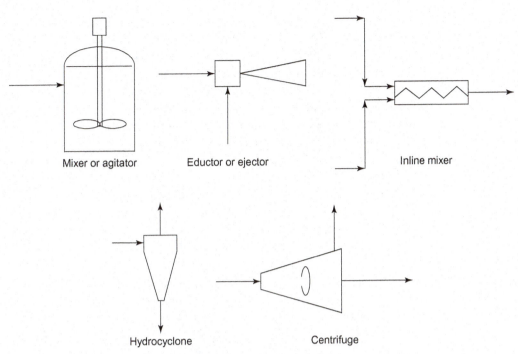

Figure 20.14 Common symbols for auxiliary equipment. These may vary somewhat from one facility to another.

Process technicians routinely monitor auxiliary equipment to ensure it is operating properly. They must always remember to look, listen, and check for the items indicated in Table 20.2.

Table 20.2 Process Technician's Role in Operation and Maintenance of Auxiliary Equipment

Look	Listen	Check
▪ Monitor system to prevent it from operating at higher or lower speeds than normal ▪ Watch for system pulling higher amps than normal ▪ Look for gases or solids leaking into the atmosphere	▪ Listen for abnormal noises such as banging or hammering ▪ Listen for normal noises at abnormal pitches	▪ Check for excessive heat in equipment components and operational areas ▪ Check for surging through the lines (e.g., normal flow, then excessively high or low flow)

Process technicians must be familiar with the appropriate procedures for monitoring and maintaining auxiliary equipment components and conditions in order to maintain a safe work environment. Typical procedures, which include startup, shutdown, emergency, and lockout/tagout, vary at each individual site and for each type of auxiliary equipment. Because of this, process technicians must always follow the standard operating procedures (SOPs) for their assigned unit(s) and speak up immediately if something does not seem right.

Summary

The auxiliary equipment described in this chapter is used in the process industries to mix, combine, or separate materials. Operators must be alert to safety and environmental hazards associated with rotating equipment and make sure all machine guards are in position. Process problems can be caused by incorrect rotational speeds or process flows outside design ranges.

Agitators are used to combine contents in a tank or vessel and can be used to create turbulence, which promotes mixing. The primary components of an agitator are the driver, the agitator

shaft and agitator blades, which produce a mixing effect to obtain the final product.

Mixers can be either dynamic (rotating) or static (inline). A dynamic mixer may rotate inside a container or the entire container may rotate to combine products. A static mixer has no moving parts, but creates a path through it that promotes the mixing of liquid streams. Mixers can be designed for continuous operation or set to run for a specific period of time. Some mixers are referred to as agglomerators.

Agglomerators feed dry and liquid products together to produce an agglomerated (clustered) product like powdered laundry detergent.

Eductors are devices that transport fluids using the Venturi effect. Eductors are reliable pieces of equipment due to their limited number of moving parts. Eductors are used in firefighting to regulate the proportion of foam distributed through the fire hose. Steam jet ejectors use a flow of steam to draw inert gases into its flow using the Venturi effect. They are used extensively in condensing steam turbines and vacuum distillation columns to create and maintain a vacuum system. Advantages of steam jet ejectors are that they are easy to install and easy to maintain. A jet pump eductor is designed to lift liquids and operates without external power.

A centrifuge is a device that uses centrifugal force to separate liquids from liquids, or liquids from solids. Heavy liquids are forced to the outer walls of an enclosed drum while the lighter liquid remains in the center.

Hydrocyclones are used in industry to separate solids or liquids from liquid streams. They have no moving parts so they are less expensive than centrifuges.

A process technician's role in operating auxiliary equipment includes being aware of potential problems when an abnormal sound is heard or equipment is not operating properly, in order to ensure the safety of people in the vicinity, the environment, and the equipment. Process technicians must always follow standard operating procedures for their unit and/or facility and be alert to potential hazards.

Checking Your Knowledge

1. Define the following terms:

   a. Agitator

   b. Centrifuge

   c. Eductor

   d. Hydrocyclone

   e. Venturi effect

2. How do agitators function?

   a. By creating a complex flow path

   b. By rotating the combined products in a clockwise direction

   c. By feeding dry products into the liquid

   d. By stirring the contents of a vessel and creating turbulence

3. (True or False) A cement truck is an example of a drum mixer.

4. Why are eductors reliable pieces of equipment?

   a. They have no moving parts.

   b. They require no maintenance.

   c. They are inexpensive.

   d. They are easily available.

5. A jet pump eductor is designed to be used in the following area(s): (Select all that apply.)

   a. Areas that need to be drained

   b. Areas that might contain high levels of debris

   c. Areas that might contain combustible fluids

   d. Areas that need more fluid added to the process

6. (True or False) In a hydrocyclone, incoming liquid enters through the bottom of the vessel.

7. List the three types of demister designs.

8. Which of the following types of auxiliary equipment can cause exposure to rotating parts? (Select all that apply.)

   a. Agitator   b. Eductor

   c. Centrifuge   d. Demister

9. Proper operation of rotating equipment such as the agitator, mixer, agglomerator, and centrifuge includes which of the following? (Select all that apply.)

   a. Having machine guards in place

   b. Running the equipment at designed speed of rotation

   c. Turning equipment on/off every shift

   d. Starting the process at 100 RPM

NOTE: Answers to Checking Your Knowledge questions are in the Appendix.

Student Activities

1. Write a report on how agitators and mixers are used in the process industries. Discuss their similarities and differences, and include a diagram of each.

2. Work with a team to prepare a presentation that discusses the theory of operation of eductors (including Venturi tube), steam ejectors, and jet pumps. Present the

material to the class. Be sure to include drawings and a description of the process.

3. Write a report describing centrifuges and hydrocyclones. Discuss their similarities and differences. In the written report, include a flow diagram that illustrates how each device works.

Appendix

Answers to Checking Your Knowledge Questions

Chapter 1	Answer	LO #
1.	See Key Terms list.	1.1 1.2
2.	True	1.2
3.	True	1.2
4.	c.	1.3
5.	a., b., c.	1.3
6.	a., c., d.	1.2
7.	Any of the following: hearing protection, gloves, hard hat, safety glasses/goggles, safety shoes, or flame-retardant clothing.	1.4
8.	Any of the following: Abnormal sounds coming from equipment; excessive vibration; leaks around equipment; faulty equipment gauges; use of incorrect tools; lack of proper personal protective equipment (PPE), such as hearing protection, safety glasses, goggles, hard hats, work gloves, safety shoes, or flame-retardant clothing; not following standard operating procedures (normal operating conditions of pressure and temperatures); or not complying with government, industry, or company regulations.	1.4
9.	d.	1.3
10.	c.	1.3
11.	b.	1.3
12.	c.	1.1

Chapter 2		
1.	See Key Terms list.	2.3 2.4 2.5 2.6
2.	b.	2.1
3.	Left to right	2.3
4.	a., b., c.	2.3
5.	False	2.4
6.	a., c., d.	2.6
7.	True	2.6
8.	b.	2.6
9.	c.	2.7
10.	a., d.	2.2
11.	True	2.2
12.	False	2.5

13.	c.	2.5
14.	NEC (National Electrical Code)	2.7

Chapter 3		
1.	See Key Terms list.	3.1 3.2 3.3
2.	a.	3.3
3.	b.	3.1
4.	c.	3.1 3.2
5.	a.	3.1
6.	True	3.1 3.2
7.	False	3.3
8.	d.	3.3
9.	c., d.	3.1 3.2
10.	True	3.4

Chapter 4		
1.	See Key Terms list.	4.1 4.2 4.3 4.4 4.6
2.	b.	4.2
3.	b.	4.1
4.	True	4.4
5.	c.	4.2
6.	a.	4.6
7.	Leakage	4.4
8.	d.	4.4
9.	a., b.	4.7
10.	c.	4.6
11.	b.	4.3
12.	I. d II. a III. b IV. c	4.4
13.	True	4.5
14.	Any of the following: exposure to chemicals, burn or injury, cross contamination, possible atmospheric release, and spill into the environment.	4.5

Chapter 5

1.	See Key Term list.	5.1
		5.2
		5.3
		5.4
2.	True	5.1
3.	a.	5.1
4.	d.	5.1
5.	a., b.	5.1
6.	True	5.1
7.	False	5.1
8.	c.	5.1
9.	True	5.1
10.	b.	5.2
11.	Causes of leakage include corrosion (internal or external), erosion (usually internal), vibration, wear (external), and damage from an external source (for example, impact from something striking the pipe).	5.2
12.	a., b.	5.3
13.	False	5.3
14.	I. f. Ball	5.1
	II. b. Butterfly	
	III. i. Check	
	IV. d. Diaphragm	
	V. e. Gate	
	VI. h. Globe	
	VII. c. Plug	
	VIII.g. Relief	
	IX. a. Safety	
15.	Any of the following: the packing, bonnets, gaskets, flanges, or threads.	5.4

Chapter 6

1.	See Key Terms list.	6.1
		6.2
		6.3
		6.4
		6.5
		6.6
		6.7
		6.8
		6.9
2.	a., b., c., d.	6.1
3.	False	6.2
4.	a., b., c.	6.2
5.	Canned	6.2
6.	False	6.3
7.	False	6.4
8.	I. b	6.6
	II. d	
	III. a	
	IV. c	
9.	b.	6.6
10.	a. C	6.2
	b. G	
	c. F	
	d. B	
	e. D	
	f. A	
	g. E	
11.	Tandem; double	6.3

12.	Total head	6.4
13.	c.	6.5
14.	True	6.7
15.	True	6.7
16.	False	6.8
17.	1. Close the pump suction and discharge so liquid does not enter the pump. 2. De-energize the driver by opening the breaker to the motor or blocking the inlet and exhaust steam valves on the turbine. 3. Drain the pump to the designated collection system. 4. Shut down any auxiliary systems, such as lube oil and seal flush systems. 5. Lock and/or tag the valves and breaker for the motor or valves on the turbine 6. Flush/purge the pump to clear the pump of the chemical contained in the pump.	6.9

Chapter 7

1.	See Key Terms list.	7.1
		7.2
		7.3
		7.4
2.	Compression	7.1
3.	They have few moving parts, are very energy efficient, and provide higher flows than similarly sized reciprocating compressors.	7.1
4.	a.	7.1
5.	a., b.	7.1
6.	Type of gas being compressed, the flow rate, and the discharge pressure	7.1
7.	a.	7.3
8.	b., c.	7.2
9.	False	7.4
10.	False	7.1
11.	True	7.2
12.	b.	7.3
13.	Any of the following: hazardous fluids/materials, excessive noise levels, slipping/tripping due to leaks, chemical exposure, skin burns, rotating equipment hazards, leaks, high or very low temperatures, high-voltage electricity, high pressure	7.5
14.	c., d.	7.5
15.	Any of the following: Wear the appropriate personal protective equipment and observe all safety rules; observe the compressor operation for signs of compressor failure or maintenance needs during normal operation; check and record pressures, temperatures, and flow rates periodically to identify a problematic trend; check rotating equipment frequently for mechanical problems, such as lack of lubrication, seal leakage, overheating of compressors and drivers, and excessive vibration; and check lubrication systems to ensure proper temperatures and lubricant levels because seals become worn and weak and because motor, turbine, and compressor bearings become worn and overheat.	7.6

Chapter 8

1.	See Key Terms list.	8.1
2.	False	8.1
3.	d.	8.1
4.	False	8.1
5.	a.	8.1

6.	a.	8.1
7.	b.	8.2
8.	a., b., c., d.	8.1
9.	a., c., d.	8.1
10.	False	8.1
11.	Sentinel	8.3
12.	d.	8.2
13.	Fouling	8.3
14.	a., d.	8.4
15.	Any of the following: inlet and exhaust steam temperatures (which could indicate dirty turbine blades) and pressures, oil levels, temperatures, flows, discoloration, foaming, filter pressure drops, speed governor lubrication, and safety interlocks and devices	8.5
16.	1. Communicate with affected personnel about the impending turbine startup. 2. Ensure that the driven equipment connected to the steam turbine is ready for startup. 3. Verify the lube oil and cooling water are in service. 4. Open drains from the turbine casing, steam lines, and steam chest. 5. Open the turbine exhaust steam valve. 6. Open the steam inlet valve slowly, to begin slow rolling the turbine. 7. Observe the steam coming out of the drains and, when it becomes dry, close the drain valves. 8. Open the steam inlet valve to adjust turbine speed as needed.	8.6

Chapter 9

1.	See Key Terms list	9.3 9.4
2.	True	9.2
3.	True	9.2
4.	False	9.2
5.	d.	9.4
6.	d.	9.5
7.	a.	9.6
8.	a.	9.6
9.	b.	9.7
10.	drivers	9.1
11.	b.	9.3
12.	Any of the following: bearing failures, loss of lubrication, overload from driven machinery, or loss of cooling water or fan airflow.	9.5
13.	True	9.7
14.	Any of the following: oil temperature, oil pressure, engine speed, vibration levels, or engine temperature.	9.8

Chapter 10

1.	See Key Terms list.	10.1 10.2
2.	b.	10.3
3.	True	10.3
4.	d.	10.4
5.	b.	10.4
6.	b.	10.5

7.	a., d.	10.5
8.	c.	10.5
9.	True	10.6
10.	True	10.6
11.	Grounding	10.1
12.	False	10.2
13.	d.	10.4
14.	Coupling	10.7
15.	Any of the following: switching or replacing oil filters, switching oil coolers or back flushing waterside oil coolers, collecting and recording vibration levels and readings, checking oil levels and adding oil if necessary, checking the grease supply and adding grease if necessary, changing oil where applicable, sampling oil for water and other contamination, checking bearing surface temperatures with a skin or surface pyrometer, or monitoring and ensuring proper operation of oil centrifuges.	10.8
16.	True	10.7

Chapter 11

1.	See Key Terms list.	11.1 11.3 11.4
2.	b.	11.1
3.	True	11.1
4.	d.	11.3
5.	a., b., c., d.	11.2
6.	b.	11.2
7.	d.	11.1
8.	a.	11.4
9.	I. d II. c III. a IV. b	11.1
10.	c.	11.5
11.	True	11.5
12.	Any of the following: Inspect equipment for external leaks, check for internal tube leaks by collecting and analyzing samples, look for abnormal pressure changes (could indicate tube plugging), check temperature gauges for high temperature readings, and inspect insulation.	11.6
13.	a., b.	11.7
14.	False	11.7

Chapter 12

1.	See Key Terms list.	12.1 12.2 12.3 12.4 12.6
2.	False	12.1
3.	False	12.1
4.	c.	12.2
5.	a.	12.2
6.	d.	12.4
7.	c.	12.4
8.	b.	12.6
9.	False	12.3

10.	True	12.3
11.	a., c., d.	12.5
12.	True	12.5

Chapter 13

1.	See Key Terms list.	13.1
		13.2
2.	a.	13.1
3.	b.	13.1
4.	a., b., c.	13.2
5.	Any of the following: inefficient burning, smoke production, high carbon monoxide (CO) emissions, or furnace explosion.	13.3
6.	False	13.3
7.	b.	13.4
8.	a., c., d.	13.5
9.	True	13.5
10.	d.	13.4
11.	1. c	13.6
	2. a	
	3. b	
12.	False	13.6
13.	True	13.7
14.	a., c.	13.7

Chapter 14

1.	See Key Terms list.	14.1
		14.2
		14.3
2.	True	14.1
3.	Any of the following: turbines, reactors, distillation columns, stripper columns, and heat exchangers.	14.1
4.	a., b., d.	14.1
5.	a., b., c.	14.1
6.	False	14.1
7.	a.	14.1
8.	d.	14.3
9.	True	14.4
10.	a., b., c.	14.5
11.	True	14.6
12.	a., b., c.	14.6
13.	fireman	14.7

Chapter 15

1.	See Key Terms list.	15.1
		15.2
2.	Any of the following: pressure and temperature requirements, product type (liquid, gas, or solid), chemical properties, corrosion factor, volume, ambient conditions, and geographic location (for example, earthquake or flood zone) of the vessel.	15.1
3.	a.	15.1
4.	c.	15.1
5.	b.	15.1
6.	d.	15.2

7.	Safety valves are intended for use in gas service whereas relief valves are used in liquid service; safety valves have a larger outlet than relief valves, because compressible gases require that a greater amount be released to reduce pressure; and safety valves open fully and quickly when set pressure is exceeded, but relief valves open more slowly and in proportion to the pressure increase.	15.2
8.	False	15.3
9.	a., c.	15.3
10.	True	15.4
11.	d.	15.5
12.	True	15.5
13.	Any of the following: monitor levels; look carefully at firewalls, sumps, and drains; inspect auxiliary equipment associated with the vessel; visually inspect for leaks (especially if associated with abnormal odor); use level gauges and sight glasses to monitor level, monitor level and pressure; and inspect for corrosion.	15.6
14.	b.	15.6
15.	I. a	15.7
	II. a	
	III. b	
	IV. a	

Chapter 16

1.	See Key Terms list.	16.4
		16.6
2.	Baffles	16.3
3.	a.	16.2
4.	b.	16.6
5.	c.	16.2
6.	a., c.	16.2
7.	a., b., d.	16.6
8.	c.	16.6
9.	c.	16.3
10.	Any of the following: addition of heat, addition of a catalyst, or increasing the number of collisions between molecules.	16.4
11.	b.	16.5
12.	True	16.7
13.	a., b., d.	16.7
14.	False	16.8

Chapter 17

1.	See Key Terms list.	17.2
2.	c.	17.1
3.	Mechanical, gravity, and vacuum	17.2
4.	b.	17.1
5.	a.	17.2
6.	a.	17.3
7.	Any of the following: stay within pressure and flow limits of the system; do not overpressurize; operate within the design limits (for example, temperatures and flows); and ensure that the correct drying agent or precoat is installed.	17.3
8.	a., b., c.	17.4

9.	True	17.5
10.	Strainers	17.6

Chapter 18

1.	See Key Terms list.	18.1
2.	False	18.1
3.	False	18.1
4.	True	18.1
5.	c.	18.1
6.	True	18.1
7.	b.	18.1
8.	c.	18.2
9.	b., c., d.	18.3

Chapter 19

1.	See Key Terms list.	19.1
2.	Switching to a fuel with a lower nitrogen content, flue gas recirculation (FGR), or the use of low NO_x burners reduces the amount of NO_x formed in combustion processes.	19.1
3.	Nitrogen oxides (NO_x), carbon monoxide (CO), and sulfur dioxide (SO_2)	19.1
4.	True	19.1

5.	c.	19.1
6.	b, c.	19.1
7.	Any four of the following: ground level ozone, carbon monoxide (CO), sulfur dioxide (SO_2), particulate matter, lead, and nitrogen dioxide (NO_2).	19.2
8.	False. Landfills must be properly lined and not allow leakage, so that contaminants do not leach into the soil and eventually into the groundwater.	19.3
9.	a.	19.4

Chapter 20

1.	See Key Terms list.	20.1
2.	d.	20.1
3.	True	20.1
4.	a.	20.1
5.	a., b., c.	20.1
6.	False. In a hydrocyclone, incoming liquid enters through the top of the vessel.	20.1
7.	A porous woven pad or a steel wool pad, a stack of screenlike material, and a zigzag vane design.	20.1
8.	a., c.	20.2
9.	a., b.	20.3

Glossary

AC power source   a device that supplies alternating current.

Activation energy   the minimum amount of energy that is required for a reaction to take place.

Adjustable pliers   pliers that have a tongue and groove joint or slot that allows the jaws to widen and grip objects of different sizes; tongue and groove also are called Channellocks®; slotted design may be called slip joint pliers.

Adjustable wrench   open-ended wrench with an adjustable jaw that contains no teeth or ridges (i.e., the jaw is smooth); also referred to as a Crescent® wrench.

Aftercooler   a heat exchanger, located after the final discharge stage of a compressor.

Agglomerator   a mixer that feeds dry and liquid materials together to produce an agglomerated (clustered together) product like powdered laundry detergent.

Agitator   a device that acts on a fluid by generating turbulence, which promotes mixing.

Agitator/mixer   a device used to mix the contents inside a vessel.

Air register   also called the burner damper, a device located on a burner that is used to adjust the primary and secondary air flow to the burner; it provides the main source of air entry to the furnace. It maintains the correct fuel-to-air ratio and reduces smoke, soot, or NO_x (nitrogen oxide) and CO (carbon monoxide) formation.

Alloys   compounds composed of two or more metals or a metal and nonmetal (e.g., carbon steel) that are mixed together in a molten solution.

Alternating current (AC)   electric current that reverses direction periodically, usually 60 times per second.

Ambient air   any part of atmospheric air that is breathable.

Ammeter   device used to measure the electrical current in a circuit.

Ampere (amp)   a unit of measure of the electrical current flow in an electrical circuit, comparable to the rate of water flow in a pipe.

ANSI   American National Standards Institute, an organization that oversees and coordinates voluntary standards in the United States.

Antisurge protection   control instrumentation designed to prevent damage to the compressor by preventing it from operating at or near an undesirable pressure and by preventing flow conditions that result in a surge.

Antisurge   recycle flow returned from the discharge of a compressor stage to a lower pressure suction; serves to prevent surging.

API   American Petroleum Institute, a trade association that represents the oil and gas industry in the areas of advocacy, research, standards, certification, and education.

Application block   the main part of a drawing that contains symbols and defines elements such as relative position, types of materials, equipment descriptions, flows, and functions.

Approach   a term that describes how closely a cooling tower can cool the water to the wet bulb temperature of the air.

Articulated drain   a hinged drain, attached to the roof of an external floating roof tank, that moves up and down with the roof as the fluid level rises and falls.

ASME   American Society of Mechanical Engineers, an organization that specifies requirements and standards for pressure vessels, piping, and their fabrication.

Assisted draft furnace   a system that uses electric motor, steam turbine-driven rotary fans, or blowers to push air into the furnace (*forced draft*) or to draw flue gas from the furnace to the stack (*induced draft*).

Atmospheric storage tanks   vessels in which atmospheric or low pressure is maintained.

Autoclave reactor   a vessel in which the reaction occurs at high pressure and temperature.

Axial compressor   a dynamic compressor that contains a rotor with contoured blades followed by a set of stationary blades (stator). In this type of compressor, the flow of gas is *axial* or parallel to the shaft.

Axial pump a dynamic pump that uses a propeller or row of blades to propel liquids along the shaft.

Backwashing (backflushing) a procedure in which the direction of flow through the exchanger is reversed to remove solids.

Baffle a metal plate, placed inside a vessel, that is used to alter the flow, facilitate mixing, or cause turbulent flow.

Bag filter a tube-shaped filtration device that uses a porous bag to capture and retain solid particles.

Baghouse air pollution control equipment that contains fabric or bag filter tubes, envelopes, or cartridges designed to capture, separate, or filter particulate matter.

Balanced draft furnace a furnace that uses both forced and induced draft fans to facilitate air flow; fans induce flow out of the firebox through the stack (*induced draft*) and provide positive pressure to the burners (*forced draft*).

Ball check valve a valve used to limit the flow of fluids to one direction; available in both horizontal and vertical designs.

Ball valve a type of valve that uses a flow control element shaped like a hollowed-out ball attached to an external handle to control flow.

Barge a flat-bottomed boat used to transport fluids (e.g., oil or liquefied petroleum gas) and solids (e.g., grain or coal) across shallow bodies of water such as rivers and canals.

Basin a reservoir at the bottom of the cooling tower where cooled water is collected so it can be recycled through the process.

Batch reaction a carefully measured and controlled process in which raw materials (reactants) are added together to create a reaction that makes a single quantity (batch) of final product.

Bearing a machine component that rotates, slides, or oscillates. Bearings reduce friction between the motor's rotating and stationary parts. They support the shaft, keep the shaft in alignment with the casing, and absorb axial and radial forces.

Belt a flexible band, placed around two or more pulleys, that transmits rotational energy.

Bernoulli principle principle stating that, as the speed of a fluid increases, the pressure of the fluid decreases.

Bin a vessel that typically holds dry solids.

Biocides chemical agents that are capable of controlling undesirable living microorganisms (such as in a cooling tower).

Blanketing the process of putting an inert gas, usually nitrogen, into the vapor space above the liquid in a vessel to prevent air leakage into it (often called a *nitrogen blanket*).

Blinds (or blanks) solid plates or covers that are installed between pipe flanges to prevent the flow of fluids and to isolate equipment or piping sections when repairs are being performed; typically made of metal.

Block flow diagram (BFD) a simple illustration that shows a general overview of a process, indicating its parts and their interrelationships.

Block valve a valve used to block flow to and from equipment and piping systems (e.g., during outage or maintenance). Block valves differ from some other types of valves, in that they should not be used to throttle flow.

Blowdown the process of removing small amounts of water (e.g., from the cooling tower or boiler) to reduce the concentration of impurities. There are two types of blowdown, continuous and intermittent.

Blower a mechanical device, either centrifugal or positive displacement in design, that has a lower ratio of pressure change from suction to discharge and is not considered to be a compressor. The resistance to flow is downstream of the blower.

Boiler a device in which water is boiled and converted into steam under controlled conditions.

Bonnet a bell-shaped dome mounted on the body of a valve where the valve disc is held when the valve is opened.

Boot a section in the lowest area of a process vessel where water or other liquid is collected and removed.

Box furnace a square, box-shaped furnace designed to heat process fluids or to generate steam.

Breakthrough the condition of any adsorbent bed or filter medium when the component being adsorbed or filtered starts to appear in the outlet stream.

British thermal unit (BTU) a measure of energy in the English (Imperial standard) system; the heat required to raise the temperature of 1 pound of water 1 degree Fahrenheit at sea level.

Bucket elevator   a continuous line of buckets attached by pins to two endless chains running over tracks and driven by sprockets.

Bullet vessel   cylindrically shaped container used to store contents at moderate to high pressures.

Burners   devices with open flames that introduce, distribute, mix, and burn a fuel (e.g., natural gas, fuel oil, or coal) in a firebox to produce heat.

Butterfly valve   a type of valve that uses a disc-shaped flow control element to control flow.

Cabin furnace   a cabin-shaped furnace designed to heat process fluids or to generate steam.

Calorie (cal)   gram calorie; the amount of heat energy required to raise the temperature of 1 gram of water by 1 degree Celsius at sea level.

Camshaft   a driven shaft fitted with rotating wheels of irregular shape (cams) that open and close the valves in an engine.

Canned pump   a sealless pump that ensures zero emissions. It is often used on EPA-regulated liquids.

Carbon ring   a seal component located around the shaft of the turbine that controls the leakage of motive fluid (typically steam) along the shaft or the entrance of air into the exhaust.

Cartridge filter   a filtration unit that uses a fine mesh that is tightly folded or pleated to remove suspended contaminants from a liquid.

Casing   a housing component of a turbine that holds all moving parts (including the rotor, bearings, and seals) and is stationary.

Catalyst   a substance used to facilitate the rate of a chemical reaction without being consumed in the reaction.

Cavitation   a condition inside a pump in which the liquid being pumped partially vaporizes because of variables such as temperature and pressure drop, and the resulting vapor bubbles then implode.

Centrifugal compressor   a dynamic compressor in which the gas flows from the inlet, located near the suction eye, to the outer tip of the impeller blades.

Centrifugal force   the energy that causes something to move outward from the center of rotation.

Centrifugal pump   a type of dynamic pump that uses an impeller on a rotating shaft to generate pressure and move liquids.

Centrifuge   a device that uses centrifugal force to separate solids from liquids or to separate fluids based on their density.

Chains   mechanical devices used to transfer rotational energy between two or more sprockets.

Channeling   formation of a path through a dryer or filter that allows feed to flow through without proper contact with the filtering medium or drying agent.

Check valve   a type of valve that allows flow in only one direction and is used to prevent reversal of flow in a pipe.

Chiller   devices used to cool a fluid to a temperature below ambient temperature. Chillers generally use a refrigerant as a coolant.

Circuit   a system of electrical components that accomplishes a specific purpose.

Circuit breaker   an electrical component that opens a circuit and stops the flow of electricity when the current reaches unsafe levels.

Circulation rate   the rate at which cooling water flows through the tower and through the cooling water system.

Closed circuit cooling towers   cooling towers that use a closed coil over which water is sprayed and a fan provides air flow, and in which there is no direct contact between the water being cooled and the water and air used to cool it.

Coagulation   a method for concentrating and removing suspended solids in boiler feedwater by adding chemicals to the water, which causes the impurities to cling together.

Coke   carbon deposits inside process tubes (*coking*), which can result in poor heat transfer to process fluid and overheating and failure of the tube metal.

Combustion chamber   a chamber between the compressor and the turbine where the compressed air is mixed with fuel. The fuel is burned, increasing the temperature and pressure of the combustion gases.

Compression ratio   the ratio of the volume of the cylinder at the start of a compression stroke compared to the smaller (compressed) volume of the cylinder at the end of a stroke.

Compressor   a mechanical device used to increase the pressure of a gas or vapor.

Condensate   condensed steam, which often is recycled back to the boiler.

Condenser   an exchanger used to convert a substance from a vapor to a liquid.

Condensing steam turbine   a device in which exhaust steam is condensed in a surface condenser. The condensate is then recycled to the boiler.

Conduction   the transfer of heat from one substance to another by direct contact.

Conductor   materials that have electrons that can break free from the flow more easily than the electrons of other materials.

Connecting rod   a component that connects a piston to a crankshaft.

Containment wall   an earthen berm or constructed wall used to protect the environment and people against tank failures, fires, runoff, and spills; also called *bund wall*, *bunding*, *dike*, or *firewall*.

Continuous reaction   a chemical process in which raw materials (reactants) are continuously fed into a reactor vessel and products are continuously being formed and removed from the reactor vessel.

Control valve   a valve through which flow is controlled by a signal from a controller.

Convection section   the upper and cooler section of a furnace where heat is transferred and convection tubes absorb heat, primarily through the method of convection; the part of the furnace where process feed and/or steam enters.

Convection   the transfer of heat as a result of fluid movement.

Convection tubes   furnace tubes, located above the shock bank tubes, that receive heat through convection.

Conveyor   a mechanical device used to move solid material from one place to another.

Coolant   a fluid that circulates around or through an engine to remove the heat of combustion. The fluid can be a liquid (e.g., water or antifreeze) or a gas (e.g., air or freon).

Coolers   heat exchangers that use cooling tower water to lower the temperature of process materials.

Cooling range   the difference in temperature between the hot water entering the tower and the cooler water exiting the tower.

Cooling tower   a structure designed to lower the temperature of water using latent heat of evaporation.

Corrosion   deterioration of a metal by a chemical reaction (e.g., iron rusting).

Countercurrent flow   flow movement that occurs when two fluids are moving in opposite directions.

Counterflow   the condition created when air and water flow in opposite directions.

Couplings   mechanical devices used to connect and transfer rotational energy from the shaft of the driver to the shaft of the driven equipment.

Crane   a mechanical lifting device used to lift and lower materials or move them horizontally.

Crankshaft   a component that converts the piston's up-and-down or forward-and-backward motion into rotational motion.

Crossflow   movement that occurs when two streams of fluid flow perpendicular (at a 90-degree angle) to each other.

Cutwater   a thick plate in the discharge nozzle of a pump that breaks the vortex.

Cycles of concentration   the maximum multiplier for the miscellaneous substances in the circulating cooling water as compared to the amount of those substances in the makeup water.

Cyclone separator   a mechanical device that uses a swirling (cyclonic) action to separate heavier and lighter components of a gas or liquid stream.

Cylinders   vessels that can hold extremely volatile or high-pressure materials; a cylindrical chamber in a positive displacement compressor in which a piston compresses and then expels the gas.

Damper   a movable plate that regulates the flow of air or flue gases.

Deadhead   also called shut-off pressure; it is the maximum pressure (head) that occurs when a pump is operating with zero flow (discharge valve shut).

Deadheading   creating a condition in which all outlet paths from a compressor's discharge line are closed.

Deaeration   removal of air or other gases from boiler feedwater by increasing the temperature using steam and stripping out the gases.

Demineralization   a process that uses ion exchange to remove mineral salts; also known as *deionization*. The water produced is referred to as deionized water.

Demister   a device that promotes separation of liquids from gases. Also called a mist eliminator.

Desiccant   a specialized substance contained in a dryer that removes hydrates (moisture) from a process stream.

Desuperheated steam   superheated steam from which some heat has been removed by the reintroduction of water. It is used in processes that cannot tolerate the higher steam temperatures.

Desuperheater   a system that controls the temperature of steam leaving a boiler by using water injection through a control valve.

Dew point the temperature at which air is completely saturated with water vapor (100 percent relative humidity).

Diaphragm pump a mechanically or air-driven positive displacement reciprocating pump that uses a flexible membrane to move liquid.

Diaphragm valve a type of valve that uses a flexible, chemical-resistant, rubber-type membrane (diaphragm) to control flow.

Die a metal plate with specifically shaped perforations designed to give materials a specific form as they pass through an extruder.

Dike a wall (earthen, metal, or concrete) built around a piece of equipment to contain any liquids should the equipment rupture or leak.

Direct current (DC) electrical current that flows in a single direction through a conductor.

Discharge static head the vertical distance between the centerline of a pump and the surface of the liquid on the discharge side of the pump (if the discharge line is submerged) or the pipe end (if the discharge line is open to the atmosphere).

Distillation the process of separating two or more liquids by their boiling points.

Dolly a wheeled cart or hand truck used to transport heavy items such as barrels, drums, and boxes.

Double-acting piston pump a type of piston pump that takes suction and discharges by reciprocating motion on every stroke.

Downcomers tubes that transfer water from the steam drum to the mud drum.

Draft fan a fan used to control draft in a boiler.

Drift carrying of water with the air stream out of the cooling tower; also known as *windage* (a force created on an object by friction when there is relative movement between air and the object).

Drift eliminators devices that prevent water from being blown out of the cooling tower; used to minimize water loss.

Drill a device that contains a rotating bit designed to bore holes into various types of materials.

Drum a specialized type of storage vessel or intermediary process vessel.

Drum mixer a screw-type device used to stir, agitate, mix, or blend the contents of a drum into a consistent composition.

Dry bulb temperature the actual temperature of the air that can be measured with a thermometer.

Dry carbon rings an easy-to-replace, low-leakage type of seal consisting of a series of carbon rings that can be arranged with a buffer gas to prevent process gases from escaping.

Dryer a device or vessel in which moisture is removed from process streams.

Dynamic compressor a compressor that uses centrifugal or axial force to accelerate a gas and then to convert the velocity to pressure.

Dynamic mixer a piece of equipment that rotates to mix products rapidly.

Dynamic pump a type of pump that converts the spinning motion of a blade or impeller into dynamic pressure to move liquids.

Economizer a type of heat exchanger that preheats a fluid with waste heat, such as boiler feedwater being preheated with flue gases exiting a boiler; the section of a boiler used to preheat feedwater before it enters the main boiler system.

Eductor a device that uses a fluid as a motive force to move another fluid using the Venturi design also called a *jet pump eductor* or *aspirator*.

Ejector a device that is used to maintain vacuum by using a Venturi design powered by a motive gas (usually steam or air) flow. The ejector helps evacuate inert gases from the vacuum atmosphere; sometimes called a *vacuum ejector*, *injector*, or *steam ejector*.

Electrical diagram illustration showing power transmission and how it relates to the process.

Electrical schematic a drawing that shows the direction of electrical current flow in a circuit, typically beginning at the power source.

Electricity the flow of electrons from one point to another along a pathway called a conductor.

Electric tool a tool operated by electrical power (either AC or DC).

Electromagnetism magnetism produced by an electric current.

Electrons negatively charged particles that orbit the nucleus of an atom.

Electrostatic precipitator a device that contains positively charged collecting plates and a series of negatively charged wires.

Endothermic a chemical reaction that requires the addition or absorption of energy.

Engine a machine that converts chemical (fuel) energy into mechanical force.

Engine block the casing of an engine that houses the pistons and cylinders.

Erosion the degradation of tower components (fan blades, wooden portions, and cooling water piping) by mechanical wear or abrasion by the flow of fluids (often containing solids). Erosion can lead to a loss of structural integrity and process fluid contamination.

Evaporation the process in which a liquid is changed into a vapor through the latent heat of evaporation.

Exchanger head device located on the end of a heat exchanger that directs the flow of fluids into and out of the tubes; also called a channel head.

Exhaust port a chamber or cavity in an engine that collects exhaust gases and directs them out of the engine.

Exhaust valve valve in the head at the end of each cylinder. It opens and directs the exhaust from the cylinder to the exhaust port.

Exothermic a chemical reaction that releases energy in the form of heat.

Expansion loop segment of pipe that allows for expansion and contraction during temperature changes.

External gear pump a type of positive displacement rotary pump in which two gears rotate in opposing directions, allowing the liquid to enter the space between the teeth of each gear in order to move the liquid around the casing to the discharge.

Extruder a device that forces materials through the opening in a die so it can be formed into a particular shape.

Fan an assembly holding rotating blades inside a motor housing or casing that cools the motor by pulling air in through the shroud.

Feeder a piece of equipment that conveys material to a processing device (e.g., an extruder) at a controllable and changeable rate.

Fill material inside the cooling tower, usually made of plastic, that breaks water into smaller droplets and increases the surface area for increased air-to-water contact.

Filter device that removes particles from a process, allowing the clean product to pass through.

Filter aid a substance applied as a slurry to a filtering medium, which coats it and aids in the filter's efficiency.

Filter cake layers of solids that are deposited on the surface of a filter medium.

Filter medium the material that actually does the filtering in the filter unit.

Filtration the process of removing particles from water or some other fluid by passing it through porous media.

Finfan coolers air coolers consisting of banks of finned tubes, connected by an external tubesheet, that transfer heat from process streams flowing through the tube banks to the atmosphere.

Firebox the area of a boiler or furnace where the burners are located and where radiant heat transfer occurs.

Fire tube boiler a device that passes hot combustion gases through the tubes to heat water on the shell side.

Fitting system component used to connect two or more pieces of piping, tubing, or other equipment.

Fixed bed dryer a device that uses a desiccant to remove moisture from a process stream.

Fixed bed filter a filter in which the fluid passes over a stationary bed of filtering medium.

Fixed bed reactor a reactor vessel in which the catalyst bed is stationary and the reactants are passed over it. In this type of reactor, the catalyst occupies a fixed position and is not designed to leave the reactor with the process.

Fixed blade a blade inside a turbine, attached to the casing, that determines the direction of the power fluid.

Fixed roof tanks storage tanks that have a roof permanently attached to the top. The roof can be dome or cone-shaped or flat.

Flare system a process safety system in which unwanted process gases are burned to prevent their release into the atmosphere.

Flaring tool a device used to create a cone-shaped enlargement at the end of a piece of tubing so the tubing can accept a flare fitting.

Flash dryer a device that forces hot gas (usually air or nitrogen) through a vertical or horizontal flash tube causing the moist solids to become suspended.

Floating roof a type of vessel covering, used on storage tanks, that floats upon the surface of the stored liquid; it is used to decrease vapor space and reduce the potential for evaporation.

Fluid bed dryer a device that uses hot gases to fluidize solids and promote the contact between dry gases and solids.

Fluidized bed reactor a reactor that uses high-velocity fluid to suspend or fluidize solid catalyst particles.

Foam chamber a fixture or piping installed on liquid storage tanks which contains fire-extinguishing chemical foam.

Foaming the formation of a froth caused by mixing of water contaminants with air; impairs water circulation and can cause pumps to lose suction or cavitate.

Forced draft cooling towers cooling towers that contain fans or blowers at the bottom or on the side of the tower to force air through the equipment.

Forced draft furnace a furnace that uses fans or blowers to force the air required for combustion into the burner's air registers.

Forklift a motorized vehicle with pronged forks used to lift and transport equipment or items placed on a pallet.

Fouling accumulation of deposits that build up on the surfaces of processing equipment.

Frame a structure that holds the internal components of a motor and motor mounts.

Fuel lines supply lines that provide the fuel required to operate the burners.

Fuel-operated tools tools powered by the combustion of a fuel (e.g., gasoline).

Furnace a piece of equipment in which heat is released by burning fuel and is transferred directly or indirectly to a fluid mass for the purpose of increasing the temperature of fluids flowing through tubes to effect a physical or chemical change; also referred to as a *process heater*.

Fuse a device used to protect equipment and electrical wiring from overcurrent.

Gantry crane a type of crane (similar to an overhead crane) that runs on elevated rails attached to a frame or set of legs.

Gasket flexible material used to seal components together so they are air- or watertight.

Gas turbine a device that uses the combustion of natural gas to spin the turbine rotor.

Gate valve a positive shutoff valve that uses a gate or guillotine that, when moved between two seats, causes a tight shutoff; it should never be used to throttle flow.

Gauge hatch an opening on the roof of a vessel that is used to check tank levels and obtain samples of the contents.

Gear a toothed wheel that engages another toothed mechanism in order to change the speed or direction of transmitted motion.

Gearbox mechanical device that houses a set of gears, connects the driver to the load, and allows for changes in speed, torque, and direction.

Generator a device that converts mechanical energy into electrical energy.

Globe valve a type of valve that uses a disc and seat to regulate the flow of fluid through the valve body.

Governor a device used to control the speed of a piece of equipment, such as a turbine.

Gravimetric feeder a device designed to convey material at a controlled rate by measuring the weight lost over time.

Grinder a device that uses the rotating force of one material against another material (e.g., a stone grinding wheel against a metal blade) to reduce or modify the shape and size of an object.

Grounding the process of using the earth as a return conductor in an electrical circuit.

Hammer a tool with a handle and a heavy head that is used for pounding or delivering blows to an object.

Hand tool a tool that is operated manually instead of being powered by electric, pneumatic, hydraulic, or other forms of power.

Handwheel the mechanism that raises and lowers a valve stem to control the flow of fluid through a valve.

Head measurement of pressure caused by the weight of a liquid, measured in feet or meters, determined by the height of the liquid above the centerline of a pump; the component of an engine on the top of the piston cylinders that contains the intake and exhaust valves.

Heat a measure of the transmission of energy from one object to another as a result of the temperature difference between them.

Heat duty the amount of heat energy a cooling tower is capable of removing, usually expressed in MBTU/hour.

Heater a term used to describe a fired furnace.

Heat exchanger a device used to transfer heat from one substance to another, usually without the two substances physically contacting each other.

Heat tracing a coil of heated electric wire or steam tubing that is wrapped around a pipe to increase the temperature of the process fluid, reduce fluid viscosity, and facilitate flow.

Heat transfer the transmission of energy from one object to another as a result of a temperature difference between the two objects.

Hemispheroid vessel a pressurized container used to store material with a vapor pressure slightly greater than atmospheric pressure (e.g., 0.5 PSI to 15 PSI). The walls of a hemispheroid tank are cylindrically shaped and the top is rounded.

Hoist a device composed of a pulley system with a cable or chain used to lift and move heavy objects.

Hopper a funnel-shaped, temporary storage container that is usually filled from the top and emptied from the bottom.

Hopper car a type of railcar designed to transport solids such as plastic pellets and grain.

Hoppers/bins vessels that typically hold dry solids.

Hose flexible tube that carries fluids; can be made of plastic, rubber, fiber, metal, or a combination of materials.

Humidity moisture content in ambient air measured as a percentage of saturation.

Hybrid flow movement of fluid in a heat exchanger that is any combination of counterflow, parallel flow, and crossflow.

Hydraulic tool a tool that is powered using hydraulic (liquid) pressure.

Hydraulic turbine a device that is operated or affected by a liquid (e.g., a water wheel).

Hydrocyclone a mechanical device that promotes separation of heavy and light liquids by centripetal force, a force that acts on a body moving in a circular path and is directed toward the center around which the body is moving (e.g., like a cyclone).

Igniter a device (similar to a spark plug) that automatically ignites the flammable air and fuel mixture at the tip of the burner.

Impact wrench a pneumatically or electrically powered wrench that uses repeated blows from small internal hammers that generate torque to tighten or loosen fasteners.

Impeller a device with vanes that spins a liquid rapidly in order to generate centrifugal force; (*in a boiler*) a fixed, vaned device that causes the air/fuel mixture to swirl above the burner; different from an impeller in a turbine or pump.

Impulse movement motion that occurs when fluid is directed against turbine blades, causing the rotor to turn.

Impulse turbine a device that uses a directed flow of high velocity fluid against rotor blades to move the rotor; an example of Newton's second law of motion.

Incinerator a device that uses high temperatures to destroy solid, liquid, or gaseous wastes.

Induced draft cooling towers cooling towers in which air is pulled through the tower internals by a fan located at the top of the tower.

Induced draft furnace a furnace that uses fans, located between the convection section and the stack, to pull air up through the furnace and induce air flow by creating a lower pressure in the firebox.

Induction motor a motor that turns slightly more slowly than the supplied frequency and can vary in speed based on the amount of load.

Inhibitor a substance used to slow or stop a chemical reaction.

Inlet guide vane a stator, located in front of the first stage of a compressor, which directs the gas into the compressor at the correct angle; also called *intake guide vane*.

Inline mixer a mixer that has no moving parts; it operates by creating a complex flow path that results in contact between fluid streams; also called a *static mixer*.

Insulation any substance that prevents the passage of heat, light, electricity, or sound from one medium to another.

Intake port an air channel that directs fuel to an intake valve.

Intake valve a valve located in the head, at the top of each cylinder, that opens, allowing air–fuel mixture to enter the cylinder.

Interchanger a process-to-process heat exchanger that uses hot process fluids on the tube side and cooler process fluids on the shell side. Also known as a cross exchanger.

Intercooler a heat exchanger used between the stages of a compressor to cool the gas and remove any liquid before it reaches the next stage.

Interlock a type of hardware or software that does not allow an action to occur unless certain conditions are met; a system for connecting mutually independent equipment that can stop upstream equipment when downstream equipment is shut down.

Internal gear pump a type of positive displacement rotary pump, called a "gear within a gear" pump, in which two gears rotate in the same direction, one inside the other, and trap liquid between the teeth of the gears to move liquid from suction to discharge.

ISA The International Society of Automation (originally known as the Instrument Society of America), a global, nonprofit technical society that develops standards for automation, instrumentation, control, and measurement.

Isometric drawing an illustration showing objects as they would appear to the viewer (similar to a three-dimensional drawing that appears to come off the page).

Jacketed pipe a pipe-within-a-pipe design that allows hot or cold fluids to be circulated around the process fluid without the two fluids coming into direct contact with each other.

Jet pump eductor a device that uses a high pressure liquid to create a Venturi effect to transfer fluid without the need for an external power source.

Jib crane a type of crane that contains a vertical rotating member and an arm that extends to carry the hoist trolley.

Knockout pots devices designed to remove liquids and condensate from the fuel gas before it is sent to the burners.

Labyrinth seal a shaft seal designed to restrict flow by requiring the fluid to pass through a series of ridges in an intricate path.

Laminar flow (also called *streamline flow*) movement of fluid that occurs when the Reynolds number is low (at very low fluid velocities).

Landfill a human-made or natural pit, typically lined with an impermeable, flexible substance (e.g. rubber) or clay, which is used to store residential or industrial waste materials.

Leaf filter a type of filter that uses a precoated screen, located inside the filter vessel, to hold the filtered material.

Legend the section of a drawing that explains or defines the information or symbols contained within the drawing.

Level indicator a gauge placed on a vessel to denote the height of the liquid level within the vessel.

Lift check valve a valve with a design similar to that of a globe valve, which allows flow in only one direction; it can be installed in both horizontal and vertical lines.

Lineman's pliers snub-nosed pliers with two flat gripping surfaces, two opposing cutting edges, and insulated handles that help reduce the risk of electrical shock.

Lining a coating applied to the interior wall of a vessel to prevent corrosion or product contamination.

Liquid buffered seal a close-fitting bushing in which oil or water are injected in order to stop the process fluid from reaching the atmosphere.

Liquid head pressure developed from the pumped liquid passing through the volute.

Liquid ring compressor a rotary compressor that uses an impeller with vanes to transmit centrifugal force into a sealing fluid (e.g., water), driving it against the wall of a cylindrical casing, to form a compression area.

Live bottom a device at the bottom of a bin or silo designed to facilitate the smooth discharge of material.

Load the amount of torque necessary for a motor to overcome the resistance of the driven machine.

Lobe pump a type of positive displacement rotary pump consisting of a single or multiple lobes; liquid is trapped between the rotating lobes and is subsequently moved through the pump.

Locking pliers pliers that can be adjusted with an adjustment screw and then locked into place when the handgrips are squeezed together; also referred to as Vise-Grips®.

Louvers movable, slanted slats that are used to adjust the flow of air.

Lubricant a substance used to reduce friction between two contact surfaces.

Lubrication the application of a substance between moving surfaces in order to reduce friction and minimize heating.

Lubrication system a system that circulates and cools sealing and lubricating oils.

Magnetic (mag) drive pump a type of pump that uses magnetic fields to transmit torque to an impeller.

Makeup water the amount of water that must be added to compensate for the water leaving the cooling water system through evaporation, drift, and blowdown.

Manual valve a hand-operated valve that is opened or closed using a handwheel or lever.

Manway an opening in a vessel or tank that permits entry for inspection or repair.

MAWP maximum allowable working pressure; a safety limit for components in a piping system; the safe pressure level determined by the weakest element in the system.

Mechanical energy energy of motion that is used to perform work.

Mechanical seals seals that typically contain two flat faces (one that rotates, and one that is stationary), that are in close contact with one another in order to prevent leaks.

Micron a unit of measure equal to 0.000001 meters. Filters are rated based on the size of the particles they are capable of filtering.

Mist eliminator a device in the top of a vessel, composed of mesh, vanes, or fibers, that collects droplets of mist (moisture) from gas to prevent it from leaving the vessel and moving forward with the gas flow (also known as a demister).

Mixer a device used to combine chemicals or other substances mechanically using a rotary motion.

Mixing system a system consisting of an agitator, circulating pump, gas spargers, and a series of baffles to provide proper mixing of reactants and a catalyst.

Monorail crane a type of crane that travels on a single runway beam.

Motive fluid a fluid whose pressure is used to drive a piece of equipment.

Motor a mechanical driver that converts electrical energy into useful mechanical work and provides power for rotating equipment; sometimes called a *driver*.

Motor control center (MCC) an enclosure that houses the equipment for motor control, including its power source, isolation power switches, lockouts, fuses, overload protection devices, ground-fault protection, and sometimes meters for current (amperes) and voltage.

Mud drum the lower drum in a boiler; also called the water drum; serves as a settling point for solids in the boiler feedwater.

Multiport valve a type of valve used to split or redirect a single flow into multiple directions or direct flows from multiple sources to the same destination.

Multistage centrifugal pump a type of pump that uses two or more impellers on a single shaft (generally used in high-volume, high-pressure applications, such as boiler feed water pumps).

Multistage compressor a device designed to compress gas multiple times by delivering the discharge from one stage to the suction of another stage.

Multistage turbine a device that contains two or more stages (sets of blades), used as a driver for high-differential pressure, high-horsepower applications, and extreme rotational requirements.

Natural draft cooling towers cooling towers in which air movement is caused by wind, temperature difference, or other nonmechanical means.

Natural draft furnace a furnace that has no fans or mechanical means of producing draft; instead, the heat in the furnace causes natural draft.

NEC National Electric Code; standard established by the National Fire Protection Agency (NFPA), which promotes the safe installation of electrical wiring and equipment.

Needle nose pliers small pliers with long, thin jaws for fine work (e.g., holding small items in place, removing cotter pins, or bending wire).

Needle valve a variation of a globe valve that controls small flows using a long, tapered plug.

Net positive suction head available (NPSH$_a$) NPSH available from the pump system, it must always be equal to or greater than the NPSH$_r$.

Net positive suction head (NPSH) the liquid head (pressure), minus the vapor pressure, that exists at the suction end of a pump.

Net positive suction head required (NPSH$_r$) pump characteristic established by manufacturer testing that indicates the amount of NPSH necessary for the pump to function properly.

Nitrogen oxides (NO$_x$) undesirable air pollution produced from reactions of nitrogen and oxygen. The primary NO$_x$ species in process heaters are nitric oxide (NO) and nitrogen dioxide (NO$_2$).

Nominal pipe size a number that is used to represent pipe size.

Noncondensing steam turbine a turbine that exhausts steam at or above atmospheric pressure, allowing it to be further used in other equipment; also called a *back pressure turbine*.

Nonrising stem valve a valve in which the stem does not rise through the handwheel when the valve is opened or closed.

Nonsparking tools tools that are manufactured from specific materials that will not produce sparks capable of igniting vapors; also referred to as *spark-resistant* or *spark-proof tools*.

Nozzle component used to drive the blades or rotor of the turbine; orients and converts the steam or motive fluid from pressure to velocity.

Ohm a measurement of resistance in electrical circuits.

Oil pan/sump a component that serves as a reservoir for the oil used to lubricate internal combustion engine parts.

OSHA Occupational Safety and Health Administration, a U.S. government agency created to establish and enforce workplace safety and health standards, conduct workplace inspections, provide worker training and education, and investigate serious workplace incidents.

Overhead traveling bridge cranes a type of crane that runs on elevated rails along the length of a factory and provides three axes of hook motion (up and down, sideways, and back and forth).

Overspeed trip mechanism an automatic safety device designed to remove power from a rotating machine if it reaches a preset trip speed.

Packing a substance such as Teflon® or graphite-coated material that is used inside the stuffing box to prevent leakage from occurring around the valve stem.

Parallel flow movement of fluid that occurs in a heat exchanger when the shell flow and the tube flow are parallel to one another.

Personnel lift a lift that contains a pneumatic or electric arm with a personnel bucket attached at the end; also referred to as a manlift, cherry picker, Condor®, or JLG®.

Pilot an initiating device used to ignite the burner fuel.

Pipe long, hollow cylinder through which fluids are transmitted; primarily made of metal, but also can be made of glass, plastic, or plastic-lined material.

Pipe clamp piping support that protects piping, tubing, and hoses from vibration and shock; also, a ring-type device used to stop a pipe leak temporarily.

Pipe hanger piping support that suspends pipes from the ceiling or other pipes.

Pipe shoe piping support that supports pipes from beneath.

Pipe wrench an adjustable wrench that contains two serrated jaws that are designed to grip and turn pipes or other items with a rounded surface; also referred to as a Stillson wrench.

Piping and instrumentation diagram (P&ID) detailed illustration that graphically represents the relationship of equipment, piping, instrumentation, and flows contained within a process in the facility.

Piston a component that moves up and down or backward and forward inside a cylinder.

Piston pump a type of positive displacement reciprocating pump that uses a piston inside a cylinder to move fluids.

Plate and frame filter a type of filter used to remove liquids from a slurry feed by using a deep filtration process that includes clarification and prefiltration.

Platform a strategically located structure designed to provide access to instrumentation and to allow the performance of maintenance and operational tasks. Platforms are usually accessed by stairs or ladders.

Pleated cartridge filter a type of cartridge filter that uses a filter medium in a pleated form to provide more surface area.

Pliers a hand tool that contains two hinged arms and serrated jaws that are used for gripping, holding, or bending.

Plot plan illustration drawn to scale, showing the layout and dimensions of equipment, units, and buildings; also called an *equipment location drawing*.

Plug valve a type of valve that uses a flow control element shaped like a hollowed-out plug attached to an external handle to control flow.

Plunger pump a type of positive displacement reciprocating pump that displaces liquid using a plunger and maintains a constant speed and torque.

Pneumatic conveyor a device used to move powdered or granular material using pressured gas flow from one point to another.

Pneumatic tool a tool that is powered using pneumatic (air or gas) pressure.

Positive displacement compressor a device that uses screws, sliding vanes, lobes, gears, or pistons to deliver a set volume of gas with each stroke or rotation.

Positive displacement pump type of pump that uses pistons, diaphragms, gears, or screws to deliver a constant volume with each stroke.

Powder-actuated tool a tool that uses a small explosive charge to drive fasteners into hard surfaces such as concrete, stone, and metal.

Power tool a tool that is powered by electric, pneumatic, or hydraulic power or is powder actuated.

Preheater a heat exchanger that adds heat to a substance prior to a process operation.

Premix burner a device that mixes fuel gas with air before either enters the burner tip.

Pressure relief device a component that discharges or relieves pressure in a vessel or piping system in order to prevent overpressurization (e.g., pressure relief valve, vacuum breaker, or rupture disc).

Pressure/vacuum relief valve valve that maintains the pressure in the tank at a specific level by allowing air flow into and out of the tank. Also known as *breather valve, conservation vent,* and *tank breather vent.*

Pressurized tanks enclosed vessels in which a greater-than-atmospheric pressure is maintained.

Priming the process of filling the suction line and casing of a pump with liquid to remove vapors and eliminate the tendency for it to become vapor-bound or to lose suction.

Process   the conversion of raw materials into a finished or intermediate product.

Process drawing   illustration that provides a visual description and explanation of the processes' equipment, flows, and other important items in a facility.

Process flow diagram (PFD)   basic illustration that uses symbols and direction arrows to show the primary flow path of material through a process.

Process industries   a broad term for industries that convert raw materials, using a series of actions or operations, into products for consumers.

Process schematic   screen display of a process in the computer control system; it shows control parameters but not piping and equipment details.

Process technician   a worker in a process facility who monitors and controls mechanical, physical, and/or chemical changes throughout a process in order to create a product from raw materials.

Psychrometer   an instrument for measuring humidity, consisting of a wet bulb thermometer and a dry bulb thermometer; the difference in the two thermometer readings is used to determine atmospheric humidity.

Pump   a mechanical device that transfers energy to move liquids through piping systems.

Pump performance curve   a specification that describes the capacity, speed, horsepower, and head needed for correct pump operations.

Radial bearing   a type of bearing designed to support and hold the rotor in place while offering minimum resistance to free rotation; prevents and offsets horizontal and vertical radial movement. Also called a *journal bearing*.

Radiant section   the lower portion of a furnace where heat transfer is primarily through radiation.

Radiant tubes   tubes located in the firebox that receive heat primarily through radiant heat transfer; (*in boilers*) tubes containing boiler feedwater that are heated by radiant heat from the burners and boiled to form steam that is returned to the steam drum.

Radiation   the transfer of heat through electromagnetic waves (e.g., warmth emitted from the sun).

Raw gas burner   a burner in which gas has not been pre-mixed with air.

Reaction turbine   a device that uses a reactive force of pressure or mass of the motive fluid to turn a rotor; this is an example of Newton's third law of motion.

Reactor   a vessel in which a controlled chemical reaction is initiated and takes place either continuously or as a batch operation.

Reboiler   a tubular heat exchanger, usually located at the bottom of a distillation column or stripper, that is used to supply the necessary column heat.

Reciprocating compressor   a positive displacement compressor that uses the inward stroke of a piston to draw (intake) gas into a chamber and then uses an outward stroke to positively displace (discharge) the gas.

Reciprocating pump   a type of positive displacement pump that uses the inward stroke of a piston or diaphragm to draw liquid into a chamber (intake) and then positively displace the liquid using an outward stroke (discharge).

Reciprocating saw   a device that uses the back-and-forth motion of a saw blade to cut materials such as wood and metal pipe.

Rectifier   a device that converts AC voltage to DC voltage.

Recuperator   a heat exchanger that preheats the waste gas feed stream with the flue gas exiting the thermal oxidizer.

Refractory lining   a form of insulation used inside high temperature boilers, incinerators, heaters, reactors, and furnaces; it is used to reflect heat back into the box and protect the structural steel; a common type is called fire brick.

Relative humidity   a measure of the amount of water in the air compared with the maximum amount of water the air can hold at that temperature.

Relief system   a system designed to prevent overpressurization or a vacuum.

Relief valve   a safety device designed to open slowly if the pressure of a liquid in a closed space, such as a vessel or a pipe, exceeds a preset level.

Reverse osmosis   method for processing water by forcing it through a membrane through which salts and impurities cannot pass (purified bottled water is produced this way).

Reynolds number   a numeric designation used in fluid mechanics to indicate whether a fluid flow in a particular situation will be laminar (smooth) or turbulent.

Riser tubes   tubes that allow water or steam from the lower drum to move to the upper drum.

Rising stem valve   a valve in which the stem rises out of the valve handwheel when the valve is opened.

Rod bearing   a flat steel ring or sleeve coated with soft metal and placed between the connecting rod and the crankshaft.

Rotary airlock valve   a valve that uses rotating pockets to meter a fixed volume of solids at a fixed rate; also called an *airlock, rotary valve,* or *rotary feeder.*

Rotary compressor   a positive displacement compressor that uses a rotating motion to pressurize and move gas through the device.

Rotary drum filter   a type of filter that contains a cloth screen on a rotating drum to hold solids being filtered.

Rotary dryer   a device composed of large, rotating, cylindrical tubes; reduces the moisture content of a material by bringing it into direct contact with a heated gas.

Rotary pump   a type of positive displacement pump that moves in a circular motion to move liquids by trapping them in a specific area of a screw or a set of lobes, gears, or vanes.

Rotating blades   components attached to the shaft of the turbine. The motive fluid causes the turbine to spin by impinging on the rotating blades.

Rotor   (*in a turbine, pump, or compressor*) a rotating assembly comprised of the shaft and the rotating blades; (*in a motor*) the rotating member of a motor that is connected to the shaft.

Runaway reaction   an uncontrolled reaction that is accelerated by the heat generated from the reaction. As the reaction rate increases, additional heat is released; the release of heat causes the reaction rate to increase and release more heat.

Safety valve   a safety device designed to open quickly if the pressure of a gas in a closed vessel exceeds a preset level.

Saltation velocity   the minimum gas velocity required to maintain particle entrainment, also known as *pickup velocity;* if gas velocity slows below the saltation velocity, particles will fall from suspension in a pipe.

Saturated steam   steam in equilibrium with water (e.g., steam that holds all of the moisture it can without condensation occurring).

Scale   dissolved solids deposited on the inside surfaces of equipment.

Schedule   piping reference number that pertains to the wall thickness, which affects the inside diameter (ID) and specific weight per foot of pipe.

Screener   a piece of equipment that uses sieves or screens to separate dry materials according to particle size.

Screwdriver   a device that engages with, and applies torque to, the head of a screw so that the screw can be tightened or loosened.

Screw pump   a type of positive displacement rotary pump that displaces liquid with a screw. The pump is designed for use with a variety of liquids and viscosities and a wide range of pressures and flows.

Scrubber   a device used to remove contaminants from a gas stream by washing the stream with water or some other neutralizing agent.

Seal   a device that holds lubricants and process fluids in place while keeping out foreign materials where a rotating shaft passes through a pump casing.

Seal flush   a small flow (slip stream) of pump discharge or externally supplied liquid that is routed to the pump's mechanical seal. This acts as a barrier liquid between the two faces of the seal to reduce friction and remove heat.

Seal system   system designed to prevent process gas from leaking out of the compressor shaft.

Sentinel valve   a spring-loaded, small relief valve that provides an audible warning (high-pitched whistle) when the turbine nears its maximum operating pressure conditions.

Separator   device used to physically separate two or more components in a mixture.

Shaft   (*in motors*) a metal rod on which the rotor resides; suspended on each end by bearings, it connects the driver to the shaft of the driven end or mechanical device; (*in turbines*) a metal rotating component (spindle) that holds the blades and all rotating equipment in place.

Sheave   a V-shaped groove centered on the circumference of a wheel designed to hold a belt, rope, or cable.

Shell   the outer casing, or external covering, of a heat exchanger.

Ships   seagoing vessels used to transport cargo across oceans and other large bodies of water.

Shock bank (shock tubes)   also called the *shield section,* a row of tubes located directly above the firebox (radiant section) in a furnace that receives both radiant and convective heat; protects the convection section from exposure to the radiant heat of the firebox.

Shroud   a casing over the motor that allows air to flow into and around the motor.

Silo   a tower-shaped container for storage of materials; typically used to store product destined for distribution.

Single-acting piston pump a type of piston pump that pumps by alternating suction and discharge actions on each piston stroke.

Single-stage compressor a device designed to compress gas a single time before discharging it.

Single-stage turbine a device that contains a set of nozzles or stationary blades and moving blades, called a *stage*.

Skirt support structure attached to the bottom of free-standing vessels.

Slurry a liquid solution that contains a high concentration of suspended solids.

Slurry dryer a device designed to convert wet process slurry into dry powder.

Socket wrench a tool that contains interchangeable socket heads of varying sizes that can be attached to a ratcheting wrench handle or other driver.

Softening the treatment of water that removes dissolved mineral salts such as calcium and magnesium, known as hardness in boiler feedwater.

Solids handling equipment equipment used to process and transfer solid materials from one location to another in a process facility; can also provide storage for those materials.

Sparger a pipe with holes drilled along its length; holes allow for the introduction and distribution of a pressurized gas into a liquid.

Spark plug a component in an internal combustion engine that supplies the spark to ignite the air–fuel mixture.

Spheres pressurized storage tanks that are used to store volatile or highly pressurized material; their spherical shape allows stress to be uniformly distributed over the entire vessel; also called Hortonspheres.

Spiders devices with a spiderlike shape that are used to inject fuel into a boiler.

Spray dryer a device that sprays a moist feed stream into a vertical drying chamber using a nozzle or rotary wheel.

Sprocket a toothed wheel used in chain drives.

Spuds devices used to inject fuel into a boiler.

Stack a cylindrical outlet, located at the top of a furnace or boiler, through which flue (combustion) gas is removed.

Stack damper a device that regulates the flow of flue gases leaving the furnace.

Static electricity electricity that occurs when a number of electrons build up on the surface of a material but have no positive charge nearby to attract them and cause them to flow.

Stator a stationary part of the motor where the alternating current supplied to the motor flows, creating a magnetic field using magnets and coiled wire.

Steam chest area where steam enters the turbine.

Steam drum the top drum of a boiler where all of the generated steam is collected before entering the distribution system.

Steam strainer a mechanical device that removes impurities from the steam.

Steam trap device used to remove condensate from the steam system or piping.

Steam turbine a device that is driven by the pressure of high-velocity steam.

Stirred tank reactor a reactor vessel that contains a mixer or agitator to improve the mixing of reactants.

Straight-through diaphragm valve a valve that contains a flexible membrane (diaphragm) that extends across the valve opening; long diaphragm movements are required to operate it; better for viscous liquids than the weir diaphragm valve.

Stress corrosion type of corrosion that results in the formation of cracks (called stress cracks).

Stuffing box the area in a pump's casing that contains the packing material.

Suction static head the vertical distance between the centerline of a pump and the surface of the liquid on the suction side of the pump.

Sump an area of temporary storage located in the bottom of a vessel from which undesirable material is removed.

Superheated steam steam that has been heated to a very high temperature so that a majority of the moisture content has been removed (also called *dry steam*).

Superheater tubes located near the boiler outlet that increase (superheat) the temperature of the steam.

Surging the intermittent flow of gas through a compressor that occurs when the discharge pressure is fluctuating, resulting in flow reversal and instability within a compressor.

Swing check valve a valve used to control the direction of flow and prevent contamination or damage to equipment caused by backflow.

Switch an electrical device used to start, stop, or otherwise reconfigure the flow of electricity in a circuit.

Symbol simple illustration used to represent a piece of equipment, an instrument, or other device on a PFD or P&ID.

Synchronous motor a motor that runs at a fixed speed synchronized with the supply of electricity.

Tank a vessel in which a feedstock or product (intermediate or finished) is stored; might be classified as atmospheric or pressurized, aboveground or underground, fixed or floating roof.

Tank car a type of railcar designed to transport liquids.

Tank trucks vehicles with containers designed to transport fluids in bulk over roadways.

Temperature a measure of the thermal energy of a substance (e.g., the "hotness" or "coldness") that can be determined using a thermometer.

Temperature control system a system that allows the temperature of a reaction to be adjusted and maintained.

Tensile strength the pull stress, in force per unit area, required to break a given specimen.

Throttling partially opening or closing a valve to restrict or regulate fluid flow rates.

Thrust (axial) bearing a type of bearing designed to support and hold the rotor in place while offering minimum resistance to free rotation; prevents and offsets back-and-forth axial movement.

Title block the section of a drawing (typically located in the bottom-right corner) that contains the drawing title, drawing number, revision number, sheet number, company information, process unit, and approval signatures.

Tool a device designed to provide mechanical advantage and make a task easier.

Torque a force that produces rotation.

Torque wrench a manual wrench that uses a gauge to indicate the amount of torque (rotational force) being applied to the nut or bolt.

Total head a measure of a pump's ability to move liquid through a pumping system.

Transformer an electrical device that takes electricity of one voltage and changes it into another voltage.

Trickle valve a valve used to continuously transfer a fixed weight of solids between two different pressure zones at a constant rate.

Trip and throttle valve a valve designed to constrict (throttle) the inlet steam to control the turbine speed and to shut down the turbine in the event of excess rotational speed or vibration.

Troubleshooting the systematic search for the source of a problem so that it can be solved.

Tube bundle a group of fixed, parallel tubes though which process fluids are circulated.

Tube leaks leaks in a tube that can result in process chemicals entering the circulating water, possibly resulting in a fire, explosion, or environmental or toxic hazard.

Tube sheet a formed metal plate with drilled holes that allow process fluids to enter the tube bundle.

Tubing hose or pipe of small diameter (typically less than 1 inch [2.5 cm]) used to transport fluids.

Tubular reactor a continuous flow vessel in which reactants are converted in relation to their position within the reactor tubes, not influenced by residence time in the reactor.

Turbine a machine that is used to produce power and rotate shaft-driven equipment such as pumps, compressors, and generators.

Turbulent flow movement of fluids that occurs when the Reynolds number is high, characterized by irregular flow patterns and high velocities.

Underground storage tank a container in which process fluids are stored underground.

Unit an integrated group of process equipment used to produce a specific product; it might be referred to by the processes it performs or be named after its end products.

Universal motor a motor that can be driven by either AC or DC power.

Utility flow diagram (UFD) illustration that provides process technicians a PFD-type view of the utilities used for a process.

Valve piping system component used to control, throttle, or stop the flow of fluids through a pipe.

Valve body the lower portion of the valve; contains the fluids flowing through the valve.

Valve cover a cover over the cylinder head that keeps the valves and camshaft clean and free of dust or debris; it also keeps lubricating oil contained.

Valve disc the section of a valve that attaches to the stem and that can fully or partially block the fluid flowing through the valve.

Valve knocker   a device usually attached to a chain operator, used to facilitate the movement (opening or closing) of an overhead valve.

Valve seat   a section of a valve where the flow control element rests when the valve is completely closed; it is designed to maintain a leak-tight seal when the valve is shut.

Valve stem   a long, slender shaft that attaches to the flow control element in a valve.

Valve wheel wrench   a hand tool used to provide mechanical advantage when opening and closing valves. These wrenches typically fit over the spoke of a valve wheel.

Vane pump   a type of positive displacement rotary pump having either flexible or rigid vanes designed to displace liquid.

Vanes   raised ribs on the impeller of a centrifugal pump designed to accelerate a liquid during impeller rotation.

Vapor recovery system   process used to capture and reclaim vapors.

Vaporizer   a device that converts a liquid into a vapor.

Venturi   a device consisting of a converging section, a throat, and a diverging section; its purpose is to create a constriction in a pipe.

Venturi effect   the drop in pressure that occurs when liquid flows through a constricted section of pipe; as the speed increases, the pressure decreases.

Vertical cylindrical furnace   a furnace design in which the radiant section tubes are laid out in a vertical configuration along the walls of the cylinder; it has a smaller footprint than a cabin furnace and is used in operations where less heat capacity is required.

Vessel   a container in which materials are processed, treated, or stored; an enclosed process container such as a tank, drum, tower, filter, or reactor.

Vessel heads   components on the top and bottom of a reactor shell that enclose it.

Viscosity   the degree to which a liquid resists flow under applied force (e.g., molasses has a higher viscosity than water at the same temperature).

Volt   the unit used to measure the difference in electrical potential between two points in a circuit. One volt is the force that will cause a current of one amp to flow through a resistance of one ohm.

Voltage   the potential energy available to push electrons from one point to another.

Voltmeter   a device that can be connected to a circuit to determine the amount of voltage present.

Volumetric feeder   a device designed to convey materials at a controlled rate by running at a set motor speed.

Volute   a widened spiral casing in the discharge section of a centrifugal pump designed to convert liquid speed to pressure without shock.

Vortex   cyclone-like rotation of a fluid.

Vortex breaker   a metal plate, or similar device, that prevents a vortex from being created as liquid is drawn out of a vessel.

Waste heat boiler   a device that uses waste heat from a process to produce steam.

Water distribution header   a pipe that provides water to a distributor box located at the top of the cooling tower so the water can be distributed evenly onto the fill.

Water tube boiler   a type of boiler that contains water-filled tubes that allow water to circulate through a heated firebox.

Watt   a unit of measure of electric power; the power consumed by a current of one amp using an electromotive force of one volt.

Wear rings   close-running, noncontacting replaceable metal rings located between the impeller and casing of a centrifugal pump; wear rings allow for a small clearance between the two components.

Weighing system   a system used to weigh material before shipping.

Weir   a flat or notched dam or barrier to liquid flow that is usually used either for the measurement of fluid flow or to maintain a given depth of fluid, as on a tray of a distillation column, in a separator, or in another vessel.

Weir diaphragm valve   a valve that contains a dam-like device, sometimes called a saddle, that functions as a seat for the diaphragm.

Wet bulb temperature   the lowest temperature to which air can be cooled through the evaporation of water.

Wind turbine   a device that converts wind energy into mechanical or electrical energy.

Wrench   a hand tool that uses gripping jaws to turn bolts, nuts, or other hard-to-turn items.

Index